Governors State University
Library
Hours:
Monday thru Thursday 8:30 to 10:30
Friday and Saturday 8:30 to 5:00
Sunday 1:00 to 5:00 (Fall and Winter Trimester Only)

DEMCO

Toxicants in Aqueous Ecosystems
A Guide for the Analytical and Environmental Chemist

T.R. Crompton

Toxicants in Aqueous Ecosystems

A Guide for the Analytical and Environmental Chemist

With 19 Figures and 134 Tables

 Springer

T. R. Crompton
Hill Cottage (Bwthyn Yr Allt)
Anglesey, Gwynedd
United Kingdom

Library of Congress Control Number: 2006927807

ISBN-10 3-540-35738-6 Springer Berlin Heidelberg New York
ISBN-13 978-3-540-35738-4 Springer Berlin Heidelberg New York
DOI 10.1007/b95924

Springer is a part of Springer Science+Business Media
springer.com
© Springer-Verlag Berlin Heidelberg 2007

Cover design: *design & production* GmbH, Heidelberg
Typesetting and production: LE-TEX Jelonek, Schmidt & Vöckler GbR, Leipzig, Germany
Printed on acid-free paper   52/3100/YL - 5 4 3 2 1 0

Preface

Pollution of the ecosystem has always occurred to some extent or other. For example, over the whole of prehistory and still, to some extent, today, the eruption of volcanoes or the occurrence of fumaroles under the ocean has resulted in large-scale contamination of the ecosystem. Since the start of the Industrial Revolution, pollution of the ecosystem has obviously increased considerably and, despite efforts to control it, is still doing so.

Such inputs of pollution obviously include discharges of industrial and other waste and sewage directly into rivers and via coastal discharges. The emission of toxic substances into the atmosphere by factory smokestack emissions, incineration plants and fires is another major source of pollution, such emissions inevitably being washed out of the atmosphere by rain and then causing pollution of the oceans and land. Another input is the dumping of industrial and sewage wastes into the seas by ships.

Pollution is defined as a change in water quality that causes deleterious effects in the organism community or that makes the aesthetic quality of the water unacceptable. Contamination refers to the presence of potentially harmful substances at concentrations that do not cause harm to the environment.

It is becoming increasingly clear that the oceans and rivers, in particular, are not an unlimited reservoir into which waste can be dumped, and that control of these emissions is necessary if complete destruction of the environment is to be avoided. Heavy metals are particular offenders in this respect, as are organometallic compounds—whether the latter are discharged directly into the environment or whether (as has been shown in recent years) they are produced by the biological conversion of inorganic metallic contaminants such as lead, mercury and arsenic.

There are also many classes of organic pollutants that are encroaching upon the aquatic ecosystem. Organic pollutants—a subject that has been increasingly discussed in the public domain in recent years via the media—include crude petroleums, polyaromatic hydrocarbons, organochlorine and organophosphorus insecticides, polychlorinated biphenyls, chlorinated dioxins, chlorinated aliphatic and aromatic compounds, and nitrosamines. However, there are many thousands of possible organic pollutants, and only some of them have been studied in detail.

Once a toxic substance enters a river, it can cause damage to animal and plant life in the river, with possible implications for the survival of fish and invertebrates and also for the health of the humans who eat these creatures. Many rivers serve as inputs to potable water treatment plants and consequently so there are further health implications for humans and animals that drink the water. River waters often carry the pollutants to the oceans, where they are added to by the pollutants in coastal discharges, atmospheric fallout and shipboard dumping. Again, the survival of animal populations and the health of humans become major considerations. Pollutants that discharge directly onto land, including sewage and domestic and industrial waste, are inevitably washed by rain to a watercourse and eventually end up in the sea.

Regulations for controlling the input of pollutants into the environment are slowly being introduced internationally, but much remains to be done.

It is the purpose of this book to describe in detail methods for the determination of all types of pollutants—inorganic, organic and organometallic— in fish, crustacea and other marine creatures, as well as in weeds, plants, phytoplankton, algae and so on (Chaps. 1–3).

The levels of pollutants that occur in these depends on the levels of pollutants that occur in the water in which they live, and in the case of bottom-feeding fish and crustacea on the pollutant levels that occur in sediments. Sediments in the beds of rivers and in the oceans adsorb many toxicants from the water in such amounts that the concentrations of toxicants in the sediment are many times—in some instances up to a million times—higher than in the surrounding water. Analysis of sediments is therefore a useful means of assessing the pollutant levels in water over a period of time, and is related to ill health or mortality of creatures living in the water. A review is given in Chap. 4 of the levels of inorganic, organic and organometallic toxicants found in such sediments in samples taken all over the world, and an attempt is made to correlate contaminant levels with the health of creatures. This aspect is fully discussed in the author's previous books [1–5].

Chapter 4 reviews the levels of metals, organometallics and organics found in the tissues of various types of fish and invertebrates as well as in phytoplankton and weeds taken at various sites throughout the world. In addition, results are reported for the levels of metals found in the organs of these creatures, as in many instances enhanced metal levels occur in particular organs, and this allows the cause of death to be identified. In particular, polyaromatic hydrocarbons, chlorinated biphenyls and 2,3,7,8-tetrachlorodibenzo-p-dioxin are discussed.

In Chaps. 5–9, examples of the effect of dissolved metallic toxicants on freshwater organisms and seawater organisms are discussed. Using published LC_{50} and maximum safe concentration (S_x) data, it is possible to draw-up 'at risk' tables for each type of creature. One can then use these to compile a list of creatures from any particular water with any particular composition that will either suffer ill health or will die. Examples of clean and dirty rivers are discussed.

These chapters discuss available toxicity data describing the effects of various types of pollutants on fish and invertebrates. These include studies on nonsaline and saline waters and cover all the toxic metal pollutants, organic pollutants and organic compounds of arsenic, lead, mercury and tin.

The exposure of creatures to known concentrations of toxicants for stipulated periods of time enables the toxicity of the pollutant to be established, as measured by the relationship between the concentration and the time taken for 50% of the creatures to die (LC_{50}), or to experience adverse effects, i.e., LE_{50}. Such water analysis-based methods for assessing the effects of pollutants are discussed in Chap. 10. A further method of assessing the toxicity of pollutants is based on relating the composition of the water in which the creatures live to the concentration of the toxicant found in the animal tissue, or, better still, in a particular organ of the animal in which the toxicant concentrates preferentially. Such data can be related to the water composition and the condition of the animal in terms of ill health or mortality. These methods are reviewed in Chap. 11.

This book is essential reading for all analytical chemists, environmentalists and toxicologists working in the field.

References

1. Crompton TR (2001) *Determination of Metals and Anions in Soils, Sediments and Sludges*, Spon Press, London, UK.
2. Crompton TR (2000) *Determination of Organic Compounds in Soils, Sediments and Sludges*, Spon Press, London, UK.
3. Crompton TR (2002) *Determination of Metals in Natural and Treated Waters*, Spon Press, London, UK.
4. Crompton TR (2000) *Determination of Organic Compounds in Natural and Treated Waters*, Spon Press, London, UK.
5. Crompton TR (2002) *Determination of Anions in Natural and Treated Waters*, Spon Press, London, UK.

Contents

1 Analysis of Fish

1.1
Cations

1.1.1
Aluminium

Harvey (Clyde River Purification Board, unpublished report) has described a method for the determination of aluminium and calcium in fish gills.

After washing with distilled water, the gill filaments were cut away from the bony gill arches, placed in acid-washed glass petri dishes and dried at 105 °C. The dry material was weighed and digested with concentrated nitric acid. The digest was filtered, the residue washed, and filtrate made up to the 25 ml. Calcium was determined by atomic absorption using background correction and an air/acetylene flame. Lanthanum chloride was added as a releasing agent to all solutions in order to give a concentration of 0.1% lanthanum in the final solutions. The calcium standards used were 2, 5 and 10 mg/l. Aluminium was determined using a nitrous oxide/acetylene flame, 309.3 nm wavelength, background correction and standards of 5, 10, and 20 mg/l for absorption. A wavelength of 396.2 nm, along with standards of 1, 3 and 5 mg/l, were used for emission. A slit of 0.3 was used and potassium chloride was added as an ionisation buffer to all solutions to give a final concentration of 0.1% potassium. Flame conditions were the same for both modes.

Heydorn et al. [1] have discussed the determination of aluminium in fish gills using neutron activation analysis and inductively coupled plasma mass spectrometry. Considerable contamination was obscured with both methods, which required the samples to be handled using a clean bench and super-pure reagents.

Ranau et al. [2] used graphite furnace atomic absorption spectrometry (AAS) after pretreatment with microwave-activated oxygen plasma to determine aluminium.

1.1.2
Arsenic

Maher [3] has described a procedure for the determination of total arsenic
in fish. The sample is first digested with a mixture of nitric, sulfuric and
perchloric acids. Then arsenic is converted into arsine using a zinc reductor
column, the evolved arsine is trapped in a potassium iodide–iodine solution,
and the arsenic determined spectrophotometrically at 866 nm as the arseno-
molybdenum blue complex. The detection limit is 0.3 mg/kg dry fish and
the coefficient of variation is 5.1% at this level. The method is free from
interferences by other elements at levels normally found in fish. Values of
9.7 ± 0.3 and 13.2 ± 0.4 mg/kg obtained for NBS reference waters SRM
1S71 and SRM 1566, respectively, were in good agreement with the nominal
values of 10.2 ± and 13.4 ± mg/kg. Spiked sampled crayfish gave 98 – 100%
recovery of arsenic by this procedure.

Agemian and Thomson [4] have described a semi-automated AAS
method for the determination of arsenic in wet homogenised fish tissue.
A combination of nitric, perchloric and sulfuric acids is used to dissolve
high fat fish tissues at 140 – 180 °C in a glass tube. Extracts are then analysed by reduction to arsine with sodium borohydride, followed by AAS
with quartz tube atomisation. Average recoveries of arsenic(III), arsenic(V),
p-arsalinic acid, benzene arsenic acid, methylarsenic, and triphenylarsine
oxide obtained using this procedure were between 90 and 102%. Arsenic
found in a NBS standard bovine liver (SRM 1577) reference sample was
500 ± 10 mg/kg against a certified value of 550 ± 5 mg/kg. Arsenic levels
found in fish samples ranged from 0.26 to 0.44 mg/kg, determined with a
coefficient of variation of 7 – 15%.

Brooke and Evans [5] described two methods for the digestion of fish
samples prior to determination of arsenic down to 0.02 mg/kg by hydride
generation AAS.

The first method involves separation of the inorganic arsenic by distilling
it from 6.6 N hydrochloric acid. The second method involved chelation and
extraction of inorganic arsenic after sample dissolution in sodium hydroxide
solution, with subsequent back-extraction and oxidation. In both methods
the arsenic concentration is measured after hydride generation by AAS with
atomisation in a flame-heated silica tube; in the first method the solution
contains arsenic(III), and in the second the solution contains arsenic(V).
Results obtained by both methods are in agreement over a range of samples.
The distillation method is favoured for reasons of efficiency and economy
in time.

Hydrochloric Acid Digestion

Weigh 5 grams of a representative wet fish sample (2 g of dry sample) into
a 125 ml pear-shaped flask. Add 5 ml of water and 1 ± 0.1 g of iron(II)
sulfate heptahydrate. Through the Bethge trap, add 50 ml of hydrochloric

acid $(3 + 2)$ and reflux the reaction mixture for ten minutes. Close the tap in the Bethge trap and collect the first 50 ml of distillate over a period of 30 minutes. Cool and transfer into a 100 ml calibrated flask, washing with water, to give 100 ml of a colourless solution free from suspended solids. Reagent blank solutions should be obtained from hydrochloric acid $(3 + 2)$ in an identical manner.

Sodium Hydroxide Digestion

Place 2 g of a representative wet fish sample (1 g of dry sample) into a 150 ml conical flask, add 10 ml of sodium hydroxide reagent and heat on a boiling water bath for 20 minutes. Cool, cautiously add 35 ml of hydrochloric acid $(1 + 3)$, and cool further. Transfer the solution into a separating funnel using 5 ml of water for washing, add 2 ml of ammonium pyrrolidone dithiocarbamate solution, and mix thoroughly. Extract with 10 ml of 4-methylpentan-2-one, shaking for 2 minutes, allow to stand for 5 minutes or until separation is complete, and run off the solvent into another separating funnel. Repeat the extraction with the addition of ammonium pyrrolidone dithiocarbamate reagent, and finally extract with 10 ml of 4-methylpentan-2-one. To the combined solvent extracts add 10 ml of nitric acid $(1 + 7)$ and shake for 2 minutes. Repeat this extraction twice and combine the extracts in a beaker. Add 5 ml of sulfuric acid $(1 + 1)$ and boil until white fumes are evolved. Cool, add 10 ml water, re-heat to fuming, and repeat. Dilute to 50 ml. Reagent blank solutions should be obtained in an identical manner.

Recoveries of inorganic arsenic in spiked fish samples were 80 – 94% for arsenic(III) and 75 – 88% for arsenic(V). Total and organic arsenic levels found by both methods in some fish samples are given in Table 1.1.

Goulden et al. [6] have described a semi-automatic procedure for the determination of arsenic in nitric–perchloric acid digests of fish. Arsenic is determined using inductively coupled argon plasma (ICAP) excitation using a plasma power of 1400 W.

Beauchemin et al. [7] identified and determined the arsenic species present in dogfish muscle reference material (DORM-1). The arsenic species present were identified using electron impact mass spectrometry (IMS), thin-layer chromatography (TLC) and high-performance liquid chromatography

Table 1.1. Arsenic determination in fish. From [5]

Fish	Total arsenic[a] mg/kg	Inorganic arsenic mg/kg		Inorganic arsenic as % of total arsenic
		HCl digestion	NaOH digestion	
Herring	1.1	0.03	0.04	3.6
Haddock	2.6	0.03	0.02	0.8
Tuna	2.9	0.17	0.13	4.5

[a] Obtained by dry ashing

(HPLC)/ICA spectrometry. Determination was by the latter technique and graphite furnace atomic absorption spectrometry (GF-AAS). Arsenobetine was the major arsenic species in the methanol/water fraction (84% of the total arsenic). Arsenic(III), arsenic(V), monomethylarsonic acid, dimethyl-arsinic acid and arsenocholine constituted 4%. The total arsenic concentration was 18.7 mg/kg. The detection limit was 0.3 ng arsenic.

High-performance liquid chromatography accompanied by ICP optical emission spectrometry and hydride generation quartz furnace AAS has been used [8] to determine six arsenic species in marine organisms.

Branch et al. [9] and Le et al. [10] directly coupled HPLC to ICP–MS to determine arsenic in fish.

Yun-Kai-Lu et al. [11] carried out a simultaneous determination of traces of arsenic and cadmium in biological samples using hydride generation–double channel atomic furnace spectrometry.

1.1.3
Cadmium

Blood and Grant (private communication) determined cadmium in fish tissue using flameless AAS at 228.8 nm using a tantalum ribbon.

Fish samples were digested by one of two techniques: either 1 ml of the acid mixture (3 parts concentrated nitric acid by volume: 1 part concentrated sulfuric acid: 1 part concentrated perchloric acid) was placed together with a weighed sample (1 to 100 mg) in a covered 3.5 ml polypropylene test tube for 2 hours at 74 °C in a water bath and, finally, diluted to 25 ml; or 1 ml of concentrated nitric acid was added to the sample, which was heated for 15 minutes at 80 – 90 °C followed by the addition of 1 ml of 10% hydrogen peroxide and heating for an additional 15 minutes.

Mean recoveries obtained on the NBS SRM 1577 reference bovine liver sample with an authenticated cadmium content of 0.27 ± 0.04 mg/kg were 96.2% (nitric–perchloric–sulfuric acid digestion) and 84.8% (nitric acid–hydrogen peroxide digestion). In general, higher results were obtained by the sulfuric-nitric acid digestion procedure. The higher mean levels (13.1 – 5.6 mg/kg) of cadmium in wet blue gill tissue were found in kidney, gut, heart, gill, and liver and the lowest levels (0.14 – 1.7 mg/kg) in muscle, skin and bone.

In a series of papers, Sperling [12–14] studied the application of flameless GF–AAS to the determination of cadmium in complex matrices resulting from the digestion of fish and other biological materials. Organic material in the sample is destroyed before atomisation by digestion with ammonium peroxydisulfate, thereby avoiding loss of volatile cadmium, which would occur in ignition methods at temperatures exceeding 420 °C [12]. Cadmium was then extracted from the digest with a saturated solution of ammonium pyrrolidone dithiocarbamate in carbon tetrachloride [13,14], and the cadmium in the lower layer determined by flameless GF-AAS.

Poldoski [15] used a molybdenum- and lanthanum-treated pyrolytically coated graphite tube for the GF–AAS determination of cadmium at 228.8 nm in nitric acid perchloric acid digests of fish tissue. Molybdenum and lanthanum help reduce chemical interferences and interference from uncompensated background signals during analyte atomisation.

Digestions were carried out on 0.6 g of dry fish using 10 ml concentrated nitric acid and 2 ml perchloric acid. After digestion was complete, the residue was dissolved in 10 ml 0.2% w/v nitric acid and stored in Nalgene bottles.

Cadmium spiking recovery experiments were carried out on fish tissue samples on an authenticated reference sample (NBS SRM 1577 bovine liver) under specified conditions of analysis. The determination of cadmium content on a NBS SRM reference fish sample (0.31 ± 0.05 mg/kg) is in good agreement with the nominal value (0.27 ± 0.04 mg/kg). In addition, 0.038 mg/kg recoveries of cadmium in the fish samples were 91 – 97%. Down to 0.2 pg cadmium could be determined in the injected portion of the sample. Cadmium contents determined by this technique were in good agreement with those obtained by anodic scanning voltammetry.

1.1.4
Cobalt

Kiriyama and Kuroda [16] applied their combined ion-exchange spectrophotometric procedure to the simultaneous determination of cobalt and vanadium in cutlass fish. In this procedure, the sample is dry ashed at 420 °C, the ash (ca. 0.5 g) is decomposed with a mixture of perchloric, nitric, and hydrofluoric acids, and is finally taken up in hydrochloric acid. The metals are adsorbed by anion exchange on an Amberlite CG 400 (SCN-) column from a dilute ammonium thiocyanate–hydrochloric acid solution. The adsorbed vanadium and cobalt are separated chromatographically by elution with 12 mol/l hydrochloric acid and 2 mol/l perchloric acid, respectively. Both fractions of vanadium and cobalt are subsequently purified by anion exchange from 0.1 mol/l hydrochloric acid–3 volume% hydrogen peroxide for vanadium and 6 mol/l hydrochloric acid for cobalt. Vanadium and cobalt in the effluents are determined spectrophotometrically with 4-(2-pyridylazo) resorcinol. A 98.2% recovery of cobalt was obtained by this procedure in the presence of appreciable excesses of elements also likely to occur in the sample, namely magnesium, calcium, aluminium, iron, copper, nickel and zinc.

1.1.5
Copper

Spark source mass spectrometry, employing the stable ^{65}Cu and ^{63}Cu isotopes of copper, has been used by Harvey to study the uptake of this element [17]. An attractive feature of this method is that both the natural copper contents of the fish organs and the concentrations of added tracers are determined on the same sample by making two measurements of

Table 1.2. Copper content and accumulation of ^{65}Cu tracer in a 30 g (wet) plaice (*Pleuronectes platessa*). From [17]

Organ	Wet weight, g	Percentage of body weight	Total natural Cu in organ, µg	Concentration of natural Cu in organ, µg/g (wet)	^{65}Cu accumulated in organ after 2-month exposure at 8 µg/l
	(A)	(B)	(C)	(D)	(E)
Blood cells	0.15	0.5	0.04 ± 0.004	0.27 ± 0.03	0.03 ± 0.003
Blood serum	0.3	1.0	0.17 ± 0.02	0.57 ± 0.06	0.14 ± 0.01
Heart	0.02	0.06	0.06 ± 0.006	3.0 ± 0.3	0.15 ± 0.02
Spleen	0.01	0.04	0.03 ± 0.003	3.0 ± 0.3	0.13 ± 0.01
Liver	0.5	1.7	0.85 ± 0.09	1.7 ± 0.2	1.0 ± 0.1
Kidney	0.12	0.4	0.08 ± 0.008	0.67 ± 0.07	0.14 ± 0.01
Gut	0.3	1.0	0.33 ± 0.03	1.1 ± 0.1	0.25 ± 0.03
Stomach	0.15	0.5	0.12 ± 0.14	0.80 ± 0.08	0.28 ± 0.03
Gill filaments	0.2	0.6	0.12 ± 0.01	0.60 ± 0.06	0.33 ± 0.03
Skin	2.2	7.2	1.4 ± 0.1	0.64 ± 0.06	0.06 ± 0.006
Muscle	17.5	58.3	3.9 ± 0.4	0.22 ± 0.02	0.02 ± 0.002
Bone	4.0	13.2	6.4 ± 0.6	1.6 ± 0.2	0.20 ± 0.02

isotopic ratio—one before and one after the addition of the standard ^{63}Cu spike. Both of these isotopes constitute ideal tracers, since they are readily available at low cost and are free from radiation hazards. A spark source mass spectrometer is an ideal way of carrying out isotopic ratio measurements.

In this procedure, the wet fish sample was weighed before and after vacuum freeze-drying, and then transferred to a Tracerlab (Richmond, CA, USA) LTA 600 low-temperature asher to remove organic matter. The residue was then digested in 0.5 mol/l hydrochloric acid–30% hydrogen peroxide, ascorbic acid being added to destroy residual hydrogen peroxide, and copper extracted from the solution with a carbon tetrachloride solution of dithizone. This extract was then evaporated directly onto graphite prior to spark source mass spectrometric evaluation. Table 1.2 illustrates the type of data obtained in this procedure for a range of biological samples.

1.1.6
Lead

The molybdenum- and lanthanum-treated pyrolytically coated GF–AAS method described by Poldoski [15] under cadmium earlier in this section has also been applied to the determination of lead in fish tissue. Lead results obtained in spiking recovery experiments, carried out on fish tissue samples and on an authenticated reference sample (NBS SRM 1577 bovine liver) under specified conditions of analysis, were reported. It is seen that the determined lead content on NBS SRM 1577 (0.33 ± 0.01 mg/kg) is in good agreement with the nominal value (0.34 ± 0.08 mg/kg). Average analytical recoveries on the fish samples are 91 – 93%. Down to 4 pg lead can be determined in the injected portion of sample. Lead contents obtained by this procedure are in good agreement with those obtained by anodic scanning voltammetry. Using this method, 0.26 mg/kg and 0.32 mg/kg of lead were found in whole catfish and blue gill samples, respectively.

Pagenkopf et al. [18] also employed G-AAS to determine lead in fish. They were able to determine down to 0.15 μg/kg lead in fish. In this procedure 1 – 5 g fish muscle tissue were removed by dissection and freeze-dried by a Thermovac (Copiague, NY, USA) lypholiser. Approximately 1 g of the dried tissue was weighed and then digested in a mixture of 7.00 ml of concentrated nitric acid and 5.00 ml of concentrated perchloric acid. The solutions were slowly heated until all foaming had stopped and dissolution was achieved. At this point, the temperature was increased so as to reduce the volume to about 1 ml. This was accompanied by copious fuming of perchloric acid. The maximum temperature was 88 °C. The colourless samples were then transferred to cleaned 100 ml volumetric flasks and diluted to volume. An Eppendorf pipette was used to transfer 20 μl of the sample into graphite cups. The cups were placed under an infrared light and heated until the solvent had evaporated. They were then placed in the furnace and peak absorbances were recorded. Spiking experiments in fish samples originally containing 0.12, 0.77 and 1.81 mg/kg lead indicated recoveries of 95 – 102%.

To overcome problems of contamination and nonreproducibility in the determination of low levels of lead in fish, Harms [19] devised a method of sample pretreatment and enrichment in which sample decomposition was performed in a closed system based on Mattinson's two-bottle system, and followed by the addition of pure nitric acid and then by neutralisation with ammonia and extraction with dithizone/toluene solution. After back-extraction into aqueous hydrochloric acid, the aqueous phase was subjected to measurement of the Pb-203 activity (recovery control) followed by electrothermal AAS for determination of stable lead. Samples of fish muscle containing less than 0.5 μg/kg could be analysed by this procedure.

May and Brumbaugh [20] used ammonium dihydrogen phosphate matrix modifier and a modified L'vov platform to overcome matrix interference effects in the determination of lead in fish tissues. The 283.3 lead line was used. They defined GF–AAS conditions to obtain maximum improvement in the slope ratio. Precisions were between 0.8 and 1.7% for fish samples.

Fish sample digestions were performed in PTFE-capped glass pressure reaction vessels in which the sample was digested with concentrated nitric acid at 70 °C for 48 hours. The digests were then made up to 50 ml with 1% hydrochloric acid. This procedure did not yield complete digestion, as lipids are not destroyed and remain as a floating white solid in the digest.

1.1.7
Mercury

Atomic Absorption Spectrometry

Various workers have discussed the application of cold vapour AAS to the determination of mercury in fish [21-33]. Various digestion procedures have been used, including concentrated nitric acid in a Teflon-lined sealed bomb [21] or glass tube [24], mixtures of nitric acid and sulfuric acid [29, 30], 50% hydrogen peroxide [28, 32] and sulfuric acid–potassium permanganate [27] in open tubes. Wickbold combustion procedures have also been used [33].

Hendzel and Jamieson [31] digested 0.1 – 0.5 g of fish sample with 5 ml nitric acid–sulfuric acid ($1:1$ v/v) in a glass digestion tube at 180 °C until white fumes appeared. After reduction with a reagent comprising hydroxylamine sulfate and stannic chloride, the elemental mercury was swept off with a stream of air and estimated at 253.7 nm by cold vapour AAS.

Louie [30] used concentrated hydrochloric acid–nitric acid–sulfuric acid in open tube digestions followed by cold vapour AAS to determine down to 0.01 mg/kg mercury in fish tissues. He claimed that this was an improvement over previous methods and that 3 g of fish was completely digested at 85 – 100 °C within 30 minutes. Using this procedure, Louie [30] obtained a mercury content on NBS Albacore Tuna Research Material 50 Reference Sample of 0.94 ± 0.05 mg/kg against a certified value of 0.95 ± 0.01 mg/kg. Levels found in various fish samples ranged between 0.1 and 0.4 mg/kg.

Davidson [28] used digestion on a hot plate with $4:1$ 50% sulfuric acid–hydrogen peroxide to digest tissue prior to the determination of mercury at 253.7 nm by cold vapour AAS.

An approximately 0.100 – 0.200 g portion (less if high mercury levels are known to exist) of homogenised, freeze-dried and ground tissue (or 0.500 – 1.00 g wet mass) was weighed into each reaction tube. Then 10 ml of 4+1 sulfuric acid was added and the tubes were covered and left to stand overnight. At this stage, 4 ml of cold (4 °C) 50% w/v hydrogen peroxide was mixed in and the tubes were placed on the hot block, set so that the sample temperature did not exceed 80 °C. When the solutions were clear and colourless, tubes were removed from the hot block. They were cooled in a cold water bath and 46.0 ml of cold (4 °C) 0.1% w/v potassium permanganate solution were added in a steady stream to ensure complete mixing. The required final volume was 60 ml. With argon flowing through the solution, a hydroxylamine sulfate-hydrazine sulfate–stannous chloride reductant was added and the elemental mercury swept into the AAS.

Nine replicate samples of NBS reference tuna (Research Material No. 50) were analysed by the 50% hydrogen peroxide method to determine the repeatability of the method. The mean and standard deviations were 1.00 µg and 0.02 µg dry mass, respectively, against the reported value of 0.95 ± 0.1 µg, and this indicated that 80 – 90% of the mercury content is present as methylmercury. Between 0.24 and 1.11 mg/kg of mercury was found in pike and lake trout samples by this method.

Konishi and Takahashi [32] have described a method for the determination of inorganic mercury in fish in the presence of organic mercury. This is based on the fact that hydrogen peroxide oxidatively liberates inorganic mercury from organic substances in strong alkali, and reduces it to the metallic state without decomposing organic materials concomitantly present. The metallic mercury, vaporised with a nitrogen stream, is trapped by gold amalgamation, and then released for electrothermal atomisation AAS. The detection limit is 1 ng of inorganic mercury, and the coefficient of variation for 40 ng of inorganic mercury is 2.8%. A 92% recovery of mercury was obtained in this procedure.

Fostier et al. [40] used microwave digestion followed by automated cold vapour AAS to determine mercury in fish.

Adeloju et al. [41] evaluated four of the most commonly used wet digestion methods for mercury in fish and found that the one based on the use of a nitric–sulfuric acid mixture was the best. Subnanogram amounts of organic and inorganic mercury have been determined by helium microwave-induced plasma atomic emission spectrometry [42]. Detection limits were around 10 pg. Organic mercury was determined as the difference between total and inorganic mercury.

Liang et al. [43] carried out a simultaneous determination of monomethylmercury, inorganic mercury and total mercury using a procedure based on ethylation, room temperature precollection, gas chromatographic separation and detection by cold vapour atomic fluorescence. The detection limit was 1 pg.

Gas Chromatography

Jones and Nickless [34] converted inorganic mercury in fish samples to its methyl derivative using 2,2'-dimethyl-2-silapentane-5-sulfonate as a reagent, prior to the determination of inorganic mercury in benzene extracts of the reaction product by gas chromatography. The highest yield was obtained by digesting the fish sample at 100 °C with 5 N nitric acid in the presence of sodium nitrite and then extracting with benzene. Between 2.8 and 8.6 mg/kg mercury were found in fish samples by this method.

Pyrolysis Ultraviolet Spectroscopy

Thomas et al. [35] described a rapid pyrolytic procedure for determining the total mercury content in fish. A weighed amount of homogenised fish tissue is combusted in a flowing air stream at 900 °C, and then over copper oxide at 850 °C to ensure complete combustion. Elemental mercury vapour is expelled into the carrier stream and, after passing through silver oxide absorbent traps to remove possible interfering gases, is detected and measured in an ultraviolet photometer at 253.5 nm. The relative error is approximately

Table 1.3. Comparison of mercury analysis in various fish specimens. From [35]

Fish	Combustion method total Hg, mg/kg	Digestion method total Hg, mg/kg	Mercury as methyl Hg, mg/kg
Carp	2.7	1.5	2.4
		1.6	2.3
Shiner	0.39	0.33	0.33
		0.28	0.35
Chub	0.19	0.10	0.16
		0.09	0.10
Buffalo	0.33	0.14	0.41
	0.53	0.12	
	0.44		
Carp	0.54	0.28	0.47
	0.64	0.29	
	0.52		
Blue Cat	0.25	0.21	0.21
	0.26	0.27	
Channel Cat	0.52	0.37	0.42
	0.47	0.55	
Carp	0.25	0.31	0.26
		0.34	0.22
Crappie	0.14	0.12	0.09
		0.12	0.11
Crappie	0.20	0.19	0.13
		0.14	0.11

$\pm 10\%$ for inorganic and organic mercury over a linear response range of 0.05 to 3.0 mg/kg.

Thomas et al. [35] compared mercury contents obtained by this method with those based on a gas chromatographic method involving the conversion of inorganic mercury to methylmercury, and with determinations of total mercury by a sulfuric acid–potassium permanganate acid digestion method.

It is seen in Table 1.3 that, whereas total mercury determinations are lower, the pyrolysis method and the gas chromatographic methods give results that are in reasonably good agreement.

Anodic Stripping Voltammetry

Nitric acid–perchloric acid digestion in a Teflon autoclave bomb has been used to prepare digests of finely powdered freeze-dried fish [36]. The extract was irradiated with ultraviolet light to complete fish sample degradation prior to the determination of mercury using a gold disc electrode. Results obtained compared well with total mercury contents obtained by neutron activation analysis.

Neutron Activation Analysis

Uthe et al. [22] found that mercury determinations in fish by digestion-flameless AAS were only slightly lower than those obtained by neutron activation analysis, but had a poorer precision.

Sivasankara-Pillay et al. [37] determined mercury in fish samples by neutron activation analysis. As a further check, the samples were wet-ashed at $120 - 160\,°C$ with sulfuric and perchloric acids in the presence of an accurately known amount of mercury carrier. A preliminary precipitation as mercury sulfide is followed by further purification, and electrodeposition or precipitation as mercuric oxide to isolate mercury. The radioactivities due to ^{196}Hg and ^{197}Hg are then measured by scintillation. The errors in this method are 5% at the 2 mg/kg mercury level and 15% at the 0.01 mg/kg level, with standard deviations of less than 5% at the 5 mg/kg level and less than 17% at the 0.01 mg/kg level, respectively. Fish samples contain both organic and inorganic mercury, predominantly organic. Sivasankara-Pillay [37] showed that freeze-drying of homogenised fish samples caused a 16 - 39% loss of organic mercury compounds, but did not cause any loss of inorganic mercury. Similarly, low-temperature ashing (Tracerlab Model 505 asher) caused an 81 - 98% loss of mercury from fish. Exposure of fish samples to X-rays or neutrons before mercury analysis, in order to convert volatile organomercury compounds to inorganic mercury, reduced mercury losses to 4.5 - 16.4% but did not eliminate them. Low-temperature ($60\,°C$) oven drying caused up to 72% losses of volatile mercury from fish. As a consequence of these findings, Sivasankara-Pillay et al. [29] decided that it was good practice before analysis not to preprocess fish samples to limit their bulk or to reduce their water content, and not to store samples in containers that adsorb mercury onto their surfaces. The procedure they adopted was to keep the samples frozen until use. They were then homogenised using a blender and/or a grinder made of stainless steel or borosilicate glass. The portion of sample used for neutron activation analysis was then vacuum-sealed in a polyethylene bag.

Table 1.4 shows the mercury contents obtained by neutron activation analysis in a survey of fish in Lake Erie.

Table 1.5 presents the results obtained in an interlaboratory comparison of methods for the determination of naturally occurring forms of mercury in fish. It is seen that, in general, the highest results are obtained by neutron activation analysis.

Lo et al. [38] digested wet fish samples with concentrated sulfuric–nitric acids until white fumes appeared, and then added excess potassium permanganate, sodium chloride and hydroxylamine hydrochloride to reduce mercury.

Mercury in the digest was then preconcentrated into a small volume of lead diethyldithiocarbamate dissolved in chloroform. The chloroform was then allowed to evaporate in an ampoule, and the ampoule sealed for neutron activation analysis and subsequent gamma spectrometry of the sensitive

Table 1.4. Mercury content of edible tissues of Lake Erie fish (1970 fall catch). From [37]

Species	Mercury content of edible tissue mg/kg		
	Western basin	Central basin	Eastern basin
Walleye	0.79 (25)[a]	0.65 (25)	0.33 (25)
Yellow perch	0.61 (25)	0.49 (25)	0.29 (25)
White bass	0.60 (25)	0.72 (25)	0.43 (25)
Channel fish	0.36 (25)	0.42 (20)	–
Freshwater drum	0.67 (25)	0.62 (20)	0.30 (25)
Carp	0.23 (25)	0.35 (17)	0.36 (14)
Coho salmon	0.69 (20)	0.58 (10)	0.51 (13)
White sucker	0.55 (24)	0.56 (8)	0.35 (25)
Gizzard shad	0.20 (25)	0.21 (15)	0.26 (18)
Smallmouth bass	–	0.55 (14)	–
Smelt[b]	–	–	0.30 (10)

[a] The numbers in the brackets refer to the number of fish samples of a particular species used when preparing the composite.
[b] Mercury content of the whole fish.

Table 1.5. Results of mercury analysis method evaluation program using fish homogenates[a]. From [37]

Analytical method used	Number of labs that participated	Range of reported values in ppm Hg		
		Sample D	Sample E	Sample G
Flameless (cold) atomic absorption	13	0.09 to 1.80	0.03 to 0.18	2.80 to 5.21
Flame atomic absorption	5	0.70 to 1.80	< 0.05 to 0.49	2.26 to 5.40
Dithizone colorimetry	1	1.31	0.05	3.98
Dithizone titration	1	0.09	< 0.03	0.09
Pyrolysis	2	0.47 to 1.52	0.04 to 0.10	2.00 to 4.25
Neutron activation analysis	6	0.95 to 1.77	0.04 to 0.19	2.83 to 4.60
Cold atomic absorption following acid digestion (Fresh Water Institute, Winnipeg, Canada).		1.46	0.04	4.53
Neutron activation analysis with post-irradiation chemical separation (Western New York Nuclear Research Center).		1.77	0.12	4.56

[a] Trace Mercury Analyses Evaluation Program sponsored by the Fresh Water Institute of the Canadian Fisheries Research Board

[197]Hg peak. As well as reducing the detection limit to 1 μg/kg in fish, precon-centration has the additional advantage of overcoming interferences from [24]Na and [82]Br which commonly occur in fish samples. Recoveries of 95% of mercury in fish samples were obtained by this procedure.

Medina et al. [39] has described a high-performance liquid chromato-graphic method for the determination of mercury speciation in fish.

1.1.8
Nickel

Pihlar et al. [44] have described a voltammetric procedure for the deter-mination of μg/kg (1 ng/l) levels of nickel in various biological materials including fish. The sample is wet-digested with nitric acid (65%) sulfuric acid (98%) in the ratio 5:1 or 2.5:1 at 150–200 °C. Alternatively, a 30-minute digestion with 30% hydrogen peroxide is carried out. The digest is then buffered at pH 9.2 using 0.1–1 M ammonia/ammonium chloride. The dimethylglyoxime complex is then formed and dc or differential pulse voltammetry at −1.25 V is applied to determine the nickel. The method gave 4.6 μg/kg nickel in a fish sample.

1.1.9
Selenium

The semi-automated atomic absorption method [4] for the determination of arsenic in fish described in Sect. 1.1.2 has also been applied to the de-termination of selenium. Average recoveries of selenium(IV), selenium(VI), selenourea, selenomethonine and selenocysteine obtained by this method were between 91 ± 10 and 100 ± 1%. Selenium found in a NBS standard bovine liver (SRM 1577) reference sample was 1106 ± 100 mg/kg against a certified value of 1020 ± 40 mg/kg. Selenium levels found in a range of fish samples ranged from 0.308 to 0.548 mg/kg, determined with a coefficient of variation of 4.5 to 6.0%.

The semi-automated inductively coupled plasma atomic spectrometric technique described by Goulden et al. [6] in Sect. 1.1.2 has also been applied to the determination of selenium in nitric–sulfuric–perchloric acid digests of fish.

Januzzi et al. [45] have reported a method for the determination of sele-nium in fish based on a slurry technique without sample preconcentration.

1.1.10
Strontium

Bagenal et al. [46] have described a method for the determination of μg/g levels of strontium in fish scales. The sample is digested with perchloric acid and the flask heated to destroy organic matter. A nitric acid solution of this digest is used for the determination of strontium by GF–AAS. In general, freshwater trout were found to contain between 76 and 142 µg/kg strontium in their scales, whilst sea trout had much higher levels present in their scales (320 – 653 µg/kg).

1.1.11
Tin

Flameless AAS [47] has been applied to the determination of tin in fish. Between 0.4 and 6.6 mg tin was reported in homogenised fish samples. Sample digestion was carried out using lumatron (a quaternary ammonium hydroxide) dissolved in isopropanol (available from H. Kurenell, Neuberg, Germany).

1.1.12
Vanadium

The combined ion-exchange spectrophotometric procedure [16] described in Sect. 1.1.4 on cobalt earlier in this chapter has been applied to the determination of vanadium in cutlass fish. A recovery of 96.3% vanadium was obtained by this procedure.

Cation exchange chromatography followed by neutron activation analysis has been used [48] to determine down to 30 µg/kg vanadium in fish.

1.1.13
Multi-cation Analysis

1.1.13.1
Atomic Absorption Spectrometry

Various workers have discussed the application of this technique to the determination of elements in fish tissue digests [49–53]. Elements determined include cadmium, lead, copper, manganese, zinc, chromium and mercury [54]; cadmium, zinc, lead, copper, nickel, cobalt and silver [50]; copper, zinc, cadmium, nickel and lead [52]; lead, cadmium, copper and zinc [52]; and lead and cadmium [53].

Various digestion systems have been studied for the decomposition of fish samples prior to analysis, including digestion with nitric acid–perchloric acid [50, 52], nitric acid–hydrogen peroxide [51, 53], all in open tubes, or decomposition with nitric acid in a closed Teflon-lined bomb [49].

Nitric Acid–Sulfuric Acid Digestion

Agemian et al. [52] have reported a simple and rapid digestion method for the simultaneous acid extraction of chromium, copper, zinc, cadmium, nickel and lead from high-fat fish tissue. Samples are digested with nitric (5 ml 16N) and sulfuric (5 ml 36N) acids at 150 °C in a modified aluminium hot-block. The method is specially set-up for fish sample sizes of up to 5 g for low-level detection of these elements. After digestion, acid extracts of the sample are analysed by direct flame AAS for copper, zinc and chromium. The other three elements, cadmium, nickel and lead, are concentrated by chelation with ammonium tetramethylene dithiocarbamate followed by solvent extraction with isobutyl methyl ketone, and determined by flame AAS.

Detection limits in whole fish tissue are 0.02 mg/kg (cadmium), 0.05 mg/kg (nickel), 0.1 mg/kg (lead) and 0.2 mg/kg (chromium, copper and zinc). Recoveries through the whole analytical procedure ranged from 90 to 110%. Precisions were in the range 9.1% to 12.1 (cadmium), 5 – 15% (nickel and copper), 4.3 – 17.0% (lead), 3.9 – 6.7% (zinc) and 7.9 – 15% (chromium).

Nitric Acid–Perchloric Acid Digestions [50]

To carry out this digestion, 0.5 – 3 g of ball mill-ground freeze-dried fish sample is digested in a silica flask with 10 – 20 ml concentrated nitric acid and then 5 – 10 ml of 1 : 1 nitric : perchloric acid to dryness. The residue is dissolved in dilute hydrochloric acid–nitric acid and adjusted to pH 8 with ammonia. This solution is extracted with a 0.02% solution of dithizone in chloroform. Metals are then back-extracted from the organic phase with 2 mol/l hydrochloric acid prior to atomic absorption spectrometry. Using this method, the following values (mg/kg) were obtained for a NBS reference kale sample (nominal values in brackets): cadmium 0.9 (0.84); zinc 29.9 (31.8); lead 2.6 (3.2); copper 4.2 (4.9); cobalt 0.05 (0.056). Concentrations (mg/kg) of metals found in whale tissues were: zinc 26 – 103; lead 0.45 – 1.37; copper 1.2 – 7.6; nickel 0.17 – 0.60; cobalt 0.07 – 0.38; silver 0.02 – 0.04; cadmium, not detected. Kale brought from Iceland contained the following concentrations: zinc 39; lead 0.89; copper 2.6; nickel 0.34; cobalt 0.14; silver 0.04 mg/kg; cadmium, not detected.

Nitric Acid–Hydrogen Peroxide Digestions

Van Hoof and Van San [51] worked on fish samples that had been calcined at 450 °C prior to digesting the ash in 2.5 : 1 v/v 14N nitric acid : 30% hydrogen peroxide. Elements determined included copper, zinc, cadmium and chromium. Low recoveries of at least some of these elements would be expected under these conditions.

Borg et al. [53] digested 10 mg freeze-dried fish livers with concentrated nitric acid at 50 °C for 2 hours in quartz tubes, and then slowly raised the temperature to 110 °C over 18 hours. Hydrogen peroxide (30%) is added to the cooled samples, which are again heated to 110 °C for six hours to digest fats completely. When made-up to a standard volume, this digest was used for the determination of copper, lead, cadmium and zinc by GF-AAS. Table 1.6 compares results for fish livers obtained by this procedure with those obtained via neutron activation analysis. The high metal concentrations found in the livers reflect the fact that the fish were taken in an area subject to heavy contamination originating from ore smelting activities.

Nitric Acid Bomb Digestion

Ramelow et al. [49] determined cadmium, lead, copper, manganese, zinc and chromium in wet fish by digesting a 0.5 – 1.0 g sample with 2 – 3 ml concentrated nitric acid in a Teflon-lined bomb at 150 °C for 1.5 hours. Elements

Table 1.6. Metal concentrations in fish liver determined by the Borg method (AAS) and by neutron activation analysis (NAA). From [53]

| | mg/kg | | | | | |
| | Zn | | Cu | | Cd | |
Sample No.	AAS	NAA	AAS	NAA	AAS	NAA
Perch						
173	120	131	13	17.80	5.1	4.79
174	120	119	12	11.70	3.8	4.31
178	100	107	11	10.70	2.0	2.45
189	120	130	6.7	8.64	6.9	8.08
191	100	115	7.2	8.30	2.8	3.30
358	150	115	6.0	8.38	8.1	7.46
361	110	112	5.7	9.96	5.2	6.73
364	120	124	5.3	8.21	6.8	9.01
368	120	107	3.7	5.69	6.2	7.34
236			22	23.9	2.1	2.1
244			23	21.8	1.5	1.6
249			55	45.0	2.2	2.6
264			48	46.2	4.0	4.0
White fish						
463			27	24.5	0.56	0.51
477			62	56.0	0.90	0.84
482			–	–	0.72	0.71
487			43	39.6	0.19	0.275
Pike			11.7 ± 0.6	10.0	0.17 ± 0.02	0.162
			$(n = 3)$		$(n = 3)$	

Table 1.7. Analytical results from the analysis of trace metals in various marine organisms (results show mg/kg fresh weight). From [49]

Species	Cd	Pb	Cu	Mn	Zn	Cr
White bream	0.04	0.61	1.11	0.51	10.6	0.58
Sardine	0.02	0.57	2.18	1.63	6.3	0.28
Gilt-head bream	0.03	0.68	1.20	–	9.5	0.49
Grey mullet	0.09	1.36	1.70	0.33	12.2	0.10
Horse mackerel	0.17	1.05	0.99	0.63	4.3	0.65
Striped mullet	0.02	0.12	0.68	0.22	6.4	0.14

were determined in the digest by flame atomisation or graphite furnace atomisation AAS. Concentrations found in whole fish in an unpolluted area are shown in Table 1.7, which should be contrasted with concentrations found in fish livers in a polluted area (Table 1.6).

Comparison of Digestion Methods

Adeloju et al. [54] evaluated four commonly used wet digestion procedures for fish and found that a method based on digestion with a mixture of nitric and sulfuric acids gave the best results.

Intercomparison Studies

The International Council for the Exploration of the Sea has arranged a series of intercomparison studies of the determination of trace elements in fish using techniques based on AAS. A summary of the test results is given in Table 1.8.

Despite the large number of participants in the fourth exercise, the results for the analysis of copper, zinc and mercury demonstrated that most analysts were continuing to produce reasonably comparable and accurate data for these metals at levels typical of those found in fish muscle and shellfish tissue. The results for mercury were particularly good in view of the relatively low concentrations present.

The analysis of arsenic appears to have posed problems for some of the analysts in view of the wide range of values reported in the fourth exercise, i.e., 6.27 – 275 µmol/kg. An independent check of arsenic in the sample by neutron activation analysis produced a mean value of 200 µmol/kg with a coefficient of variation of 6%. With the exception of one analyst, who used x-ray fluorescence (mean arsenic concentration of 216 (µmol/kg), all analysts employed a similar, but individually modified, procedure for the analysis of arsenic: following destruction of the organic matter by wet digestion or dry ashing, the arsenic was liberated from the resultant matrix as arsine and then measured by either flame and flameless AAS or colorimetry. If it is assumed that the results produced by x-ray fluorescence and neutron activation analysis represent the true concentration of arsenic in the reference material, then the low results produced by some participants are incorrect. It follows that the methods used by these analysts may suffer from some form of matrix interference.

From an analysis of the arsenic methodology, it appears that the root of the analytical problem may lie with the choice of technique used for the destruction of organic matter. This is suggested by the fact that all methods incorporating a dry ashing step produced high values (> 133 µmol/kg), whereas some methods employing a wet digestion step produced very low values, in the range 6.7 – 119 µmol/kg. Some of the wet digestion procedures which produced high values appear to have overcome the effects of matrix interference through either the addition of nickel salts to the digest before

Table 1.8. Summary of the results from the analysis of reference materials distributed in the ICES metals intercomparison exercise (1971 – 1980) (from author's own files)

Elements	Exercise	No. of partici-pants	Range of mean values submitted, µmol/kg	Grand mean, µmol/kg	SD	CV	Outliers (or qualifications)
Copper	4a	36	< 6 – 63	28	11	39	< 6 as 6
Zinc	4a	36	199 – 566	352	61	18	None
Mercury	4a	33	0.25 – 1.9	1.05	0.35	33	None
	4b	34	< 0.05 – 1.3	0.3	0.15	50	All < values (two) and two high values (1.05 and 1.25) omitted
Cadmium	4a	35	0.05 – 8.8	0.29	0.24	87	All < values (five) and four high values (2.5, 2.8, 3.5 and 8.8) omitted
Lead	5a	52	4.7 – 9.9	7.1	4.8	17	None
	4a	32	0.96 – 36.0	1.0	0.7	71	
	5b	52	1.06 – 37.4	13.0	6.1	47	Two high values (29.3 and 37.4) omitted
	5c	32	0.53 – 15.4	3.6	2.5	71	One high value (15.4) omitted
Arsenic	4a	16	6.7 – 275	196	56	28	Three low values (6.7, 8.4 and 21.3) omitted

Exercise	(Year)	Raw Material	Brief description of preparation of reference material	Elements under study
4a	(1978)	Fish fillet (cod, skinned)	Wet tissue cut into small pieces (3 cm × 3 cm); blast-frozen, freeze-dried and re-peatedly ground in a hammer mill to a fine flour	Cu, Zn, Hg, Cd, Pb and As
4b		Fish fillet (cod, skinned)	Chopped wet tissue washed with dilute acid to reduce Hg content. Freeze-dried and ground into flour as above	Hg only
5a	(1980)	White meat of edible crab	As for 4a	Cd only
5b		Commercial fish meat	As for 4a	Pb only
5c		Hepato-pancrease of lobster	Prepared in the form of acetone powder	Pb only

SD: Interlaboratory standard deviation
CV: Interlaboratory coefficient of variation

measurement by flameless AAS or the utilisation of a much stronger reducing agent at the arsine generation stage. It appears that some component(s) of the matrix, which is destroyed or eliminated during dry ashing but not during wet digestion, suppresses the release of arsenic as arsine and also suppresses the arsenic signal in flame and flameless AAS unless nickel salts are added to the digest prior to measurement.

The results from the fifth exercise show that the majority of participants can produce comparable (i.e., interlaboratory coefficient of variation of 10%) and accurate data for cadmium but not for lead.

In conclusion, over the nine years of the intercomparison exercise, a progressive improvement was shown in the determination of copper, zinc and mercury. Difficulties were still being encountered in relation to producing comparable data for lead and cadmium at low tissue concentrations in the range 0.05 – 0.9 µmol/kg, but at higher tissue concentrations (2 – 12 µmol/kg) there have been few problems producing accurate data for cadmium, although somewhat greater difficulties in the case of lead.

Ramelow et al. [49] digested fish samples with concentrated nitric acid in a Teflon-lined bomb for 1.5 hours at 150 °C prior to the determination of mercury by reduction to elemental mercury with stannous chloride and determination by cold vapour AAS.

1.1.13.2
Hydride Generation Atomic Absorption Spectrometry (HG-AAS)

Welz and Melcher [55] decomposed fish tissue with nitric–sulfuric–perchloric acids in a Teflon-lined bomb to decompose arsenic, selenium and mercury. Nitric acid alone gave low recoveries for arsenic and selenium but quantitative recovery for mercury. The final determination of down to 0.3 mg/kg arsenic, 0.2 mg/kg selenium and 0.005 mg/kg mercury was carried out by hydride generation and cold vapour AAS.

1.1.13.3
Isotope Dilution Coupled Plasma Mass Spectrometry (IDC–PMS)

Buckley and Ihnat [56] determined trace elements in fish samples by isotope dilution ICP–MS.

1.1.13.4
Inductively Coupled Plasma Atomic Emission Spectrometry (ICP–AES)

The ICPES [57] procedure has been applied to the determination of arsenic, antimony and selenium in fish. Sample digestion was carried out in open vessels at room temperature using nitric acid, followed by heating with a mixture of nitric, perchloric and sulfuric acids on a hot plate. Accurate determinations (mg/kg) were obtained by this procedure. On NBS Reference Sample NBS 1566 (oyster tissue), for arsenic the method gave (certified values in brackets) 11.1 ± 1.1 (13.4+0.9), antimony 0.42 ± 0.3, and selenium 1.7 ± 0.2 S ((2.1 ± 0.5) and for NBS 1571 (orchard leaves) it gave for arsenic 11.9 + 0.6 (10.2 ± 2.0), and antimony 2.8 + 0.02 (0.08 ± 0.01).

Sakai and May [58] used ICPAES, AAS and hydride generation AAS to determine cadmium, arsenic, boron, chromium, mercury, molybdenum, nickel, lead and selenium in common carp. The highest concentrations found were: arsenic 1.5 mg/kg, boron 20 mg/kg, cadmium 0.27 mg/kg, chromium 2.2 mg/kg, mercury 2.9 mg/kg, molybdenum 3.6 mg/kg, nickel 2.2 mg/kg, lead 2.3 mg/kg and selenium 5.5 mg/kg.

1.1.13.5
Differential Pulse Anodic Stripping Voltammetry (DPASV)

Adeloju et al. [59] used this technique to determine selenium, copper, lead and cadmium in fish tissues. Detection limits were in the µg/kg range. Samples were first digested with concentrated nitric acid and 80% magnesium nitrate solution, and then dry ashed at 500 °C. The ash was dissolved in boiling 6 mol/l hydrochloric acid. This solution was analysed for selenium on a hanging mercury drop polarographic analyser, and copper, lead and cadmium were determined in the anodic scanning voltammetry mode using the peaks appearing at −0.20, −0.5 and −0.7 V versus SCE, respectively. Results (mg/kg) obtained by this method for crayfish are in good agreement with certified values (reference values in brackets): selenium 0.17 (0.16), copper 3.46 (3.10), lead 0.48 (0.48) and cadmium 0.10 (0.05). Relative standard deviations in determinations of selenium, copper, lead and cadmium were 12, 5, 15 and 20%, respectively.

1.1.13.6
Neutron Activation Analysis (NAA)

This technique has been applied to the determination of cobalt, chromium, selenium, silver, rubidium, nickel and zinc [60], and aluminium, gold, bromine, calcium, chlorine, cobalt, chromium, copper, iron, iodine, potassium, magnesium, manganese, sodium, rubidium, scandium, vanadium and tungsten [61] in fish.

Neutron activation analysis has been used to determine miscellaneous elements in fish at sub-ng/g concentrations [62].

1.1.13.7
Secondary Ion Mass Spectrometry and X-Ray Spectrometry (SIMS/XS)

This technique can be used [63] to provide simultaneous morphological and chemical identifications in histological sections of fish, molluscs and crustaceans.

1.1.13.8
Miscellaneous

Topping (private communication) has reviewed methodology for the determination of copper, zinc, mercury, cadmium and lead in fish flesh, fish flour and shellfish, and has organised intercomparison tests. Various techniques were applied, including neutron activation analysis, x-ray fluorescence and AAS.

Throughout this study, the participants, particularly those who took part in all of the exercises, showed a progressive improvement in the analysis of copper, zinc and mercury. On the basis of these results, it is concluded that the analytical data produced by these participants for these metals in a fish and shellfish monitoring programme are comparable.

The identification of significant differences in the concentrations of working standards and the subsequent adoption of a common procedure for the preparation of these solutions are considered to be important factors in the achievement of this improved comparability for the above metals.

The study revealed that the participants were unable to produce comparable, and in most cases accurate, data for lead and cadmium at low tissue concentrations, i.e., in the range 0.024 – 4.8 µmol/kg and 0.009 – 0.89 µmol/kg, respectively. However, at relatively high tissue concentrations (2.5 – 12.0 µmol/kg and 10 pmol/kg, respectively), the majority of analysts experienced little difficulty in producing accurate data for cadmium, but the analysis of lead presented some problems for a minority of the participants. On the basis of these results, it is considered that the participants in ICES fish and shellfish monitoring programmes can produce comparable data for cadmium in shellfish tissue but not for cadmium in fish muscle or lead in both fish muscle and shellfish tissue.

Das [64] has reviewed the trace metal status of methods used in marine biological samples.

Arslan et al. [65] have shown that Toyo Peasil AF-Chelate 650M gives a 25-fold preconcentration in the determination of metals in juvenile bluefin tuna fish in the Pacific Ocean.

1.2
Organic Compounds

1.2.1
Hydrocarbons

Farrington et al. [66] used column chromatography and thin-layer chromatography to isolate hydrocarbons (arising from marine contamination) in fish lipids. The hydrocarbon extracts were then examined to select those that could be determined by gas chromatography mass spectrometry, by combinations of spectrophotometric methods, or by wet chemistry. As a

Table 1.9. Reproducibility of pristane analyses. From [66].

Analyst	Technique	Pristane concentration, mg/kg lipid
Cod liver sample A	Saponification – TLC, GC	30.1
Quinn, Wade	Saponification – TLC, GC, 3 analyses by GC	35.8 ± 1.6
Farrington	Saponification – CC, 2 analyses by GC	39.4 ± 2.3[a]
	No saponification – CC, GC, Subsample A-1	36.4 ± 1.6[a]
	No saponification – CC, GC, Subsample A-2	37.3 ± 1.0[a]
		Mean $= 35.7 \pm 3.5$
Cod liver sample B		
Farrington	No saponification – CC, GC,	
	Subsample 1	38.7 (n-C_{14})[b]
		40.3 (n-C_{28})
	Subsample 2	37.1 (n-C_{14})
		39.7 (n-C_{28})
		Mean $= 39.0 \pm 1.2$
IDOE-5		
Quinn, Wade	Saponification – TLC, GC, 2 analyses 259, 268	Mean $= 264$
Teal, Burns	Saponification – CC, 1 analysis	276
Farrington	Saponification – CC, GC, 6 analyses, range 225-308	Mean $= 272$
	Mean for 3 laboratories	271 ± 4.5
Tuna meat		
Quinn, Wade	Saponification – TLC, GC, 2 analyses 3.0, 3.6	Mean $= 3.3$
Teal	Saponification – CC, GC, 1 analysis	2.0
Farrington	Saponification – CC, GC, 2 analyses, 1.9, 2.0	Mean $= 2.0$
	Mean for 3 laboratories	2.4 ± 1.5

TLC: Thin-layer chromatography
CC: Column chromatography
GC: Gas chromatography
[a] Mean $\pm 1\sigma$ estimated from 2 or 3 analyses by GC of hydrocarbons isolated from the same sample
[b] Internal standard used to calculate concentration. Both n-C_{14} and n-C_{28} addedto subsamples as internal standard

screening method, gas chromatography was shown to be fairly accurate and precise for hydrocarbons boiling in the range 287–450 °C and of suitable polarity.

Farrington et al. [66] also described a gas chromatographic method for the determination of hydrocarbons such as petroleum cuts, fuel oil and lubricating oils in tuna meat and cod liver lipid extracts. The cod liver oil samples used in this study were spiked with known amounts of various commercial hydrocarbon products and subjected to analysis in order to check

Table 1.10. Results from hydrocarbon analyses of intercalibration sample IDOE-5: $x =$ 372 mg petroleum per kg cod liver lipid (µg hydrocarbons per g cod liver lipid). From [66]

Analyst	Petroleum hydrocarbon peaks and unresolved complex mixture	Peaks[a]	Unresolved complex mixture	Pristane
Subsample date, January 1972. Analyses, January to October 1972				
Quinn, Wade[a]	373	85.5	288	264
Teal, Burns[a]	438	87.7	350	276
Farrington	407	64.4	343	272
Mean standard deviation	406 ± 26	79.2 ± 10.5	327 ± 27.7	271 ± 4.5
Subsample date, August 1972. Analysis, February 1974				
Medeiros[b]	455	71	384	267
Subsample date, October 1972. Analysis, February 1974				
Medeiros[b]	426	59	367	270
Subsample date, June 1974. Analysis, November 1974				
Quinn, Wade	474	78	396	262
	493	108	385	256
	676	63	613	272

[a] Does not include pristine or squalene
[b] taken from [67]

analytical recoveries. In one recovery procedure, the cod liver oil sample was refluxed with methanolic 1 N potassium hydroxide to saponify esters. An ether extract of the digest was prepared, the ether was removed, and the residue dissolved in chloroform. This extract was subjected to TLC and gas chromatography. Further cod liver oil samples were chromatographed on a column consisting of layers of alumina and silica gel, elution of hydrocarbons being carried out with a 5% solution of benzene in pentane. Solvent was then removed from the eluate in vacuo, and the residue dissolved in a small volume of carbon disulfide. This extract was then gas chromatographed to give the distribution of hydrocarbons present.

Table 1.9 gives the results from naturally occurring pristane determinations in unspiked cod liver oil and tuna meat samples analysed by different methods. Agreement and precision are good for cod liver oil but not so good for tuna meat.

Table 1.10 shows the results obtained for a sample of cod liver [66,67] spiked with a known amount of crude oil. The measured concentration of petroleum hydrocarbons is in fair agreement with the actual concentration spiked in the sample: 372 mg/kg lipid.

Law [68] and Chesler et al. [69] have discussed the application of GC–MS to the determination of traces of hydrocarbons in Chesler et al. [6] dynamic headspace analysis of an alkaline digest of sample to prepare extracts for gas chromatography.

Picer [70] has reported a spectrofluorimetric method for the determination of petroleum hydrocarbons in benthic organisms.

Polyaromatic Hydrocarbon

Vassilaros et al. [71] have described a method for determining polyaromatic hydrocarbons down to 0.2 µg/kg in fish tissue. The analytical procedure includes the following steps: aqueous alkaline digestion, acidification of the digest with glacial acetic acid, extraction with methylene chloride, liquid–liquid partitioning with water and then a 10% potassium hydroxide solution, adsorption chromatography on basic alumina using hexane, benzene and chloroform sequentially, gel permeation chromatography on BioBeads with methylene chloride, capillary gas chromatography using nitrogen- and sulfur-specific detectors, and GC–MS. An average recovery of 72% of spiked ^{14}C-labelled anthracene was obtained.

Figures 1.1 and 1.2 show capillary gas chromatograms obtained using this procedure for the polyaromatic hydrocarbon (PAH) and polycyclic aromatic sulfur heterocycle (PASH) fractions of a brown bullhead catfish, obtained from the Black River in Ohio (heavily polluted), and from a similar fish obtained from Buckeye Lake, a relatively unpolluted area.

Table 1.11 lists the peak numbers, compound names, retentive indices and intensities of the peaks numbered in Figs. 1.1 and 1.2. The fraction obtained from the fish from the Black River (Fig. 1.1) consisted primarily of polyaromatic hydrocarbons ranging from 2-methyl-naphthalene to benzo[*ghi*]perylene, with fairly high levels of polycyclic aromatic sulfur heterocycles. The major components are acenaphthylene, dibenzofuran, fluorene, phenanthrene, fluoranthene and pyrene. The lower detection limit for this sample was about 0.5 µg/kg.

A bullhead catfish from a pristine area (resort lake, no industry, agricultural) (Fig. 1.2) had a total polyaromatic hydrocarbon content of about 38 µg/kg.

Birkholz et al. [72] have described a method for the extraction, clean-up and high-resolution gas chromatographic analysis of polycyclic aromatic hydrocarbons, polycyclic aromatic sulfur heterocycles and basic polycyclic aromatic nitrogen heterocycles in fish. Soxhlet extraction of fish tissue with methylene chloride was followed by gel permeation chromatography using Bio-beads SX-3.

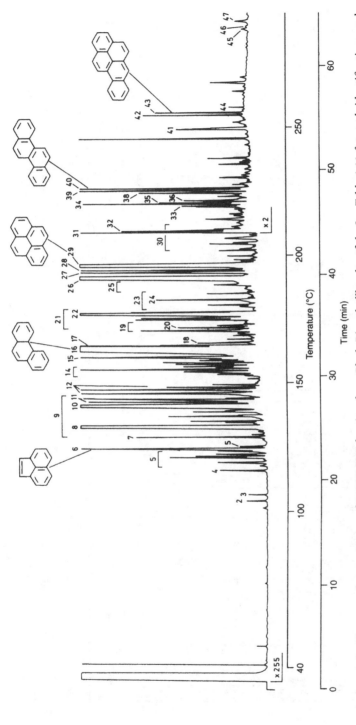

Figure 1.1. Capillary gas chromatogram of PAH/PASH fraction from Black River bullhead catfish. See Table 1.11 for peak identifications and text for chromatography conditions. From [71]

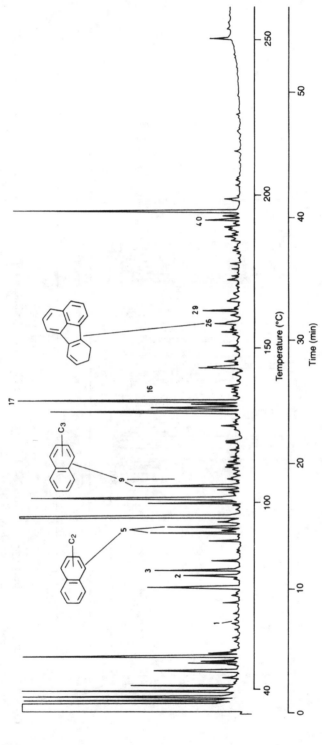

Figure 1.2. Capillary gas chromatogram of PAH/PASH fraction from Buckey Lake bullhead catfish. See Table 1.11 for peak identifications and text for chromatography conditions. From [71]

Table 1.11. PAH and PASH determined in two fish samples. From [71]

Peak no.	Black River	Buckeye Lake	Retention Index	Compound name
1			200.000	Naphthalene
2	6	5	220.400	2-Methylnaphthalene
3			223.240	1-Methylnaphthalene
4	14		236.237	Biphenyl
5	100	17		C_2-Naphthalenes
6	270		246.557	Acenaphthylene
7	39		252.792	Acenaphthene
8			258.546	Dibenzofuran
9				C_3-Naphthalenes
10			269.476	Fluorene
11				Methylbiphenyls or methylacenaphthenes
12				Methyldibenzofurans
13				C_2-Biphenyls
14				Methylfluorenes
15	270		295.323	Dibenzothiophene
16	2700	2	300.000	Phenanthrene
17			301.162	Anthracene
18				Naphtho [2, 3−b] thiophene
19	17			Methyldibenzothiophenes
20			316.350	1-Phenylnaphthalene
21				Methylphenanthrenes
22			321.789	4 H-Cyclopenta[def]phenanthrane
23				C_2-Dibenzothiophenes
24			333.516	2-Phenylnaphthalene
25				C_2-Phenanthrenes
26	1800	4	344.372	Fluoranthene
27			347.669	Acephenanthrylene
28	78		348.644	Phenanthro [4, 5-bcd] thiophene
29	1500	4	351.263	Pyrene
30				Methylfluoranthenes and methylpyrenes
31			366.811	Benzo[a]fluorine
32			369.458	Benzo[b]fluorine
33	6		389.063	Benzo[b]naphtho[2, 1-d]thiophene
34			389.768	Benzo[ghi]fluoranthene
35			391.245	Benzo[c]phenanthrene
36			392.546	Benzo[b]naphtho[1, 2−d]thiophene
37			395.594	Benzo[b]naphtho[2, 3-d]thiophene
38			396.341	Cyclopenta[cd]pyrene
39	22		398.782	Benz[a]anthracene
40	61	6	400.000	Chrysene
41				Benzofluoranthenes
42	14		452.294	Benzo[e]pyrene
43	7		453.939	Benzo[a]pyrene
44	8		457.128	Perylene
45			493.250	Indeno[1,2, 3-cd]pyrene
46				Dibenzanthracenes
47			500.227	Benzo[ghi]perylene

For polyaromatic hydrocarbons and polycyclic aromatic sulfur hetero-cycles, clean-up was by absorption chromatography on Florisil and elution with hexane. Basic polycyclic aromatic nitrogen heterocycle clean-up was by partitioning with hydrochloric acid and chloroform, increasing the pH to 11 with sodium hydroxide, and extracting with chloroform. Samples were anal-ysed using capillary gas chromatography with flame ionisation detection or with mass spectrometry. Similar results were obtained with the two meth-ods. Average recoveries for these three classes of compounds were 87, 70 and 97% from fish tissue fortified at 0.24 – 1.1, 0.024 – 0.11 and 1.2 – 1.4 mg/kg, respectively.

Ariese et al. [73] applied chemical derivativisation with methyl iodide and Shpol'skii spectrofluorimetry to the determination of benzo[a]pyrene in amounts down to 0.005 mg/l in flounder fish bile.

Synchronous fluorescence spectroscopy has been used as a screening method for biomonitoring of polyaromatic hydrocarbons in fish. An analysis could be completed in 3 minutes [74].

Ariese [75] has reviewed the application of fluorescence spectroscopy to the determination of polyaromatic hydrocarbons and their metabolites in fish and crustacea.

Subcritical water extraction has been used to extract polyaromatic hy-drocarbons from fish [72].

1.2.2
Phthalate Esters

Giam et al. [76] determined phthalate esters in amounts down to less than 5 μg/kg in marine fish using capillary column gas chromatography with an electron capture detector. Chlorinated insecticides and chlorinated biphenyls interfere in this chromatographic analysis and consequently must first be removed by column chromatography on water deactivated Florisil.

The tissue was macerated with acetonitrile, then diluted with methylene chloride-petroleum ether (1:5) and extracted with saltwater. The dried or-ganic phase was concentrated and diluted with *iso*-octane and subjected to clean-up in a Florisil column. Elution of the Florisil column with 6%, then 15%, then 20% diethyl ether in petroleum ether provided three fractions containing: (i) chlorinated insecticides and chlorinated biphenyls, (ii) di-ethylhexylphthalate and dibutylphthalate, and (iii) dibutylphthalate.

Extreme precautions are necessary in this procedure to avoid contami-nation due to phthalates present as impurities in commonly used laboratory materials, e.g. aluminium foil contains 300 mg/kg phthalate.

Between 2 and 20 μg/kg of diethylhexyl phthalate was found by this procedure in various types of fish taken in the Gulf of Mexico.

1.2.3
Chlorine Compounds

1.2.3.1
Aliphatic Chlorohydrocarbons

Various workers have discussed the application of gas chromatography to the determination of chlorinated aliphatics in fish [77–83]. Compounds that have been determined include trichloroethane, tetrachloroethylene, chloroform and carbon tetrachloride [84, 85], and 1,1,1- trichloroethaner-trichloroethylene, perchloroethylene, 1,1,1,2-tetrachloroethane, 1,1,2,2-tetrachloroethane, pentachloroethane, hexachloroethane, pentachlorobuta-diene, hexachlorobutadiene, chloroform and carbon tetrachloride [86–89].

Hiatt [90] has described a GC–MS system for the determination of a wide range of volatiles, including haloparaffins, in vacuum distillates of fish samples. Linde et al. [91] determined organohalogen compounds as halogens in fish samples. The samples were steam distilled with cyclohexane for halogen-containing nonpolar compounds, and hexane extracts of oils from all species were treated with concentrated sulfuric acid. Total amounts of halogens in the original oils, in the volatile compounds in the cyclohexane distillate, and in the sulfuric acid-treated hexane extracts were determined by neutron activation analysis. The total level of organic chlorine ranged from 30 to 240 ppm: 2 – 10% of the chlorinated compounds were volatile, and from 5 to 50% of the chlorinated compounds remained after acid treatment. This chlorine exceeded the amount of chlorine in polychlorinated biphenyl by a factor of 1.5 to 5, and most of the chlorine in untreated and acid-treated lipids could not be accounted for as known compounds.

Coelhan [92] determined short-chain (C_{10}-C_{13}) chlorinated paraffins in fish using short-column gas chromatography electron capture negative ion chemical ionisation mass spectrometry.

After a clean-up procedure with a silica gel minicolumn and gel perme-ation chromatography, detection was achieved by short-column gas chro-matography/electron capture negative ion mass spectrometry. The quan-tification was performed by reintegration of selected ions from full-scan spectra. Without chromatographic separation, all of the short-chain poly-chlorinated paraffins eluted from the column as only one peak. This leads to better sensitivity and makes it easier to survey the spectra. A great number of C_{10}, C_{11}, C_{12} and C_{13} polychlorinated alkanes with different chlorine con-tents were used as standards. Detection limits in the full-scan mode varied between 10 and 100 pg depending on the chlorination grade of the alkanes. The results show that polychlorinated decanes are the dominant residues in most of the investigated fish samples.

Tomy et al. [93] published a method in which, in contrast to other studies, a high-resolution mass spectrometer was utilised. Noninterferences between chlorinated paraffin (CP) and other organochlorine compounds were observed at a resolution of 12,000, significant below a resolution of

1000. The detection of [M–Cl] ions from C_{13} CP was performed in seven retention windows. Commercially available C_{10}–C_{13} polychlorinated paraffins with chlorine contents of 60 and 70% were used as standards. The method has a linear range of 0.5 – 500 ng and a detection limit of 60 pg.

De Boer et al. [94] determined tris(4-chlorophenyl) methanol and tris(4-chlorophenyl)methane in aquatic organisms using a method based on Soxhlet extraction and gel permeation chromatographic fractionation over silica gel followed by GC–MS.

Vetter et al. [95] separated chiral organochlorine compounds in fish using modified cyclodextrin phases.

1.2.3.2
Polychlorinated Styrenes

Polychlorinated styrenes have been determined in fish [96–98]. Steinwandter and Zimmer [97] used analysis by GC–MS and mass fragmentography after liquid chromatographic fraction in silica gel to identify 14 isomeric polychlorinated styrenes in Rhine fish via negative ion chemical ionisation mass spectrometry. Ramdahl et al. [98] determined polychlorinated styrenes in fish samples. The fish muscle and liver samples were homogenised and extracted with equal parts of cyclohexane and isopropanol. Concentrated extracts were treated with concentrated sulfuric acid before undergoing GC–MC analysis. Separation of chlorinated styrenes was achieved using gas chromatography columns of fused silica with the injector and detector temperature at 250 °C, and the temperature was programmed at 60 °C for two minutes and then 4 °C per minute to 250 °C. All eight possible chlorostyrenes with fully chlorinated aromatic nuclei were identified in codfish samples by this method. Gas chromatography with electron capture detection could be used for quantification.

1.2.3.3
Chlorophenols

Stark [99] has described a gas chromatograph method for the determination of pentachlorophenol as the trimethyl silylether in amounts down to 0.5 mg/kg in fish. Rudling [100] determined pentachlorophenol in fish and water by an electron capture gas chromatographic method. In this method, a sample of fish tissue (1 g) in water is transferred to a centrifuge tube with 5 ml water, 1 ml 6 mol/l sulfuric acid and 5 ml 1 : 5 isopropyl alcohol-hexane. The tube is centrifuged and cooled in ethanol-solid carbon dioxide. The organic layer is decanted and extracted with 0.1 mol/l $Na_2B_4O_7$. Hexane (0.5 ml) and fresh acetylation reagent (pyridine (2 ml) plus acetic anhydride (0.8 ml) (40 µl) are added to the combined aqueous extracts. The hexane phase is analysed by gas chromatography on a glass column (1 m × 1.5 mm) packed with 5% of QF-1 on Varaport 30 (100- to 120-mesh).

Renberg [101] has used an ion exchange technique for the determination of chlorophenols and phenoxyacetic acid herbicides in fish tissue. The sample (5 g) is homogenised in a mixture of hexane and acetone (5 + 10 ml). The liquid is dropped into a separatory funnel containing 1.0 mol/l hydrochloric acid (5 ml). The funnel is shaken and the upper phase transferred into a centrifuge tube. Sodium sulfate (100 – 300 mg) is added to bind any water present. After centrifugation, the extract is transferred into a weighed flask, the sodium sulfate is rinsed with diethylether (2 ml), and the solvents gently evaporated on a water bath in a nitrogen stream. The flask is reweighed and the fat content calculated; the fat is then dissolved in benzene (about 1 ml per 25 mg fat). After treatment with ion exchange resin, the chlorophenols in the extract are converted to their methylethers using diazomethane and determined by gas chromatography. Detection limits for a 10 g fish sample are 0.1 to 1 µg/kg.

Hoben et al. [102] have described a gas chromatographic technique for determining 0.1 ppb pentachlorophenol in fish tissues. Confirmation of the identity of the chlorophenol was provided by GC–MS. In this method, the pentachlorophenol is extracted from the acidified sample with n-hexane and then re-extracted into a borax solution. It is then acetylated by extracting with n-hexane containing acetic acid anhydride and pyridine. The resulting pentachlorophenyl acetate is analysed by gas chromatography using an electron capture detector.

This extraction procedure gave 83 – 91% recovery of pentachlorophenol from fish. The method was successfully used to determine pentachlorophenol at the 0.15 – 3 mg/kg level in fish. Confirmation of the identity of the chlorophenol was established by a combined gas chromatographic–mass spectrometric analysis.

Thin-layer chromatography and gas chromatography have been used to determine microgram levels of pentachlorophenol, trichlorophenol isomers and 2,4-dichlorophenol in fish tissue [103].

1.2.3.4
Hexachlorobenzene

Residues of hexachlorobenzene in fish have been determined at the µg/kg level using gas chromatography combined with mass spectrometry [104].

1.2.3.5
Chlorinated Fatty Acids

Chlorinated fatty acids have been determined in fish lipids [105]. These account for 90% of the extractable organically bound chlorine in fish.

Laramee and Deinzer [410] determined hexachlorobenzene and Aroclor in fish muscle using a method based on trochoidal electron monochromater mass spectrometry.

1.2.3.6
PCB Polychloroterphenyls

Thin-Layer Chromatography

Sackmauerova et al. [106] determined polychlorobiphenyls (PCB) in water
and fish by TLC on silica gel plates impregnated with 8% paraffin oil. A
mixture of acetonitrile, acetone, methanol and water (20 : 9 : 20 : 1) was used
as a mobile phase. A solution of silver nitrate and 2-phenoxyethanol followed
by irradiation with UV light was used for detection. The detection sensitivity
for Aroclor 1242 is 0.5 – 1.0 μg.

Szelewski et al. [107] claim that some loss of PCB homologues occurs
during the chromium trioxide extraction of fish tissue. The biphenyl-free
PCB extract is then perchlorinated using antimony pentachloride at 200 °C.
Following acidification and toluene extraction, the aqueous phase remaining is extracted with hexane, and this extract is passed down an anhydrous
sodium sulfate microcolumn and concentrated in a Kuderna-Danish evaporator prior to gas chromatography. Comparison of the gas chromatograms
of the polychlorobiphenyls thus obtained with those obtained for the per-
chlorination product of an authentic sample of PCB (e.g., Aroclor 1260)
enables the types of PCBs in the sample extract to be identified.

Tausch et al. [109] determined PCB in fish from the Danube River by capillary gas chromatography and mass spectrometry. Altogether, 40 separate
peaks representing various PCB isomers were identified, the total PCB concentration in the sample amounting to 6.1 mg/kg, with much lower amounts
of pesticide residues and DDT breakdown products.

Bush and Barnard [109] analysed fish eggs by gas chromatography for
78 PCB congeners, hexachlorobenzene, octachlorostyrene, p,p'-DDE and
Mirex. Concentrations of the substances could be detected at 1 mg/kg accurately using 10 mg samples. The precision was better than ± 5% relative
standard deviation.

A gel permeation chromatography clean-up procedure has been used
to determine specified PCBs in fish [110] prior to analysis by capillary gas
chromatography. Gel permeation chromatography gave identical results to
those obtained by the conventional saponification clean-up procedure.

Gaskin et al. [111] have described a gas chromatographic method for
determining DDT, Dieldrin, and PCBs in the organs of whales and dolphins.
Total DDT in blubbers ranged from 1.25 to 7.4 ppm, Dieldrin in blubber from
0.007 to 0.04 ppm, and PCB in blubber from 0.69 to 5.0 ppm.

Luckas et al. [113] have described a method for determining PCBs and
chlorinated insecticides in environmental samples by the simultaneous use
of electron capture gas chromatography and derivatisation gas chromatography. The method is based on the different stabilities of chlorinated insecticides and PCBs towards magnesium oxide in a microreactor. Extracts of
samples are injected twice, first into a regular gas chromatograph and then
into a gas chromatograph equipped with a microreactor for derivatisation. A

Table 1.12. Stability of chlorinated insecticides and PCB. From [112]

Substance	Treatment with conc. H_2SO_4	Treatment with ethanolic KOH
Aldrin	+	+
Dieldrin	–	+
Endrin	–	+
Endosulfan	–	–
HCH isomers	+	–
PCB	+	+
p,p'-DDT	+	→ *p,p'*-DDE
o,p-DDT	+	→ *o,p*-DDE
p,p'-DDE	+	+
o,p-DDE	+	+
p,p'-DDD	+	→ *p,p'*-DDMU
o,p-DDD	+	→ *o,p*-DDMU
p,p'-DDMU	+	+
o,p'-DDMU	+	+

+ Unchanged
– Decomposed (products of decomposition are not detected)
→ Dehydrochlorination to the olefin

'basic' chromatogram and a 'derivatisation' chromatogram are obtained, and the combination of the two chromatograms provided a satisfactory solution.

Chemical derivatisation of sample extracts is very convenient. The extracts containing insecticides and PCBs, after the first injection into the gas chromatograph, are treated with derivatisation reagents, the insecticides being converted into derivatives while the PCBs remain unchanged. Table 1.12 demonstrates the stability of chlorinated insecticides and PCBs towards reagents used for chemical derivatisation.

Based on these considerations, Luckas et al. [112] developed their microreactor gas chromatographic technique in which derivatisation is carried out in situ. Preheated magnesium oxide affects the rapid quantitative dehydrochlorination of saturated DDT metabolites to the corresponding DDT olefins [113]. The derivatisation products immediately obtained in the gaseous phase by means of the microreactor (with nitrogen as the carrier gas and magnesium oxide as the catalyst) are comparable with the products of chemical derivatisation with an alkali in the liquid phase, and substances that are stable to treatment with alkali are also not decomposed in the microreactor (Table 1.12). Two gas chromatographs with an all-glass system and an electron capture detector were used. One chromatograph was equipped with a microreactor for the derivatisation gas chromatography. Luckas et al. [112] used sample extracts obtained by digestion with perchloric acid–acetic acid extraction with *n*-hexane and clean-up with sulfuric acid [114, 115].

In the chromatogram of an extract of a fish sample obtained by this method, the peaks of *γ*-HCH and DDT metabolites appear, but the background suffers from interference from peaks of PCBs. After derivatisation, the peaks of γ-HCH and the saturated DDT metabolites disappear. The saturated DDT metabolites (p,p'-DDT, p,p'-DDD, and o,p-DDT) are converted quantitatively into the corresponding DDT olefins (p,p'-DDE, p,p'-DDMU, and o,p-DDE). The main peak in the 'derivatisation' gas chromatogram represents the sum of p,p'-DDT and p,p'-DDE from the 'basic' chromatogram, and is often sufficient to determine the total DDT content. The content of PCBs can be calculated in the 'derivatisation' gas chromatogram without interference effects due to saturated DDT metabolites.

Rapid derivatisation in the gaseous phase for the determination of DDT metabolites has been carried out by means of catalytic reduction (carbon skeleton chromatography with hydrogen as carrier gas). This method has the disadvantage that a flame ionisation detector is used, which has insufficient sensitivity [116, 117].

Prescott and Cooke [118] have described the application of carbon skeleton gas chromatography to the analysis of environmental samples containing residues of organochlorine insecticides, PCBs and polychlorinated naphthalenes. Their results suggest that extraction by steam distillation followed by carbon skeleton gas chromatography with either a flame ionisation or mass spectrometric detector is a practical method of determining organochlorine compounds in the environment. In this procedure, sulfur, nitrogen, oxygen and halogens in the sample are replaced by hydrogen and the unsaturated bonds formed are saturated by hydrogenation over a palladium–platinum catalyst prior to passing the vapours into the gas chromatograph.

A gas chromatograph fitted with dual flame ionisation detectors was employed. The catalyst was packed into the part of the column that passed through the injection port heater and was thus maintained at the required temperature. Hydrogen was used as the carrier gas. Products were identified by a mass spectrometer linked to the gas chromatograph. At low temperatures hydrogenation of the aromatic rings tended to occur. As the temperature of the catalyst was increased from 140 to 305 °C, there was a progressive decrease in the formation of bicyclohexyl and phenylcyclohexyl and an increase in the yield of biphenyl. At 305 °C biphenyl was the only product from Aroclors. It is likely that at low temperatures loss of chlorine is followed by or coupled with hydrogenation of the aromatic rings. At higher temperatures a secondary reaction involving the dehydrogenation of the cyclic system is also present.

Dechlorination of polychloronaphthalenes was much easier. At catalyst temperatures of less than 280 °C hydrogenation of the rings was less pronounced, although gas chromatographic–mass spectrometric studies indicated that some tetrahydronaphthalene was present. At 305 °C polychloronaphthalenes were quantitatively converted into naphthalene. Using this technique, polychlorinated terphenyls were converted into a mixture of o-, m-, and p-terphenyl at 305 °C. Using a 5% platinum catalyst, conversion of polychloronaphthalenes was poor. At 205 °C, naphthalene gave two compounds, which, from mass spectrometric studies, were suggested to be tetra- and decahydronaphthalene. As the temperature increased the peak heights decreased, and at 305 °C no peaks remained; presumably the naphthalene skeleton was completely destroyed at this temperature. The results for polychloronaphthalenes were similar to those for naphthalene. PCBs were also completely destroyed at 305 °C. At 280 °C small amounts of bicyclohexyl, phenylcyclohexyl and biphenyl were eluted. As the temperature was decreased, the amount of bicyclohexyl increased. At 180 °C, conversion into bicyclohexyl was quantitative, and polychloronaphthalenes were only converted into decahydronaphthalene. Initially a 5% SE-30 column was used by Prescott and Cooke [118] to separate the biphenyl and naphthalene after catalysis. Later gas–solid chromatography with a rubidium chloride column was employed, as inorganic salts give excellent separations of hydrocarbons and also very good reproducibility over a long period of time.

Ling et al. [119] used a matrix solid-phase extraction procedure with gas chromatography–electron capture detection to determine PCB in fish.

Ling and Teng [120] showed that PCB and chlorinated insectides could be simultaneously determined in fish by simultaneous extraction and clean-up followed by GC with electron capture detection, with GC–MS used to confirm identities.

Ewald et al. [121] showed that for the extraction of both PCB and total lipids from fish, the Bligh and Dyer method recovered a greater amount of total lipid, but a lesser quality of total PCB then Soxhlet extraction.

Hajslova et al. [122] employed two-dimensional GC for the GC–electron capture determination of PCB in fish. The number of chromatographically unresolved sample components was then significantly reduced.

Wiberg et al. [123] carried out enantioselective GC–MS of methyl/sulfonyl PCB.

Four different commercially available cyclodextrin (CD) capillary gas chromatography columns were tested for the enantioselective separation of nine environmentally persistent atropisomeric 3- and 4-methylsulfonyl PCBs. The selected columns contained cyclodextrin with various cavity diameters (β- or γ-CD), which were methylated and/or tert-butyldimethylsilylated in the 2,3,6-O-positions. The β-CD column with tert-butyldimethylsilylated substituents in all of the 2,3,6-O-positions was by far the most selective column for the methylsulfonyl PCBs tested. Enantiomers of congeners with 3-methylsulfonyl substitution were more easily separated than those with 4-methylsulfonyl substitution. The separation also seemed to be enhanced for congeners with the chlorine atoms on the ring that did not contain methylsulfonyl and were clustered on one side of the same ring. The 2,3-di-O-methyl-6-O-tert-butyldimethylsilylated-β-CD was found to give somewhat better selectivity than the corresponding γ-CD, upon comparing the two columns, which were identical in all other respects. Enantioselective analysis of arctic ring seal (*Phoca hispida*) and polar bear

(*Ursus maritimus*) adipose tissue revealed a strong dominance of certain enantiomers. For example, the enantiomer ratio of 3-methylsulfonyl was 0.32 and <0.1 in ringed seal blubber and polar bear fat, respectively. These low enantiomer ratio values are indicative of highly enantioselective formation, enantioselective metabolism, enantioselective transport across cell membranes, or a combination of the three in both species. Comparable results for the enantiomeric analysis of methylsulfonyl PCBs in biotic tissue extracts were obtained using two highly selective MS techniques, ion trap MS/MS and electron capture negative ion low-resolution MS.

High-Performance Liquid Chromatography (HPLC)

Echols et al. [124] described an automated HPLC method utilising a porous graphitic carbon column for the fractionation of PCB and chlorinated dibenzo-p-dioxins in fish tissue.

Miscellaneous

Buser et al. [125] determined up to 20 tetra- to heptachloro PCB or their aryl sulfone derivatives in grey seal adipose fat and liver by mass spectrometric techniques.

De Voogt and Haeggberg [126] have also discussed the determination of methyl sulfone PCB and chlorodibenzo-p-dioxins in fish tissue.

Yu Ma and Bayne [127] determined PCB in fish using linear discrimination by electron capture negative ion chemical ionisation mass spectrometry.

Ling and Huang [128] described an effective multiresidue method for PCB and other organochlorine compounds in fish.

An interlaboratory study has been carried out on the determination of chlorobiphenyl congeners in marine media [129].

Wong et al. [130] carried out food web studies of chiral PCB in river and riperious biota.

Blanche et al. [131] quantified the enantiomeric ratios of chiral PCB.

Jansson et al. [408] determined polychlorinated naphthalenes in fish livers using a method based on carbon column chromatography and mass fragmentography using the negative ions formed during chemical ionisation. Isomers containing 4 – 6 chlorine atoms dominated the mass pattern. Fish livers examined contained from 3 to 6.2 mg/kg of lipid.

This study revealed a small enantiomeric bias for PCB mass 132 while polychlorobiphenyl masses 95 and 149 were present in racemic or almost racemic form.

Wong et al. [132] has carried out enantioselective bioaccumulation measurements of PCB as well as α-hexachlorocyclohexane and *trans*-chlordane in rainbow trout.

Various workers [133–135] have discussed isomer-specific determinations of selected PCB in fish.

Wirth et al. [136] have described a novel microextraction technique for determining PCB in fish, which involves extraction of µg dry marine benthic copepod tissue with 100 µl of extraction solvent.

Various workers [137–139] have reviewed methods for the determination of PCB in fish.

1.2.3.7
Polychlorodibenzo-p-Dioxins and PCB

Polychlorinated dibenzo-p-dioxins, polychlorinated dibenzofurans, and other substituted polychlorinated biphenyls, are three structurally and tox-icologically related families of anthropogenic chemicals that have, in recent years, been shown to have the potential to cause serious environmental con-tamination. These substances are trace-level components or byproducts in several large-volume and widely used synthetic chemicals, principally poly-chlorinated biphenyls and chlorinated phenols, and can also be produced during combustion processes and by photolysis. In general, polychlorinated dibenzo-p-dioxins, polychlorinated dibenzofurans, and non-*ortho* polychlo-rinated biphenyls are classified as highly toxic substances, although their toxicities are dramatically dependent on the number and positions of the chlorine substituents. About ten individual members out of a total of 216 polychlorinated dibenzo-p-dioxins, polychlorinated dibenzofurans and non-*ortho* polychlorinated biphenyls are among the most toxic manmade or nat-ural substances to a variety of animal species. The toxic hazards posed by these chemicals are exacerbated by their propensity to persist in the envi-ronment and to readily bioaccumulate, and although the rate of metabolism and elimination is strongly species-dependent, certain highly toxic isomers have been observed to persist in the human body for more than ten years. Work on the determination of these classes of compounds in fish is dis-cussed below.

High-Performance Liquid Chromatography (HPLC)

Lamparski et al. [140] have developed a procedure for the determina-tion of 10 – 100 µg/kg quantities of 2,3,7,8-tetrachlorodibenzo-p-dioxin in fish. The technique involves digestion with alcoholic potassium hydrox-ide and extraction of the matrix with hexane, followed by a series of absorbent and chemically modified adsorbent liquid column chromato-graphic clean-up steps involving the use of silica–sulfuric acid, alumina and silica–silver nitrate. A final 'residue polishing' step via elevated tem-perature reversed-phase HPLC is applied prior to detection by multi-ple ion mode gas chromatography–mass spectrometry. Using ^{13}C-labelled 2,3,7,8-tetrachlorodibenzo-p-dioxin as an internal standard and carrier, the procedure has been validated for rainbow trout from approximately 10 to 100 ng/kg 2,3,7,8-tetrachlorodibenzo-p-dioxin. Relative to this range,

2,3,7,8-tetrachlorodibenzo-p-dioxin recovery is 75% ± 25%, and the precision of a single determination at the 95% confidence level (2σ) is ± 20% relative standard deviation at 50 ng/kg concentration.

Results obtained using this procedure in spiking experiments in which 10 – 100 ng/kg quantities of 2,3,7,8-tetrachlorodibenzo-p-dioxin were added to trout gave recoveries of between 80 and 120%. Other anthropogenic compounds in the trout did not interfere.

In a method for determining polychlorodibenzo-p-dioxin and polychlorodibenzo furans in fish, Sherry et al. [141] blended the fish sample with solid anhydrous sodium sulfate. The blend was packed into a glass column and the lipid fraction eluted with methylene dichloride. Lipids were removed from the extract by size exclusion chromatography followed by acid–base silica chromatography and HPLC on basic alumina and activated carbon.

Gas Chromatography (GC)

Phillipson and Puma [142] identified chlorinated methoxybiphenyls as contaminants in fish, and recognised them as potential interferers in their gas chromatography–mass spectrometric method for the determination of chlorinated benzo-p-dioxins. These workers used a similar sample digestion–clean-up procedure to that used by Lamparski et al. [140] involving digestion of the fish with alcoholic potassium hydroxide followed by hexane extraction, extraction with concentrated sulfuric acid, and clean-up on alumina and Florisil.

Gas chromatography–mass spectrometry of the solvent extracts of carp samples revealed the presence of a group of xenobiotics (Cl_3, Cl_4 and Cl_5 ring-substituted methoxybiphenyls). These compounds gas chromatographed in the region of Cl_3, Cl_4 and Cl_5 polychlorinated dibenzo-p-dioxins and produced intense molecular ions with the same nominal masses and chlorine isotopic abundances as those observed in the molecular ion clusters from trichloro-, tetrachloro- and pentachlorodibenzo-p-dioxins. Synthesis of 3,3',4',5-tetrachloro-4-methoxybiphenyl and 2,3',4,4'-tetrachloro-3-methoxybiphenyl provided model compounds which came through the polychlorinated dibenzo-p-dioxin clean-up procedure, and had gas chromatographic and mass spectrometric properties consistent with those of the residues recovered from the fish. The finding of chlorinated methoxybiphenyls as contaminants in fish, combined with the potential for their molecular ions to be mistaken for those from polychlorinated dibenzo-p-dioxins, indicates a need for the reappraisal of reported identifications of polychlorinated dibenzo-p-dioxin residues in environmental samples by selected-ion monitoring gas chromatography–mass spectrometry methods based on monitoring exclusively for the isotopic molecular ions from polychlorinated dibenzo-p-dioxins. For reliable identification and quantitation of polychlorinated dibenzo-p-dioxins by GC low-resolution MS, it might be necessary to examine a sufficient segment of the mass spectrum of the suspect residue to rule out the presence of polychloro methoxybiphenyls (by

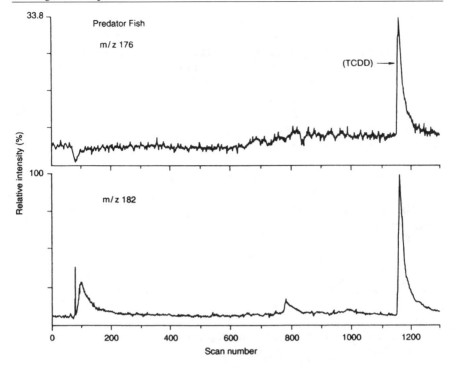

Figure 1.3. GC/NIAPI/MS response for a fish (predator) extract spiked with 1 ppb TCDD-^{13}C. The result corresponds to a level of 230 ppt TCDD in the sample. From [143]

the absence of fragment ion clusters resulting from losses of CH_3, CH_3CO and $CH_3CO + 2Cl$ from the M^+ cluster) and other potential interferences not yet observed.

Mitchum et al. [143] determined $10 - 30\,\mu g/kg$ levels of 2,3,7,8-tetrachlorodibenzo-p-dioxin in various fish samples by a procedure involving the use of a capillary column gas chromatograph interfaced directly to an atmospheric pressure ionisation mass spectrometer. Isolation of 2,3,7,8-tetrachlorodibenzo-p-dioxin from tissue samples was accomplished via multistep HPLC on silica gel incorporating stable label isotope dilution. Polychlorinated biphenyls were not found to interfere with 2,3,7,8-tetrachlorodibenzo-p-dioxin analysis at low ng/kg levels. No mention is made of potential interference by chlorinated methoxy biphenyls, as discussed by Phillipson and Puma [142]. Again, ethanolic potassium hydroxide was used to digest fish samples, followed by hexane extraction and clean-up on charcoal column.

Figure 1.3 shows a selected ion chromatograph obtained from injection onto the capillary gas chromatographic–negative ion atmospheric pressure mass spectrometry system of a hexane extract of a fish sample spiked with

230 ng/kg of 2,3,7,8-tetrachloro-dibenzo-p-dioxin. Between 7 and 230 μg/kg of this dioxin have been found in river water samples by this method.

Lawrence et al. [144] detailed equipment and the procedure used for the determination of dioxins in fish tissues and sediment samples, involving isolation and extraction by acid digestion, gel permeation chromatography, trisodium phosphate treatment, microalumina chromatography, and carbon-fibre column chromatography followed by determination of tetrachlorodibenzo-p-dioxins by GC–electron capture detection screening, and confirmation by high-resolution GC–MS.

Harless et al. [146] also used high-resolution GC–MS to determine 2,3,7,8-tetrachlorodibenzo-p-dioxin in fish tissues.

Niemann et al. [409] determined 2,3,7,8-tetrachlorodibenzo-p-dioxins in extracts of fish by electron capture capillary GC. The components were separated on size exclusion C_8 and C_{18} LC and OV 101 capillary GC columns. Precision is less than 20% relevant standard deviations.

King et al. [147] have reported a rapid screening method for determining polychlorodibenzo-p-dioxins and polychlorodibenzo furans in fish. The method is based on saponification of the sample, extraction with hexane and gel permeation chromatography, as well as clean-up with sulfuric acid and HR–GC–LR–MS.

At concentrations above the detection limit, concentrations obtained by HR and LR–MS were comparable.

Clement et al. [145] compared results obtained by various techniques for the determination of chlorinated dibenzo-p-dioxins and chlorinated dibenzofurans in fish in a round robin experiment involving GC–MS as well as mass spectrometry–mass spectrometry and HR–MS. They commented that the analysis of these complex samples was difficult without extensive clean-up procedures. However, determination by GC and HR–MS was possible with reduced clean-up compared with GC and LR–MS.

Taguchi et al. [148] used HR–MS to determine polychlorinated dibenzo-p-dioxins and polychlorinated dibenzo furans in fish. They used mixtures of the solid aromatic hydrocarbons, coronene, tetraphenylcyclopentadiene and decacyclene as lockmasses in the analysis of polychlorinated dibenzodioxins and polychlorinated dibenzofurans with between four and eight chlorine atoms. Enhanced sensitivity, easier control of the lockmass concentration in the ion source, and greater resolution between lockmass and sample ion signals were obtained for this system compared to the conventional perfluorokerosene lockmass system. Both systems exhibited a trend of decreasing signal strength with increasing chlorine substitution and stronger signals obtained for polychlorinated dibenzo furans than for polychlorinated dibenzo-p-dioxins. Crummelt [149] has reviewed methods for the determination of polychlorinated dibenzo-p-dioxins and polychlorinated dibenzo furans in fish.

1.2.3.8
Polychlorinated Paraffins

Because of the complexity of polychlorinated paraffins, it is extremely difficult to analyse them. This is certainly one of the reasons that they have only
rarely been determined. The HR–GC electron capture detection of polychlorinated paraffins with chlorine contents above 60% is easy, but unspecific.
Since these compounds elute over a wide range, unequivocal identification
is not possible even with different stationary phases of different polari-
ties. Therefore, electron capture negative ion mass spectrometry is generally
favoured. The possibility of using this technique for the determination of
chlorinated paraffins was demonstrated for the first time by Gyøs and Gustavsen [415]. They studied two Cl0–C13 polychlorinated paraffins with 59
and 70% chlorine content. A mass spectrometer with a direct inlet was used
as the detection system. Fish was also analysed after having been exposed
to both polychlorinated paraffins. A heavy fuel oil prepared with CP10–
CP13:70 was used to simulate the residue peak pattern of environmental
samples. However, a method developed on the basis of spiked samples that
contained only a few selected polychlorinated paraffins in an artificial matrix and without the background of the whole range of chlorinated residues
(which make separation and identification so very difficult) is not suitable
for grown samples.

1.2.3.9
Miscellaneous

Passivirta et al. [150] reported on the occurrence of polychlorinated aromatic ethers in salmon and fish liver oil. They described an improved clean-
up technique that separated polychlorinated diphenyl ethers, anisoles and
veratroles, phenoxyanisoles and biphenylanisoles, and dioxins and dibenzofurans into different fractions. Compounds from all of these groups of
substances were detected and, in many cases, identified.

Marquis [151] has described the methods used by the US Environmental Protection Agency in their surveys of polychlorodibenzo-p-dioxins and
polychlorodibenzofurans in fish.

A non-statistically based laboratory data scoring system has been developed to evaluate interlaboratory performance, and it has been tested for the
analysis of polychlorodioxins and furans in fish tissues [152].

Smith et al. [153] have discussed the determination of ng/kg quantities of polychlorinated dibenzofurans and dioxins in fish. They point out
that polychlorinated dibenzofurans may commonly occur at comparable
or greater levels than the dioxins, and could generally pose a greater hazard than polychlorinated dibenzo-p-dioxins. The latter are often found as
cocontaminants in, and are readily produced from the pyrolysis of, polychlorinated biphenyls. Most important, the polychlorinated dibenzo-p-dioxins
produced from the pyrolysis of polychlorinated biphenyls are predomi-

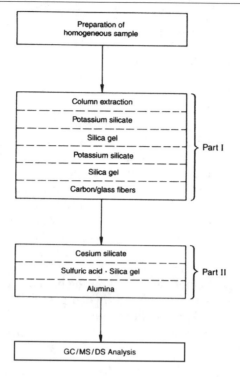

Figure 1.4. Flow chart of total procedure. From [153]

nantly the most toxic isomers, and they are those having a 2,3,7,8-chlorine substitution pattern.

In this procedure, the fish samples (spiked with an isotopic compound) are processed as shown in Figure 1.4. Table 1.13 shows some recovery data obtained by this procedure for various polychlorinated dibenzo-p-dioxins (PCDD) and polychlorinated dibenzofurans (PCDF) spiked into fish samples. These results demonstrate the effectiveness of the recovery of a large proportion of these compounds, especially those tetra-, penta- and hexachloro isomers possessing the critical 2,3,7,8-chlorine substitution pattern (Column (a), Table 1.13). The reduced precision for tetrachloro-dibenzo-p-dioxins (TCDD) tetrachlorodibenzo furans (TCDF) and octachloro-dibenzo-p-dioxins (OCDD) is evident (column (a) Table 1.13).

Compounds that do not interfere in this procedure include DDE, polychlorinated biphenyls, methoxy polychlorinated biphenyls, polychlorinated diphenyl ethers, and methoxy polychlorinated biphenyl ethers.

Determinations of 2,3,7,8-tetrachlorodibenzo-p-dioxin in fish carried out by this and other methods of analysis are compared in Table 1.14. The agreement, in terms of identification and quantitation, between the results obtained by Smith et al. [153] and those of the other laboratories was con-

Table 1.13. Recoveries of selected PCDD and PCDF from spiked samples of homogenised whole fish using the unabbreviated enrichment procedure (from [153])

Sample	Recoveries of selected compounds							
	$2,3,6,8$-Cl_4-PCDF	$2,3,7,8$-Cl_4-PCDF and PCDD	$1,2,4,7,8$-Cl_5-PCDF	$1,2,4,6,7,9$-Cl_6-PCDF	$1,2,3,4,7,8$-Cl_6-PCDD	$1,2,3,4,6,7,9$-Cl_7-PCDF	OCDD	OCDF
a) 100 g of grass carp and 10 ng each of PCDD and PCDF (100 mg/kg)	81 (1)	92 (30)	94 (3)	98 (6)	104 (4)	95 (8)	99	91 (16)

Sample	Recoveries of selected compounds		
	$[^{13}C]$-$2,3,7,8$-TCDD	$[^{37}Cl]$-$2,3,7,8$-TCDF	$[^{37}Cl]$-$1,2,7,8$-TCDF	$[^{37}Cl]$-OCDD
b) samples spiked at 25 – 50 mg/kg	82 ± 27	58 ± 16	75 ± 18	83 ± 30

Sample	Recoveries of selected compounds							
	Cl_4 PCDF	Cl_5 PCDF	Cl_6 PCDF	Cl_7 PCDF	OCDF	Cl_5 PCDD	Cl_6 PCDD	Cl_7 PCDD	Cl_4 biphenylene
c) fish spiked at 20 mg/kg	58 ± 10	64 ± 6	64 ± 7	63 ± 10	59	41	49	58	52
Fish spiked at 100 mg/kg	52 ± 7	55 ± 4	53 ± 6	56 ± 4	52	84	60	51	59

OCDD: octachlorodibenzo-p-dioxin
OCDF: octachlorodibenzo furan
TCDD: tetrachlorodibenzo-p-dioxin
TCDF: tetrachlorodibenzo-p-furan

Table 1.14. Results from interlaboratory studies and comparisons of the determination of 2,3,7,8-TCDD in fish and birds (from [153])

Study number	CNFRL	No.1	No.2	No.3	No.4	No.5	No.6	No.7	Reported average
USFDA									
Sample 1	9					6	5		
Sample 2	47	67			77	89	67		
Sample 3	22	25			57	42	34		
Sample 4	117	113			128	99	188		
Sample 5	56	45			38	53			
Sample 6	96	100			107	199	178		b
H & WC/USFDA									
Sample 7	58	104	58	49, 58	< 5	72	70	60	61
Sample 8	< 1	< 10	< 1	< 2, < 2	< 5	< 2	< 5	37	3.6
Sample 9	34	35	37	23, 32	51	25	33	26	30
Sample 10	38	45	33	19, 31	55	32	27	32	32
USEPA									
Sample 11	37	52	45	55					
Sample 12	36	39							
Sample 13	19	15	25						
Sample 14	< 1	< 9	< 5	< 25					

	Independent laboratories		
	CNFRL	Swiss Federal Reserve[a]	National Center Toxicology Research[b]
Carp, Lake Huron	22, 27	29	10
Carp, Lake Erie	< 1	5	< 10
Lake trout, Lake Ontario	56, 58		54
Ocean herring, control	< 1		< 10
Lake trout, Lake Huron	39		32
Rainbow trout, Lake Ontario	38		31
Carp, Saginaw Bay	94		75
Carp, Tittabawassee River, MI	81		65

[a] analysed by HRGC/high-resolution electron impact MS
[b] analysed by HRGC/chemical ionisation atmosphere pressure MS
USFDA: United States Food and Drug Administration
HRWC: Health and Welfare, Canada
CNFRL: Colombia National Fisheries Research Laboratory, USA

sistently good, and no false positive results were indicated in any of the determinations made with this procedure.

1.2.4
Bromine-Containing Compounds

Natural and Anthropogenic Bromine Compounds

The detection of brominated compounds has attracted increasing interest in environmental science. Gas chromatographic separation followed by ECNI–MS detection in the SIM mode using m/z 79 and m/z 81 is a highly sensitive and selective method for the determination of brominated compounds in environmental samples [154]. These studies identified brominated flame-retardants in several environmental compartments with a tendency toward increasing global concentrations [155]. These studies have further suggested that the brominated trace chemicals observed in the tissues of higher organisms originate from the use of brominated flame retardants.

The major brominated flame retardant pollutants are polybrominated diphenyl ethers, which have been detected in numerous environmental samples [154]. Relevant polybrominated diphenyl ether (PBDE) congeners occurring in a variety of samples include PBDE 47 (2,2',4,4'-tetrabromodiphenyl ether) and to a lesser extent PBCD 99 (2,2',4,4',5-pentabromodiphenyl ether) and PBCD 100 (2,2',4,4',6-pentabromodiphenyl ether) [154,156,157]. PBDE 47 and 99 are found in the technical formulation pentabromodiphenyl ether (Bromkal 70-5 DE) at ~ 40% [154, 158]. Lower quantities are also found for PBDE 100 [159]. Another class of nonpolar brominated flame retardants is polybrominated biphenyls (for an example, see below), but polybrominated biphenyl concentrations in environmental samples were generally lower than those of the PBDEs [154].

In a recent study of marine mammals from Australia, several other previously unknown brominated compounds were detected.

Vetter [160] has described a gas chromatography ECNI mass spectrometric method for the identification of lipophilic anthropogenic and natural brominated compounds in the blubbers of marine samples from various locations.

The residues from these brominated flame retardants PBDE 47, 99 and 100 were included in the study were several potentially naturally occurring brominated compounds recently identified in dolphins from Australia (BC1, BC2, BC3, BC10, BC11).

ECNI–MS full scan spectra were obtained for BC-3 and BC-10. A natural mixed halogenated compound (MHC-1) and an unknown brominated compound (UBC-1) were investigated as well. Evidence for the natural production of these secondary metabolites and their bioaccumulation in higher organisms as well as analytical protocols for their detection in the environment are presented. Some of these naturally occurring compounds may be misinterpreted as anthropogenic brominated compounds. In ECNI–MS, brominated compounds are usually identified by the detection of the fragment ions m/z 79 ($[^{79}Br]^-$) and m/z 81 ($[^{81}Br]^-$). Vetter [160] showed that the monitoring of additional ion traces corresponding to $[Br_2]^-$ (160 type), $[HBr_2]^-$ (161 type), $[BrCl]^-$ (116 type) and $[HBrCl]^-$ (117 type) fragment ions allows different classes of brominated compounds to be distinguished. This technique was used to demonstrate that UBC-1 is neither a PBDE nor a PBB congener, whereas a second mixed halogenated compound (MHC-2) was identified as a result of the ECNI–MS response at m/z 114/116. Studies on blubber extracts of marine mammals from four continents resulted in the detection of significant differences in the global distributions of brominated compounds, and it is suggested that naturally occurring organobromines are more abundant than anthropogenic brominated compounds in several regions of the world.

1.2.5
Nitrogen-Containing Compounds

Nitrogen Bases

Methods are available for the determination of trimethylamine, trimethylamine oxide and hypoxanthin in fish.

Trimethylamine. Amines and basic substances that are not rendered unreactive by formaldehyde are released from an extract at room temperature by potassium hydroxide and extracted into toluene. The toluene phase is separated and dried, and picric acid reagent is added in order to form a coloured complex with trimethylamine. The absorbance of this solution at 410 nm is a measure of the concentration of trimethylamine. Other tertiary amines and bases not rendered completely unreactive by formaldehyde under the conditions of the method may also give coloured complexes, but normally this interference is small.

Methylamines. The volatile amines in the sample are extracted by means of perchloric acid. The perchloric acid is made alkaline with sodium hydroxide, and the amines liberated are steam-distilled into hydrochloric acid. The concentration of amines, including trimethylamine, is determined in the hydrochloric acid solution by gas–liquid chromatography.

Trimethylamine oxide. Trimethylamine oxide in an extract of the sample is reduced with titanium(III) chloride to trimethylamine, which is then determined. The value obtained for the trimethylamine content of the sample before reduction is subtracted from this value to give the trimethylamine oxide.

Hypoxanthine. Hypoxanthine is extracted from the sample by macerating it with perchloric acid. After neutralisation, the extract is treated with an enzyme that converts the hypoxanthine quantitatively into uric acid, which is determined by measuring its absorbance at 290 nm. Both the enzyme and the extract absorb at this wavelength, and so it is necessary to carry out measurements on blanks. The method is not specific for hypoxanthine, as similar results will be given by xanthine, which, although not normally present, should be looked for when unusual samples are being examined.

Richling et al. [161] have reviewed methods for the determination of heterocyclic aromatic amines in fish.

Niculescu et al. [162] have discussed the use of biosensors based on redox hydrogel-based amperometric bienzyme electrodes for fish freshness monitoring.

Rapid evaluation of fish and meat quality is permanently required in the food industry, motivating a continuous search for freshness biomarkers and efforts to develop simple and inexpensive methods for their determination. Among these biomarkers, inositol monophosphate, hypoxanthine and xanthine, which are intermediate degradation products of nucleic acids [163, 164], and some biogenic amines such as histamine [165–169], putrescine [170, 171] and cadaverine [170, 172], produced by microbial decarboxylation of the amino acids histidine, ornithine and lysine, respectively, have been proposed. The biogenic amine contents of various foodstuffs have been intensively studied due to their potential toxicity [173]. Histamine is the most biologically active compound from that class, affecting the normal functions of heart, smooth muscle, motor neurons and gastric acid secretion [174]. Other biogenic amines, such as putrescine and cadaverine, may amplify the effects caused by histamine intoxication, inhibiting the enzymes involved in histamine biodegradation: diamine oxidase and histamine-N-methyl transferase [175]. Numerous countries have adopted maximum levels for histamine in food, especially in fish products; e.g. Italian laws fixed this level at 100 mg/kg food [165], and similar limits have been adopted by EEC regulations [168].

Classical methods for the analysis of biogenic amines generally involve chromatographic techniques, such as gas chromatography [176], thin-layer chromatography [177], reversed phase liquid chromatography [177, 178], and liquid chromatography with derivatisation techniques [179–181]. However, they often require sample pretreatment steps and skilled operators, and their relatively long analysis times and high costs make these methods unsuitable for routine use.

Enzymic determination of biogenic amines has been carried out and this represents an alternative that can solve the problems mentioned above. In this context, amperometric [171–182], spectrophotometric [183–186], fluorimetric [187] or chemiluminometric detection methods have been used. Amperometric electrodes using purified amine oxidase as the biological recognition element have been reported.

Niculescu [162] presents the design and optimisation of amperometric biosensors for the determination of biogenic amines (e.g. histamine, putrescine, cadaverine, tyramine, cystamine, agmatine, spermidine) commonly present in food products, and their application to freshness monitoring in fish samples. The biosensors were used as the working electrodes of a three-electrode electrochemical cell of wall-jet type, operated at -50 mV vs. Ag/AgCl, a flow injection system. Two different bioenzyme electrode designs were considered, one based on the two enzymes (a newly isolated and purified amine oxidase and horseradish peroxidase) simply adsorbed onto graphite electrodes, and one when they were crosslinked to an Os-based redox polymer. The redox hydrogel-based biosensors showed better biosensor characteristics, i.e. sensitivities of 0.194 A M^{-1} cm^{-2} for putrescine and 0.073 A M^{-1} cm^{-2} for histamine and detection limits (calculated as three times the signal-to-noise ratio) of $0.17\,\mu$ M for putrescine and $0.33\,\mu$ M for histamine. The optimised redox hydrogel-based biosensors were evaluated in terms of stability and selectivity, and were used to determine the total amine content in fish samples kept for ten days in different conditions.

Nitrobenzenes

Gas chromatography mass spectrometry has been used to determine nitrobenzenes in fish [188].

Aromatic Amines

Okumura et al. [189] determined anilines in samples by GC–MS after liquid/liquid extraction and steam distillation. In addition to finding aniline, trace levels of its methyl, methoxy and chloro derivatives were detected.

Polychlorinated Nitrobenzenes

Procedures have been described for the determination of polychlorinated nitrobenzene compounds in fish [190]. The method comprised liquid chromatographic separation in silica gel, selective fractionation on Florisil, and then GC, with mass spectroscopy and mass fragmentography as methods of identification. In all, 12 separate nitrocompounds were detected, together with other polychlorinated pesticide derivatives.

1.2.6
Sulfur-Containing Compounds

Dichlorodimethylsulfone

Lindström and Schubert [191] used GC combined with multistage mass spectrometry and direct inlet multistage mass spectrometry to determine 1,1-dichlorodimethylsulfone in aquatic organisms in waters. The results obtained using the multistage mass spectrometry technique are reported, using extracts from flounders. The method is shown to be sensitive, selective and rapid, and does not require selective workup procedures.

Linear Alkyl Benzene Sulfonates

Linear alkyl benzenesulfonate is the most widely used synthetic surfactant, with an annual production rate of 1.8 million tons [206]. It is a mixture of n-(p-sulfophenyl) alkanes, and the individual constituents are abbreviated as C_n-m-linear alkyl benzenesulfonate, with n specifying the number of Cnatoms in the alkyl chain [200–203], and m identifying the C-atom at which the p-sulfophenyl moiety is attached to the alkyl chain. In technically produced linear alkyl benzene sulfonates, m can assume values between 2 and 7. Since it is the 'workhorse' surfactant in laundry detergents, linear alkyl benzene sulfonate is discharged with household wastewater. Although efficiently removed by wastewater treatment [192–198], the fraction remaining in the wastewater eventually reaches surface waters [197–203], therefore exposing aquatic organisms to linear alkyl benzenesulfonate. The bioaccumulation potential of linear alkyl benzenesulfonate required evaluation in the course of an environmental risk assessment of the surfactant [204,205].

Tolls et al. [207] have described a procedure for the determination in fish of linear alkyl benzenesulfonate and its biotransformation products, sulfophenylcarboxylic acids.

Matrix solid-phase dispersion extraction with subsequent ion-pair liquid–liquid partitioning of the extract was a time-efficient sample preparation method for the analysis of linear alkyl benzene sulfonate. The recovery of parent linear alkyl benzene sulfonate from spiked fish exceeded 70%, and the limit of quantitation was around 0.2 mg/kg, corresponding to 0.6 mol/kg. In a simultaneous determination of linear alkyl benzene sulfonate and sulfophenyl carboxylic acids, the analytes were matrix solid-phase dispersion-extracted in different fractions. The target compounds were separated from the sample matrix by protein precipitation and subsequent isolation of (a) sulfophenylcarboxylic acids by graphitised carbon black solid-phase extraction of the supernatant and (b) parent linear alkyl benzene sulfonate by ion-pair liquid–liquid partitioning of the pellet obtained after protein precipitation. The recoveries of the model compounds, C_{12}-2-linear alkyl benzene sulfonate and C_4-3-sulfophenylcarboxylic acids were 84 ± 6 and 65 ± 11%, respectively. The use of C_3-3 sulfophenylcarboxylic acids as an internal standard corrected for the loss of the biotransformation product during sample workup. The suitability of both methods was demonstrated by analyzing fish containing linear alkyl benzene sulfonate and sulfophenylcarboxylic acids incurred during aqueous exposure.

Tolls et al. [407] have studied a method based on measurements of total radioactivity for determining linear alkyl benzene sulfonates in fish. Unfortunately, this method does not distinguish between the parent surfactant and its biotransformation products.

1.2.7
Phosphorus-Containing Compounds

Triarylphosphate Esters

Murray [208] has described a gas chromatographic method for the determination in fish tissues of triarylphosphate esters (1 mol S-140, tricresyl phosphate, cresol diphenyl phosphate). These substances are used commercially as lubricating oil and plastic additives, hydraulic fluids and plasticisers. The method involves extraction from the samples, hydrolysis, and measurement of the individual phenols by GC as the trimethylsilyl derivatives. The lower detection limit was about 3 mg/kg of fish.

1.2.8
Chlorine-Containing Pesticides and Insecticides

Gas chromatography has been used extensively for the determination of chlorinated insecticides in extracts of fish tissue [209–216]. Solvent extraction and clean-up procedures are summarised in Table 1.15.

Luckas et al. [112] have described a method for determining polychlorobiphenyls (PCBs) and chlorinated insecticides in fish by the simultaneous use of electron capture gas chromatography and derivatisation gas chromatography (See Sect. 1.2.3).

Norheim and Oakland [218] studied the distribution of persistent chlorinated insecticides such as DDT, polychlorinated biphenyls and hexachlorobenzene in cod samples taken along the Norwegian coast down to 0.001 mg/kg.

In their analytical procedure, a 0.5 g sample of fish was digested with 6 ml concentrated sulfuric acid for four hours at 60 °C. The cooled digest was shaken with 1 ml hexane and this extract examined by electron capture gas chromatography. Recoveries of 95% of added hexachlorobenzene and octachlorostyrene were obtained in spiking experiments. Good interlaboratory comparisons were obtained in determinations of hexachlorobenzene, octachlorostyrene and pentachlorobenzene present at 0.001–4 mg/kg levels in fish.

Neely [219] determined PCBs in fish and water in Lake Michigan whilst Frederick [220] measured, by gas chromatography, the comparative uptake of PCBs and dieldrin by the white sucker. Jan and Malservic [221] determined PCBs and polychlorinated terphenyls in fish using acid hydrolysis of the fish tissue and destructive clean-up of the extract. Olsson et al. [222] studied the seasonal variation of PCB levels in roach. Szelewski et al. [108] have also determined PCBs in fish samples.

Ludke and Schmitt [223] have reported on pesticide and polychlorobiphenyl concentrations found in fish in the US National Pesticide Monitoring Programme.

Sackmauerova et al. [106] have described a gas chromatographic method for the determination of chlorinated insecticides in fish. In this method,

Table 1.15. Extraction and clean-up procedures used in the determination of chlorinated insecticides in fish (from author's own files)

Extraction solvent	Sample clean-up	Gas chromatography	Recovery %	Reference
Miscellaneous chlorinated insecticides				
Mix fish with granular sodium sulfate and sand and perform Soxhlet extraction with *n*-hexane	–	Electron capture	*p,p′*-DDE 97.8 *p,p′*-DDT 91.2 *p,p′*-TDE 94.6 *p,p′*-DDT 89.6 Dieldrin 89.3	[210,216]
Or				
Fish mixed with deactivated Florisil and extracted in a column with 1:4 v/v dichloromethane–hexane	–	Electron capture		
Blend fish with sodium sulfite and solid carbon dioxide. Extract with cyclohexane	Florisil column	–	95-100	[211]
Dieldrin				
Light petroleum–acetonitrile partitioning or direct homogenisation with acetonitrile	Florisil column	Flame ionisation and capture detection	Dieldrin 100	[212]
Photodegradation products of Endrin	Florisil column	Flame ionisation of silylated or acetylated derivatives	–	[213]
Dieldrin and polychlorobiphenyls	–	GC-MS	–	[214]
BHC isomers, DDE, DDT hexachlorobenzene. Homogenised fish mixed with standard anhydrous sodium sulfate extracted with light petroleum. Extract concentrated prior to clean-up on celite-oleum column.	Celite-oleum column or Florisil for Aldrin- (unstable on celite-oleum)	Electron capture	BHC isomers 93-103 DDT, DDE 90-93	[215]
Organochlorine pesticides	Solid-phase extraction	GC with ECD	–	[217]

100 g of fish sample are weighed and homogenised. From the homogenate, 10 g are weighed and rubbed with cleaned sea sand. Water contained in the sample is bound by adding some anhydrous sodium sulfate to the sample in order to obtain a homogeneous powdery mixture. The mixture is shaken three times with portions of light petroleum (200, 100 and 100 ml) for one-hour periods. The separate extracts are filtered into 500 ml flasks through a layer of anhydrous sodium sulfate and concentrated using a vacuum rotary evaporator. The concentrated extract is quantitatively transferred into a 250 ml separation funnel using 20 ml petroleum ether saturated with acetonitrile. The mixture is shaken three times with 40 ml acetonitrile saturated with petroleum ether. The acetonitrile extracts are combined in a one-litre separating funnel, 500 ml of 5% sodium chloride in water solution is added, and the insecticides are extracted from the samples twice with 100 ml petroleum ether. The mixed ether extracts are concentrated using the vacuum rotary evaporator to a small volume and purified on a chromatographic column filled with 4 g Celite and a mixture of 8 g Celite with 6 ml oleum. The upper layer of the column consisted of a 15 mm layer of anhydrous sodium sulfate. The thickened extract is quantitatively transferred to the top of the washed column and the insecticides eluted with 250 ml petroleum ether. The eluate is reduced to a volume of 1 ml and used for gas chromatography [224].

The working conditions of the gas chromatographic column were as follows; temperature of the column 180 – 200 °C, temperature of the injection port 210 °C, temperature of the electron capture detector (^{63}Ni) 200 – 225 °C, nitrogen flow rate 60 – 80 ml/min, EC detector voltage 20 – 70 V.

A column filled with 1.5% siliconeOV-17 plus silicone oil (fluoral-chylsiloxane on Chromosorb W (80 – 100 mesh) is used to separate the BHC alpha, beta, gamma and delta isomers (hexachlorocyclohexane) o,p'-DDT, p,p'-DDE, p,p'-DDD and p,p'-DDT. α-BHC and hexachlorobenzene (HCB) have a common peak. They can be separated on a column filled with 2.5% Silicone XE-60 (β-cyanoethyl–methylsilicone) on Chromosorb W (80 – 100 mesh).

Gas chromatography has been applied to the determination of endosulfan in fish [225].

A gas chromatograph with a nitrogen/phosphorus-specific detector has been used to determine carbaryl in rainbow trout liver. A freeze-out column and a Florisil column were used in the workup procedure [226].

Miscellaneous

Galceran [227] has reviewed methods for the determination of toxaphene in fish.

Lott and Barker [228] compared matrix solid-phase dispersion and classical extraction methods for the determination of organochlorine pesticides in fish muscle.

Hughes and Lee [271] has used GC to determine toxaphene in fish samples.

Musial and Uthe [272] described a simple procedure using a combination of chromatography and fuming nitric–concentrated sulfuric acid cleanup followed by capillary GC for the estimation of toxaphene residues in marine fish. Wet weight concentrations were 1.1 mg/kg in cod liver and 1.0 – 0.4 mg/kg in herring fillet from the Gulf of St. Lawrence and Halifax respectively.

Vetter et al. [273] used cogener-specific methods for the determination of toxaphene in Antarctic Seals [275]; the same method has been used to determine toxaphene in fish.

Alder and Vieth have reported a method for the determination of toxaphene in fish.

Polychloroterphenyls have been determined in fish [276].

Sackmauerova et al. [224] used thin-layer chromatography on silica plates to confirm the identity of chlorinated insecticides previously identified by GC. The compounds can be separated by single or repeated one-dimensional development in n-heptane or in n-heptane containing 0.3% ethanol. The plate is dried at 65 °C for ten minutes and detected by spraying with a solution of silver nitrate plus 2-phenoxyethanol. Thereafter, the plate was dried at 65 °C for ten minutes and illuminated with an ultraviolet light ($\lambda = 254$ nm) until spots representing the smallest amounts of standards were visible (10 – 15 minutes). The pesticide residues may be evaluated semi-quantitatively by simple visual evaluation of the size and the intensity of spot colouration, and by comparing extracts with standard solutions.

In this method, the recoveries of both DDT and DDE were from fish. Purification on a Florisil column was used when determining chlorinated insecticides unstable at low pH (aldrin, dieldrin). The type and activity of Florisil influence the yield and accuracy of the method.

The average content of λ-BHC in samples of herbivorous fishes (*Abramis ballerm* L., *Cyprinus carpio* L., *Chrondrosroma nastus* L.) was 0.054 mg/kg, that of β-BHC 0.009 mg/kg, and that of the remaining BHC isomers and HCB 0.049 mg/kg. The average DDE content was 0.133 mg/kg and that of DOT 0.094 mg/kg. From the analyses of 78 samples of carnivorous fishes (*Esox indus* L., *Lepomis gibbosus* L., *Aspius fluviatilis* L.), these workers found the average content of λ-BHC 0.062 mg/kg, that of β-BHC 0.023 mg/kg, and that of the remaining BHC isomers plus HCB 0.060 mg/kg. The average DDE content was more than ten times higher (1.53 mg/kg) and likewise that of DDT (1.175 mg/kg) in comparison with the herbivorous fishes. Besides the BHC isomers, DDE and DDT, the concentration of hexachlorobenzene has also been studied in waters and fish since 1973. Its concentration in waters varied between 0.001 and 0.03 µg/l, while in fish it was from 0.001 to 0.26 mg/kg. Comparing the average contents of the BHC isomers of DDT and metabolites in fish, it can be seen that DDT and DDE levels greatly exceed the levels of BHC, mainly as far as carnivorous fish are considered. Upon comparing the contents of the BHC isomers and DDT and its metabolites in waters and fish, a 1000 – 10,000-fold higher concentration was detected in fish.

1.2.9
Nitrogen-Containing Insecticides and Pesticides

Steinwandter [229] determined nitrogen in fish by using silica gel clean-up and capillary GC with an electron capture detector.

Aminocarb and a metabolite have been extracted from whole fish, derivatised with heptafluorobutyric anhydride prior to determination by GC with an electron capture detector [230,414].

1.2.10
Phosphorus-Containing Insecticides and Pesticides

Duesch et al. [231] determined Dursban in fish. After a preliminary clean-up, the extract is chromatographed on a column packed with 3% Carbowax 20 M on Gas-Chrom (60 to 80-mesh), which gives excellent separation of Dursban from other organophosphorus insecticides. Both thermionic and flame photometric detectors are satisfactory. Recoveries range from 75 to 105% (i.e. 90 ± 15%) depending on the nature of the sample. This procedure will detect as little as 0.5 ng of Dursban, corresponding to a level of 0.01 mg/kg in a 10 g sample of fish.

Pirimiphos methyl (O-(2-diethylamino)-6-methyl-4-pyrimidyl-O,O-dimethyl phosphorothioate) has been determined in fish by a procedure involving GC of a hexane–acetone extract of the sample [232].

Lores et al. [233] described a method for the determination of Fenthion in amounts down to 0.01 mg/kg in fish. The method involved solvent extraction followed by a silica gel clean-up procedure, then determination by gas liquid chromatography with thermionic detection. The clean-up procedure required relatively little time. Recovery of Fenthion exceeded 85% with adequate sensitivity.

Szeto et al. [234] have described a gas chromatographic method for the determination of acephate and methamidophos residues in fish.

1.2.11
Miscellaneous Insecticides and Herbicides

Trihalomethrin, deltamethrin and related compounds have been determined in fish tissue at the ppb and ppt levels by silica gel HPLC with radiometric detection [235].

Bonwick et al. [236] have determined synthetic pyrethroids in fish.

1.2.12
Ciguatoxins

Ciguatera (fish poisoning) is a major economic and social problem throughout tropical and subtropical waters, with an estimated 25,000 persons poisoned annually. The disease is characterised by neurological and gastrointestinal disorders which typically appear from 1 to 24 hours following the consumption of contaminated fish [237, 238, 253]. The toxins involved are potent sodium channel activator toxins known as ciguatoxins that are produced by the benthic dinoflagellate *Gambierdiscus toxicus* [239]. The ciguatoxins and structurally related brevetoxins (e.g. PbTx-2[4]) [240] compete in the voltage-sensitive sodium channel. Two related families of Pacific ciguatoxins (P-CTX) have been identified in Pacific Ocean fish [241–246]. A third family of Caribbean ciguatoxins (C-CTX) has been identified in fish of the Caribbean Sea [247–251]. The ciguatoxins are heat-stable polyether toxins of 1023 – 1157 Da. P-CTX-1 is the major toxin in the flesh of carnivorous fish of the Pacific, contributing to ∼ 90% of the total lethality and posing a health risk at levels above 0.1 ppb [241, 242, 252, 253]. The minimum risk level for C-CTX-1 has not been determined [250].

The traditional method of detecting the presence of ciguatoxins in fish involves testing lipid extracts by mouse bioassay [254–257]. More recently, cytotoxicity [258] and radio ligand binding [248, 249, 259] methods have been discussed.

Lewis et al. [252, 414] developed a method based on high-performance liquid chromatography–tandem electrospray spectrometry (HPLC/MS/MS) to determine sub-ppb levels of Pacific and Caribbean ciguatoxins in crude extracts of fish.

This method gave a linear response to pure Pacific and Caribbean ciguatoxins (P-CTX-1 and C-CTX-1) and the structurally related brevetoxin (PbTx-2) spiked into crude extracts of fish. Levels equivalent to 40 ppt P-CTX-1, 100 ppt C-CTX-l and 200 ppt PbTx-2 in fish flesh could be detected by HPLC/MS/MS. Using P-CTX-1 as an internal standard, the analysis of extracts of 30 ciguateric fish from the Caribbean Sea (8 toxic, 12 borderline, and 10 nontoxic by mouse bioassay) confirmed the reliability of the method and allowed an estimated risk level of > 0.25 ppb C-CTX-1 to be established. High-performance liquid chromatography–tandem electrospray spectrometry provides a sensitive analytical approach for the determination of Pacific and Caribbean ciguatoxins present at sub-ppb levels in fish flesh.

1.2.13
Miscellaneous Organic Compounds

Geosmin and 2-Methylisoborneol

Martin et al. [260] have described a method for isolating and quantifying 2-methylisoborneol in fish flesh. Samples of channel catfish flesh were cooked in a microwave oven under a nitrogen stream, and the condensate trapped at –80 °C. Hexane extracts of the condensate were then concentrated to 100 – 200 µl and analysed by gas chromatography. Concentrations as low as 5 ng methylisoborneol per gram of fish were detectable. About 80% of the organoleptically off-flavoured fish had elevated 2-methylisoborneol concentrations (5.0 – 815.5 ng 2-methylisoborneol/g), but other musty odorants (geosmin, pyrazine isomers) were not detected. On average, the 2-

methylisoborneol levels in channel catfish from off-flavour ponds were approximately 34 and 28 times higher than those in mud and water, respectively.

Lovell et al. [261] showed in a survey that a high proportion of catfish samples taken in ponds in Alabama had an off flavour due to the presence of geosmin. Persson [262] investigated threshold odour concentrations of geosmin and 2-methylisoborneol in fish.

Conte et al. [411] have described a microwave distillation solid-phase adsorbent trapping device for the determination of geosmin and isobornol in catfish tissue. Ethyl acetate was used to trap the two compounds which were then determined by GC in the sub-µg/kg range.

Fluridone

A method has been described [263] for the simultaneous determination of underivatised fluridone herbicide and its major metabolite, 1-methyl-3-(4-hydroxyphenyl)-5-(3-trifluoromethyl)phenyl-4(1H)-pyridinone, in fish and crayfish tissues by liquid chromatography. Compounds are extracted with methanol, followed by evaporation, acid hydrolysis to release conjugated residues, liquid–liquid partitioning purification using Florisil Sep-Pak, and liquid chromatographic analysis. In the absence of interfering peaks the detection limit was 0.04 – 0.05 mg/kg for either compound. In method validation studies, overall recoveries of fluridone and its metabolite averaged 84 and 70% in edible crayfish tissues, 74 and 67% in inedible crayfish tissues, 111 and 103% in edible fish tissues and 109 and 76% in inedible fish tissues, i.e. slightly high recoveries.

Squoxin (1,1′-Methylene-2-Naphthol)

Kiigemagi et al. [264] developed a method for the determination of 0.1 mg/kg residues of this piscicide in fish using derivitisation GC and spectrophotometric methods. Fish were homogenised by grinding with dry ice. The pulverised mixture was poured into a plastic bag which was lightly sealed and placed in a –10 °C freezer overnight to allow the carbon dioxide to sublime. The samples were stored in this condition prior to extraction. A benzene extract of the fish was treated with diazomethane to produce the methylated derivative of the piscicide. After concentration, the residue was dissolved in acetonitrile then extracted with hexane prior to Florisil clean-up and GC, or conversion to a diazo blue B complex and spectrophotometric evaluation at 552 nm.

Chiral Hexachlorobornane

Enantioselective bioaccumulation measurements including measurements of chiral hexachlorobornane have been carried out on a number of minnow-like fish [265].

Vitamin E (Tocopherol)

Schulz [266] has described a high-performance liquid chromatographic method using fluorimetric detection that permitted the simultaneous determination of four tocopherol isomers. The four tocopherols were completely separated within 15 minutes on a 5 μm silica gel column with a silanised stationary phase. Fluorescence detection, with excitation at 206 nm and measurement at 340 nm, permitted recoveries averaging 95% from spiked samples, based on measurement of the peak areas.

Eulan WA (Polychloro-2-(Chloro Methyl Sulfonamide) Diphenylethers)

Wells and Cowan [267] described a gas liquid chromatographic method for determining this mothproofing agent in fish tissue down to 0.005 mg/kg, in which Eulan was extractively methylated using tetrabutylammonium ion, which forms ion pairs with sulfonamide at pH $10-12$. This ion pair was subsequently back-extracted into the organic phase and methylated using methyl iodide. The methyl derivatives of the Eulan were quantified by gas liquid chromatography using electron capture detection. The coefficient of variation ($p > 0.05$) for perch liver was 4.82 mg/kg $\pm$ 5.6% for between-batch extractions.

To prepare the fish digest, 5 g of fish tissue is ground with sufficient anhydrous sodium sulfate to obtain a free-flowing powder. The sample is then extracted for two hours with 100 ml hexane with a Soxhlet extractor. The extract is then passed down a column containing acidic and basic alumina and eluted with hexane to remove chlorinated insecticides, then with diethyl ether–glacial acetic acid to remove Eulan. Addition of tetrabutyl ammonium hydroxide and methyl iodide to the second extract methylates the Eulan ready for gas chromatographic analysis.

Concentrations of Eulan were determined in perch muscle and liver, with coefficients of variation of 38% and 5.6%, respectively, at the 0.2 and 4.8 mg/kg levels.

Neutral Priority Pollutants

An extraction procedure [268] utilising sonication with acetonitrile and clean-up using aminopropyl and/or C_{18}-bonded silica phases prior to GC has been applied to the analysis of fish extracts. Using this procedure the following values were obtained for chlorinated insecticides and polychlorobiphenyls in a range of Environmental Protection Agency Reference fish samples:

	Found mg/kg	Normal
DDE	18.6 ± 2.7	20.0 ± 1.0
DDD	6.8 ± 2.7	8.0 ± 0.4
DDT	4.2 ± 1.1	7.5 ± 0.3
PCB 1254	3.12 ± 1.32	2.2 ± 0.08

Analytical recoveries from fish were variable, ranging from 7% (naphthalene) to 60 – 70% (isopherone, fluorene, hexachlorobenzene, anthracene, pyrene, dimethyl phthalate, diethyl phthalate, di-*n*-butyl phthalate, *n*-butyl benzylphthalate, bis(2-ethylhexyl) phthalate and *n*-octyl phthalate). Crysene was exceptional in giving a recovery of 76%.

α,α,α-Trifluro-4-Nitro-*m*-Cresol

This pesticide has been determined gas-chromatographically in fish in amounts down to 0.01 ppm [269]. The fish sample is homogenised with hexane–ethyl ether (7:3). The phenol is extracted into 0.1 M sodium hydroxide, back-extracted into hexane–ether (7:3) after acidification, and methylated with diazomethane. The product formed is analysed by GC at 140 °C using a glass column (1.8 m × 4 mm) packed with 3% of OV-1 on 80 – 100 mesh Chromosorb W, with nitrogen as carrier gas (60 ml/min) and electron capture detection.

B$_2$ Vitamins

A sensitive method has been described [270] for the determination of riboflavin, flavin mononucleotide and flavin adenine dinucleotide in sera of different fish using reversed-phase HPLC with fluorimetric detection. Trichloroacetic acid was used to isolate B$_2$ vitamins from the serum, and an aliquot of this solution was analysed by HPLC using a Zorbax-NH$_2$ column with methanol-0.2 mol/l phosphate buffer (1 + 9) as the mobile phase. The detection limits of riboflavin, flavin mononucleotide and flavin adenine dinucleotide in the sera were 4.89, 9.13 and 73.1 ng/ml, respectively.

Samples were prepared for analysis either by acid hydrolysis in 1 N hydrochloric acid at 120 °C for one hour or by enzyme hydrolysis at pH 4.2. Proteins were then precipitated with trichloroacetic acid to provide a solution ready for HPLC. Typical concentrations in fish sera were: riboflavin 0.31 – 0.37 µg/ml, flavin mononucleotide LD 0.05 µg/ml, and flavin adenine dinucleotide 0.25 – 0.44 µg/ml. Recoveries were in the range 92 to 97%.

Saraflaxaxin

Schilling et al. [412] used liquid chromatography combined with tandem mass spectrometry to confirm the presence of this antibacterial agent in catfish.

Mirex (Dechlorane, C$_{10}$H$_{12}$)

Markin et al. [209] have discussed the possible confusion between Mirex and PCBs in the analyses of crabs, shrimp, fish and fish products. In their method, the samples were thoroughly scrubbed to remove mud, algae and other residues; they were ground whole and mixed in a Waring blender to make a composite sample. Samples were prepared and analysed on a whole-body basis as received. A 20-g subsample of the composite was removed and analysed as follows. The homogenised sample was extracted with a mixture of hexane and isopropanol, and the extract subjected to a concentrated sulfuric acid clean-up. The sulfuric acid destroys dieldrin, endrin and organophosphorus insecticides, but the improvement in sensitivity by this clean-up was considered more than adequate compensation for the loss of these other insecticides. The final extract was cleaned upon a Florisil column and concentrated to the desired level for analysis. If PCBs were suspected in the first analysis (their presence usually being indicated by a series of characteristic peaks), the sample was reprocessed to separate the PCBs from the insecticides as described by Armour and Burke [277], Gaul and Leuz La Grange [278], and Markin et al. [279]. After concentrating to the appropriate volume, the extract from both methods of clean-up were chromatographed on a gas chromatograph equipped with dual electron capture detection. Each sample was analysed on two different columns: the first column was a mixture of 1.5% OV-17 and 1.95% QF-1 on Gas Chrom Q. The second column was 2% DC-200 on Gas Chrom Q. Levels of detection were 0.001 mg/kg for DDT and its metabolites, 0.005 mg/kg for Mirex and 0.01 mg/kg for Aroclor 1260.

Markin et al. [209] found Mirex in only a minority of the samples they analysed, contrary to results obtained by earlier workers. All samples containing Mirex were from around Savannah, Georgia, an area with a history of concentrated Mirex use among the most extensive in the United States. The recovery of Mirex in only 12% of the samples, all from one area, could indicate that Mirex is not so general nor widespread as a contaminant of seafood as are PCB and DDT. This does not correspond to earlier seafood studies [280–282] which reported that Mirex occurred much more frequently and densely in many of these same collection sites. Probably the reason for the discrepancy between their study and earlier studies is the confusion of Aroclor 1260 with Mirex. The retention time for the last peaks of Aroclor 1260 is almost identical to the retention time for Mirex on most columns routinely used for analysing Mirex [289]. Unless extensive additional clean-up procedures are employed such as those used by Markin et al. [209] it is almost impossible to separate these two peaks. In their study, if the problem of PCB confusion had not been recognised and the special clean-up procedure used, the PCB peaks would probably have been reported as Mirex.

Kaiser [283] used gas chromatography–mass spectrometry to identify Mirex (dechlorane, $C_{10}H_{12}$) in fish in Lake Ontario, Canada. Under standard gas chromatographic conditions, the peak due to this substance is superimposed on that from the PCBs, and, as a result, the presence of Mirex may have been unrecognised and it may therefore have been misinterpreted as a PCB isomer by previous workers. The fish samples were digested with sulfuric acid. The purified extracts were analysed for their PCB contents by two parallel means:

Table 1.16. Sample data for PCB and Mirex residues from two fishes from the Bay of Quinte, Lake Ontario, Canada (from [283])

	Weight, g	PCB as Aroclor			Mirex, ppm
		1242	1254	1260	
Northern longnose gar (*Lepistosteus osseus* (L)) 902 g					
Gonads	32	2.09	1.18	0.44	0.020
Viscera, fat	62	3.14	1.95	0.90	0.041
Liver	17	3.68	2.31	1.08	0.047
Northern pike (*Esox lucius* (L)) 2930 g					
Pectoral to pelvic fin	950	ND	ND	ND	0.025
Post-anal fin	280	0.89	1.01	0.48	0.050

ND: not determined

(1) quantitative determination of PCBs by gas chromatography with electron capture detectors (Chau ASY, Wilkinson WJ, personal communication; [284–286]) and
(2) quantitative investigation of the gas chromatographic peaks by computerised gas chromatography–mass spectrometry [287]. The sample and analytical data obtained are summarised in Table 1.16.

One of the PCB peaks was found to have a different mass spectrometric fragmentation pattern to those of known PCB isomers. The base peak of this compound had a mass-to-charge ratio (m/e) of 272 with an isotope cluster centred on this peak, unambiguously indicating a (C_5Cl_6) + moiety. Mass spectrometric fragmentations show that this cluster is derived from compounds containing perchlorocyclopentadiene units in their molecule structures or, for a very few cases, from similar, highly chlorinated hydrocarbons. Compounds of this kind include insecticides such as aldrin, chlordane, dieldrin, endrin, endosulfan, heptachlor, kepone, Mirex, pentac and toxaphene. The identification of the unknown in the fish samples as Mirex was established by a combination of gas chromatographic and mass spectrometric techniques. For several gas chromatographic conditions the retention volumes of the compounds aldrin, chlordane, endrin, dieldrin, heptachlor and toxaphene were considerably smaller than that of Mirex. The retention volume of Mirex was identical to that of the observed compound, and endosulfan cannot be chromatographed under these conditions. The differentiation between Mirex, kepone and pentac was achieved by combined gas chromatography–mass spectrometry with a computer-controlled system. Thus a constant retention time for the unknown compound was assured. The mass spectrometer was set to observe seven small mass ranges, five of which are relevant to Mirex or kepone fragments, or both. The m/e ranges were as follows: 220–225, 235–241, 253–258, 270–278, 353–361, 451–463, and 505–517, with integration times for each atomic mass unit of 30 ms.

Table 1.17. Mass spectra of Mirex observed in fish samples and of Mirex and kepone standards (from [283])

		Relative intensities	
M/e	Fish sample	Mirex	Kepone
235	50	28	24
236	7	4	3
237	77	55	38
238	8	6	4
239	49	30	25
240	4	4	3
271	15	10	9
272	54	52	50
273	8	5	5
274	100	100	100
275	10	7	7
276	80	76	80
277	6	5	6
278	34	36	35
353	2	3	3
355	9	9	8
356	2	2	6
357	5	4	11
358	2	1	2
359	6	5	12
360	3	1	2
361	4	3	7
451	3	1	4
453	2	1	3
455	0	0	2
457	0	0	7
459	0	0	8
461	0	0	6
463	0	0	4
507	1	2	2
509	1	3	1
511	1	4	0
513	1	2	0

Only major peaks with $m/e \geq 235$ are presented

Table 1.17 lists the observed intensities together with those of Mirex and kepone standards. For $m/e < 360$, the mass spectra of Mirex and kepone are quite similar. Both compounds have base peaks of m/e 272 due to the $(C_5Cl_6)^+$ ion. Mirex, however, has a different fragmentation pattern at the high mass end, which is demonstrated by the mass spectra of the fish samples: peaks due to kepone (m/e 451 – 463) are absent, and those due to Mirex (m/e 505 – 517) are observed. The molecular ions of Mirex (m/e 540 – 544) are of very low relative intensities and were not investigated. All gas chromatographic and gas chromatographic–mass spectrometric data for the compound observed were in *good* agreement with those of authentic Mirex.

Kaiser [283] points out that the analytical conditions employed in many laboratories do not allow for the separation of or the differentiation between PCBs and Mirex. In fact, in the usual analytical procedure, the Mirex peak is exactly superimposed on one of the major Aroclor 1260 peaks. Since it is impossible to differentiate between a PCB and coeluting Mirex using the electron capture detector, there is a strong likelihood that the presence of Mirex in many environmental samples has not been recognised.

Laseter et al. [288] used a combination of GC and mass spectrometry to determine Mirex in fish.

Acid digestions of fish samples are rapidly achieved using microwave oven heating with nitric acid at 105 °C [289,290].

Klingston et al. [291] showed that a vacuum drying procedure compares favourably with sodium sulfate drying of fish samples, but is more automated and/or less labour-intensive.

Secondary ion mass spectrometry, laser ablation inductively coupled plasma mass spectrometry (ICP–MS) and high- resolution plasma mass spectrometry (HRP–MS) followed by ICP–MS have been used to deconstruct fish migratory patterns and contamination of fish habitats [292].

Wells and Hess [293] have reviewed the separation, clean-up and recovery of persistent organic contaminants from biological matrices.

Lima et al. [294] found that ultrasonic-assisted extraction gives twofold better precision and better detection limits than microwave-assisted digestion in biological methods.

Dodo and Knight [295] have described a clean-up method for the removal of lipids from fish extracts. Liquid chromatographic columns of 100% poly (divinylbenzene) were used to determine semi-volatile organics in fish.

Solid-phase microextraction has been used to sample the headspace above fish tissue extracts. Approximately 170 compounds were identified by GC–MS [296].

Tyurin [297] has discussed the choice of bio-tests and bio-indicators available for evaluating the quality of the marine environment.

1.2.14
Methods Based on Supercritical Fluid Extraction (SCFE)

Both Moralez-Munos et al. [301] and Jaevenpaece et al. [298] have applied this technique to the determination of polyaromatic hydrocarbons in fish.

Jaevenpaeae et al. [298] isolated polyaromatic hydrocarbons from smoked and broiled fish by SCFE and then quantitated by reverse-phase HPLC. For the SCFE with carbon dioxide, methanol was needed as a modifier because larger polyaromatic hydrocarbon molecules were only weakly extracted by pure carbon dioxide.

Supercritical fluid chromatography is an excellent method for the analysis of fish oils [299]; no pretreatment of the fish oil sample was needed, unlike gas chromatographic or HPLC methods.

Wells [300] reviewed the application of SCFE to the determination of organics in fish.

1.3
Organometallic Compounds

1.3.1
Organoarsenic Compounds

Fishman and Spencer [302] used an ultraviolet radiation or an acid persulfate digestion procedure to decompose organoarsenic compounds. The automated methods of Agemian and Cheam [303] use hydrogen peroxide and sulfuric acid for the destruction of organic matter, combined with permanganate–persulfate oxidation for the complete recovery of organoarsenic compounds from fish. An automated system based on sodium borohydride reduction with atomisation in a quartz tube is used to determine the inorganic arsenic thus produced.

Agemian and Cheam [303] found that, in the sodium borohydride reduction of inorganic arsenic to arsenic, concentrations of 0.5–1.5 mol/l of hydrochloric acid gave the highest sensitivity; both As(III) and As(V) were detected equivalently. When the hydrochloric acid concentration was increased from 2 to 6 mol/l, the sensitivity for both species decreased, particularly for As(V). Replacement of the hydrochloric acid line with a sulfuric acid line reduced the sensitivity for As(III) by about 30%, and As(V) gave a sensitivity of about 50% As(III). Replicate determinations at a level of 0.10 μg As/g in a fish sample gave a relative standard deviation of 15%.

Maher [304] has described a method for determining inorganic arsenic, monomethylarsenic acid and dimethylarsinic acid in fish. The procedure involves the use of solvent extraction to isolate the arsenic species which are then separated by ion exchange chromatography, and determined by arsine generation.

Beauchemin et al. [305] identified and determined organic arsenic species in a dogfish muscle reference sample using HPLC coupled with ICP–MS, thin-layer chromatography and electron impact mass spectrometry–graphite furnace AAS. The major species (84% of the total arsenic) was arseno betaine, present at 16 μg/g (as arsenic) in the dogfish sample.

Branch [9] applied HPLC with ICP–atomic emission detection to the determination of organoarsenic compounds in fish.

Le et al. [306] coupled HPLC to ICP–MS for the determination of organic and inorganic arsenic species in fish. These workers noted the changes in arsenic separation that occurred in samples stored at 4 °C for nine months.

Hanumura et al. [307] applied thermal vapourisation and plasma spectrometry to the determination of organoarsenic compounds in fish.

1.3.2
Organolead Compounds

Chau et al. [308, 309] have described a simple and rapid extraction procedure for extracting the five tetraalkyl lead compounds (Me_4Pb, Me_3EtPb, Me_2Et_2Pb, $MeEt_3Pb$ and Et_4Pb) from fish samples. The extracted compounds are analysed in their authentic forms by a gas chromatographic–atomic adsorption spectrometry system (GC-AAS). Other forms of inorganic lead do not interfere. The detection limit for fish (2 g) was 0.025 mg/kg. The recovery of the five alkylated lead compounds from fish tissue averaged 74%, and the coefficient of variation was 7.4%. Whilst this method would be applicable to the determination of tetraalkyl lead compounds in fish, the main interest of Chau et al. [308, 309] was in the determination of organically bound lead produced by biological methylation of inorganic and organic lead compounds in the aquatic environment by microorganisms. The gas chromatographic–atomic absorption system used by Chau et al. (without a sample injection trap) for this procedure has been described previously [308]. The extract was injected directly into the column injection port of the chromatograph. In this method, the frozen fish tissue was homogenised in a Hobart grinder and a Polytron homogeniser, and 2 g of the fish homogenated with 5 ml of EDTA reagent and 5 ml of hexane were immediately placed in a 25 ml test tube with a Teflon-lined screw-cap. The contents were shaken rigorously for two hours in a reciprocating shaker and centrifuged to facilitate phase separation. A suitable aliquot, 5 – 10 µl of the hexane phase, was withdrawn and injected into the gas chromatographic–atomic absorption system. Chau et al. [308, 309] pointed out that, since the authenticity of the compounds to be analysed must be preserved, any digestion method with acid or alkali is not suitable, and that extraction seemed to be the method of choice for removing these compounds from samples. For this extraction, they adopted hexane, octanol or benzene for the quantitative extraction of tetramethyl lead and tetraethyl lead from fish homogenates suspended in aqueous EDTA solution. Although ionic forms of lead such as Pb(II), diethyl lead dichloride, and trimethyl lead acetate do not extract in the organic phase, any lead compounds that distribute into this phase as tetraalkyl lead will be determined.

Chau et al. [308, 309] found that tetraalkyl lead compounds have high vapour pressures and are not stable in water. It was observed that water containing 4.2 µg l/l tetramethyl lead decreased to 2.8 and 3.9 µg l/l tetramethyl lead when stored respectively at room temperature and at 4 °C overnight.

Results obtained in measurements of the accumulation of tetramethyl lead in rainbow trout indicated that the tetramethyl lead content of dead trout tissue increased from 0.43 mg/kg (one day of exposure) to 2.09 mg/kg (three days of exposure). Trout exposed for different periods of time to water containing 3.5 µg/l tetramethyl lead were found to contain tetramethyllead.

Chau et al. [28] have also described a method for the determination of dialkyl lead and trialkyl lead compounds in fish. This method involves use of a tissue solubiliser to digest the sample followed by chelation extraction with sodium diethyldithiocarbamate, followed by n-butylation using butyl magnesium chloride to their corresponding tetraalkyl forms, $R_n Pb \, Bu_{(4-n)}$ and R_4Pb, respectively (R = methyl and ethyl). The method determines tetraalkyl lead, ionic alkylmlead (R_2Pb^{2+} and R_3Pb^+) and divalent inorganic lead simultaneously in one sample, all of which are determined by gas chromatography using an atomic absorption detector.

In this method, the fish samples were homogenised a minimum of five times. About 2 g of the homogenised paste was digested in 5 ml of tetramethylammonium hydroxide solution in a water bath at 60 °C for 1 – 2 hours until the tissue had completely dissolved to a pale yellow solution. After cooling, the solution was neutralised with 50% hydrochloric acid to pH 6 – 8. The mixture was extracted with 3 ml of benzene for two hours in a mechanical shaker after addition of 2 g sodium chloride and 3 ml of sodium diethyldithiocarbamate. After centrifugation of the mixture, a measured amount (1 ml) of the benzene was transferred to a glass-stoppered vial and butylated with 0.2 ml of butyl magnesium chloride with occasional mixing for ca. 10 minutes. The mixture was washed with dilute sulfuric acid to destroy the excess Grignard reagent. The organic layer was separated in a cupped vial and dried with anhydrous sodium sulfate. Suitable aliquots (10 – 20 µL) were injected into the gas chromatographic atomic absorption system for analysis.

The recoveries of trialkyl lead and dialkyl lead species at different levels obtained by this procedure are shown in (Table 1.18). The relatively low recovery of dimethyl lead is in agreement with the results of other investigators. Chau et al. [308, 309] noticed that there was a large Pb(II) peak in the fish sample containing spiked dimethyl lead, but this was not found in the standard run in parallel but without the sample. They attributed this

Table 1.18. Recovery and reproducibility of alkyl lead and lead(II) compounds from fish[a]. From [309]

Amount of Pb added[a], µg	Recovery[b], %				
	Me_3Pb	Et_3Pb	Me_2Pb	Et_2Pb	Pb(II)
1	75 (5)	102 (5)	79 (4)	93 (0)	
5	88 (4)	88 (3)	89 (5)	103 (2)	
10	93 (2)	88 (2)	56 (10)	92 (2)	
20	91 (2)	81 (2)	62 (6)	114 (2)	
Average	86	92	71	101	
%RSD ($n = 6$) a[+] 5 mg/kg level	15	7	18	20	14[c]

[a] Fillet, 2 g; spiked compounds expressed as Pb

[b] Average of two results with average deviation in brackets

[c] The fish fillet contained 142 mg/kg of Pb(II) which was used to evaluate the reproducibility. No Pb(II) was added to the sample

Table 1.19. Analysis of environmental samples (St. Lawrence River near Maitland, Ontario, Canada)[a]. From [309]

Sample	Me$_4$Pb	Me$_3$EtPb	Me$_2$Et$_2$Pb	MeEt$_3$Pb	Et$_4$Pb	Me$_3$Pb$^+$	Me$_2$Pb^{2+}	Et$_3$Pb$^+$	Et$_2$Pb^{2+}	Pb^{2+}
Carp	137	–	–	–	780	2735	362	906	707	1282
	–	–	96	142	7475	162	–	1215	1310	4133
Pike	–	–	–	169	1018	215	–	–	–	1040
	–	–	–	146	1125	205	–	53	–	1187
White sucker	–	–	–	–	48384	196	–	3433	4268	3477
	–	–	–	293	2984	95	–	2171	2196	3610
Small mouth bass	–	–	57	187	1204	–	–	223	92	254
	–	–	71	252	1834	–	–	660	275	305
Sediment	–	–	–	142	1152	–	–	187	22	10000
	–	–	–	–	309	–	–	–	–	5582
Macrophytes, mixed										
Surface	–	–	–	–	68	–	–	132	–	4327
4 m deep	–	38	1501	3613	16515	–	–	558	113	59282

[a] Data expressed in µg/kg as Pb, wet weight; whole fish for fish samples
– Not detectable

Table 1.20. Recovery of tetraalkyllead compounds from cod liver homogenate (from [310])

Compound	Amount added, µg Pb	Amount added, ng/kg	Total Pb present prior to spike, µg	Total Pb present after spike, µg	Amount of spike found, µg	Recovery, %
Tetramethyl lead	0.10	20	0.25	0.38	0.13	130
	0.10	20	0.27	0.40	0.13	130
	0.50	100	0.06	0.575	0.515	103
	0.50	100	0.06	0.625	0.565	113
Tetraethyl lead	0.10	20	0.14	0.21	0.07	70
	0.10	20	0.16	0.26	0.10	100
	0.50	100	1.056	1.548	0.492	98
	0.50	100	0.053	0.65	0.542	119
	0.50	100	0.045	0.42	0.375	75

low recovery to the decomposition of dialkyl lead in the fish matrix. Diethyl lead, however, did not decompose significantly and was recovered at near-quantitative levels. For the first time, the occurrence of triethyl and diethyl lead compounds was detected in fish samples and in other environmental materials (Table 1.19).

Chau et al. [308, 309] have also applied derivatisation with butylmagnesium halides followed by GC to the determination of mono-, di-, tri- and tetraalkyl lead compounds in biological samples. Detection was achieved by an atomic absorption detector.

Table 1.21. Concentrations of total lead and tetraalkyl lead in various marine tissues (from [310])

Tissue	Concentration of total Pb, mg/kg	Concentration of total PbR$_4$, mg/kg	%Tetraalkyl lead from total lead
Frozen cod (liver homogenate)	0.39 ± 0.04	0.37 ± 0.003	9.5
		0.010 ± 0.001	
Large, freshly killed cod (liver homogenate)	0.52 ± 0.05	0.125 ± 0.005	24
	A 0.21 ± 0.04[a]	0.028	13.3
Small, freshly killed cod (2 separate lobes analysed)	B	0.044	20.9
Lobster digestive gland (homogenate)	0.20 ± 0.02	0.162 ± 0.004	81
Frozen mackerel muscle (homogenate)	0.14 ± 0.02	0.054 ± 0.005	38.6
Flounder meal	5.34 ± 1.02	4.79 ± 0.32	89.7

[a] Total lead determination for both lobes

Sirota and Uthe [310] have described a fast, sensitive atomic absorption procedure for determining tetraalkyllead compounds in biological materials such as fish tissue. Tissue homogenates were extracted by shaking with a benzene/aqueous EDTA solution, a measured portion of the benzene was removed, and after digestion the residue was defatted if necessary. The resultant Pb^{2+} was determined by flameless atomic absorption spectroscopy using a heated graphite atomiser. Using a sample weight of 5 g, 10 µg/kg of lead as PbR$_4$ can be determined with a relative standard deviation of 5%. No other forms of lead that were tested (e.g. PbR$_3$X, PbR$_2$X$_2$) were found to partition into the benzene layer under these conditions.

The recovery and selectivity of the method was evaluated by adding known amounts of different lead compounds to previously analysed tissue samples. The results obtained are summarised in Table 1.20 and indicate satisfactory recoveries and selectivities for tetraalkyl lead compounds. Various marine tissues were sampled for total lead and tetraalkyl lead. Results are summarised in Table 1.21. Di- and tri-substituted alkyl leads were also evaluated in this system; the results were satisfactory.

Birnie and Hodges [311] have given details of a procedure for the determination of down to 0.01 mg/kg of ionic species of alkyl lead in marine organisms by solvent extraction and differential pulse anodic stripping voltammetry. The sample is homogenised in the presence of a mixture of salts (lead nitrate, sodium benzoate, potassium iodide, sodium chloride, EDTA), which effectively releases the di- and trialkyl lead species present and facilitates their transfer into toluene before back-extraction into dilute

nitric acid in preparation for differential pulse anodic stripping voltammetry. Recoveries were in the range $70-90\%$ (Et_3Pb^+, Et_2Pb^{2+}, Me_3Pb^+) to
$10-40\%$ (Me_2Pb^{2+}).

Lobinski and Adams [312] reviewed methods for the determination of
alkyl lead compounds in fish.

1.3.3
Organomercury Compounds

Over the last few decades, extensive damage has been caused to the environment by mercury and organomercury compounds in commonly used
products such as paints, wood preservatives, paper and pesticides. However,
the toxicities of these compounds vary considerably, necessitating the determination of individual species in order to accurately assess environmental
impact. Methylmercury is one of the most toxic of these mercury species
and is commonly found in the marine environment.

Interest in methylmercury as an environmental contaminant first arose
in the 1960s with reports of alkylmercury poisoning of marine life and
people in Japan and of birds and marine life in Sweden. Most alkylmercury
poisoning incidents have been due to short-term exposure to high levels of
these compounds. Acute poisoning effects include kidney damage, damage
to the central nervous system, and death. Chronic poisoning symptoms
include tremors, constriction of visual field, lack of coordination, damage
to the central nervous system and kidney damage, and it can also be fatal.
Little is known of the long-term effects of low-level alkylmercury exposure,
however. Much concern has been expressed over the potential danger to
humans exposed to low levels of alkylmercury on a daily basis.

The most common mode of human exposure to alkylmercury, predominantly as methylmercury, is through the ingestion of fish and bivalves.
Methylmercury is present at extremely low levels in lower forms of marine
life such as algae and small fish due to constant exposure to methylmercury in ocean water. These lower life forms are consumed by higher life
forms, and these higher life forms are consumed by marine mammals and
large fish. Over time, methylmercury bioaccumulates in the tissue and fats
of organisms due to the inability of most species to efficiently eliminate it
from the body. Methylmercury concentrations increase as they move up the
food web, often resulting in dangerously high levels in some marine species
consumed by humans. Therefore, it is important to monitor methylmercury
concentrations in marine materials to ensure the safety and welfare of both
humans and the marine ecosystem.

Fish frequently have $80-100\%$ of the total mercury in their bodies in
the form of methylmercury, regardless of whether the sites at which they
were caught were polluted with mercury or not [314]. Methylmercury in the
marine environment may originate from industrial discharges or can be syn-
thesised by natural methylation processes [313]. Fish do not methylate inorganic mercury themselves [318–320] but they can accumulate methylmer-

cury from both seawater [320] and food [319]. It has been found in some sediments but only at very low concentrations, mainly in areas of known mercury pollution. It usually represents less than 1% of the total mercury in the sediment, and frequently less than 0.1% [317,322,323]. Microorganisms within the sediments are considered to be responsible for the methylation [317], and it has been suggested that methylmercury may be released by the sediments to the seawater, either in a dissolved form or attached to particulate material and thereafter rapidly taken up by organisms [316,319,321, 325,327,328]. Davies et al. [324] set out to determine the concentrations of methylmercury in seawater samples much less polluted than Minamata Bay; namely The Firth of Forth in Scotland. They described a tentative bioassay method for determining methylmercury at the 0.06 ng/l level. Mussels from a clean environment were suspended in cages at several locations in the Firth of Forth. A small number were removed periodically, homogenised, and analysed for methylmercury by solvent extraction–gas chromatography, as described by Westhoo [326]. The rate of accumulation of methylmercury was determined, and by dividing this by the mussel filtration rate, the total concentration of methylmercury in the seawater was calculated. The methylmercury concentration in caged mussels increased from low levels (less than 0.01 µg/g) to 0.06 – 0.08 µg/g in 150 days, giving a mean uptake rate of 0.4 ng/g/day; i.e., a 10 g mussel accumulated 4 ng/g/day. The average percentage of total mercury in the form of methylmercury increased from less than 10% after 20 days to 33% after 150 days. Davies et al. [324,343] calculated the total methylmercury concentration in the seawater as 0.06 µg/l, i.e., 0.1 – 0.3% of the total mercury concentration, as opposed to less than 5-32 ng/l methylmercury found in Minamata Bay, Japan. These workers point out that a potentially valuable consequence of this type of bioassay is that it may be possible to obtain estimates of the relative abundance of methylmercury at different sites through the exposure of 'standardised' mussels (as used in their experiment) in cages for controlled periods of time, and through comparisons of the resultant accumulations of methylmercury.

1.3.3.1
Sample Digestion Procedures

Sodium Hydroxide Digestion

Magos [331] has described a simple method for the determination of total mercury in biological samples contaminated with inorganic mercury and methylmercury. The method is based on the rapid conversion of organomercurials, first into inorganic mercury and then into atomic mercury suitable for aspiration through the gas cell of a mercury vapour concentration meter, by a combined tin(II) chloride–cadmium chloride reagent. It was found that, if 100 mg of tin(II) chloride alone were added instead of the tin(II) chloride–cadmium chloride reagent, only the release of inorganic mercury influenced the peak deflection of the potentiometer, thus permitting the selective determination of inorganic mercury, and then (after re-acidification of the reaction mixture) methylmercury, by adding the tin(II) chloride–cadmium chloride reagent and sodium hydroxide. When total mercury and inorganic mercury were determined separately, the difference between the results gave the methylmercury content of the sample.

Sulfuric Acid–Nitric Acid–Hydrogen Peroxide Digestion

In this method [341], the fish sample is treated with boiling sulfuric acid, nitric acid and hydrogen peroxide, with excess peroxide being removed by boiling and the addition of potassium permanganate. A portion of the digest is reduced with hydroxylammonium chloride solution and stannous chloride, aerated through a cell in an atomic absorption spectrometer fitted with a cold cathode mercury lamp, and the absorption measured at 253.7 nm. Some results obtained by this and other methods in round robin determinations of total mercury and methylmercury in tuna are shown in Table 1.22. Total mercury was determined by the atomic absorption method discussed above, whilst methylmercury was determined by the gas chromatographic method discussed below [341].

Table 1.22. Determination of total mercury in tuna: collaborative tests (from [341])

Laboratory	By method discussed in [341]	Mean	Blank
	All figures are concentrations of mercury in mg/kg, corrected for the blank		
1	0.53, 0.56	0.55	0.02
2	0.49, 0.52, 0.51, 0.62, 0.62, 0.71, 0.75, 0.60, 0.66	0.61	–
4	0.54	0.54	–
6	–	–	< 0.01
7	–	–	–
8	0.41, 0.45	0.43	0.01
9	0.49, 0.47	0.48	< 0.02
10	0.55, 0.57, 0.56, 0.55, 0.57, 0.58, 0.56	0.56	< 0.01
11	–	–	–

Mean: 0.56
Standard deviation: ± 0.08

Determination of methylmercury in tuna: second round of collaborative tests

Laboratory	By method of [341]
	All figures are concentrations of mercury in mg/kg, corrected for the blank
1	0.49, 0.48
2	0.44, 0.45, 0.43
4	0.48, 0.52
8	0.39, 0.42
9	0.47, 0.48
11	–

Mean: 0.46
Standard deviation: ± 0.04

Hydrochloric Acid Digestion

In this procedure [340], the fish sample is digested with 2 mol/l hydrochloric acid, and then organic mercury is extracted into benzene. Organic mercury is then extracted from the benzene phase with 0.0003% glutathione in 0.1 mol/l ammonia solution.

To a gas washing bottle, 150 ml of water, 10 ml of 10 mol/l sodium hydroxide, 2 ml of 1000 ppm copper solution and 5 ml of 5% tin(II) chloride dihydrate solution are added. Nitrogen gas at a flow rate of 1.4 l/min is passed for six minutes to eliminate any mercury in the reagent solutions. Then the aqueous back-extract from the sample is added. Mercury is concentrated on 1.5 grams of gold granules (about 1 mm diameter) packed in a glass tube (4 mm id) by passing nitrogen gas for six minutes. The gold granules are heated in a boat to 500 °C in a furnace for two minutes, and the absorbance at 253.7 nm is measured by passing nitrogen gas at a flow rate of 1.2 l/min. Between 0.1 and 1.0 mg/kg mercury in rockfish samples were determined by this method.

Hydrobromic Acid Digestion–Toluene Extraction

Capelli et al. [346] have described a procedure for the determination of down to 0.7 ng/kg mercury in fish.

A sample of 5.00 – 10.00 g of fish flesh is homogenised with 10 ml of water in a mortar and then transferred into a centrifuge tube with water; 14 ml of 47% m/m hydrobromic acid is then added (i.e., 2 N acid concentration in the total volume).

Then 50 ml of toluene are added and the tube is shaken for five minutes. After centrifuging, 25 ml of the organic phase are transferred into another 100 ml test-tube; 6 ml of 1% cysteine solution are added and the solution is shaken for two minutes and then centrifuged, and 5 ml of the aqueous phase is diluted with water to 50 ml. Mercury is determined in this solution by atomic absorption spectroscopy with the cold vapour technique, using the standard additions method. Then 5 ml of the solution is placed in the bubbler and 2 ml of 45 wt% sodium hydroxide solution and 1 ml of the reducing solution [tin(II) chloride–cadmium chloride] are added. A stream of air or argon at 0.3 l/min is used to strip the mercury vapour and to convey it into the silica-walled cell across the spectrophotometer beam.

Table 1.23 compares determinations of total mercury in fish obtained by the above method with those of methylmercury obtained by a gas chromatographic method [340]. It would appear that up to 77% of the mercury is present as the methylmercury derivative.

Acid Cupric Bromide Digestion-Toluene Extraction

Shum et al. [345] have described a procedure for determining down to 0.8 mg/kg methylmercury in fish, in which an acetone extract of the sample is

Table 1.23. Determination of total mercury and methyl mercury in lyophilised samples of tuna (from [346])

Sample	Mercury as total mercury, µg/g, Capelli [346]			Mercury as methyl mercury, µg/g, Capelli [346]		
	Uthe, GLC [347]	Average[a], AAS	95% confidence interval	Uthe, GLC [347]	Average[a], AAS	95% confidence interval
Lyophilised 29-63 (Liver of white tuna fish from the Atlantic)	0.50	0.52	±0.3	0.34	0.334	±0.012
Lyophilised 78-31 (Muscle of red tuna fish from the Mediterranean Sea)	3.00	3.14	±0.19	1.90	1.89	±0.025
Lyophilised CG-56 (Muscle of red tuna fish from the Mediterranean Sea)	11.50	10.7	±0.6	8.20	8.29	±0.12

[a] Average values from the results of five determinations

Table 1.24. Determination of methylmercury (MeHg) in fish and shellfish (in µg/g Hg) by the heated graphite furnace and gas chromatography techniques (from [345])

Sample	MeHg in toluene extract	MeHg in Na$_2$S$_2$O$_3$ extract	Total Hg	MeHg by GLC	Total Hg	Inorganic	Total Hg
Freeze-dried							
Oyster	ND	0.06, 0.06	0.18 ± 0.02[a]	0.02	0.25	–	–
Halibut	5.37, 5.68	5.49, 5.55	5.79 ± 0.07	4.55	4.77	0.200	5.85
Canned tuna	0.76 ± 0.05 (6)	0.78, 0.74	0.92 ± 0.03	0.60	0.80	0.034	0.95
Scallop muscle	ND	0.06, 0.06	0.06 ± 0.02	0.05	0.11	–	–
Lobster tomalley	ND	0.05, 0.06	0.74 ± 0.02	0.09	0.64	–	–
Swordfish muscle 1	0.43 ± 0.06	0.40, 0.42	0.41 ± 0.02	–	–	0.022	0.42
Swordfish muscle 2	2.37 ± 0.06	2.52, 2.48	2.54 ± 0.07	–	–	0.109	2.45
Swordfish muscle 3	3.00 ± 0.10	–	3.34 ± 0.06	2.75	3.03	0.143	3.31
Canned tuna 1	0.40, 0.40	0.41, 0.42	0.50 ± 0.01				
Canned tuna 2	1.90, 1.95	–	2.23 ± 0.19				
Frozen							
Swordfish liver 1	0.41, 0.42		1.05, 1.00				
Swordfish liver 2	8.67, 8.25	9.14, 8.39	19.1 ± 0.72				
Swordfish liver 3	1.50, 1.62	1.38, 1.81	2.27 ± 0.12				
Swordfish muscle 1	1.96 ± 0.15 (4)	–	2.02, 1.99				
Swordfish muscle 2	3.98, 4.22	–	4.40, 4.41				
Swordfish muscle 3	1.60, 1.50	–	1.71, 1.70				
Swordfish muscle 4	0.87, 0.87	–	0.92 ± 0.12				

[a] Figures are the mean and standard deviations of triplicate analyses except where otherwise indicated

ND: not detected

treated with acidic cupric bromide in order to release methylmercury, which is then extracted into toluene. Addition of dithizone to the extract allows the determination of mercury by graphite furnace atomic absorption spectrometry. The average recovery of methylmercuric chloride added at 2.00 and 4.00 μg levels is 97.7 + 5.5%. Determination of methylmercury in aqueous sodium thiosulfate, after partitioning from toluene, permits an autosampler to be used. A number of fish samples were analysed for methylmercury using this procedure (Table 1.24). Total mercury contents of these samples were also determined for comparison samples. The average recovery was 97.7 + 5.5%. An average of 92.2% of the mercury in muscle samples was methylmercury. This showed that prewashing the samples with acetone did not remove appreciable amounts of tightly bound organic mercury. The swordfish liver samples, however, contained a much lower percentage of methylmercury than muscle. This was not surprising, because the liver had been suggested as one of the sites associated with the demethylation of methylmercury.

Gutierrez et al. [351] have reported improvements in digestion procedures for inorganic mercury and methylmercury in fish.

1.3.3.2
Atomic Absorption Spectrometry (AAS)

Various workers have applied cold vapour atomic absorption spectrometry to the determination of organomercury in compounds in fish [303, 329, 329–349, 379–382]. As reported above, various methods have been used to decompose organomercury compounds prior to AAS, including digestion with acidic potassium permanganate [348], sodium hydroxide [331], sulfuric acid–hydrogen peroxide [336, 338–341], sulfuric acid–nitric acid [330, 342], hydrochloric acid digestion–benzene extraction [340], hydrochloric acid digestion–toluene extraction, and acidic cupric bromide digestion–toluene extraction [299]. Steam distillation has also been used to isolate organomercury compounds from fish prior to determination by cold vapour AAS [348].

Stuart [342] used ^{203}Hg-labelled methylmercury chloride to label fish in vivo in order to study the efficacy of various wet ashing procedures, and obtained 93% recovery of activity in digestions of fish with concentrated sulfuric acid and fuming nitric acid.

1.3.3.3
Plasma Atomic Emission Spectrometry (PA–ES)

Palmeri and Leonel [352] determined methylmercury in fish at sub-ppt levels by employing a microwave-induced PA-E spectrometer.

Fukushi et al. [42] performed subnanogram determination of organic and inorganic mercury in fish by helium microwave–IPA-ES. Detection limits were around 10 pg, and organic mercury was determined as the difference between total and inorganic mercury.

1.3.3.4
Gas Chromatography (GC)

This technique is essential if it is necessary to obtain an unequivocal identification of the type of organomercury compound present in a biological
material as opposed to the total organic plus inorganic mercury content
that is provided by AAS. An ideal combination is to use GC for separation
of the organomercury compounds in combination with a flameless atomic
absorption or an ICP–MS spectrometer used as a detector system. Much of
the original work on the application of GC to the identification and determination of organomercury compounds in biological materials was performed
by Westhoo [353–355]. In view of the comparatively high mercury contents
of fish found in Swedish lakes and rivers, Westhoo et al. embarked on an
extensive survey of the nature and the concentration of mercury in fish from
these waters.

He describes a combined gas chromatographic and thin-layer chromatographic method [353–355] for the identification and determination
of methylmercury compounds in fish, in animal foodstuffs, egg yolk, meat
and liver. He has also used a combination of GC and mass spectrometry to
identify and determine methylmercury compounds in fish [356].

To extract organically bound mercury from muscle tissue of fish, Westhoo homogenised the fish with water and acidified with concentrated hydrochloric acid (to a fifth of the volume of the suspension). Organomercuric
compounds were then extracted in one step with benzene using the method
described by Gage [357]. Methylmercury which was either originally present
or had been added to the fish could only be extracted with difficulty when
only a small amount of acid was added (e.g. at pH 1). Organomercury could
be extracted from an aliquot of the benzene solution with ammonium or
sodium hydroxide solution saturated with sodium sulfate to eliminate lipids.
The yields were low and variable, but could be improved as described below.

Several workers have found that a clean-up procedure is required to
remove fatty acid and amino acids which could otherwise poison the gas
chromatography column. The clean-up is achieved by adding a reagent such
as sulfide [357], cysteine [353–355], sodium thiosulfate or glutathione to the
organic phase, which forms a strong water-soluble alkylmercury complex
to extract the mercury complex into the aqueous phase. A halide is added
to the aqueous phase, and the alkylmercury halides formed are re-extracted
into an organic phase. Aliquots of this phase are finally injected into the
gas chromatograph.

The mercury compound in the shellfish that caused the Minimata disease
(Japan) was methyl(methylthio)mercury. Westhoo concluded that it is reasonable to assume that methylmercury, if present in Swedish fish, should be
a methylthio derivative, at least to some extent. The Hg–S bond is stronger
than the Hg–NH bond or the Hg–OH bond. Accordingly, it prevents the
formation of these bonds, which should be produced by the ammonium hy-

droxide solution and increase the solubility in water. Any methylthio group present should therefore be removed before the extraction with alkali.

Distillation of the benzene extract at reduced pressure at room temperature or at 760 mm Hg pressure at 80 °C to 1/10 of the original volume removed the factor that prevented an acceptable extraction by ammonium or sodium hydroxide solution (probably methanethiol and perhaps hydrogen sulfide). After distillation and subsequent extraction with ammonium hydroxide solution, the extract was acidified with hydrochloric acid and the organomercury compound was extracted once with benzene. After drying with anhydrous sodium sulfate, the benzene solution was ready for GC and, after concentration, also ready for thin-layer chromatography.

About 30% of the methylmercury was lost in the above procedure, mainly due to unfavourable partition coefficients. In a model experiment of the benzene extraction of methylmercury from a hydrochloric acid solution, 14% (for example) of the methylmercury was left in the water layer. The losses by partition are, however, characteristic of the compounds involved and reproducible. Consequently, they can be included in the calibration curve, thus disturbing the results only slightly. The yields can be increased by repeated extractions, but good results are obtained with the above simple procedure. The calibration curve is based on the partition laws for methylmercury chloride, through some methylmercaptide and perhaps sulfide are probably present in fish. However, when hydrogen sulfide or methanethiol was added (30 µg per 5 µg mercury as methylmercury) to the aqueous phase before the first extraction, the 5 µg point was unaltered on the calibration curve. Large amounts of these sulfur compounds disturbed the analysis because they were not completely removed by the distillation.

When known amounts of methylmercury dicyandiamide were added to saltwater fish (frozen cod (*Gadus morrhua*), or haddock (*Gadus aeglefinus*)), 82 – 95% of the additions were recovered.

Westhoo [353–355] used an electron capture detector and 150 cm × 3 mm (60 in × 8 in) stainless steel columns filled with Carbowax 1500 (10%) on Teflon 6 and washed DMCS. Nitrogen was used as carrier gas, and column temperatures were 130 – 145 °C. He identified methylmercury chloride in pike caught in the Baltic Ocean at concentrations of between 0.07 and 4.4 mg/kg of fish.

Westhoo [355] pointed out that if methylmercury attached itself to a sulfur atom by reaction with a thiol or hydrogen sulfide, then the nonvolatile HgS compound produced would not be included in the determination. He has developed a modification to this method in order to render it applicable to a wider range of foodstuffs (egg yolk and white, meat, liver or fish) by binding interfering thiols in the benzene extract of the sample to mercuric ions added in excess or by extracting the benzene extract with aqueous cysteine to form the cysteine methylmercury complex.

Westhoo et al. [356] reported results obtained by gas chromatography with electron capture and with mass spectrometric detection on a range of

Table 1.25. Comparison between results for mercury levels in fish flesh, determined by combination gas chromatography–mass spectrometry, gas chromatography with electron capture detection and activation analysis (from [356])

	Methylmercury (mg Hg/kg fish flesh)		Total mercury (mg/kg fish flesh)
	GLC-MS measurement of $^{202}Hg^+$	GC-ECD	NAA
Pike 1	0.14	0.17	ND
Pike 2	0.55	0.54	0.59
Pike 3	2.53	2.57	2.70
Pike 4	0.43	0.41	0.39
Pike 5	0.49	5	0.54
Pike 6	0.75	0.66	0.63
Pike 7	0.72	0.70	0.66
Pike 8	3.19	3.29	3.12

samples of fish (Table 1.25). Total mercury was also determined on these samples by neutron activation analysis. Results obtained by the three methods agree within ±10% of the average value.

It was mentioned above that in the Westhoo method [355] for organomercury compounds in fish, low recoveries are obtained unless the benzene extract of the fish homogenate is boiled to remove volatile mercaptans prior to extraction with ammonia. This distillation procedure was assumed to remove volatile thio compounds that bind some of the methylmercury and prevent its uptake into ammonia.

Longbottom et al. [359] used the Westhoo clean-up procedure to detect down to 10 μg/kg of methylmercury. They improved the Westhoo clean-up procedure by replacing cysteine with the more stable sodium thiosulfate when forming the methylmercury adduct. For the GC of methylmercury iodide, these workers recommend the use of a ^{63}Ni electron capture detector, as it does not form an amalgam at 280 °C, the temperature at which it is used.

Kamps and McMahon [360] determined methylmercury in fish by GC. The method involves the partitioning of methylmercury chloride in benzene and analysis with electron capture detection. Down to 0.02 ppm of methylmercury chloride were detected in a 10 g sample.

Bache and Lisk [361] determined methylmercury compounds in benzene extracts of fish by chromatography on a 60 cm glass column of Chromsorb 101 or 20% 1:1 OV-17/QF-1. Detection of the separated organomercury compounds was achieved by measurement of the emission spectrum of the 253.7 nm atomic mercury line, which gave a linear response over the range 0.1 – 100 mg of injected methylmercury chloride. Average recoveries of methylmercury chloride in fish were 62% at the 0.3 mg/kg level.

Newsome [362] has described a method for the determination of methylmercury in fish, in which the sample (10 g) is homogenised for ten minutes with 1 N hydrobromic acid and 2N potassium bromide (60 ml) and filtered through glass wool. The combined filtrate is extracted twice with benzene. The combined benzene layers are extracted with a cysteine acetate solution, an aliquot of which is acidified with 48% hydrobromic acid and extracted with benzene. The benzene extracts are submitted to GC on a glass column (40 cm × 4 mm) packed with 2% of butanediol succinate on Chromosorb W (AW–DCMS) (100–120 mesh) operated at 120 °C with nitrogen as carrier gas (80–100 ml/min) and a ^{3}H foil electron capture detector. The sensitivity of the method is in the range 0.01–0.90 mg Hg/kg. Mean recovery generally exceeds 95%. When direct gas chromatographic methods are used in the determination of alkylmercury compounds, interferences are often a problem, especially with the electron capture detector, which is sensitive to other halogen compounds.

Uthe et al. [350] have described a rapid semi-micro method for determining methylmercury in fish. The procedure involves extracting the methylmercury into toluene as methylmercury(II) bromide, partitioning the bromide into aqueous ethanol as the thiosulfate complex, re-extracting methylmercury(II) iodide into benzene, followed by GC on a glass column (4 ft × 0.25 in) packed with 7% of Carbowax 20 M on Chromosorb W and operated at 170 °C with nitrogen as carrier gas (60 ml/min), as well as electron capture detection. Down to 0.01 mg/kg of methylmercury in a 2 g sample could be detected.

The application of GC with electron capture detection has also been discussed by Zeleuko and Kosta [383].

Longbottom [363] cooled the gases from the flame ionisation detector and led the gases through an atomic absorption spectrometer, but reported that it was less sensitive than the electron capture detector for dialkyl mercury compounds. Bye and Paus [364] solved this problem by leading the effluent from the gas chromatographic column through a steel tube in a furnace at a temperature at which the organic mercury molecules are cracked. The products are then led through a 10 cm quartz cuvette placed in the beam from a hollow-cathode lamp in an atomic absorption spectrometer. These workers state that for many of the earlier methods, the calibration curves are obtained from measurements of peaks from pure standard solutions of organic mercury compounds. They doubt the correctness of such a procedure, because it does not take into account the fact that appreciable amounts of mercury may be lost during the many extraction steps used in the analysis, especially in work with small samples and small volumes. They state that a standard addition procedure should be used for calibration, and the standard organic mercury solution should be added as early as possible in the procedure.

A Perkin–Elmer model 800 gas chromatograph was used. The following operating conditions were satisfactory: column, 10% SP2300 on Chromosorb W 80-100 mesh; oven temperature 145 °C; inlet temperature 200 °C; nitrogen carrier gas at a pressure of 3.5 kPa/cm^2 measured at the GC inlet; 90 ml/min flow rate.

The Perkin–Elmer model 303 atomic absorption spectrometer was run at the 254 nm mercury line. Deuterium background correction was essential. Portions (0.5 g) of frozen fish were transferred to a tissue grinder, 0.5 µl of 1 M copper sulfate solution was added to each, and 50 – 100 µl of the standard mercury solution were added to two of the samples. A fairly detailed workup procedure in which the sample is treated successively with bromine, sodium thiosulfate and potassium iodide is followed. A final benzene extract is obtained for GC.

Bye and Paus [364] detected methylmercury (not ethyl or phenylmercury) in fish samples. Ranges up to 10 mg/kg mercury as methylmercury and ethylmercury chloride in mixtures was measured. Fish samples were found to contain 2.2 mg/kg of mercury as methylmercury.

Ealy et al. [365] discussed the determination of methyl-, ethyl- and methoxymercury halides in water and fish. The mercury compounds were separated from the samples by leaching with M-sodium iodide for 24 hours and then the alkylmercury iodides were extracted into benzene. These iodides were then determined by GC of the benzene extract on a glass column packed with 5% of cyclohexane–succinate on Anakron ABS (70 – 80 mesh) and operated at 200 °C with nitrogen (56 ml/min) as carrier gas and electron capture detection. Good separation of chromatographic peaks was obtained for the mercury compounds as either chlorides/bromides or iodides. The extraction recoveries were monitored using alkyl mercury compounds labelled with ^{208}Hg.

In an official procedure [341], 5 g of the homogenised sample is placed in a centrifuge tube and 25 ml of sodium hydroxide added. The tube is heated at 100 °C for 30 minutes, and then 8 ml of hydrochloric acid (specific gravity 1.18), 25 ml of freshly distilled toluene and 1 ml of 1 mol/l copper(II) sulfate are added and the mixture shaken and centrifuged. The toluene layer is siphoned into a 125 ml separating funnel.

The combined toluene extracts are shaken successively with 3 and 2 ml portions of cysteine hydrochloride reagent solution, and the extracts combined in a 25 ml separating funnel. Then 1 ml of hydrochloric acid (specific gravity 1.18) is added prior to extraction with 10 ml of toluene. The cysteine solution is run into a second 25 ml separating funnel, retaining the toluene layer in a separate container. The toluene extracts are combined and made up to 25 ml. This solution is now ready for gas chromatography on a Carbosorb 20 M Chromosorb G column at 160 °C using a ^{63}Ni electron capture detector and nitrogen as carrier gas.

Cappon and Crispin Smith [366] has described a method for the extraction, clean-up and gas chromatographic determination of alkyl and arylmercury compounds and inorganic mercury in fish, crustacea and other biological samples. Detection limits for fish and crustea were 0.02 and 0.001 mg/kg, respectively. Methyl-, ethyl- and phenylmercury are first extracted as the chloride derivatives. Inorganic mercury is then isolated as methylmercury upon reaction with tetramethyltin. The initial extracts are subjected to thio-

Table 1.26. GC–AA intercomparison study (from [366])

		Mg/l Hg	
	GC	AA (as MeHg)	GC-AA
Fish	1.0	1.06	1.04
Muscle	0.27	0.70 as EtHg	1.03
Kidney	0.66	0.68 as inorganic	0.97
Fish	0.08	0.07	1.14
Sediment	0.17	0.19	0.89

sulfate clean-up, and the organomercury species are isolated as the bromide derivatives.

Total mercury recovery ranges between 75 and 90% for both forms of mercury, as assessed by using appropriate ^{203}Hg-labelled compounds for liquid scintillation spectrometric assay. Specific gas chromatographic conditions permit the detection of mercury concentrations of 1 ng/kg. Mean deviation and relative accuracy average 3.2 and 2.2%, respectively. The accuracy and precision of this procedure was evaluated by analysing different sample types fortified with mercuric chloride and methylmercuric chloride. Results were cross-checked by an atomic absorption procedure. Results obtained on samples by both methods are given in Table 1.26. There is good agreement between the two methods for samples containing methyl, ethyl and inorganic mercury, and this is expressed in terms of gas chromatographic/atomic absorption ratios.

Callum et al. [367] used the proteolytic enzyme subtilisin Carlsberg Type A for the breakdown of fish tissues prior to the release of methylmercury. The finely chopped tissue is homogenised with 1 mol/l tri(hydroxymethyl)-amino methane–hydrochloric acid buffer (pH 8.5) and then incubated with the subtilisin for an hour at 50 °C. Then 2 ml of 40% w/v sodium hydroxide solution and 1 ml of 1% w/v cysteine hydrochloride solution were added and the samples stirred for five minutes at 50 °C. When cool, 1 ml of 0.5 mol/l copper(II) bromide and 10 ml of acidic sodium bromide were added. The methylmercury(II) bromide was then extracted with two 5 ml portions of toluene. In each extraction, the mixtures were shaken for two minutes then centrifuged at 6000 g for ten minutes. The two toluene extracts were removed and combined, and the methylmercury was extracted twice with 1 ml of ethanolic sodium thiosulfate solution (a 1 : 1 mixture of 95% ethanol and 0.005 mol/l sodium thiosulfate solution). During each extraction, the solutions were vortex-mixed and centrifuged at 4000 g for two minutes. The lower aqueous layers were removed and combined, and then 0.5 ml of 3 mol/l potassium iodide was added to these combined aqueous extracts followed by 0.5 ml of benzene (pesticide grade, distilled in glass) containing ethylmercury(II) iodide as an internal standard. These solutions were shaken and then centrifuged at 3000 g for one minute. Standard solutions of methylmercury(II) iodide were prepared in the benzene containing

Table 1.27. Analysis of tuna and dry fish (from [367])

	Method	Methylmercury (mg/kg)	Range (mg/kg)	Coefficient of variation (%)
Tuna	Enzyme	0.693	0.661 – 0.734	3.7
	Sodium bromide	0.912	0.812 – 1.006	5.5
Fish homogenate	Enzyme	0.057	0.043 – 0.069	15.4
	Sodium bromide	0.346	0.329 – 0.362	3.6

the internal standard. Samples were analysed by GC on a column comprising 5% w/w ethylene glycol adipate polyester on 80 – 100 mesh Gas-Chrom G at 155 °C with electron capture detection.

Methylmercury recoveries of 96 – 97% from tuna were obtained by this procedure. The coefficient of variation was in the range 3.6–5.5%.

Table 1.27 shows a comparison of the results obtained by this procedure with the values obtained by a method involving extraction with acid sodium bromide alone.

Decadt et al. [368] determined methylmercury in bird liver and kidney samples using a headspace injection system coupled to a gas chromatograph–microwave-induced plasma system. Vapour concentrations decreased in the following order; $CH_3HgI > CH_3HgBr > CH_3HgCl$. The methylmercury compounds in the sample were transformed into methylmercury iodide by reaction with iodoacetic acid:

$$(CH_3)_2Hg + ICH_2COOH = CH_3HgI + CH_3CH_2COOH$$

The detection limit in fish organ samples was 1.5 µg/L of homogenate.

Workers at the Society for Analytical Chemistry, London [369] utilised GC with an electron capture detector for the determination of down to 0.25 mg/kg of methylmercury chloride in swordfish, shark, shrimp, oyster, clams and tuna fish.

Hight [370, 371] extracted methylmercuric chloride from homogenised, acetone-washed, acid-digested (hydrochloric acid) fish tissue using toluene. Toluene extracts were analysed by GC with electron capture for the detection, using a 5% DEGS-PS column pretreated with mercuric chloride. Samples of swordfish, shark, shrimp, oysters, clams and tuna were analysed for methylmercury. The detection limit for the method was 0.25 mg/kg.

Panaro et al. [372] determined methylmercury in fish by GC–direct current plasma atomic emission spectrometry (DCP–AES).

The sample was dissolved in methanolic potassium hydroxide and then ethylated by derivitisation with sodium methyl borate. The reaction products were trapped cryogenically and then analysed by GC. Down to 4 pg/g (as Hg) of $CH_3 Hg^+$ could be determined.

Cai and Bayona [374] determined methylmercury in fish in situ using sodium tetraethylborate derivitisation followed by GC–mass spectrometry.

Derivitisation was carried out in a vial followed by solid-phase microextraction and aqueous solid-phase microextraction were studied.

Fisher et al. [373] has described a rapid method for the determination of methylmercury in fish involving dissolution of the sample in methanolic potassium hydroxide, aqueous-phase ethylation, cryogenic trapping on a packed gas chromatographic column, and GC with an atomic adsorption spectrometric detector. Low-level detection limits have been reported [375, 378, 379] in determinations of alkylmercury species in biological marine materials using GC.

1.3.3.5
Microwave-Induced Plasma Atomic Emission Detection (MIP–AED)

This method includes a solid-phase extraction procedure with preparative gel permeation chromatography clean-up.

Donais et al. [375] developed a chromatographic method for the quantification of alkylmercury species using MIP–AED. The column conditioning and analyte derivatisation required for previous methods were not found to be necessary for stable, accurate and sensitive element-specific detection using MIP–AED. Chromatographic and detection parameters such as stationary phase type, stationary phase film thickness, gas chromatographic column dimensions, helium mobile phase column head pressure, detector make-up gas flow rate and detector reagent gas type and flow rate were found to significantly affect analyte response. The detection limit for the optimised MIP–AED conditions was 0.8 pg (0.1 pg/s) of methylmercury chloride (as mercury). A solid–liquid extraction procedure with preparative gel permeation chromatography clean-up and MIP–AED analysis was used to quantify methylmercury in a variety of complex matrix marine materials. The methylmercury quantification method was validated with four marine certified reference materials (CRMs). The method was then applied to 13 standard reference materials, CRMs and control materials for which no certified reference values for methylmercury have been determined. Four National Institute of Standards and Technology Standard Reference Materials and one control material, which were analysed using the MIP–AED method, were also analyzed by two other laboratories using independent methods to further validate the method.

Jiminez and Sturgeon [384] and Swan [385] used ethylation of fish extracts with sodium tetraethylborate followed by GC in studies of the speciation of methylmercury. Low picogram detection limits were achieved.

Liang et al. [410] reported the simultaneous determination of monomethylmercury, inorganic mercury and total mercury in fish by using a procedure based on ethylation, room temperature precollection, gas chromatographic separation and detection by cold vapour atomic fluorescence. Absolute detection limits for each analyte were on the order of 1 pg.

Cai et al. [43] carried out an aqueous-phase microextraction of methylmercury in fish samples followed by solid-phase microextraction and GC with atomic fluorescence detection.

Holak [386] has developed a method that uses a simplified sample preparation procedure and atomic absorption or electrochemical detection of column eluents for the determination of methylmercury in fish. Methylmercury is isolated from the blended sample by chloroform elution from a diatomaceous earth–hydrochloric acid column. The organomercury compound is then extracted into a small volume of 0.01 N sodium thiosulfate solution. An aliquot of this solution is injected onto a Zorbax ODS column and eluted with methanol–ammonium acetate solution (3:2) buffer, pH 5.5 containing mercaptoethanol. Detection can be accomplished by atomic absorption spectrophotometry with the aid of a specially designed apparatus for the generation of mercury vapour. Alternatively, a commercially available electrochemical detector equipped with a dropping-mercury electrode may be used.

Linearity was maintained for up to 5 µg/ml solution of methylmercury(II) chloride (when 100 µl were injected). The reproducibility of multiple injections of 2.95 µg/ml of methylmercury(II) chloride was 3.2% in terms of relative standard deviation. The sensitivity, i.e. the amount of methylmercury(II) chloride that gave an adsorption of 1% (0.0044 absorbance units), was 0.0037 µg. The detection limit was 0.6 ng.

Table 1.28 shows the results and the recoveries for a number of spiked fish samples analysed by this proposed method for methylmercury. The recoveries ranged from 96 to 106%, (i.e., $101 + 5\%$). The precision was, in terms of the relative standard deviation, 4.1%. In the majority of instances, atomic absorption detection was used. When determining mercury compounds, this is the preferred mode of detection because of the ease with which mercury vapour can be generated.

Table 1.28. Analysis of marine samples (from [386]). Atomic absorption detection was used for samples 1–7 and electrochemical detection was used for samples 8 and 9[a]

| Sample | Mercury contents (µg/g) | | | Recovery, % |
	Found (as CH_3HgCl)	Added (as CH_3HgCl)	Total found	
1 Tuna	0.54[a]	1.0	1.50	96
2 Shrimp	0.01	0.18	0.19	100
3 Sardines	0.03	0.59	0.61	93
4 Swordfish	0.57	0.59	1.12	93
5 Whiting	0.08	0.27	0.35	100
6 Turbot fillets	0	1.18	1.16	98
7 Octopus	0	1.18	1.26	106
8 Swordfish	1.01	1.18	2.26	106
9 Squid	0	0.47	0.46	98

[a] Average of 6 determinations: standard deviation – 0.022 µg/g, relative standard deviation – 4.1%

1.3.3.6
High-Performance Liquid Chromatography (HPLC)

Rezende et al. [387] extracted methylmercury from fish as bromide using chloroform as a solvent. Mercury was determined by HPLC [39].

A method for the fully automated determination of both organomercury and inorganomercury species in fish has been reported [388]. This method is based on solid-phase extraction, flow injection, HPLC separation, reduction combined with thermolysis, and detection by cold vapour.

1.3.3.7
Column Chromatography

Liquid chromatography using differential pulse electrochemical detection has been used to determine organomercury cations in tuna fish and shark meat [389]. The differential pulse mode of detection offers a substantial increase in selectivity over amperometry.

Following alkaline hydrolysis, the sample (1 g) is acidified with hydrochloric acid. The organomercury cations can then be extracted from the aqueous solution with toluene as the neutral chloride complexes. The aqueous back- extraction solution used was 0.01 mol/L disodium thiosulfate buffered to pH 5.5 with 0.05 mol/L ammonium acetate. This extraction solution was compatible with the column chromatographic separation, and the determination was performed directly on this aqueous extract after filtering through a 0.2 µm syringe filter. In all cases, a standard addition procedure was used for the determination with known amounts of diluted $CH3Hg^+$ solution added to the solid material before the hydrolysis step. The recovery was checked by comparison to a standard curve and found to be about 95%. Various interferences on the determination of organomercury compounds and how they are overcome are discussed by these workers.

Table 1.29 shows the results obtained when the method was applied to standard NBS fish samples. The sample chromatograms were characterised by a single response for methylmercury with high signal-to-noise ratio. Ethyl and phenylmercury were not detected in these samples.

The results obtained (see Table 1.29) for the methylmercury content of the fish samples were in fairly close agreement to the total mercury (as measured by an alternate technique such as atomic absorption and neutron activation analysis). MacCrehan and Durst [389] achieved a detection limit

Table 1.29. Methylmercury contents of fish samples (from [389])

Sample	Mercury species (ng/kg)			
	$MeHg^+$	$EtHg^+$	$PhHg^+$	Total Hg
RM 50 Albacore tuna	0.93 ± 0.1	ND	ND	0.95 ± 0.1
Japanese shark paste	8.41 ± 0.1	ND	ND	7.4

ND: not detected

of 0.002 mg/kg for methylmercury compounds in fish utilising HPLC with an electrochemical detector.

1.3.3.8
Thin-Layer Chromatography

Thin-layer chromatography was carried out by Westhoo [355], either on the original methylmercury chloride- containing fish extract, or on derivatives prepared from this extract, such as dithizonate, bromide, iodide or cyanide. Light petroleum:diethylether (70 : 330) was used as developing solvent, and aluminium oxide or silica gel plates were used. Separated organomercury compounds were detected with a saturated ethanolic solution of Michler's thioketone in ethanol.

Methylmercury dithizonate and phenylmercury dithizonate could be separated from each other in the fish extracts by thin-layer chromatography on aluminium oxide (limit of detection 0.2 µg). Methylmercury cyanide, chloride, bromide and iodide were separated by thin-layer chromatography on silica gel (limit of detection of the chloride and bromide: 0.02 µg).

1.3.3.9
Capillary Electrophoresis

Medina et al. [391] speciated organomercury compounds in fish using capillary electrophoresis.

The combination of capillary zone electrophoresis with ICP-M spectrometric detection enabled absolute detection limits of a few picograms to be achieved in the determination of various organomercury species in fish [392]. Sample pretreatment consisted of only digestion and pH adjustment neutron activation analysis.

1.3.3.10
Neutron Activation Analysis

Vasankara-Pillay et al. [37] have used this technique to determine down to 0.01 mg/kg organomercury as total mercury in frozen homogenised fish samples after neutron activation; the samples were wet ashed with concentrated sulfuric–70% perchloric acid at 120 – 160 °C with mercury carrier. A preliminary precipitation as sulfide in acidic medium was followed by a further precipitation as sulfide, and electrodeposition or precipitation was used to isolate mercury. The radioactivities of ^{197}Hg and ^{197m}Hg were measured by scintillation spectrometry using a thin sodium iodide detector. These techniques were used to carry out a survey of mercury levels of edible fish taken in Lake Erie. In general, the fish from the Western Basin of Lake Erie had elevated levels of mercury in their edible tissues, 0.2 – 0.79 mg/kg, when compared with similar species caught from the Central (0.2 – 0.65 mg/kg) and Eastern (0.26 – 0.51 mg/kg) basins.

1.3.3.11
Spectrophotometry

Jones and Nickless [413] have described a dithizone spectrophotometric procedure for the measurement of trace concentrations of methylmercury salts in fish tissue. The application of a simple equation using absorbance measurements taken at two wavelengths cancels out small differences in excess dithizone arising between blank and sample, thus ensuring good precision in the range 0.1 – 4.0 µg/ml. Dithizone reacts with most organomercury salts of the type RHgX, where X is any anion. The extraction procedure used is based on that described by Westhoo [353–355] in which 40 g of fish is homogenised with water and then concentrated hydrochloric acid and sodium chloride are added. A benzene extract is centrifuged to provide a clear phase, which is then treated with aqueous 1% cysteine solution. After acidification, the aqueous phase is then again extracted with benzene, and a benzene solution of dithizone is then added to the extract. This extract is evaluated spectrophotometrically at 628 and 475 nm. Extraction efficiencies were low (24 – 32%) but reproducible. Methylmercury contents found in tuna fish flesh ranged from 0.15 to 0.69 mg/kg.

1.3.4
Organosilicon Compounds

Wanatabe et al. [393] have described a method for the separation and determination of siloxanes in water, sediment and samples of fish tissues using ICP–ES. Petroleum ether extracts of the organosilicone are evaporated to dryness. The damp residue is dissolved in methyl isobutyl ketone and aspirated into the plasma. The detection limit is 0.01 mg/kg. Recoveries are about 50% with coefficients of variation of about 13%.

1.3.5
Organotin Compounds

Smith [394] discussed the determination of tin in fish. McKie [395] determined total tin and tributyltin in fish by graphite furnace atomic AAS following extraction by digestion with nitric acid for total tin, and by *n*-hexane after treatment with hydrochloric acid for tributyltin.

Short [396] has compared two methods for the determination of tributyltin in salmon. One method was a simple screening procedure, determining tin by flameless AAS, while the other method was specific for tributyltin and involved separation of tributyltin by GC, its reduction to metallic tin, and determination by AAS. The screening method tended to overestimate tributyltin in fish flesh, but could be useful for identifying samples requiring more detailed examination.

Pannier et al. [397] converted butyltin compounds into volatile hydrides using sodium tetrahydroborate. The volatile hydrides were trapped cryogenically and then determined by selective volatilisation onto an on-line quartz furnace atomic absorption spectrometer.

Sasaki et al. [398] determined tri-n-butyltin and di-n-butyltin in fish by GC with flame photometric detection. The method involved extraction with acidified solvent, gel permeation chromatography clean-up, methyl derivatisation with Grignard reagent and gas chromatographic analysis using flame photometric detection. Recoveries from fish samples spiked at 0.2 and 1.0 mg/kg were 80 – 104% (i.e. 92 ± 12%) tri-n-butyltin and 92 – 105% (i.e. 98 ± 6%) for di-n-butyltin. The detection limit for both compounds was 0.2 mg/kg. The levels determined in reared yellowtail fish were similar for both flame photometric detection GC and GC–MS.

Shawky et al. [399] reported a method for the speciation of organotin compounds in fish based on aqueous ethylation gas chromatographic separation and on-line quartz furnace AAS. Digestion with 0.5 mol/l methanolic acetic acid was assisted by microwave sonication.

Kumar et al. [400] determined organotin compounds in fish by supercritical fluid extraction followed by liquid chromatography on a column directly connected to an ICP–mass spectrometric detector. Reproducibility extractions were coupled within 15 m, although recoveries were only 44% for tributyltin and 23% for triphenyltin.

1.4
Nonmetallic Elements

1.4.1
Total Nitrogen

In the Kjeldahl method [401, 402] for the determination of total nitrogen in fish, the sample is digested with concentrated sulfuric acid, with copper(II) sulfate added as a catalyst, in order to convert organic nitrogen into ammonium ion; alkali is added and the ammonia liberated is distilled. The ammonia is adsorbed in boric acid and titrated with hydrochloric acid.

Nitrogen Derived from Total Volatile Bases

This includes the nitrogen content of those volatile bases that do not react with formaldehyde [402]. Extracts or solutions are made alkaline with sodium hydroxide in a suitable semi-micro stream distillation apparatus. The bases, including trimethylamine, liberated in this way are steam-distilled quantitatively into standard acid, and the excess of acid is then back-titrated with standard alkali. Formaldehyde is added to the neutralised mixture in order to render amines other than trimethylamine unreactive, and the acid released (equivalent to the non-reactive amines) is titrated with standard alkali.

1.4.2
Phosphorus

Total phosphorus in fish has been determined [402] by ashing the sample at 550 – 600 °C in the presence of magnesium acetate as an ashing aid, the ash being dissolved in dilute sulfuric acid and the phosphorus determined spectrophotometrically as molybdovanadophosphoric acid.

Kirkpatrick and Bishop [403] determined total phosphorus due to organophosphonates in fish. The sample was digested with 1.5 ml of a 98 : 230 : 1200 : 120 v/v mixture of sulfuric acid : water : nitric acid : perchloric acid for 1.5 hours at 225 °C. Orthophosphate was then determined in the digest molybdophosphate by ultramicrospectrophotometry at 830 nm. Recoveries from fish of 2-aminoethyl phosphoric acid, N-methyl-2-aminoethyl phosphoric acid, N,N-dimethyl-2-aminoethyl-phosphoric acid, 2-amino-3-phosphoropropionic acid and phosphoserine were in the range 95 – 101%, and relative standard deviations were in the range 1.4 – 2.4%.

Elementary phosphorus has been determined in benzene or isooctane extract of fish by gas chromatography with a flame photometric detector [404].

1.4.3
Halogens

White [405] developed methods for the detection of trace amounts of organic halogens in organisms. Three fractions (lipid-soluble material, cationic water-soluble molecules, and large macromolecules) were assayed for each organism, the inorganic halogens being monitored by means of radioactive chloride and radioactive iodide. Fractions were burned in an oxygen combustion tube, and the resulting adsorbed halides were assayed on two automatic analytical instruments: chloride and bromide on one instrument and iodide on the other [406].

Linde et al. [401] determined organohalogen compounds, as halogen, in fish samples. The samples were steam-distilled with cyclohexane for halogen-containing nonpolar compounds, and hexane extracts of oils from all species were treated with concentrated sulfuric acid. Total amounts of halogens in the original oils, in the volatile compounds in the cyclohexane distillate, and in the sulfuric acid-treated hexane extracts were determined by neutron activation analysis. The total level of organic chlorine ranged from 30 to 240 ppm: 2 – 10% of the chlorinated compounds were volatile, and from 5 to 50% of the chlorinated compounds remained after acid treatment. This chloride exceeded chlorine in polychlorinated biphenyl by a factor of 1.5 – 5, and most of the chlorine in untreated and acid-treated lipids could not be accounted for as known compounds.

1.5
Detection Limits for the Analysis of Fish

1.5.1
Cations

Available information is listed in Table 1.30.

Generally speaking, achievable detection limits are in the range 0.1–1 mg/kg (100 – 100 µg/kg) of fish sample (arsenic, chromium, copper, lead, mercury, nickel, selenium, strontium and zinc).

Slightly greater sensitivity is available in the case of cadmium (0.02 mg/kg) and nickel (0.05 mg/kg).

Hydride generation AAS is capable of achieving detection limits of 0.005 mg/kg for mercury and 0.02 mg/kg for arsenic.

Table 1.30. Detection limits for metals in fish (from author's own files)

Element	Method	Limit of detection, mg/kg unless otherwise stated	Reference
Arsenic	Spectrophotometric	0.3	[3]
Arsenic	Hydride atomic absorption spectrometry	0.02	[5]
		0.3	[55]
Arsenic	HPLC	0.3 ng absolute	[7]
Cadmium	Graphite furnace atomic absorption spectrometry	2 pg absolute	[15]
Cadmium	Atomic absorption spectrometry	0.02	[52]
Chromium	Atomic absorption spectrometry	0.2	[52]
Copper	Atomic absorption spectrometry	0.2	[52]
Lead	Graphite furnace atomic absorption spectrometry	4 pg absolute	[15]
Lead	Atomic absorption spectrometry	0.1	[52]
Mercury	Atomic absorption spectrometry	1 ng absolute	[32]
Mercury	Helium microwave-induced plasma atomic emission spectrometry	10 pg absolute	[42]
Mercury	Hydride atomic absorption spectrometry	0.005	[55]
Mercury	Neutron activation analysis	1	[38]
Nickel	Voltammetry	µg/kg	[44]
Nickel	Atomic absorption spectrometry	0.05	[52]
Selenium	Hydride atomic absorption spectrometry	0.2	[55]
Strontium	Graphite furnace atomic absorption spectrometry	1	[46]
Vanadium	Neutron activation analysis	0.03	[48]
Zinc	Atomic absorption spectrometry	0.2	[52]

1.5.2
Organic Compounds

Typical values are listed in Table 1.31.

Table 1.31. Detection limits for organic compounds in fish (from author's own files)

Organic compound	Method	Limit of detection, µg/kg unless otherwise stated	Reference
Polyaromatic hydrocarbons	Capillary GC (NP detection)	0.2 – 0.5	[71]
Alkyl phthalates	Capillary GC	5	[76]
Chloroparaffins	GC with ECD	10 – 100 pg absolute	[92]
Chloroparaffins	High-resolution MS	60 pg absolute	[93]
Chlorophenols	Methylation GC with ECD	100 – 1000	[102]
Pentachlorphenols	Derivatisation GC	500	[99]
Hexachlorobenzene	GC-MS	~ 1	[104]
Polychlorobiphenyls	TLC	0.5 µg absolute	[106]
Polychlorobiphenyls	GC	1000	[109]
Polychlorobiphenyls	GC	10	[209]
2,3,7,8-Tetrachloro dibenzo-p-dioxin	HPLC	0.1	[140]
2,3,7,8-Tetrachloro dibenzo-p-dioxin	GC-MS	10	[143]
Linear alkyl benzene sulfonate	Ion-pair liquid–liquid partitioning	200	[207]
Phenoxyacetic acid herbicides	Methylation GC	100 – 1000	[102]
Chloro insecticides	GC with ECD	1	[218]
DDT	GC	100	[329]
Dursban	Thermionic GC	40	[202]
Fluridone	LC	5	[209]
Mirex	GC	1	[209]
Ciguatoxin	HPLC	Sub 1	[248,249]
Geosmin 2-methyl isoborneol	GC	5 ng absolute	[260]
Eulon WA (polychloro-2-chlor-methyl sulphonamide diphenyl ether)	GC	5	[206]
α,α,α-Trifluoro-4-nitro-m-cresol	GC with ECD	10	[269]
B_2 vitamins, including riboflavin,	HPLC	4.9	[270]
flavin mononucleotide,		9	
flavin adenine dinucleotide		7.3	

Sensitivity ranges from 0.01 – 1 µg/kg (dioxins, polyaromatic hydrocarbons, chlorinated insecticides) to the range 100 – 1000 µg/kg (chlorophenyls, organophosphorus, phenoxyacetic acid derivatives, linear alkyl benzene sulfonates).

1.5.3
Organometallic Compounds

Exceptionally low detection limits of 0.1 µg/kg, 4 µg/kg and 0.2 µg/kg were obtained, respectively, for organic compounds of arsenic, mercury and tin; otherwise detection limits were usually in the range 1 – 20 µg/kg of fish (Table 1.32).

Table 1.32. Detection limits for organometallic compounds in fish (from author's own files)

Organometallic compound	Method	Limits of detection, µg/kg unless otherwise stated	Reference
Arsenic type	Hydride atomic absorption spectrometry	< 0.1	[303]
Lead type	Gas chromatography	25	
Lead type	Anodic stripping voltammetry	10	[311]
Mercury type	Microwave-induced plasma atomic emission spectrometry	Sub ppt 10 pg absolute 0.08 pg absolute	[352] [42]
Mercury type	Gas chromatography	20 10 20 250 4 0.6 ng absolute	[350] [366] [369] [372] [386]
Mercury type	Ethylation gas chromatography	Low pg absolute	[43, 384]
Mercury type	Thin-layer chromatography	0.02 µ absolute	[355]
Mercury type	Inductively coupled plasma mass spectrometry	Low pg absolute	[391]
Silicon type	Ethylation gas chromatography	10	[393]
Tin type	Ethylation gas chromatography	0.2	[398]

References

1. Heydorn K, Alfassi Z, Damsgaard E, Rietz B, Solgaard P (1995) *J Radioanal Nucl Chem* **192**:321.
2. Ranau R, Oehlenschläger J, Steinhart H (1999) *Fresen J Anal Chem* **364**:599.
3. Maher WA (1983) *Analyst* **108**:939.
4. Agemian H, Thomson R (1980) *Analyst* **105**:902.
5. Brooke PJ, Evans WH (1981) *Analyst* **106**:514.
6. Goulden PD, Anthony DHJ, Austen KD (1981) *Anal Chem* **53**:2027.
7. Beauchemin D, Bednas ME, Berman SS, McLaren JW, Siu KWM, Sturgeon RE (1988) *Anal Chem* **60**:2209.
8. Amran MB, Lagarde F, Le Roy MJF (1997) *Microchim Acta* **127**:195.
9. Branch S, Ebdon L, O'Neill P (1994) *J Anal Atom Spectrosc* **9**:33.
10. Le SXC, Cullen WR, Reimer KJ (1994) *Environ Sci Tech* **28**:1598.
11. Lu Y-K, Sun H-W, Yuan C-G, Yan X-P (2002) *Anal Chem* **74**:1525.
12. Sperling ER (1977) *Fresen Z Anal Chem* **287**:23.
13. Sperling ER (1980) *Fresen Z Anal Chem* **301**:294.
14. Sperling ER (1980) *Fresen Z Anal Chem* **310**:254.
15. Poldoski JE (1982) *Anal Chem* **52**:1147.
16. Kiriyama T, Kuroda K (1982) *Analyst* **107**:505.
17. Harvey BR (1978) *Anal Chem* **50**:1866.
18. Pagenkopf GK, Neuman DR, Woodriff R (1972) *Anal Chem* **44**:2248.
19. Harms SU (1985) *Fresen Z Anal Chem* **323**:53.
20. May TW, Brumbaugh WG (1982) *Anal Chem* **54**:1032.
21. Holak W, Frinitz B, Williams JC (1972) *J AOAC* **55**:741.
22. Uthe JF, Armstrong FAJ, Tam KC (1971) *J AOAC* **54**:866.
23. Saha JG, Lee YW (1972) *Bull Environ Contam Toxicol* **7**:301.
24. Cumont G (1971) *Chim Analyt* **53**:634.
25. Fabbrini A, Modi G, Signorelli L, Siniani G (1971) *Bull Lab Chim* **22**:339.
26. Lidums V (1972) *Chem Scripta* **2**:159.
27. Uthe JF, Armstrong FAJ, Stainton MP (1970) *J Fish Res Board Canada* **27**:805.
28. Davidson JW (1979) *Analyst* **104**:683.
29. Tong SL, Leow WK (1980) *Anal Chem* **52**:581.
30. Louie HW (1983) *Analyst* **108**:1313.
31. Hendzel MR, Jamieson DM (1976) *Anal Chem* **48**:926.
32. Konishi T, Takahashi H (1983) *Analyst* **108**:827.
33. Kunkel E (1972) *Fresen Z Anal Chem* **258**:337.
34. Jones P, Nickless G (1974) *J Chromatogr* **89**:201.
35. Thomas RJ, Hagstrom RA, Kuchar EJ (1972) *Anal Chem* **44**:512.
36. Gustarsson J, Golimowski J (1981) *Sci Total Environ* **22**:85.
37. Sivasankara-Pillay KK, Thomas CC Jr, Sondel JA, Hyche CM (1971) *Anal Chem* **43**:1419.
38. Lo JM, Wei JC, Yang MH, Yeh SJ (1982) *J Radioanal Chem* **72**:571.
39. Medina I, Rubi E, Mejuto C, Casais C, Cela R (1993) *Analisis* **21**:215.
40. Fostier AH, Ferreira JR, Oesterrer de Andrado M (1995) *Quim Nova* **18**:425.
41. Adeloju SB, Dhindsa HA, Tendon RK (1994) *Anal Chim Acta* **285**:359.
42. Fukushi K, Willie SN, Sturgeon RE (1993) *Anal Lett* **26**:325.
43. Liang L, Bloom NS, Horvat H (1994) *Clin Chem* **40**:602.
44. Pihlar B, Valenta P, Nürnberg HW (1981) *Fresen Z Anal Chem* **307**:337.
45. Januzzi GS, Krug FJ, Arruda MAZ (1997) *J Anal Atom Spectrosc* **12**:375.

46. Bagenal TB, Mackereth FJH, Heron J (1973) *J Fish Biol* **5**:555.
47. Dogan S, Haerdi W (1980) *Int J Environ Anal Chem* **8**:249.
48. Blotcky AJ, Falcone C, Medina VA, Rack EP, Hobson DW (1979) *Anal Chem* **51**:178.
49. Ramelow G, Ozkan MA, Tuncel G, Saydam C, Balkas TI (1978) *Int J Environ Anal Chem* **5**:125.
50. Armannsson H (1979) *Anal Chim Acta* **110**:21.
51. Van Hoof F, Van San M (1981) *Chemosphere* **10**:1127.
52. Agemian H, Sturtevant DP, Austen KD (1980) *Analyst* **105**:125.
53. Borg H, Edin A, Holm K, Sköld E (1981) *Water Res* **15**:1291.
54. Adeloju SB, Dhindsa HS, Tandon RK (1994) *Anal Chim Acta* **285**:359.
55. Welz B, Melcher M (1985) *Anal Chem* **57**:427.
56. Buckley WT, Ihnat M (1993) *Fresen J Anal Chem* **345**:217.
57. De Olveira E, McLaren JW, Berman SS (1983) *Anal Chem* **55**:2047.
58. Saikai MK, May W (1988) *Sci Total Environ* **74**:199.
59. Adeloju SB, Bond AM, Hughes HC (1983) *Anal Chim Acta* **148**:59.
60. Greig RA, Jones J (1976) *Arch Environ Contam Toxicol* **4**:420.
61. Awadallah RM, Mohamed AE, Gaber SA (1985) *J Radioanal Nucl Chem Lett* **95**:145.
62. Kucera J, Soulal L (1993) *J Radioanal Nucl Chem* **168**:185.
63. Chassard-Bouchard (1987) *Anal Chim Acta* **195**:307.
64. Das AK (2000) *Int J Environ Pollut* **13**:208.
65. Arslan Z, Paulson AJ (2002) *J Anal Bioanal Chem* **372**:776.
66. Farrington JW, Teal JM, Quinn JG, Wade T, Burns K (1973) *Bull Environ Contam Toxicol* **10**:129.
67. Medeiros GC, Farrington JW (1974) In: *Marine Pollution Monitoring (Petroleum)* (Special Publication 409 29), National Bureau of Standards, Gaithersburg, MD, USA, p. 167.
68. Law RJ (1982) *Anal Procs* **19**:248.
69. Chesler SN, Gump BH,Hertz HS, May WE, Wise SA (1978) *Anal Chem* **50**:805.
70. Picer M (1998) *Chemosphere* **37**:607.
71. Vassilaros DL, Stoker PW, Booth GM, Lee ML (1982) *Anal Chem* **54**:106.
72. Birkholz DA, Coutts RT, Hrudey SE (1988) *J Chromatogr* **449**:251.
73. Ariese F, Kok SJ, Verkaik M, Hoornweg GP, Gooijer C, Velhorst HN, Hofstraat JW (1993) *Anal Chem* **65**:1100.
74. Ariese F, Kok ST, Verkaik M, Gooijer C, Velhorst N, Hofstraat JW (1993) *Aquatic Toxicol* **26**:273.
75. Ariese F, Gooijer C, Velhorst N (1993) *Tech Instrum Anal Chem (Environ Anal)* **13**:449.
76. Giam CS, Chan HS, Neff GS (1975) *Anal Chem* **47**:2225.
77. Itoh K, Chikuma M, Tanada H (1988) *Fresen Z Anal Chem* **330**:600.
78. Sinex SA, Cantillo AY, Helz GB (1980) *Anal Chem* **52**:2342.
79. Cantillo AY, Sinex SA, Helz GB (1984) *Anal Chem* **56**:33.
80. Suhr NH, Ingamells CO (1968) *Anal Chem* **38**:730.
81. Sturgeon RE, Desaulniers JAH, Berman SS, Russell DS (1982) *Anal Chim Acta* **134**:283.
82. McQuaker NR, Kluckner PD, Chang GN (1979) *Anal Chem* **51**:888.
83. Walsh JN, Howie RA (1980) *Min Management* **43**:967.
84. Murray AJ, Riley JP (1973) *Anal Chim Acta* **65**:261.
85. Murray AJ, Riley JP (1973) *Nature* **242**:37.
86. Deetman AA, Demeulemeester P, Garcia M, Hauck G, Hollies JI, Krockenberger D, Palin DE, Prigge H, Rohrschneider L, Schmidthammer L (1976) *Anal Chim Acta* **82**:1.

87. Parejko R, Keller R (1975) *Bull Environ Contam Toxicol* **14**:480.
88. Solomon J (1979) *Anal Chem* **51**:1861.
89. De Leon IR, Mayberry MA, Overton EB, Roschke CK, Remele PC, Steele CF, Warren VI, Leister JL (1980) *J Chromatogr Sci* **18**:85.
90. Hiatt MH (1983) *Anal Chem* **55**:506.
91. Linde G, Gether J, Steinnes E (1976) *Ambio* **5**:180.
92. Coelhan M (1999) *Anal Chem* **71**:4498.
93. Tomy G, Stern GA, Muir DCG, Fisk AT, Cymbalisty CP, Westmore JB (1997) *Anal Chem* **69**:2762.
94. De Boer J, Wester DG, Evers EHG, Brinkmann UAT (1996) *Environ Pollut* **93**:39.
95. Vetter W, Klobes U, Hummert K, Luckas BJ (1997) *J High Res Chromatogr* **20**:85.
96. Kuchl DW, Koppermann HL, Veith GD, Glass GE (1976) *Bull Environ Contam Toxicol* **16**:127.
97. Steinwandter H, Zimmer L (1983) *Fresen Z Anal Chem* **316**:705.
98. Ramdahl T, Carlberg GE, Kolsaker P (1986) *Sci Total Environ* **48**:147.
99. Stark A (1969) *J Agric Food Chem* **17**:871.
100. Rudling L (1970) *Water Res* **4**:533.
101. Renberg L (1974) *Anal Chem* **46**:459.
102. Hoben HJ, Ching SA, Casarett LJ, Young RA (1976) *Bull Environ Contam Toxicol* **15**:78.
103. Sackmasserova-Vennigerova M, Uhnak J (1981) *Vodni Hospodarstvi Ser B* **31**:133.
104. Johnson JL, Stalling DL, Hogan J (1974) *Bull Environ Contam Toxicol* **11**:393.
105. Mu H, Wesen C, Novak T, Sundin P, Skramstad J, Oldham G (1996) *J Chromatogr A* **731**:225.
106. Sackmauerova M, Pal'usova O, Szokolay A (1977) *Water Res* **11**:551.
107. Szelewski MJ, Hill DR, Spiegel SJ, Tifft EC (1979) *Anal Chem* **51**:2405.
108. Tausch H, Stehlik G, Wihlidal H (1981) *Chromatographia* **14**:403.
109. Bush B, Barnard EL (1982) *Anal Lett* **15**:1643.
110. Tuinstra LGMT, Drieson JJM, Kenkens HJ, Vna Munsteren TJ, Rees AH, Traag WA (1983) *Int J Environ Anal Chem* **14**:147.
111. Gaskin DE, Smith GJD, Arnold PW, Louisy MV, Frank R, Moldrinet M, McWade JW (1974) *J Fish Res Board Canada* **31**:1235.
112. Luckas B, Pscheidl H, Haberland P (1978) *J Chromatogr A* **147**:41.
113. Luckas B, Pscheidl H, Haberland P (1976) *Nahrung* **20**:K-K2.
114. Murphy PG (1965) *J AOAC* **48**:666.
115. Wenzel H, Luckas B (1977) *Nahrung* **21**:347.
116. Zimmerli b, Marek B, Sulzer H (1973) *Mitt Geb Lebensm Hyg* **64**:70.
117. Zimmerli B (1974) *J Chromatogr A* **88**:65.
118. Prescott AM, Cooke M (1979) *Proc Anal Div Chem Soc* **16**:10.
119. Ling Y-C, Chang M-Y, Huang I-P (1994) *J Chromatogr A* **669**:119.
120. Ling Y-C, Teng H-C (1997) *J Chromatogr A* **790**:153.
121. Ewald G, Bremle G, Karlsson A (1998) *Mar Pollut Bull* **36**:222.
122. Hajslova J, Schoula R, Holadova K, Poustka J (1995) *Int J Environ Anal Chem* **60**:163.
123. Wiberg K, Letcher R, Sandau C, Duffe J, Norstrom R, Haglund P, Bidleman T (1998) *Anal Chem* **70**:3845.
124. Echols K, Gale R, Tillit D, Schwartz T, O'Laughlin J (1997) *Environ Toxicol Chem* (1997) **16**:1590.
125. Buser HR, Zook DR, Rappe C (1992) *Anal Chem* **64**:1176.
126. De Voogt P, Häggberg L (1993) *Chemosphere* **27**:271.

127. Yu Ma C, Bayne CK (1993) *Anal Chem* **65**:772.
128. Ling Y-C, Huang I-P (1995) *Chromatographia* **40**:259.
129. De Boer J, Duinker JC, Calder JA, Van der Meer J (1992) *J AOAC Int* **75**:1054.
130. Wong CS, Garrison AW, Smith PD, Foreman WT (2001) *Environ Sci Technol* **35**:2448.
131. Blanche GP, Glausch A, Schurig MJ, Serrana R, Gonzalez MJ (1996) *J High Res Chromatogr* **19**:392.
132. Wong CS, Lau E, Clark M, Mabury SA, Muir DCG (2002) *Environ Sci Tech* **36**:1257.
133. Wells DE, Echarri I (1994) *Anal Chim Acta* **286**:431.
134. Ford CA, Muir DCG, Norstrom RJ, Simon M, Mulvihill MJ (1993) *Chemosphere* **26**:1981.
135. Rahman MS, Bowadt S, Larsen B (1993) *J High Res Chromatogr* **16**:731.
136. Wirth EF, Chandler GT, DePinto LM, Bidleman TF (1994) *Environ Sci Tech* **28**:1609.
137. Andrews P, Newsome WH, Boyle M, Collins P (1993) *Chemosphere* **27**:1865.
138. Bruns GW, Birkholz D (1993) *Chemosphere* **27**:1873.
139. Fowler B, Hoover D, Hamilton MC (1993) *Chemosphere* **27**:1891.
140. Lamparski LL, Nestrick TJ, Stehl BH (1979) *Anal Chem* **51**:1453.
141. Sherry JP, Carron L, Leger D, Kohli J, Wilkinson R (1993) *Chemosphere* **27**:651.
142. Phillipson DW, Puma BJ (1980) *Anal Chem* **52**:2328.
143. Mitchum RK, Moler GF, Korfmacher WA (1980) *Anal Chem* **52**:2278.
144. Lawrence J, Onuska F, Wilkinson R, Afghan BK (1986) *Chemosphere* **15**:1085.
145. Clement RE, Bobbie B, Taguchi V (1986) *Chemosphere* **15**:1147.
146. Harless RL, Oswald EO, Lewis RG, Dupuy, Jr., AE, McDaniel DD, Tai H (1982) *Chemosphere* **11**:193.
147. King TL, Uthe JF, Musial CJ (1993) *Analyst* **118**:1269.
148. Taguchi VY, Reiner EJ, Wang DT, Meresz O, Hallas B (1988) *Anal Chem* **60**:1429.
149. Crummett WB (1983) *Chemosphere* **12**:429.
150. Paasivirta J, Tarhanen I, Soikkeli J (1986) *Chemosphere* **15**:1429.
151. Marquis PJ, Hackett M, Holland LG, Larsen ML, Butterworth B, Kuehl DW (1994) *Chemosphere* **29**:495.
152. Ramamoorthy S, Clement R (1993) *Chemosphere* **26**:1679.
153. Smith LM, Stalling DL, Johnson JL (1984) *Anal Chem* **56**:1830.
154. De Boer J, De Boer K, Boon JP (2000) In: Passivirta J (ed) *The Handbook of Environmental Chemistry, Vol 3, Part K*, Springer, Berlin Heidelberg New York, p.61.
155. Noren K, Meironyte D (1998) *Organohalogen Compounds* **38**:1.
156. Van Bavel B, Sundelin E, Lillback J, Lindstrom G (1999) *Organohalogen Compounds* **40**:359.
157. Ikonomou MG, Fischer M, He T, Addison RF, Smith T (2000) *Organohalogen Compounds* **47**:77.
158. Sellstrom U (1999) *Doctoral Thesis*, University of Stockholm.
159. Strandberg B, Dodder NG, Basu I, Hites RA (2001) *Environ Sci Tech* **35**:1078.
160. Vetter W (2001) *Anal Chem* **73**:4951.
161. Richling E, Haering D, Herderich M, Schreir P (1998) *Chromatographia* **48**:258.
162. Niculescu M, Nistor C, Frebort I, Pec P, Mattiasson B, Csoregi E (2000) *Anal Chem* **72**:1591.
163. Volpe G, Mascini (1996) *Talanta* **43**:283.
164. Gruger E (1972) *J Agric Food Chem* **20**:781.
165. Tombelli S, Mascini M (1998) *Anal Chem* **358**:277.
166. Male KB, Bouvrette P, Luong JHT, Gibbs BF (1996) *J Food Sci* **61**:1012.
167. Bouvrette P, Male KB, Luong JHT, Gibbs BF (1997) *Enzymol Microbiol Technol* **20**:32.

168. Chemnitius GC, Bilitewski U (1996) *Sens Actuators* **B32**:107.
169. Ohashi M, Nomura F, Suzuki M, Otsuka M, Adachi O, Arakawa N (1994) *J Food Sci* **59**:519.
170. Yano Y, Yokoyama K, Tamiya E, Karube I (1996) *Anal Chim Acta* **320**:269.
171. Chemnitus GC, Suzuki, Isobe K, Kimura K, Karube I, Schmid RD (1992) *Anal Chim Acta* **263**:93.
172. Gasparini R, Scarpa M, Di Paolo ML, Stevanato R, Rigo A (1991) *Bioelectroch Bioener* **25**:307.
173. Taylor SL (1986) *Crit Rev Toxicol* **17**:91.
174. Stratton JE, Hutkins RW, Taylor S (1991) *J Food Prot* **54**:460.
175. Hui JY, Taylor SL (1985) *J Appl Pharmacol* **8**:241.
176. Bachrach U, Plesser YM (1986) *Anal Biochem* **152**:423.
177. Yang X, Rechnitz GA (1995) *Electroanal* **7**:105.
178. Hauschild MZ (1993) *J Chromatogr A* **630**:397.
179. Hui JY, Yatlor SL (1983) *J AOAC Int* **66**:853.
180. Yen GC, Hsieh CL (1991) *J Food Sci* **56**:158.
181. Veciana-Nogues MT, Hernandez-Jover T, Marine-Font A, del Carmen Vidal-Carou M (1995) *J AOAC Int* **78**:1045.
182. Xu CX, Marzouk SA, Cosofret VV, Buck RP, Neuman MR, Sprinkle RH (1997) *Talanta* **44**:1625.
183. Matsumoto T, Suzuki O, Katsumata Y, Oya M, Suzuki T, Nimura Y, Hattori T (1981) *J Cancer Res Clin Oncol* **100**:73.
184. Stevanato R, Mondovi B, Sabatini S, Rigo A (1990) *Anal Chim Acta* **273**:391.
185. Rinaldi AC, Sanjust E, Rescigno A, Finazziagro A, Rinaldi A (1994) *Biochem Mol Biol Int* **34**:699.
186. Toul Z, Macholan L (1975) *Collect Czech Chem Commun* **40**:2208.
187. Hungerford JM, Arefyev AA (1992) *Anal Chim Acta* **261**:351.
188. Nishihawa Y, Okumura T (1995) *Anal Chim Acta* **312**:45.
189. Okumura T, Imamura K, Mishikawa Y (1996) *J Chromatogr Sci* **34**:190.
190. Steinwandter H (1987) *Fresen Z Anal Chem* **326**:139.
191. Lindstrom K, Schubert R (1984) *J High Res Chromatogr* **7**:68.
192. Brunner PH, Capri S, Marcomini A, Giger W (1988) *Water Res* **22**:1465.
193. Gledhill WE, Huddleston RL, Kravetz L, Nielsen AM, Sedlak RI, Vashon RD (1989) *Tenside Surf Deterg* **26**:276.
194. Prats D, Ruiz F, Vazuez B, Zarzo D, Berna JL Moreno A (1993) *Environ Toxiocol Chem* **12**:1599.
195. Feijtel TCJ, Matthijs E, Rottieres A, Rijs GBJ, Kiewiet A, De Nijs A (1995) *Chemosphere* **30**:1053.
196. Moreno A, Ferrer J, Ruiz Bevia F, Prats D, Vazquez B, Zarzo D (1994) *Water Res* **28**:2183.
197. Rapaport RA, Eckhoff WS (1990) *Environ Toxicol Chem* **9**:1245.
198. Schoberl P (1996) *Proc 4th World Surfactant Congr*, 3–7 June 1996, Barcelona, Spain, p. 87.
199. Tabor CF, Barber LB II (1996) *Environ Sci Technol* **30**:161.
200. Kikuchi M, Tokai A, Yoshida T (1986) *Water Res* **20**:643.
201. Gonzalez-Mazo, Honing M, Barcelo D, Gomez-Parra A (1997) *Environ Sci Technol* **31**:504.
202. Di Corcia A, Marchetti M, Samperi R, Marcomini A (1991) *Anal Chem* **63**:1179.
203. Schoster M (1993) PhD Thesis, Universitat Dusseldorf, Julich, Germany.

204. Feijtel T, Kloepper-Sams P, den Haan K, van Egmond R, Comber M, Heusel R, Wierich P, Ten Berge W, Gard A, de Wolf W, Niessen H (1997) *Chemosphere* **34**:2337.
205. Kloepper-Sams PJ, Cowan CE, Larson RJ, Versteeg DJ (1996) In: *Proc 4th World Surfactant Congr*, 3–7 June 1996, Barcelona, Spain, p. 212.
206. Berth P, Jescke P (1989) *Tenside Surf Deterg* **26**:75.
207. Tolls J, Haller M, Sijm DTHM (1999) *Anal Chem* **71**:5242.
208. Murray DAJ (1975) *J Fish Res Board Canada* **32**:457.
209. Markin GB, Hawthorne JC, Collins HL, Ford JH (1974) *Pest Monit J* **7**:139.
210. Frank R, Armstrong AF, Boeleus RG, Braun HH, Douglas CN (1974) *Pest Monit J* **7**:165.
211. Hesselberg RJ, Johnson JL (1972) *Bull Environ Contam Toxicol* **7**:115.
212. Simal J, Crous Vidal D, Maria-Chareo A, Arras A, Roado MA, Diaz R, Vilas D (1971) *An Bromat* (Spain) **23**:1.
213. Chau ASY (1972) *J AOAC* **55**:519.
214. Kuem DW (1977) *Anal Chem* **49**:521.
215. Sackmauerova H, Pal'Usova O, Szokolay A (1977) *Water Res* **11**:551.
216. Langlois RE, Stemp AR, Liska BJ (1954) *Milk Food Technol* **27**:202.
217. Kohler PW, Su SY (1986) *Chromatographia* **21**:531.
218. Norheim G, Oakland MO (1980) *Analyst* **105**:990.
219. Neely WB (1977) *Sci Total Environ* **7**:117.
220. Frederick LL (1975) *J Fish Res Board Canada* **32**:1705.
221. Jan J, Malservic S (1978) *Bull Environ Contam Toxicol* **6**:722.
222. Olsson M, Jenson B, Reutergard L (1978) *Ambio* **7**:66.
223. Ludke JL, Schmitt CJ, (1980) *Proc 3rd USA–USSR Symp on the Effect of Pollutants upon Aquatic Ecosystems–Theoretical Aspects of Aquatic Toxicology*, 2–6 July 1979, US Environmental Protection Agency, Duluth, MN, p. 97–100.
224. Sackmauereva M, Pal'Usova O, Hluckan E (1972) *Vodni Hospodarsevi* **10**:267
225. Zoun PEF, Spierenburg TJ, Baars AJ (1987) *J Chromatogr A* **393**:133.
226. Nakamoto RJ, Page M (1986) *Bull Environ Contam Toxicol* **37**:415.
227. Galceran MT, Santos MJ, Caixach J, Ventura F, Rivera J (1993) *J Chromatogr A* **643**:399.
228. Lott HM, Barker SA (1993) *Environ Monit* **28**:109.
229. Steinwandter H (1987) *Fresen Z Anal Chem* **327**:363.
230. Richardson GM, Qadri SU (1987) *J Agric Food Chem* **35**:877.
231. Duesch ME, Westlake WE, Gunther FA (1970) *J Agric Food Chem* **18**:178.
232. Zakitis LH, McCray, Jr., EM (1982) *Bull Environ Contam Toxicol* **28**:334.
233. Lores EM, Moore JC, Knight J, Forester J, Clark J, Moody D (1985) *J Chromatogr Sci* **23**:124.
234. Szeto SY, Yee J, Brown MJ, Oloffs PC (1982) *J Chromatogr A* **240**:526.
235. Mao J, Erstfield KM, Fackler PH (1993) *J Agric Food Chem* **41**:596.
236. Bonwick GA, Yasin M, Hancock P, Baugh PJ, Williams JHH, Smith CJ, Armitage R, Davies DH (1996) *Food Agric Immun* **8**:185.
237. Gillespie NC, Lewis RJ, Pearn J, Bourke ATC, Holmes MJ, ourke JB, Shields WJ (1986) *Med J Australia* **145**:584.
238. Glaziou P, Legrand A-M (1994) *Toxicon* **32**:863.
239. Lewis RJ, Holmes MJ (1993) *Comp Biochem Physiol Part C* **106**:615.
240. Lin Y-Y, Risk M, Ray SM, Van Engen D, Clardy J, Golik J, James JC, Nakanishi K (1981) *J Am Chem Soc* **103**:6773.
241. Murata M, Legrand AM, Ishibashi Y, Fukui M, YAsumoto T (1990) *J Am Chem Soc* **112**:4380.

242. Lewis RJ, Sellin M, POli MA, Norton RS, MacLeod JK, Sheil MM (1991) *Toxicon* **29**:1115.
243. Lewis RJ, Norton RS, Brereton IM, Eccles CD (1993) *Toxicon* **31**:637.
244. Satake M, Murata M, Yasumoto T (1993) *Tetrahedron* **34**:1975.
245. Satake M, Ishibashi Y, Legrand A-M, Yasumoto T (1996) *Biosci Biochem Biotech* **60**:2103.
246. Legrand A-M, Teai T, Cruchet P, Satake M, Murata K, Yasumoto T (1998) In: Reguera B, Blanco J, Fernandez ML, Wyatt T (eds) *Harmful Algae*, Xunta de Galicia and Intergovernmental Oceanographic Commission of UNESCO, Santiagode Compostela, Spain, p. 473.
247. Vernoux JP (1988) *Oceanol Acta* **11**:37.
248. Crouch RC, Martin GE, Musser SM, Granade HR, Dickey RW (1995) *Tetrahedron Lett* **36**:6827.
249. Poli MA, Lewis RJ, Dickey RW, Musser SM, Buckner CA, Carpernter LG (1997) *Toxicon* **35**:733.
250. Vernoux J-P, Lewis RJ (1997) *Toxicon* **35**:889.
251. Lewis RJ, Vernoux J-P, Brereton IM (1998) *J Am Chem Soc* **120**:5914.
252. Lewis RJ, Sellin M (1992) *Toxicon* **30**:915.
253. Legrande AM, Fukui M, Cruchet P, Ishibashi Y, Yasumoto T (1992) In: Toteson TR (ed) *Proc Third Int Conf on Ciguatera Fish Poisoning*, 30 April–5 May, Puerto Rico, Polyscience Publ., Quebec, Canada, p. 25–32.
254. Hoffman PA, Granade HR, MacMillan JP (1983) *Toxicon* **21**:363.
255. Lewis RJ, Sellin M (1993) *Toxicon* **31**:1333.
256. Vernoux J-P (1994) *Mem Qld Museum* **34**:625.
257. Lewis RJ (1995) In: Hallegraeff GM, Anderson DM, Cembella AD (eds) *Manual on Harmful Marine Microalgae*, IOC Manuals and Guides 33, UNESCO, Paris, France, p. 135.
258. Manger RL, Leja LS, Lee SY, Hungerford JM, Hokama Y, Dickey RW, Granade HR, Lewis R, Yasumoto T, Wekell MM (1995) *J AOAC* **78**:521.
259. Pauillac S, Blechaut J, Cruchet P, Lotte C, Legrande A-M (1995) In: Lassus P, Arzul G, Erad E, Gentien P, Marcaillou C (eds) *Marine Algal Blooms*, Lavoisier, Paris, France and Intercept Ltd., Andover, UK, p. 801–808.
260. Martin JF, McCoy CP, Greenleaf W, Bennett L, (1987) *Canadian J Fish Aquatic Sci* **44**:909.
261. Lovell RT, Lelana IY, Boyd CE, Armstrong MS (1986) *Trans Am Fish Soc* **115**:485.
262. Persson PE (1980) *Water Res* **14**:1113.
263. West SD, May W (1986) *J AOAC* **69**:856.
264. Kiigemagi U, Burnard RJ, Terriere LC (1975) *J Agric Food Chem* **23**:717.
265. Vetter W, Smalling KM, Maruya KA (2001) *Environ Sci Toxicol* **35**:4444.
266. Schulz H (1985) *Fresen Z Anal Chem* **320**:725.
267. Wells DE, Cowan AA (1981) *Analyst* **106**:862.
268. Ozretich RJ, Schroeder WR (1986) *Anal Chem* **58**:2041.
269. Allen JL, Sills JB (1974) *J AOAC* **57**:387.
270. Ichinose N, Adochi I, Schwedt G (1985) *Analyst* **110**:1505.
271. Hughes RA, Lee GF (1973) *Environ Sci Toxicol* **7**:934
272. Musial CJ, Uthe JF (1983) *Int J Environ Anal Chem* **14**:117.
273. Vetter W, Krock B, Luckas B (1997) *Chromatographia* **44**:65.
274. Cederberg T, Fromberg A, Soerenson MK (1997) *Organohalogen Compounds* **31**:64.
275. Alder L, Vieth B (1996) *Fresen J Anal Chem* **354**:81.
276. Wester PG, de Boer J, Brinkman UAT (1996) *Environ Sci Toxicol* **30**:473.

277. Armour JA, Burke JA (1970) *J AOAC* **53**:761.
278. Gaul J, Leuz La Grange P (1971) *Separation of Mirex and PCBs in Fish*, Laboratory Information Bulletin, Food and Drug Administration, New Orleans, LA, USA.
279. Markin GP, Ford JH, Spence JH, Davies J, Leftis D (1972) *Environmental Monitoring for the Pesticide Mirex*, APHIS, US Department of Agriculture, Riverdale, MD, USA.
280. Butler PH (1969) *Biol Sci* **19**:889.
281. McKenzie MD (1970) *Fluctuations in the Abundance of the Blue Crab and Factors Affecting Mortalities,* Technical Report No. 1, South Carolina Wildlife Resources Division, Charleston, SC, USA.
282. Mahoad RK, McKenzie MD, Middough D-P, Bellar SJ, Davis JR, Spitsbergen D (1970) *A Report on the Cooperative Blue Crab Study in South Atlantic States,* Project Nos. 2-79-R1, 2-81-R-1, 2-82-R-1, Bureau of Commercial Fisheries, US Department of the Interior (now National Marine Fisheries Service, Silver Spring, MD, USA).
283. Kaiser KLE (1974) Science **185**:523.
284. Reynolds LM (1971) *Res Rev* **34**:27.
285. FDA (1971) *Pesticide Analytical Manual*, Department of Health, Education and Welfare, Food and Drug Adminisatration, Washington.
286. Sawyer LD (1973) *J AOAC* **56**:1015.
287. Bonelli EJ (1972) *Anal Chem* **44**:603.
288. Laseter JL, DeLeon IR, Remele PC (1978) *Anal Chem* **50**:1169.
289. Niazi SB, Littlejohn D, Halls DJ (1993) *Analyst* **118**:821.
290. Soto-Ferreirio RM, Bermeja-Barrera P (1993) *Analusis* **21**:197.
291. Klingston DW, Henry MA, Aldrin KJ, Pryde SD (1994) *J Chromatogr Sci* **32**:383.
292. Arslan Z, Paulson A (2002) *Anal Bioanal Chem* **372**:776.
293. Wells DE, Hess F (2000) *Tech Instrum Anal Chem* **21**:73.
294. Lima EC, Barbosa Jr., F Krug FJ, Silva MM, Vale MGR (2000) *J Anal Atom Spectrosc* **15**:995.
295. Dodo GH, Knight MM (1999) *J Chromatogr A* **859**:235.
296. Grimm CC, Lloyd SW, Batisa R, Zimba PV (2000) *J Chromatogr Sci* **38**:289.
297. Tyurin AN (2000) *Int J Environ Pollut* **13**:45.
298. Jaevenpaeae E, Huopalahti R, Tapanaineh R (1996) *J Liq Chromatogr R T* **19**:1473.
299. Staly A, Borsch-Jensen C, Balchem S, Mollerup J (1994) *J Am Oil Chem Soc* **71**:355.
300. Wells DE (1993) *Tech Instrum Anal Chem* **13**:79.
301. Moralez-Munos S, Luque-Garcia JL, Luque de Castro MD (2002) *Anal Chem* **74**:4213.
302. Fishman M, Spencer R (1977) *Anal Chem* **49**:1599.
303. Agemian H, Cheam V (1978) *Anal Chim Acta* **101**:193.
304. Maher WA (1981) *Anal Chim Acta* **126**:157.
305. Beauchemin D, Bednas ME, Berman SS, McLaren JW, Siu KWM, Sturgeon RE (1988) *Anal Chem* **60**:2209.
306. Le SXC, Cullen WR, Reimer KJ (1994) *Environ Sci Tech* **28**:1598.
307. Hanamura S, Smith BW, Winefordner JD (1983) *Anal Chem* **55**:2026.
308. Chau YK, Wong PTS, Goulden PD (1976) *Anal Chim Acta* **85**:421.
309. Chau YK, Wong PTS, Bengert GA, Dunn JL (1984) *Anal Chem* **56**:271.
310. Sirota CR, Uthe JF (1977) *Anal Chem* **49**:823.
311. Bernie SE, Hodges DJ (1981) *Environ Tech Lett* **2**:433.
312. Lobinski R, Adams FC (1994) In: Kicenuik JW, Ray S (eds) *Analysis of Contaminants in Edible Aquatic Resources: General Considerations, Metals, Organometallics, Tainting and Organics,* VCH, New York.
313. HMSO DoE (1976) *Pollution Papers, No. 10*, HMSO Department of the Environment, Central Unit of Environmental Pollution, London.

314. Holden AV (1973) *J Food Technol* **8**:1.
315. Jernelov A (1970) *Limnol Oceanogr* **15**:958.
316. Langley DG (1973) *J Water Pollut Cont Fed* **49**:44.
317. Olsen BH, Cooper RC (1976) *Water Res* **10**:113.
318. Pennacchioni A, Marchetti R, Gaggino GF (1976) *J Environ Qual* **5**:451.
319. Pentreath RJ (1976) *J Exp Mar Biol Ecol* **25**:51.
320. Pentreath RJ (1976) *J Exp Mar Biol Ecol* **25**:103.
321. Shin E, Krenkel PA (1976) *J Water Pollut Cont Fed* **48**:473.
322. Andren AW, Harris RC (1973) *Nature* **245**:256.
323. Bartlett PO, Craig PJ, Morton SF (1977) *Nature* **267**:606.
324. Davies IM, Graham WC, Pirie JM (1979) *Mar Chem* **7**:111.
325. Windom H, Gardner W, Stephens J, Taylor F (1976) *Est Coast Mar Sci* **4**:579.
326. Westoo G (1968) *Acta Chem Scand* **22**:2277.
327. Egawa H, Tajima S (1977) In: *Proc 2nd US/Japan Experts Meeting*, 1976, Tokyo, Japan.
328. Gillespie DC (1972) *J Fish Res Board Canada* **29**:1035.
329. Jones P, Nickless G (1978) *Analyst* **103**:1121.
330. Stainton MP (1971) *Anal Chem* **43**:625.
331. Magos L (1971) *Analyst* **96**:847.
332. Kopp JF, Longbottom MC, Labring LB (1972) *J Am Waterworks Assoc* **64**:20.
333. US EPA (1972) *Mercury in Water, Provisional Method*, Environmental Protection Agency, Analytical Quality Control Laboratory, Cinncinnati, OH, USA.
334. Yamanaka S, Ueda K (1975) *Bull Environ Contam Toxicol* **14**:409.
335. Schultz CD, Crear D, Pearson JE, Rivers JB Hylin JW (1976) *Bull Environ Contam Toxicol* **15**:230.
336. Analytical Methods Committee of the Chemical Society, London (1976) *Analyst* **92**:403.
337. Analytical Methods Committee of the Chemical Society, London (1976) *Analyst* **101**:62.
338. Friend MT, Smith CA, Wishart D (1977) *Atomic Absorption Newsletter* **16**:46.
339. Agemian H, Chau ASY (1975) *Anal Chim Acta* **75**:297.
340. Matsunaga K, Tahahashi S (1975) *Anal Chim Acta* **87**:487.
341. Analytical Methods Committee of the Chemical Society, London (1977) *Analyst* **102**:769.
342. Stuart DC (1978) *Anal Chem* **96**:83.
343. Davies IM (1978) *Anal Chem* **102**:189.
344. Aspila KI, Carron JM (1979) *Total Mercury in Sediments*, Interlaboratory Quality Control Study No.1, Report Series Inland Waters Directorate, Water Quality Branch, Department of Fisheries and Environment, Burlington, Canada,
345. Shum GTC, Freeman HC, Uthe JF (1979) *Anal Chem* **51**:414.
346. Capelli R, Fezia C, Franchi A, Zanicchi G (1979) *Analyst* **104**:1197.
347. Collett DL, Fleming DE, Taylor GE (1980) *Analyst* **105**:897.
348. Abo-Rady MDK (1979) *Fresen Z Anal Chem* **299**:187.
349. Holden AV (1973) *Pest Sci* **4**:399.
350. Uthe JF, Soloman J, Grift B (1972) *J AOAC* **55**:583.
351. Gutierrez J, Travieso H, Pubillones MA (1993) *Water Air Soil Poll* **68**:315.
352. Palmeri HEL, Leonel LV (2000) *Fresen J Anal Chem* **366**:466.
353. Westoo G (1967) *Acta Chem Scand* **21**:1790.
354. Westoo G (1968) *Acta Chem Scand* **22**:2277.
355. Westoo G (1966) *Acta Chem Scand* **20**:2131.

356. Westoo G, Johannsen B, Ryhage R (1970) *Acta Chem Scand* **24**:2349.
357. Gage JC (1961) *Analyst* **86**:457.
358. Sjostrand B (1964) *Anal Chem* **36**:814.
359. Longbottom JE, Dressman RC, Lichtenberg JJ (1973) *J AOAC* **56**:1297.
360. Kamps LR, McMahon L (1970) *J AOAC* **18**:351.
361. Bache CA, Lisk DJ (1971) *Anal Chem* **43**:950.
362. Newsome WH (1971) *J Agric Food Chem* **19**:567.
363. Longbottom JE (1972) *Anal Chem* **44**:1111.
364. Bye R, Paus PE (1979) *Anal Chim Acta* **107**:169.
365. Ealy JA, Sculz WD, Dean DA (1973) *Anal Chim Acta* **64**:235.
366. Cappon CJ, Smith JC (1977) *Anal Chem* **49**:365.
367. Callum GI, Ferguson MM, Lenihan JMA (1981) *Analyst* **106**:1009.
368. Decadt G, Baeyens WB, Bradley D, Goeyens L (1985) *Anal Chem* **57**:2788.
369. Analytical Methods Committee and the AOAC (1977) *Analyst* **102**:769.
370. Hight SC (1987) *J AOAC* **70**:667.
371. Hight SC (1987) *J AOAC* **70**:24.
372. Panaro KW, Erickson D, Krull IA (1987) *Analyst* **112**:1097.
373. Fischer R, Rapsomanikis S, Andreae MO (1993) *Anal Chem* **65**:763.
374. Cai Y, Bayona JM (1995) *J Chromatogr A* **696**:113.
375. Donais MKB, Uden PC, Schantz MM, Wise SA (1996) *Anal Chem* **68**:3859.
376. Kiyoura R (1964) In: Pearson EA (ed) *Advances in Water Pollution Research*, Pergamon, New York, USA, p. 291.
377. Johnels AG, Westermark T (1969) In: Miller MW, Berg G (eds) *Chemical Fallout*, Charles C Thomas, Springfield, IL, USA.
378. Peakall DB, Lovett RJ (1972) *Bioscience* **22**:20.
379. Horvat M, Liang L, Bloom NS (1993) *Anal Chim Acta* **282**:153.
380. Dressman RC (1972) *J Chromatogr Sci* **10**:472.
381. Bzezinska A, Loon JCV, Williams D, Oguma K, Fuwa K, Haraguchi IH (1983) *Spectrochim Acta* **38B**:1339.
382. Robinson JW, Wu JC (1985) *Spectrosc Lett* **18**:47.
383. Zelenko V, Kosta L (1973) *Talanta* **20**:115.
384. Jiminez MS, Sturgeon RE (1997) *J Anal Atom Spectrosc* **12**:597.
385. Swan HB (1998) *Bull Environ Contam Toxicol* **60**:511.
386. Holak W (1982) *Analyst* **107**:1457.
387. Rezende M de Carmo R, Campos RC, Curtius AJ (1993) *J Anal Atom Spectrosc* **8**:247.
388. Yin XF, Frech W, Hoffmann E, Lüdke C, Skole J (1998) *Fresen J Anal Chem* **36**:761.
389. MacCrehan WA, Durst RA (1978) *Anal Chem* **50**:2108.
390. Cai Y, Monsalud S, Furton KG, Jaffe R, Jones RD (1998) *Appl Organometall Chem* **12**:565.
391. Medina I, Rubi E, Mejuto MC. Cela R (1993) *Talanta* **40**:1631.
392. Tu Q, Qvarnström J, Frech (2000) *Analyst* **125**:705.
393. Watanabe N, Yasuda Y, Kato K, Makamura T, Funasaka R, Shimokaura K, Sato E, Ose Y (1984) *Sci Total Environ* **34**:169.
394. Smith JD (1970) *Nature* **225**:103.
395. McKie JW (1987) *Anal Chim Acta* **197**:303.
396. Short JW (1987) *Bull Environ Contam Toxicol* **39**:412.
397. Pannier F, Astruc A, Astruc M (1994) *Anal Chim Acta* **287**:17.
398. Sasaki K, Ishisaka T, Suzuki T, Saito Y (1988) *J AOAC* **71**:360.
399. Shawky S, Emons H Dürbeck HW (1996) *Anal Comm* **33**:107.

400. Kumar UT, Vela NP, Dorsey JG, Caruso JA (1993) *J Chromatogr A* **655**:340.
401. Linde G, Gether J, Steinnes E (1976) *Ambio* **5**:180.
402. Analytical Methods Committee of the Society for Analytical Chemistry, London (1979) *Analyst* **104**:434.
403. Kirkpatrick DS, Bishop SH (1971) *Anal Chem* **43**:1707.
404. Addison RF, Ackman RG (1970) J *Chromatogr A* **47**:421.
405. White RH (1968) PhD Thesis, University of Illinois, IL, USA.
406. Sorcac M, Baraj B, Celo V, Babi D (1996) *Fresen Environ Bull* **5**:661.
407. Tolls J, Kleoepper-Sams P, Sijm DTHM (1994) *Chemosphere* **29**:693.
408. Jansson R, Asplund L, Olsson M (1984) *Chemosphere* **13**:33.
409. Niemann RA, Brumley WC, Firestone D, Sphon JA (1983) *Anal Chem* **55**:1497.
410. Laramee JA, Deinzer ML (1994) *Anal Chem* **66**:719.
411. Conte ED, Shen C-Y, Miller DW, Perschbacher DW (1996) *Anal Chem* **68**:2713.
412. Schilling JB, Cepa SP, Menacherry SD, Bavda LT, Heard BM, Stockwell BL (1996) *Anal Chem* **68**:1905.
413. Jones P, Nickless G (1978) *Analyst* **103**:1121.
414. Lewis RJ, Jones A, Vernoux J-P (1999) *Anal Chem* **71**:247.
415. Gyos N, Gustavsen KO (1982) *Anal Chem* **54**:1316.

2 Analysis of Invertebrates

2.1
Cations

2.1.1
Arsenic

Maher [1] has described a procedure for the determination of total arsenic in crustaceans. The sample is first digested with a mixture of nitric, sulfuric and perchloric acids. Then arsenic is converted into arsine using a zinc reductor column, the evolved arsine is trapped in a potassium iodide–iodine solution, and the arsenic determined spectrophotometrically at 866 nm as the arseno-molybdenum blue complex. The detection limit is 0.05 mg/kg dry sediment and the coefficient of variation 5.1% at this level. The method is free from interferences by other elements at levels normally found in crustacea. Recoveries of 5–10 µg arsenic added to crayfish with a basal arsenic content of 168 ± arsenic were between 98 and 100%.

Brooke and Evans [2] described two methods for the digestion of samples of crustacea prior to determination of arsenic down to 0.02 mg/kg by hydride generation atomic absorption spectrometry (HG–AAS).

The first method involves separation of the inorganic arsenic by distilling it from 6.6 N hydrochloric acid. The second method involved chelation and extraction of inorganic arsenic after sample dissolution in sodium hydroxide solution, with subsequent back-extraction and oxidation. In both methods the arsenic concentration is measured after hydride generation by AAS with atomisation in a flame-heated silica tube; in the first method the solution contains arsenic(III) and in the second the solution contains arsenic(V). Results obtained by both methods are in agreement over a range of samples. The distillation method was favoured for reasons of efficiency and economy in time.

Hydrochloric Acid Digestion

Weigh 5 g of a representative wet sample (2 g of dry sample) into a 125-ml pear-shaped flask. Add 5 ml of water and 1 ± 0.1 g of iron(II) sulfate heptahydrate. Through the Bethge trap, add 50 ml of hydrochloric acid (3+2) and reflux the reaction mixture for 10 minutes. Close the tap in the Bethge trap and collect the first 50 ml of distillate over a period of 30 minutes. Cool and transfer into a 100-ml calibrated flask, washing with water, to give 100 ml of a colourless solution free from suspended solids. Reagent blank solutions should be obtained from hydrochloric acid $(3 + 2)$ in an identical manner.

Sodium Hydroxide Digestion

Place 2 g of a representative wet sample (1 g of dry sample) into a 150 ml conical flask, add 10 ml of sodium hydroxide reagent and heat on a boiling water-bath for 20 minutes. Cool, cautiously add 35 ml of hydrochloric acid $(1 + 3)$ and cool further. Transfer the solution into a separating funnel and, using 5 ml of water for washing, add 2 ml of ammonium pyrrolidone dithiocarbamate solution and mix thoroughly. Extract with 10 ml of 4-methylpentan-2-one, shaking for 2 minutes, allow to stand for 5 minutes or until separation is complete, and run off the solvent into another separating funnel. Repeat the extraction with the addition of ammonium pyrrolidone dithiocarbamate reagent, and finally extract with 10 ml of 4-methylpentan-2-one. To the combined solvent extracts add 10 ml of nitric acid $(1 + 7)$ and shake for 2 minutes. Repeat this extraction twice and combine the extracts in a beaker. Add 5 ml of sulfuric acid $(1 + 1)$ and boil until white fumes are evolved. Cool, add 10 ml water, reheat to fuming and repeat. Dilute to 50 ml. Reagent blank solutions should be obtained in an identical manner.

Some results obtained by these procedures on samples of crustacea are listed in Table 2.1.

Uthus et al. [3] also used hydride generation atomic absorption spectrometry to determine arsenic in oyster tissue. Arsine generated from dry combusted biological samples was measured using an atomic absorption spectrophotometer equipped with a graphite furnace. Innovations of the method included the introduction of arsine into the interior of the graphite

Table 2.1. Comparison of results for the determination of total inorganic arsenic following acid and alkaline digestion (from [2])

Sample	Total arsenic determined, mg/kg	Inorganic arsenic determined, mg/kg		Inorganic arsenic as percentage of total arsenic, %
		Hydrochloric acid digestion	Sodium hydroxide digestion	
Canned crab	1.5	0.10, 0.06	0.08, 0.08	5.3
Whelks	3.2	0.06, 0.08	0.06, 0.08	2.2
Canned lobster	3.6	0.08, 0.08	0.06,0.08	1.9
King prawns	14	0.02, 0.02	0.04, 0.04	0.3
Whelks	26	0.18, 0.14	0.10, 0.10	0.4

tube via one internal purge gas pot only, the use of three traps to remove water from the generated arsine, and the use of Erlenmeyer flasks in the generation of arsine. EDTA was added to the sample mixture to prevent interference from copper, iron and nickel cations. With the described procedure, the arsenic found in NBS 1566 standard oyster tissue (13.17 ± 0.34 mg/kg) agreed well with the certified value of 13.4 ± 1.9 mg/kg. Sensitivity and absolute detection limits of the method were 0.11 ng and 0.14 ng respectively.

Siu et al. [4] determined arsenic in standard reference materials including lobster hepatopancreas and oyster tissue. Biological samples were analysed after digestion with concentrated acid and derivatisation with 2,3-dimercaptopropanol using GC with electron capture detection. Results from the analysis compared favourably with certified values. A detection limit of 10 pg was reported with an analytical precision of 10%.

Brzezinska-Paudyn et al. [5] compared detection limits for arsenic in various standard references samples (oyster, lobster, scallop) for five different analytical techniques. The results obtained by GF–AA, combined furnace–flame atomic absorption, nondestructive neutron activation analysis, conventional inductively coupled plasma atomic emission spectroscopy (ICP–AES) and flow injection/hydride generation ICP–AES showed that all of these methods were appropriate for arsenic determinations at concentrations higher than 5 mg/kg. Graphite furnace atomic absorption, with a L'vov platform and nickel matrix modifier, was the most suitable method for analysing arsenic in biological materials with a detection limit of 0.5 – 1.0 mg/kg.

Liquid Chromatography

Ultraviolet irradiation generation ICP–MS has been used to study the speciation of arsenic in mussels [6].

2.1.2
Cadmium

Freeze-dried crab tissue was digested in open tubes with nitric and perchloric acids. Spectrometric evaluation was carried out using the cadmium 226.502 nm line, which is not subject to arsenic interference as is the cadmium 228.803 nm line, but does need a two-point background correction. Very good agreement was obtained in determinations of cadmium in crab by three different methods of analysis, these being ICP–AES 0.76 ± 0.6 mg/kg, IDSSMS 0.83 ± 0.08 mg/kg and GFAAS 0.71 ± 0.08 mg/kg.

Mazzucotelli et al. [11] pointed out that interference by inorganic elements frequently occurs in the determination of cadmium in mussels by methods based on electrothermal AAS and ICP–AES. Electrothermal AAS of cadmium in solutions containing 50 µg/kg plus increasing amounts (0.5 – 500 mg/kg) of interfering elements showed that sodium, potassium and calcium acted as enhancing agents, whereas iron and magnesium did not. In similar experiments using ICP–AES (wavelengths 228.802 and 214.438 nm), calcium and iron acted as enhancing agents at both wavelengths, whereas sodium and potassium acted as enhancing agents at 228.802 nm but depressive agents at 214.438 nm. Liquid anion exchange extraction was suggested as a way of overcoming metal interaction (only applicable to electrothermal atomic absorption spectrometry), but separation was necessary when an absolute cadmium value was required.

Greenberg [12] has developed a radiochemical neutron activation procedure for the determination of cadmium in oysters. The procedure is based on irradiation of the sample in a quartz tube with neutrons. Then, following a three-day decay period, the sample is digested with concentrated nitric and sulfuric acids in a sealed PTFE-lined bomb at 140 °C for two hours, followed by treatment with hydrofluoric acid to remove silica and hydrogen peroxide to destroy nitrogen oxides. Zinc nitrate is added as a holdback carrier, and a chloroform solution of nickel diethyldithiocarbamate added to extract mercury into the organic phase (which can also be determined by this procedure—see under Sect. 2.1.5 Mercury). The remaining aqueous fraction is extracted with a chloroform solution of zinc diethyldithiocarbamate. Back-extraction of this organic phase with aqueous hydrochloric acid provides an extract containing cadmium. The hydrochloric acid solution is allowed to decay for 24 hours to establish the equilibrium between cadmium-115 and its daughter, indium-115m. The 336-keV line from indium-115m and the 528 keV line from cadmium-115 were both used to evaluate the cadmium content of the sample.

Table 2.2 shows the determined cadmium contents of NBS reference samples contents and some oyster homogenisates obtained by this procedure.

Atomic absorption spectrometry has been applied to the determination of cadmium in mussels [7,8] and clam tissue [9].

Poldoski [9] used a molybdenum- and lanthanum-treated pyrolytically coated graphite tube for the graphite furnace atomic absorption spectrometric determination of cadmium at 228.8 nm in nitric acid–perchloric acid digests of clam tissue. Molybdenum and lanthanum help reduce chemical interferences and interference from uncompensated background signals during analyte atomisation.

Table 2.2. Cadmium contents of reference samples and oyster homogenisates (from [12])

Sample	mg/kg ± 2SD	
	Cadmium found Cd 115-Cd/115m-In	Cadmium certified
NBS SRM 1571 orchard leaves	116 ± 13	110 ± 10
NBS SRM 1577 bovine liver	288 ± 29	270 ± 40
International Atomic Energy Agency Oyster homogenisate MA-M-1	2.49 ± 0.15	2.30 ± 0.2

Digestions were carried out on 0.6 g of dry tissue using 10 ml concentrated nitric acid and 2 ml perchloric acid. After digestion was complete, the residue was dissolved in 10 ml 0.2% w/v nitric acid and stored in Nalgene bottles. An average value of 1.3 mg/kg was found in clam tissue. The determined cadmium content (0.31 ± 0.05 mg/kg) of NBS SRM 1577 bovine liver standard is in good agreement with the nominal value (0.34 ± 0.04 mg/kg). Average analytical recovery of cadmium in the clam sample is $104 \pm 10\%$.

Cadmium determinations in clam tissue digests obtained by the above procedure agreed well with those obtained by anodic scanning voltammetry in the range 1.0 to 2.3 mg/kg.

Ashworth and Farthing [8] have described a procedure for extracting cadmium from common mussels prior to analysis by AAS. The individual whole mussels were dehydrated to constant weight at 50 °C, digested under simple reflux in nitric acid, and the solution buffered to pH 5 with sodium hydroxide and sodium citrate. The cadmium was extracted with dithizone–methyl isobutyl ketone and the organic layer stored in polyethylene containers for analysis by AAS.

Mussels collected from the same region of Port Phillip Bay were found to have a cadmium concentration of approximately 0.5 mg/kg dry weight. A surprisingly high variability of $\pm 0.4\,\mu g/g$ was found in a group of 100 individuals, ranging in size from 2 to 5 g dry weight, with no correlation in cadmium concentration with size. A correction was necessary to overcome interference from the 214.445 nm and 226.505 nm lines of iron which interfere with both the cadmium 214.438 nm and the cadmium 226.502 nm lines.

2.1.3
Cobalt

Van Raaphorst et al. [13] have investigated the loss of cobalt-60 during the dry ashing of marine mussels. They observed no loss by volatilisation in porcelain crucibles when ashing was carried out at temperatures of up to 1000 °C. After ashing at 450 – 550 °C, the cobalt could be removed from the crucible with hydrochloric acid prior to counting with a thallium-activated sodium iodide crystal corrected to a single channel analyser.

2.1.3.1
Gadolinium

A method has been described for the determination of this element in crabs, based on solvent extraction followed by GF–AAS using a tantalum boat.

2.1.4
Lead

The atomic absorption spectrometric method described under cadmium (see Sect. 2.1.2) [9] has been applied to the determination of lead in clams. Lead results on clam tissues obtained in a spiking recovery experiment carried out on an authenticated reference sample (NBS 1577 bovine liver) gave a lead content of 0.33 ± 0.01 mg/kg, which is in good agreement with the nominal values ($0.34 + 0.08$ mg/kg). Average analytical recovery of lead in the clam sample is $100 \pm 6\%$. A value of 0.83 mg/kg lead was obtained on a clam sample.

Lead determinations in clam tissue digests obtained by the above procedures are in excellent agreement with those obtained by anodic scanning voltammetry in the concentration range 0.9–2.4 mg/kg. Relative standard deviations obtained by flame atomic absorption spectrometry in this concentration range are between 18 and 42%.

The wavelength modulation inductively coupled plasma echelle spectrometric technique [10], described in Sect. 2.1.2 for the determination of cadmium in crab tissue, has also been applied to the determination of lead in crab tissue. Freeze-dried crab tissue was digested in open tubes with nitric and perchloric acids. Spectrometric evaluation was carried out using the lead -220.353 nm line. Very good agreement was obtained in determinations of lead in crab tissue by three different methods of analysis, namely ICP–AES 3.0 ± 0.5, IDS-SMS 2.9 ± 0.1, and GF-AAS 2.4 ± 0.3.

Mikac et al. [14] have discussed the determination of lead in mussels. Digestion of mussel tissue with tetramethylammonium hydroxide is recommended. This digestion should be performed at room temperature to avoid the decomposition of some alkyl lead compounds. Cleaning of the extracts through a silica column was recommended before GC–AA spectrometric determination.

2.1.5
Mercury

Nondispensive AAS has been applied to the determination of down to 0.04 mg/kg mercury in shrimps. The mercury is reduced to its elemental form with acidic stannous chloride solution and swept with argon into the fluorimeter.

Various digestion procedures for mussels and oysters, including wet oxidation with nitric–sulfuric acids [16], digestion with concentrated nitric acid in a PTFE-lined bomb [17], Wickbold combustion [18], and digestion with concentrated nitric, sulfuric and nitric acids [19] have been used to digest these materials prior to the determination of mercury by cold vapour AAS. Recoveries of between 90 and 105% are claimed, with a detection limit of 0.01 mg/kg.

In the method described by Louie [19], the sample was digested with concentrated hydrochloric acid–nitric acid–sulfuric acid followed by cold vapour AAS in order to determine down to 0.01 mg/kg of mercury in oysters. A 97% recovery of mercury was obtained in spiking experiments with oysters following open tube digestion with hydrochloric–nitric and sulfuric acids at 70–95 °C.

The radiochemical neutron activation procedure [12] described in Sect. 2.1.2 for the determination of cadmium in oysters has also been applied to the determination of mercury in oysters. The final hydrochloric acid extract containing nickel diethyldithiocarbamate and mercury obtained in this procedure was counted immediately after separation via the 67.0 keV Au X-ray and the 77.5 keV combination gamma ray, and the Au X-ray produced by the decay of mercury-197 and/or (after decaying for several weeks) the 279 keV gamma ray from mercury-203.

Determined mercury contents obtained on an IAEA oyster homogenisate sample MAM-1 of $0.15 \pm 0.012\,\mu g/kg$ obtained by this procedure are in reasonably good agreement with the certified values of $0.20 \pm 0.02\,\mu g/kg$.

Lo et al. [20] digested oyster samples with concentrated sulfuric acid–nitric acid until white fumes appeared, and then added excess potassium permanganate solution as well as sodium chloride and hydroxylamine hydrochloric to reduce mercury. Mercury in the digest was then preconcentrated into a small volume of lead diethyldithiocarbamate dissolved in chloroform. The chloroform was allowed to evaporate in an ampoule, and then the ampoule sealed for neutron activation analysis and subsequent gamma spectrometry of the selective mercury-197 peak. As well as reducing the detection limit to $1\,\mu g/kg$ of oyster, the preconcentration has the additional advantage of overcoming interferences from sodium-24 and bromine-82, which commonly occur in crustacea. Recoveries of 95% were attained.

2.1.6
Selenium

Ahmed et al. [21] determined selenium(IV) in oysters by cathodic stripping voltammetry; arsenic(III), copper(II), lead(II), iron(III) and zinc(II) did not interfere. The oyster sample was digested at 50 °C with aqueous Lumatom (Hans Kurner, Neuberg, Germany), and then methanol was added and the solution acidified with hydrochloric acid prior to polarography. Pre-electrolysis was carried out at $-0.05\,V$ for 120 seconds and the solution was then cathodically polarised and quantified by standard addition and measurement of the peak heights of selenium at $E_p = -0.47\,V$ vs SCE:

$$Se(IV) + 4e + Hg \rightarrow HgSe \qquad (-0.05\,V)$$
$$HgSe + 2H^+ + 2e \rightarrow HgSe + H_2 \quad (-0.47\,V).$$

Good agreement was obtained by this procedure for the determination of selenium in NBS reference SRM 1577 oyster tissue, for which a value of $2.26 \pm 0.24\,mg/kg$ was obtained against a certified value of $2.1 \pm 0.5\,mg/kg$.

Maher [22] has reported on selenium levels in prawn (4.01 mg/kg) and scallop (1.24 mg/kg), and has pointed out that the selenium is predominantly associated with soluble protein and is not present as inorganic selenium.

Arruda et al. [23] used electrothermal AAS to determine selenium in shellfish tissue. The sample was digested in a microwave oven.

2.1.7
Tin

Thin film anodic stripping voltammetry has been applied to the determination of tin in distillates of sulfuric acid–hydrobromic acid digests of marine organisms [24].

Brown speculated methyl and butyl tin compounds and inorganic tin in amounts down to $11 - 25\,\mu g/kg$ in oysters using hydride generation AAS [25].

2.1.8
Vanadium

Blotcky et al. [26] have described a pre-irradiation chemistry neutron activation analysis procedure employing cation-exchange chromatography for the determination of trace-level vanadium in marine biological specimens, including shrimps, crabs and oysters. The procedure, utilizing a low-power nuclear reactor $(1 \times 10^{11}\,n/cm^2/s)$, consists of wet digestion of the sample with concentrated nitric acid at $65\,°C$, cation exchange chromatography employing a nitric acid wash to remove the major radioactivatable contaminants (sodium and chloride ions), ammonium hydroxide elution to remove vanadium from the resin, and neutron irradiation and radioassay for ^{52}Va. The limit of detection of the method is $30\,\mu g/kg$.

Between 600 and $2000\,\mu g/kg$ vanadium were found in crustacea taken at Galveston Island by this method.

2.1.9
Zinc

It has been reported [13] that no loss of zinc occurs when mussels are dry-ashed at $450 - 550\,°C$ in porcelain crucibles and the zinc subsequently dissolved in hydrochloric acid.

2.1.10
Multi-cation Analysis

2.1.10.1
Atomic Absorption Spectrometry (AAS)

Ramelow et al. [27] determined cadmium, lead, copper, zinc and chromium in wet shrimp, crab, oyster and mussels by digestion of $0.5 - 1\,g$ sample with $2 - 3\,ml$ concentrated nitric acid in a Teflon-lined bomb at $150\,°C$ for 1.5 hours. Elements were determined in the digest by flame atomisation or graphite furnace atomisation AAS.

Table 2.3. Trace metal concentrations of Mediterranean mussel vs. size (from [27])

Size	Total weight, g	Fresh weight, g	Shell length, cm	μg/g (fresh weight)			
				Hg	Cd	Cu	Pb
Very small (I)	6.2	1.0	4.0	0.03	–	–	–
Very small (II)	7.2	1.7	4.4	–	–	0.75	0.48
Small (I)	11.8	2.3	5.5	0.05	–		
Small (II)	17.9	5.8	5.8	–	0.07	0.89	0.54
Medium	27.7	5.9	7.3	0.02	0.24	1.27	0.61
Large	67.8	7.4	8.7	0.05	0.40	2.65	0.57

Four arbitrarily chosen sizes of Mediterranean mussel, collected near Gemlik on the Sea of Marmara in Western Turkey, were analysed for mercury, cadmium, copper and lead. The analytical results, together with the weights and shell lengths, are given in Table 2.3.

The ability of shellfish to concentrate many elements is well-known. It is thus expected that the mussel, which gets its nourishment by effectively filtering the surrounding water, might show an increase in the concentrations of some elements with size (and thus age). Such a trend is indicated in Table 2.3 for cadmium and copper, although more data is needed to be conclusive.

Topping [28] has reported on interlaboratory comparison studies on atomic absorption spectrometric and other methods for the determination of lead and cadmium in crab and lobster. He demonstrated that the majority of the participants in this study can produce comparable (i.e. interlaboratory CV of 17%) and accurate data for cadmium at a tissue concentration of around 10 μmol/kg, which is typically encountered in shellfish monitoring programs. Unfortunately, the results from the analysis of lead indicate that difficulties are experienced in producing comparable and accurate data at concentrations which cover the range of values encountered in shellfish (coefficients of variation 4.7–71%); see Table 2.4.

A breakdown of the analytical results for lead in lobster sample B and in pancreas sample C in relation to the analytical procedures used is presented in Table 2.5. Data in this table shows that the analysts incorporating a chelation–extraction step (e.g. APDC–MIBK or dithizone–CHCl$_3$) in their atomic absorption procedure produced more comparable results (i.e. lower CV results) than those employing the more commonly adopted wet digestion–atomic absorption procedure. It is worth noting that two analysts in the former group, who were the only ones in this group to receive sample C, produced a mean value for lead in sample C which was similar to that obtained by the analyst employing isotope dilution solid source mass spectrometry.

Schlemmer and Welz [29] investigated the determination of arsenic, lead, cadmium, copper and selenium in lobster and mussels by acid extraction–

Table 2.4. Intercomparison study of the determination of lead and cadmium in crab and lobster (from [28])

Sample	Sample preparation	Number of participants	Elements determined	Results obtained by Topping [28]					Analysis by alternative methods[a]		
				Range of values determined[a], µmol/kg	Mean	SD	CV, %	Nominal value mean	CV	Mean	CV
Crabmeat (Sample a)	Chopped pieces blast-frozen, freeze-dried and ground to fine flour	52	Cd	4.7 – 9.9	7.1	1.2	17	7.0	4	–	–
Lobster (Sample b)	Prepared in the form of acetone powder	32	Pb[b]	0.53 – 15.4	3.6	2.5	71	12	4	12.8	17
Pancreas (Sample c)			Pb[b]	1.44 – 14.4	3.36	1.6	4.7	2.5		1.6	18[b]

[a] AAS, ICPAES and isotope dilution solid source mass spectrometry

[b] If the reported lead values are taken as true lead content, then 9 out of 31 produced mean values which fell within the limits for the true concentrations

Table 2.5. Results for the analysis of lead in samples B and C (fifth exercise) in relation to analytical technique (from [28])

Technique	No of ana-lysts	Sample B (lobster), mean value, μmol/kg	CV[a]	No of ana-lysts	Sample C (pancreas)[a] mean value, μmol/kg	CV[a]
Wet digestion/AAS[c]	30	13.0	58	24	4.42	86
Dry ashing/AAS	9	13.6	24	3	3.36	76
Wet digestion/chelation/AAS	7	12.3	14	2	1.44	2
IDSSMS[d]	1	13.9	5[b]	1	1.68	18[b]
Nominal value		12.8	17	–	16	18

[a] Interlaboratory coefficient of variation based on mean values submitted by analysts
[b] Interlaboratory coefficient of variation based on six replicate analyses
[c] AAS: Atomic absorption spectrometry
[d] IDSSMS: Isotope dilution solid source mass spectrometry

Zeeman AAS. Factors investigated were the importance of Zeeman background compensation and the relative merits of two different methods, one an extraction under pressure with sulfuric acid and the other the complete combustion of the sample in a stream of oxygen. The sulfuric acid extraction under pressure appeared to give more acceptable results, except for arsenic and very low levels of cadmium, for which the Trace-O-Mat combustion process was preferable. The advantages of the Zeeman background correction were confirmed.

Welz and Melcher [30] investigated three different decomposition procedures for lobster and scallops prior to the determination of arsenic, selenium and mercury using hydride generation and cold vapour AAS. These procedures involved the following:

(i)   Decomposition with nitric acid under pressure in a PTFE bomb. This resulted in low values for arsenic and selenium but was adequate for the subsequent determination of mercury.
(ii) Decomposition with nitric, sulfuric and perchloric acids. This method gave the highest values for arsenic and selenium, whereas mercury was partly lost under these conditions.
(iii) Combustion in a stream of oxygen, which could be applied for all three elements and gave results that were in good agreement with the mean values of an intercalibration.

Pressure decomposition with nitric acid is recommended for mercury, followed by a sulfuric and perchloric acid treatment for the subsequent determination of arsenic and selenium (Table 2.6). Detection limits under routine conditions are 0.3 mg/kg for arsenic, 0.2 mg/kg for selenium and 0.005 mg/kg for mercury.

Table 2.6. Determination of arsenic, selenium and mercury in marine biological tissue samples using different composition procedures (from [30])

Samples	Accepted mean	Nitric acid under pressure ($n = 3$)	Perchloric acid at 310 °C, mg/kg ($n = 4$)	Oxygen combustion ($n = 6$)
		Arsenic		
Lobster	25 ± 5	15.1 ± 0.4	25.5 ± 0.5	24.6 ± 0.7
Scallops	7.1 ± 2.1	3.5 ± 0.4	7.7 ± 0.2	7.0 ± 0.3
		Selenium		
Lobster	6.4 ± 0.4	6.4 ± 0.2	6.7 ± 0.2	6.2 ± 0.4
Scallops	0.78 ± 0.08	0.71 ± 0.04	0.87 ± 0.01	0.71 ± 0.08
		Mercury		
Lobster[a]		0.30 ± 0.00	0.17 ± 0.04	0.31 ± 0.02
Scallops[b]		0.09 ± 0.01	0.07 ± 0.02	0.10 ± 0.01

[a] accepted mean 0.256 ± 0.049
[b] accepted mean 0.081 ± 0.012

Table 2.7. Maximum tolerable elemental concentrations in solution used in an interference-free determination of arsenic and selenium in 10 ml of 0.5 mol/l hydrochloric acid (from [30])

Element or sample to be determined or analysed	Maximum interferent concentration, µg/10 ml				
	Cu	Fe	Ni	As	Se
As	1000	1000	2		0.01
Se	0.5 (20)[a]	5000	2.5 (25)[a]	1	
1 mg of lobster	0.01	0.2	0.2	0.025	0.006
2 mg of scallops	0.01	< 0.2	< 0.02	0.015	0.002

[a] Values in brackets are for 10 ml of 5.0 mol/l hydrochloric acid medium. Also shown are the approximate elemental concentrations in the aliquots of sample solution analysed.

Table 2.7 gives the maximum tolerance levels for several elements that allow interference-free determination of arsenic and selenium.

Solchaga and De La Guardia [31] have proposed a method for acid pressure extraction, using nitric acid, of cadmium, copper, iron, lead and zinc in stoppered borosilicate glass, followed by flame atomic absorption spectroscopy. The five metals were determined in a 300 mg single sample of mussel meat. Detection limits were 0.3 mg/kg for cadmium, 0.7 mg/kg for copper, 33.0 mg/kg for iron, 0.7 mg/kg for lead, and 6.0 mg/kg for zinc. Precision was estimated from the coefficient of variation for 20 independent analyses, and was 7, 7, 6, 14 and 8%. Recoveries were between 90 and 107% (i.e. 98.5 ± 8.5%).

Amiard et al. [32] applied Zeeman AAS to the determination of silver, cadmium, chromium, copper, manganese, nickel, lead, and selenium in oyster and lobster. Aliquots (100 mg powdered sample) were digested in 1 ml concentrated nitric acid at 95 °C for one hour, the volume adjusted to 4 ml with deionised water, and then they were analysed using a graphite furnace coated with tantalum carbide. Detection levels were less than 1 μg/kg for silver, cadmium and manganese, about 1 μg/kg for chromium and lead, 5 μg/kg for copper and nickel, and 15 μg/kg for selenium. Variation coefficients were 5 – 10% for two series of six determinations, and experimental values agreed with certified values.

Ridout et al. [33] used ICP–AES to determine various elements in nitric acid digests of lobster hepatopancreas.

A continuous hydride generation ICP echelle spectrometer has been applied to the determination of 0.1 μg/kg of arsenic, antimony and selenium in oysters [34]. The following wavelengths were used: arsenic 193.696 nm, antimony 206.833 nm and selenium 196.026 nm.

Procedure for Antimony and Arsenic

The oyster sample (0.5 g) was digested with 4 g potassium hydroxide in a nickel crucible. The crucible was placed in a furnace and heated at 500 °C for 30 minutes. The crucible was then cooled and the contents dissolved in 50-ml of 1 mol/l hydrochloric acid; precipitated silicic acid was allowed to settle before analysis.

Procedure for Selenium

The same procedure as described above was used for antimony and arsenic but 4 g potassium hydroxide was replaced with 4 g sodium hydroxide [35] or by fusion with nitric–perchloric–hydrofluoric acid [36,37].

Heating to 50 °C for 50 minutes along with the addition of potassium bromide and concentrated hydrochloric acid converts selenium to the Se(IV) state, leaving arsenic and antimony in the pentavalent state. Reduction with sodium borohydride then converts all three metalloids to their hydrides.

Raith [38] has reported the use of ultraviolet laser ablation ICP–MS for the determination of minor and trace elements in shellfish.

The development of this system facilitated the achievement of spatial resolution of craters of 10 μm diameter on the shell. This permitted measurements to be performed over many years by measuring changes between growth bands of the shell.

High-resolution ICP–MS has been used to examine the partitioning of ^{66}Cu, ^{111}Cd and ^{207}Pb between seawater and the organs of *Mylitus galloprovincialis* mussel [39].

Brewer and Sacks [40] studied the atomisation characterisation and the direct determination of manganese and magnesium in a biological sample (oyster liver) using a magnetically attained thin-film plasma.

Chisela et al. [41] used epithermal and thermal neutron activation analysis to determine arsenic, bromine, cadmium, iron, manganese, molybdenum,

Table 2.8. Results for the analysis of NBS SRM 1566 (oyster tissue) µg/g by instrumental NAA with ENAA and RNAA (from [41])

Element	Isotope	Epithermal NAA [b]			Thermal NAA [c]			NBS value
		χ±s	CV	LD	χ±s	CV	LD	
As	^{76}As	11.96 ± 0.56	4.1	0.16	15.87 ± 3.5	22.0	1.80	13.49 ± 1.9
Br	^{82}Br	51.70 ± 7.1	13.7	0.62	50.57 ± 0.45	0.9	0.26	(55)[a]
Co	^{60}Co	0.39 ± 0.06	15.4	0.08	0.44 ± 0.07	15.9	0.02	(0.4)[a]
K	^{42}K	8600 ± 300	3.5	165	8200 ± 700	8.5	72.4	9700 ± 500
Mn	^{56}Mn	16.57 ± 0.97	5.8	0.16				17.50 ± 1.2
Na	^{24}Na	4200 ± 300	4.1	11.2	4700 ± 200	4.2	3.0	5100 ± 300
Sr	^{87}Sr^m	0.99 ± 0.76	6.9	1.15				10.36 ± 0.56
Zn	^{69}Zn^m	848.5 ± 4.5	5.3	25.6				852.0 ± 14.0
Irradiation time			1.0 h			1.0 h		
Cooling time			1.0 – 2.0 h			4 – 6 d		
Counting time			3600 s			3600 s		
Ag	^{110}Ag^m	0.86 ± 0.09	10.5	0.024	0.93 ± 0.06	6.4	0.016	0.89 ± 0.09
Br	^{82}Br	52.9 ± 3.3	6.2	1.6				(55)[a]
Co	^{60}Co	0.42 ± 0.07		0.010	0.34 ± 0.01	2.9	0.001	(0.4)[a]
Cr	^{51}Cr				0.75 ± 0.10	13.3	0.036	0.69 ± 0.27
Fe	^{59}Fe	212.5 ± 37	16.7	6.2	218.9 ± 9.0	4.1	2.8	195.0 ± 34
Mo	^{99}Mo – ^{99}Tc	0.16 ± 0.04	25.0	0.016				(≤ 0.2)[a]
Ni	^{58}Co	0.98 ± 0.10	10.2	0.01				1.03 ± 0.19
Rb	^{86}Rb	5.08 ± 0.10	2.0	0.07	4.27 ± 0.19	4.4	1.2	44.45 ± 0.09
Sc	^{46}Sc				0.015 ± 0.002	13.3	0.0004	–
Se	^{75}Se	2.04 ± 0.04	2.0	0.041	2.21 ± 0.08	3.6	0.02	2.10 ± 0.5
Zn	^{65}Zn	887.6 ± 0 10	1.1	0.56	884.60± 17	19.2	0.18	852.0 ± 14
Irradiation time			48 h			48 h		
Cooling time			15 – 21 d			50 – 70 d		
Counting time			7200 s			7200 s		

[a] Values in brackets are recommended values
[b] Arithmetic mean (χ) ± standard deviation from five parallel determinations
[c] Arithmetic mean (χ) ± standard deviation from eight parallel determinations
All values are arithmetic mean (χ) ± standard deviation from ten parallel determinations.

nickel, rubidium, selenium, strontium and zinc in lobster and oysters; see Table 2.8.

The reliability of the concentrations determined by two irradiation techniques can be evaluated in terms of the precision and accuracy achieved, defined as the difference between the mean and the certified value. For many elements, good precision is achieved with conventional thermal neutron activation. This is simply because many elements have essentially the same response to thermal neutrons, and the activation yields are generally high with little or no discrimination. In epithermal activation, however, selectivity is more pronounced and good precision is usually obtained for those elements that exhibit favourable resonance cross-section characteristics. However, in both instances, the matrix composition of the sample may

Table 2.9. Trace elements and minor constituents of the oyster hepatopancreas reference materials and NBS 1555 and NBS 1566, as determined by instrumental photon activation analysis (from [44])

Element	IPAA, NBS 1555	Certified value, NBS 1555	Detection limits, NBS 1555	Oyster tissue, NBS 1566
		Trace elements, mg/kg		
Mn	22.5 ± 2.7	23.4 ± 1.0	4	17.5 ± 1.2
Ni	2.4 ± 0.3	2.3 ± 0.3	0.6	19.4 ± 1.0
Cu	363 ± 32	439 ± 22	34	63.0 ± 3.5
Zn	175 ± 9	177 ± 10	20	852 ± 14
As	28.7 ± 4.2	24.6 ± 2.2	0.3	13.4 ± 1.9
Sr	110 ± 9	113 ± 5	12	84 ± 9
Cd	26.8 ± 3.4	26.3 ± 3.1	2	3.5 ± 0.4
Pb	11.6 ± 1.3	10.4 ± 2.0	3	12.4 ± 0.6
		Minor constituents, %		
Na	3.79 ± 0.25	3.67 ± 0.20	0.02	0.51 ± 0.03
Mg	0.267 ± 0.020	0.255 ± 0.025	0.0004	0.128 ± 0.009
Cl	5.47 ± 0.20	5.58 ± 0.10	0.01	785 ± 20
Ca	0.9606 ± 0.126	0.895 ± 0.058	0.0003	0.15 ± 0.02

influence the precision. The accuracy of the determinations, on the other hand, is comparable for many elements in both activation techniques.

This technique has been used to determine trace elements in crustacea [42].

Zeisler et al. [43] determined 44 elements in digest of marine bivalve tissue using X-ray fluorescence spectrometry. He also used neutron capture, gamma activation and neutron activation analysis.

Photon activation analysis (PAA) has been used to determine a wide range of elements in two reference oyster hepatopancreas samples (NBS 1555 and TORT 1) [44]. Photon activation analysis was carried out at the National Research Council of Canada's electron linear accelerator using the bremsstrahlung produced by the impact of a focused electron beam on a tungsten converter.

The reference samples were homogenised, spray-dried, acetone-extracted, vacuum-dried, screened, blended, bottled, and finally irradiation-sterilised. All samples were dried at 105 °C to remove residual water, and photon activation analysis was performed on the samples in aluminium vials.

Table 2.9 shows determined and expected concentrations of elements found in NBS reference oyster tissue, together with detection limits.

2.1.10.2
Gamma-Ray Spectrometry

This technique has been used to determine trace elements in mussels and oysters [45, 46]. Fourie and Peisach [46] studied the loss of traces of chromium, manganese, iron, cobalt, zinc, arsenic, selenium, cadmium, antimony and lead from oyster tissue during the dehydration of wet samples. They used radioactive tracers in these studies and, following an oven-drying (at 50 – 120 °C) or freeze-drying dehydration procedure, they estimated unvolatilised residual elements by gamma-ray spectrometry. Live oysters were fed with various radioactive elements and then subjected to various dehydration procedures in order to establish whether element loss occurred.

The results indicate the following: that for the elements chromium, manganese, iron, cobalt and zinc, where possibly no losses occurred during drying, the existing techniques are applicable and reliable; that for the elements selenium, cadmium and lead, where appreciable losses were detected, the application of existing techniques without additional precautions or corrections would probably lead to inaccurate results; that antimony and arsenic were not studied owing to 100% mortality among the oysters; and that although these results are applicable only to *C. gigas*, this study points to the need to reinvestigate the analytical validity of dehydration processes in other biological systems.

2.1.10.3
The Mussel Watch Programmes

Perhaps the most extensive set of data relating metal concentrations in mussels and oysters to those present in the environment have come from the Mussel Watch Programmes [47]. Both geographical and temporal trends in seawater concentrations are sought through soft-tissue analyses. In the US programme [48], bivalves (oysters and mussels) were collected at over 100 stations along the East, West and Gulf coasts of the United States. Animals of uniform size (approximately 5 – 8 cm long) were sought, where possible, – although oysters were slightly larger. Elements determined in this program include lead, cadmium, silver, zinc, copper, nickel, plutonium and uranium.

Two laboratories carried out the metal analyses, usually by atomic absorption spectrometry: the Scripps Institution of Oceanography and the Moss Landing Marine Laboratories, both in California.

The general picture that emerges for most of the metals is a distribution pattern that repeats itself year after year. Such a situation can result from long biological half-lives of metals in the organisms (half lives of the order of a year or more), or from uniform levels of the metals in the seawater, or a combination of the two.

Lead

The high lead concentrations in seawaters adjacent to urban areas result from the combustion of lead alkyls as anti-knock agents in gasolines [49]. Both direct atmospheric input and sewage, storm and river runoff contribute to the anthropogenic lead burden of surface waters. It has been estimated that the annual lead inputs to the southern California coastal waters are 310 metric tons from the atmosphere, 200 tons from sewage, 190 tons from storm runoff, and 40 tons from natural sources [50]. As lead alkyls are phased out from use in gasolines, it is expected that lead concentrations in the waters, and in the bivalves, will decrease with time.

Mussel analyses led to the identification of 'hot spots', where the lead concentrations in the mussels, and presumably in their environmental waters, are raised over adjacent areas as a consequence of fluxes from highly populated industrial areas. The regional variations may be seen in the data from the US west coast, from its northern boundary in the state of Washington to its more southerly parts in California. Low lead concentrations were found in mussels taken in the central California stations, San Mimeon to San Francisco, with the exception of those samples taken from the Farallon Islands, where the levels were high (3.3 – 9.3 ppm, dry weight). In contrast, mussels from southern California stations had high lead concentrations, greater than 2.5 ppm dry weight, with the exception of Point Arguello and Rincon Cliffs. Levels were especially high at Point La Jolla (6.5 – 10.0 ppm), Point Fermion (7.9 – 8.0 ppm), Santa Catalina (5.2 – 6.4), and San Pedro Harbor (8.8 – 17.7). These high mussel concentrations are attributed to high influxes of lead in the Los Angeles/San Diego/Santa Barbara region, primarily from automotive exhausts. The lead is transported principally through the atmosphere, and accommodated in the seawaters following wet and dry fallout. The subsequent uptake by the mussels is evidenced by their unusual concentrations.

A similar situation was noted in the mussel lead concentrations on the East coast. Highest values were found in animals living adjacent to highly populated areas. For the three-year sampling period, elevated levels were observed at Cape Newagen (4.4 – 9.5), Portland (4.6 – 5.3), Cape Ann (8.7 – 15.6), Boston (5.9 – 14.2), and Cape Cod (3.6 – 6.5). Relatively low lead concentrations were observed in mussels from the northernmost (Blue Hill Falls, Sears Island) and the southernmost (Atlantic City to Assateague) stations, where concentrations of less than 2 ppm were recorded. There were no trends in the lead concentrations of oysters collected along the southeast coast of the United States.

Silver

In comparison with lead, the sources of silver in the coastal waters of the US have not been clearly identified. The photographic industry is, perhaps, the largest consumer of silver, and its discards probably enter the oceans via sewage. Inputs from the plating industry to sewage are a secondary source.

There appears to be no significant atmospheric input of silver into the marine environment. Thus, elevated levels of lead in mussels living adjacent to urban areas may or may not be accompanied by complementary increased concentrations of silver. Such appears to be the case in two west coast stations, the Farallon Islands and the San Pedro Harbor. Both have high lead concentrations in mussels and low silver values for the three-year sampling period. Neither of these stations receive sewage. The San Pedro Harbor area does not receive outflow from the Los Angeles River, which undoubtedly carried anthropogenic lead, washed into it from storm run-off.

In contrast, mussels from stations exposed to sewage show elevated amounts of both lead and silver. West coast sample sources include Point La Jolla, San Diego Harbor, Point Fermin and Santa Catalina Island. Of these four stations, the first three are located in the vicinity of major urban sewer outfalls, and the elevated levels of silver and lead are not unexpected. However, the Santa Catalina station is located well offshore (40 km across the sea) and the high values there are puzzling. In comparison with their west coast counterparts, the east coast mussels had consistently low silver levels. This may relate to lower levels of silver in the waters, to different bioaccumulating abilities of mussels of different species, or to differences in the silver contents of the food consumed by the mussels. It is suspected that the primary reason is the species difference between the east coast *M. edulis* and the west coast *M. californianus*, since there is a lower silver concentration in *M. edulis* in San Francisco Bay, where both species are taken from the same environment.

Cadmium and Plutonium

In addition to lead, plutonium serves as an example of the usefulness of isotopic composition for the identification of sources. The Pu-238/Pu-239 + 240 ratio resulting from the entry of nuclear weapons debris ranges between 0.03 and 0.08 in the byssal threads of mussels taken from waters where there are no localised nuclear point sources. On the other hand, the ^{238}Pu / ^{239}Pu and ^{238}Pu / ^{240}Pu ratio in the byssal threads of mussels taken near the site of nuclear reactors in San Onofre, California, had values of 0.21 and 0.16. The ^{238}Pu is used as a fuel and probably leaked into the marine environment in the cooling water discharges. Thus, the plutonium burden of these coastal waters has been increased by a factor of three or four over that of the background fallout of plutonium on this basis. Monitoring of byssal threads, in which the plutonium is enriched, is a far simpler task than the monitoring of the waters themselves.

Metal concentrations in mussels can achieve unusual values from natural processes. Such appears to be the case with cadmium and plutonium. Mussel cadmium levels are generally higher on the west coast than the east. The high values on the west coast are most probably the result of upwelling. This can be illustrated with the mussels from Diablo Canyon and Soberanes Point, California, located near Point Sur at the midsection of the US west

Table 2.10. Cadmium and plutonium in mussels from the US West Coast (from [48])

Location	mg/kg					
	Year 1		Year 2		Year 3	
	Cd	Pu	Cd	Pu	Cd	Pu
Point La Jolla, CA, USA	1.7	1.0	0.8	0.7	1.8	
San Pedro Harbor, CA, USA	2.3	0.3	1.3	0.4	2.0	
Diablo Canyon, CA, USA	7.7	3.7	5.9	5.7	9.2	2.1
Soberanes Point, Ca, USA	9.4	5.5	20.2	11.2	9.7	3.9
				13.9		
Columbia River, WA, USA	1.4	1.0	1.0	1.0	3.8	
Puget Sound, WA, USA	2.6	0.7	0.8	0.5	2.7	0.3

coast, an area in which upwelling occurs during much of the year. The upwelling process brings cadmium-rich waters to the surface.

Normally cadmium in surface waters exhibits a depletion as a consequence of its transfer to greater depths by fast sinking biogenous particles. The higher concentrations in mussels from these two sites can be compared with those to the north (Columbia River and Puget Sound) and those to the south (Pt La Jolla and San Pedro Harbor) (Table 2.10). The latter values are subsequently low. Similarly, the transuranic element plutonium, introduced to the marine environment in fallout from nuclear weapons testing, shows surface depletion and mid-depth enrichment in the Pacific. There is a strong covariance between plutonium and cadmium in mussels from the Pacific coast. The silver, lead, cadmium and plutonium data quoted above illustrate the importance of bivalve monitoring programmes. First of all, they provide evidence of metal pollution along parts of the conterminous US.

Clean environments can be defined, without actual measurements, within the water column. The US Mussel Watch suggests a lead baseline of 1.0 mg/kg, a west coast silver baseline (*Mytilus californianus*) of 0.1 mg/kg, and an east coast silver baseline (*M. edulis*) of 0.05 mg/kg for organisms inhabiting a clean environment.

Secondly, without expensive and time-consuming water analyses, natural phenomena influencing metal concentrations in seawater can be identified. Clearly, there is a crucial need for confirmation of such hypotheses through actual water studies. Without systematic surveys, elevated lead and cadmium might have been interpreted as the result of a localised anthropogenic input rather than a natural physical phenomenon such as upwelling.

Other techniques used to determine metals in invertebrates include magnetron dc arc plasma [51], ion chromatography [52], X-ray fluorescence spectroscopy, prompt activation analyses and neutron activation analysis [43].

2.2
Organic Compounds

2.2.1
Aliphatic Hydrocarbons

Morgan [53] has described a gas chromatographic method for the determination of Bunker C fuel oil in marine organisms at the 0.5 mg/kg level.
Pentane–methanol extraction of tissues, using a blender, is followed by adsorption chromatography.

Meyers [54] investigated the occurrence of nonbiogenic hydrocarbons
in shrimps occurring in the vicinity of offshore drilling and petroleum exploration in the Gulf of Mexico. Samples were stored in glass at $-20\,°C$
prior to analysis. The crushed sample was treated with 0.5 N potassium hydroxide in $1:1\ v/v$ benzene:methanol, and, following the addition of water,
unsaponifiable lipids were extracted with petroleum ether. Column chromatography using alumina over silica gel 50/50 separated saturated from
unsaturated plus aromatic hydrocarbons. Gas–liquid chromatography with
a flame ionization detector was employed to resolve and to quantify the
various components of each hydrocarbon fraction. Both a nonpolar column
and a polar column were used. The nonpolar column was $4\,m \times 2.1\,mm$ id
of 3% SP-2100 on 100–120 mesh Supelcoport and was operated from 150
to $325\,°C$ at $4\,°C/min$ using nitrogen carrier gas at 15 ml/min. The polar
column was $2.5\,m \times 2.1\,mm$ ID 10% SP-1000 mesh Supelcoport and was
operated from 150 to $250\,°C$ at $8\,K/min$ using nitrogen at a flow rate of
15 ml/min.

Although the organisms are from different orders of Crustacea, and were
collected during different sampling periods, their traces are very similar. Few
normal alkanes are found in any of the samples, and the saturated hydrocarbon compositions of these animals appear to be composed mostly of
branched compounds. The unsaturated hydrocarbon compositions of these
organisms also display a fairly simple pattern. Usually 4–6 peaks dominate
the chromatograms obtained from both nonpolar and polar columns. The
major peaks of chromatograms of the saturated and unsaturated hydrocarbon fractions from representative samples of the five species are listed in
Table 2.11 in terms of Kovats Retention Indices and weight percent contributions of each peak to the total fraction. Peaks from both polar and nonpolar
columns are listed. As shown in this tabulation, hydrocarbon compositions
are dominated by only a few peaks, and some peaks with the same Kovats
Indices are common to all five crustaceans. The largest peak from the nonpolar chromatograms of the saturated fraction has an index of 2506–2510
in all five samples. However, a peak with an index of around 2500 is not
a major contributor to polar chromatographs of this fraction. Instead, the
most common major peak in these latter distributions has a Kovats Index of 2140–3144. Major peaks comprising the unsaturated fractions of the
samples are grouped between indices of 1900–2500 on the nonpolar chro-

Table 2.11. Hydrocarbon content of shrimp (from [54])

Organism	Saturated hydrocarbons				Unsaturated hydrocarbons			
	Nonpolar Column		Polar Column		Nonpolar Column		Polar Column	
	KI[a]	wt%	KI[a]	wt%	KI[a]	wt%	KI[a]	wt%
Penaeus setiferus	2248	15	2967	19	1935	14	2237	16
	2271	15	3013	13	2114	20	2453	14
	2508	23	3140	19	2143	11	2475	16
					2274	18	2843	14
Penaeus duoraroum	2506	34	2343	29	1950	19	2224	11
	2894	8	3145	26	2133	15	2244	11
					2300	18	2461	27
					2477	11	2810	16
Trachypenaeus similis	2506	31	2495	7	1921	14	2222	16
	2894	7	2798	8	2099	20	2430	11
			3144	26	2125	10	2455	18
					2256	18	2826	13
Squilla empusa	2244	11	2043	35	1933	20	2234	23
	2249	11	2253	27	2103	13	2438	12
	2507	16			2133	11	2463	11
					2263	13	2833	10
					2445	10		
Squilla chydea	1703	15	1699	8	1911	10	2219	14
	1803	8	1800	5	2005	8	2352	11
	1977	10	2202	11	2087	13	2449	11
	2510	16	3144	11	2237	8	2824	8

KI: Kovat's Index
[a] Peaks labelled by Kovat's Index

matograms. A shift to indices of 2200–2850 on polar chromatograms is indicative of the relatively polar nature of these unsaturated hydrocarbons.

Clearly, a combination of GC and mass spectrometry would provide more useful information in studies of hydrocarbons in crustacea and would allow a clearer distinction to be made between contaminant hydrocarbons and nonbiogenic naturally occurring hydrocarbons.

Chesler et al. [55–57] have described a headspace sampling–gas chromatographic method for the determination of petroleum hydrocarbons in mussels, oysters and clams. This procedure utilises dynamic headspace sampling of an aqueous caustic tissue homogenate to extract and collect volatile organic components. Interfering polar biogenic (nonanthropogenic) components are removed by normal-phase HPLC. Quantitation and identification of the individual compounds are accomplished using GC and GC–MS. The nonvolatile polynuclear aromatic hydrocarbons which remain in the homogenate after headspace sampling are solvent-extracted and then analysed by reversed-phase liquid chromatography. The crustacea samples were kept at a low temperature ($-10\,^\circ$C) between sampling and analysis. Approximately 30 g tissue, 500 ml hydrocarbon free water, and 50 g of sodium hydroxide were combined in a flask, together with aliphatic or aromatic hydrocarbon internal standards, and the mixture homogenised. The tissue homogenate was heated to 70 °C, and the headspace sampled for 18 hours at a nitrogen flow rate of 150 ml/min. The headspace vapours were passed into a Tenax GC packed stainless steel column. The homogenate solution remaining in the flask after headspace sampling was extracted with pentane to remove nonvolatile polyaromatic hydrocarbons. This extract was concentrated by nitrogen purge and devolved in 1 ml acetonitrile for HPLC.

Following headspace sampling and drying, the Tenax GC was connected as part of the injection loop of a liquid chromatograph and the organic compounds eluted with pentane onto a µBondepak NH_2 clean-up column, the first 15 ml of eluate containing the hydrocarbons. This fraction was reduced to 300 µl by nitrogen purge and the residue washed onto a Tenax GC column, the contents of which were thermally purged onto a gas chromatographic column for analysis.

Recovery data for the aromatic and aliphatic compounds used as internal standards in the tissue analyses are given in Table 2.12. Using an 18-hour headspace sampling period, recoveries from water for the higher molecular weight aromatic and aliphatic components (i.e. trimethyl naphthalene, phenanthrene, MeC_{16} and MeC_{18}) were nearly 100%. Aliphatic hydrocarbon recoveries were found to be much lower than aromatic hydrocarbon recoveries in the headspace sampling of the tissue homogenate. Using caustic digestion, recoveries from mussel tissue homogenate approached 100% for the higher aromatics but were only 30% for the aliphatic components. It is assumed that the aliphatic hydrocarbons were being retained in the lipophilic portion of the tissue homogenate and that the partition coefficient for these hydrocarbons between the headspace sampling gas and the lipophilic fraction was quite unfavourable. Recovery data for the complete analytical scheme indicate that some losses of the internal standards also occur during the liquid chromatographic clean-up and concentration step. The losses that occur during the concentration step amount to 25% for mesitylene, 30% for 2-methylundecane, 40% for naphthalene, 11% for 5-methyltetradecane, 5% for trimethylnaphthalene, and less than 1% for 7-methylhexadecane, 2-methylnaphthalene, phenanthrene and hydrocarbons of higher molecular weight.

Since quantitation in these analyses was dependent upon an internal standard added at the beginning of the analytical scheme, it was important to know whether the internal standard components were recovered to the same extent as these components would be if incorporated into the tissue matrix. In a series of experiments with live *Mytilus* (mussels) exposed to ^{14}C-naphthalene and then analyzed using the four-hour headspace sampling procedure and no HPLC clean-up, a ^{14}C recovery of 78 ± 12% was observed. In comparison, the recovery of unlabelled naphthalene added as an internal standard was found to be 66 ± 8% for the same four-hour headspace sampling procedure (see Table 2.12). Therefore, the indications are that, at least in the case of naphthalene, an internal standard added to the mussel tis-

Table 2.12. Internal standard recovery, in % (from [55])

	Mesitylene	Naphthalene	Trimethylnaphthalene	Phenanthrene	MeC$_{11}$[d]	MeC$_{14}$	MeC$_{16}$	MeC$_{18}$
Water (4 h headspace sampled)	8 ± 2[a] (6)[b]	29 ± 4 (6)		12 ± 4 (6)	17 ± 5 (6)	62 ± 4 (6)	74 ± 8 (6)	57 ± 8 (6)
Water (18 h headspace sampled)	6 ± 1 (3)	52 ± 9 (3)	95 ± 5 (3)	92 ± 4 (3)	31 ± 8 (3)	84 ± 3 (3)	97 ± 3 (3)	94 ± 3 (3)
Mussels (4 h)	20 ± 3 (7)	66 ± 8 (8)	44 ± 6 (5)	17 ± 4 (5)	12 ± 3 (3)	11 ± 2 (4)	4 ± 0.3 (4)	2 ± 0.5 (4)
Mussels 2 mol/l KCl (4h)	5 ± 1 (2)	40 ± 5 (2)	20 ± 3 (2)	7 ± 1 (2)	8 ± 3 (2)	11 ± 1 (2)	6 ± 0 (2)	3 ± 1 (2)
Mussels 2 mol/l KCl (4h)		83 ± 18 (2)	88 ± 6 (2)	40 ± 6 (2)	4 ± 1 (2)	7 ± 3 (2)	2 ± 1 (2)	2 ± 1 (2)
Mussels 2 mol/l KCl + 4 M KCl (18h) No HPLC	30 ± 2 (2)	69 ± 6 (2)	111 ± 2 (2)	119 ± 1 (3)	28 ± 11 (2)	33 ± 1 (2)	32 ± 1 (2)	26 ± 4 (2)
HPLC clean-up	...[c]	39 ± 7 (3)	74 ± 28 (3)	91 ± 6 (9)	20 ± 6 (3)	29 ± 6 (3)	32 ± 8 (3)	26 ± 8 (3)
Mussels 2.5 mol/l NaOH (18 h) No HPLC clean-up	...	62 ± 1 (4)	90 ± 8 (3)	96 ± 6 (3)				
HPLC clean-up	...	54 ± 5 (9)	73 ± 2 (10)	82 ± 2 (9)				

[a] Data reported as the standard deviation (1σ) of a set of replicate values from the mean of the replicate values

[b] Denotes number of samples analysed

[c] Indicates no recovery of internal standard

[d] MeC$_{11}$ = 2-methylundecane, MeC$_{14}$ = 5-methyltetradecane, MeC$_{16}$ = 7-methylhexadecane, and MeC$_{18}$ = 2-methyloctadecane

Table 2.13. Comparison of volatile hydrocarbon levels obtained with and without HPLC clean-up, in μg/kg (from [55])

	No HPLC	HPLC
Mytilus (mussels, Northeastern Gulf of Alaska)	1406 ± 98 (2)	540 ± 46 (3)
Oysters (Middle March, SC, USA)	1834 (1)	652 (1)
Clams A (control)	509 ± 11 (2)	377 ± 88 (2)
Clams B (1 μg crude oil/g water)[a]	1421 ± 114 (2)	491 ± 88 (3)
Clams C (10 μg crude oil/g water)[b]	1704 (1)	1413 ± 398 (2)

[a] Exposed to 1 μg crude oil/g water
[b] Exposed to 10 μg crude oil/g water
Numbers in brackets denote number of samples analysed

sue solution can be recovered essentially to the same extent as naphthalene incorporated into live mussels.

These data from various tissue samples (mussels and clams) indicate that HPLC removal of the non-hydrocarbon components is necessary for effective determination of low hydrocarbon levels in tissue. Of particular interest in Table 2.13 are the results obtained with various clam samples with and without exposure to 1 and 10 μg/g of crude oil in water. A comparison of the data for the control clams with and without HPLC clean-up reveals that the six most abundant components (∼ 100 μg/kg total) in the sample without HPLC clean-up are non-hydrocarbon. A comparison of the results obtained after HPLC and after excluding the control level (i.e. 400 μg/kg) for the clams exposed to 1 μg/g of crude oil in water shows a difference of a factor of 10 in petroleum uptake. The data in Table 2.13 support the applicability of the above method for the determination of hydrocarbons in marine organisms exposed to toxic levels, as well as those from unpolluted environments.

The headspace sampling procedure for the analysis of hydrocarbons in marine biota offers several advantages over solvent extraction procedures. The headspace sampling technique requires minimal sample handling, few sample transfers, and only a minimal amount of organic solvent, thereby reducing the risk of contamination (a system blank for the headspace sampling method results in a value of only ∼ 5 μg/kg based on a sample of 600 ml of water [56, 59]. In addition, only one solvent concentration step is involved, thereby reducing the losses of the more volatile components. When compared to solvent extraction procedures, the analyst's time is greatly reduced by using the headspace sampling technique. During the lengthy headspace sampling period, the system is left to run unattended.

Berthou et al. [58] used GC to determine weathered aliphatic and aromatic hydrocarbons in oyster samples.

Mason [59] studied the feasibility of using fluorescence spectroscopy to determine aromatic compounds in mussel tissues and compared the results with those obtained by GC. There were significant correlations between the concentrations of aromatic hydrocarbons found by fluorescence spectroscopy and both aliphatic and aromatic hydrocarbon concentrations obtained by GC. Analysis of the aliphatic fraction by GC and of the aromatic fraction by fluorescence spectroscopy would give a reasonable estimation of the relative degree of contamination of mussels by petroleum hydrocarbons.

Polyaromatic Hydrocarbons (PA-HCs)

Bjorseth et al. [60] described a capillary gas chromatographic method for determining PA-HCs in mussels. Up to 34 PA-HCs were identified. Dunn and Stich [61] have described a monitoring procedure for PA-HCs, particularly benzo(a)pyrene in marine organisms in coastal waters. The procedures involve the extraction and purification of hydrocarbon fractions from the sediments or organisms, and the determination of compounds by thin-layer chromatography and fluorimetry, or GC.

To avoid possible photodecomposition of PA-HCs, all extraction and purification procedures were carried out under subdued yellow tungsten light. Between 20 and 40 g of tissue were placed in a 300 ml flask and 150 ml of ethanol, 7 g of potassium hydroxide, boiling chips, and an aliquot of radioactive benzo(a)pyrene (either around 5 µg 1,000 dpm ^{14}C-benzo(a)pyrene or around 0.1 ng 25,000 dpm ^{3}H-benzo(a)pyrene) were added. The tissue was digested by refluxing gently for 1.5 hours with occasional swirling. The digest was added while hot to 150 ml of water in a two-litre separator funnel, and the digestion flask rinsed out with an additional 50 ml of ethanol. The water–ethanol mix was extracted three times with 200 ml of iso-octane, and the iso-octane extracts were combined and washed with 4×200 ml warm (60 °C) water. This extract was then passed down a Florisil clean-up column. Polycyclic aromatic hydrocarbons were eluted from the column with 3×100 ml benzene. The combined eluate was reduced to 5 ml by rotary evaporation, 50 ml of iso-octane were added, and the volume again reduced to 5 ml to remove the benzene.

Polyaromatic hydrocarbons were extracted from the iso-octane with 3×5 ml dimethyl sulfoxide. The dimethyl sulfoxide extracts were combined with 30 ml of water, and the PA-HCs extracted into 2×10 ml iso-octane. The iso-octane extracts were combined, washed with water, and dried by passage through 10 g of sodium sulfate in a 15 ml coarse-fritted glass Buchner funnel. This extract was used for thin-layer chromatography, with benzo(a)pyrene being detected under long-wavelength ultraviolet light.

The adsorbent at the position of the benzo(a)pyrene band was scraped off the plate while still damp, and placed in a fine-fritted Buchner funnel. The benzo(a)pyrene was removed from the cellulose acetate by washing with 4×4 ml hot (65 °C) methanol, using gentle suction. The methanol was added to 10 ml of a solution of 20% hexadecane in iso-octane, and the methanol and iso-octane were removed by rotary evaporation to leave the benzo(a)pyrene in 2 ml of hexadecane, ready for fluorimetry.

Mussels (*Mytilus edulis*) taken from the outer Vancouver harbour showed lower benzo(*a*)pyrene levels in the summer than in the winter, perhaps a result of seasonal discharges of sewage and storm drain water into the harbour. Elevated levels of benzo(*a*)pyrene in mussels growing near creosoted timbers or piling suggested that creosote may be a significant source of this substance in the marine environment. Direct evidence for this suggestion was obtained by comparison of GC profiles of polycyclic aromatic hydrocarbons isolated from mussels with those from creosoted wood.

Benzo(*a*)pyrene was measured fluorimetrically in hexadecane using the baseline technique of Kunte [62]. Samples and standards of 10 – 200 ng/ml benzo(*a*)pyrene in hexadecane were excited at 365 nm in an Aminco-Bowman spectrophotofluorimeter, and the emission spectrum was recorded from 375 to 500 nm. An artificial baseline was drawn between minima in the fluorescence spectrum occurring at 418 and 448 nm, and the height of the peak at 430 nm above this baseline was measured. Where necessary, highly fluorescent samples were diluted with hexadecane to bring their fluorescence within the range of the standards used.

After fluorimetry, the amount of radioactive benzo(*a*)pyrene internal standard in each sample was determined by scintillation counting. The recovery of benzo(*a*)pyrene was calculated by comparing the amount of radioactivity added at the beginning of the digestion procedure with the amount recovered in the fluorimetry sample. The amount of benzo(*a*)pyrene determined by fluorimetry was then corrected if necessary for the contribution of radioactive tracer (this correction is negligible if the ^{3}H-benzo(*a*)pyrene, which has higher specific activity, is used), and the net amount of benzo(*a*)pyrene originally present in the sample was then expressed as µg/kg benzo(*a*)pyrene wet weight of tissue or dry weight of sediment.

The overall recovery of benzo(*a*)pyrene was generally 60 – 80% for tissue samples (mussels, clams, oysters).

Estimations of the precisions of benzo(*a*)pyrene determinations in mussels at the 10 – 20 ng/kg level ranged from a standard deviation of 0.4 to 1.45. Shoreline mussel samples had a mean benzo(*a*)pyrene content of 0.55 mg/kg net weight with a standard deviation of 0.11. Samples stored for 12 weeks at −10 °C showed no significant change in benzo(*a*)pyrene content, suggesting that this is an adequate method of sample storage.

Uthe and Musal [63] carried out an intercomparison study on the determination of polynuclear aromatic hydrocarbons in lobster. Intercomparative kits comprising lobster digestive gland acetone powder and lobster digestive gland oil were sent to participants in Europe, the USA and Canada. The participants were requested to measure a suite of non-alkylated PA-HC and to analyse each material. The methods used were either liquid chromatography with UV absorption–fluorescence detection or GC–MS on cleaned-up extracts. Interlaboratory relative standard deviations for PA-HC concentrations in oil ranged from 4.3 to 24.1%. Interlaboratory relative standard deviation ranged from 39 to 96%. Laboratories using GC–MS reported a greater number of compounds, whereas those using liquid chromatography–ultraviolet spectroscopy reported higher concentrations.

Giam et al. [64] have reported on the uptake and depuration of benzopyrene, hexachlorobenzene and pentachlorophenol in marine organisms. Methods capable of determining down to 0.2 µg/kg of these substances are discussed.

Iosifidou et al. [136] gave details of a gas chromatographic method for the determination of PA-HCs in Greek Gulf waters.

2.2.2
Phthalate Esters

Giam et al. [65] determined phthalate esters in amounts down to less than 5 µg/kg in shrimps and crab using capillary column GC with an electron capture detector. Chlorinated insecticides and chlorinated biphenyls interfere in this chromatographic analysis and consequently need to be removed first by column chromatography on water-deactivated Florisil.

The tissue was macerated with acetonitrile, then diluted with methylene chloride–petroleum ether (1 : 5) and extracted with saltwater. The dried organic phase was concentrated and diluted with iso-octane and subjected to clean-up in a Florisil column. Elution of the Florisil column with 6%, then 15%, then 20% diethyl ether in petroleum ether provided three fractions containing, respectively, (i) chlorinated insecticides and chlorinated biphenyls, (ii) diethylhexylphthalate and dibutylphthalate, and (iii) dibutylphthalate.

Extreme precautions are necessary in this procedure to avoid contamination due to phthalates present as impurities in commonly used laboratory materials; e.g. aluminium foil contains 300 mg/kg phthalate.

Between 3 and 20 µg/kg of diethyl hexyl phthalate was found in crab taken in the Gulf of Mexico, while dimethyl, diethyl and dibutyl phthalates occurred at concentrations less than the detection limit of the method (1 µg/kg).

2.2.3
Chloro Compounds

Chlorinated Aliphatic Compounds

Murray and Riley [66, 67] described gas chromatographic methods for the determination of trichloroethylene, tetrachloroethylene, chloroform and carbon tetrachloride in marine organisms. These substances were separated and determined on a glass column (4 m × 4 mm) packed with 3% of SE-52 on Chromosorb W (AW DMCS) (80- to 100-mesh) and operated at 35 °C, with argon (30 ml/min) used as carrier gas. An electron capture detector was used, with argon–methane (9 : 1) as quench gas. A limitation of this procedure is that compounds which boil considerably above 100 °C could not be determined.

Chlorinated Insecticides and Polychlorobiphenyls (PCBs)

The determination of chlorinated insecticides in crustacea has been discussed by several workers [68–76]. Mills et al. [68] dehydrated oyster samples by mixing them with a 9 : 1 mixture of anhydrous sodium sulfate and
Quso (a micro-fine silica). They could be held at room temperature for up to
15 days without loss or degradation of chlorinated insecticides. The tissues
of oysters were homogenised. Approximately 30 g of the homogenate was
added to a second Mason jar and blended with a 9 : 1 mixture of sodium
sulfate and Quso. By alternately chilling and blending, a free-flowing powder was obtained. The blended sample was wrapped in aluminium foil and
shipped to the laboratory. Upon receipt of the sample, it was weighed and
extracted in a Soxhlet apparatus for four hours with petroleum ether. The
extracts were then purified by concentrating and transferring the extract to
separator funnels. The extracts were diluted to 25 ml with petroleum ether
and partitioned with two 50 ml portions of acetonitrile previously saturated
with petroleum ether. The acetonitrile was evaporated to dryness and the
residue eluted from a Florisil column [68]. In this technique, increasing
proportions of ethyl ether to petroleum ether were used to elute fractions
containing increasingly polar insecticides. The extracts were analysed by GC.
Recoveries of DDE, DDD and DDT were between 79 and 96%. The detection
limit for a 30 g oyster sample was 10 µg/kg.

Arias et al. [69] have described a method for the determination of
organochlorine insecticide residues in molluscs. The method involves extraction, Florisil column clean-up and analysis of the extract by thin-layer
chromatography on silica gel G or alumina with hexane or hexane–acetone
(49 : 1) as solvent, or GC on a polar column of 10% of DC-200 on Chromosorb W HMDS and on a semipolar column of 5% of DC-200 plus 7.5%
of QF-1 on Chromosorb W, with electron capture detection.

Ernst et al. [70,71] have determined, by GC–MS, residues of DDT, DDE,
DDD and polychlorinated biphenyls in scallops from the English Channel.

Neudorf and Khan [72] investigated the uptake of ^{14}C-labelled DDT,
dieldrin and photodieldrin by *Ankistrodesmus amalloides*. The results of
liquid scintillation spectrometric analyses show that the total pick-up of DDT
during a 1 – 3 hour period was 2 – 5 times higher than that of dieldrin, and
ten times higher than that of photodieldrin. The algae metabolised 3 – 5%
of DDT to DDE, and 0.8% to DDD. The metabolism of DDT by *Daphnia
pulex* was also monitored by exposing 100 organisms to 0.31 ppm of the
labelled pesticide for 24 hours without feeding. The metabolites were then
extracted and separated by thin-layer chromatography, and the R_f values of
radioactive spots were compared to R_f values for non-radioactive DDD and
radioactive DDE. The results show a conversion of DDT to DDE of about
13.6%.

Teichman et al. [73] have discussed the determination of chlorinated
insecticides and PCBs in oysters.

Table 2.14. Percentage recoveries of insecticides eluted from neutral alumina (from [73])

Compound	First fraction				Second fraction
	0 – 15 ml	15 – 20 ml	20 – 25 ml	25 – 30 ml	30 – 60 ml
Lindane			10		
Heptachlor	100				
Aldrin	100				
Heptachlor epoxide					100
p,p'-DDE	100				
Dieldrin					100
p,p'-DDD			50	50	
p,p'-DDT	100				
PCB	100				
γ-Chlordane	10	80	10		
α-Chlordane	80	20			

Table 2.15. Percentage recoveries of insecticides eluted from charcoal (from [73])

	First fraction: 90 ml acetone–diethyl ether (25:75)				Second fraction: 60 ml benzene
	0 – 30 ml	30 – 30 ml	60 – 90 ml	0 – 30 ml	30 – 60 ml
Lindane	30	40	30		
Heptachlor	100				
Aldrin	100				
p,p'-DDE	50	50			
p,p'-DDT	80	20			
PCB				80	20
Clordane	100				

These workers used GC coupled to mass spectrometry. PCBs were separated from DDT and its analogues and from the other common chlorinated insecticides by adsorption chromatography on columns of alumina and charcoal. Elution from alumina columns with increasing fractional amounts of hexane first isolated dieldrin and heptachlor from a mixture of chlorinated insecticides and PCBs. When added to a charcoal column, the remaining fraction could be separated into two fractions, one containing the chlorinated insecticides, the other containing the PCBs, by eluting with acetone–diethyl ether (25:75) and benzene, respectively. The PCBs and the insecticides were then determined by GC on the separate column eluates without cross-interference.

A summation of the elution of the chlorinated organic insecticides and the PCBs from the alumina column is given in Table 2.14. Heptachlor epoxide and dieldrin were removed from the column by extending the elution solvent beyond the 30 ml volume with an additional, but separated, elution volume of 30 ml. The PCBs remained an integral part of the mixture containing the insecticides in the first 30 ml of eluate. The elution pattern of alumina column fraction 1 on the charcoal column, see Table 2.15, shows that the insecticides were separated from the PCBs by means of acetone–diethyl ether eluent. The PCBs were subsequently removed from the charcoal column with benzene. Known amounts of insecticides and PCBs (Aroclor 1254) were added to oyster samples; the samples were analysed as described above to check the efficency of the analytical procedure. Recoveries were in the range 68.2% (lindane PCBs) to 102% (heptachlorepoxide).

Polychlorobiphenyls

Markin et al. [74] have also discussed the possible confusion between Mirex and PCBs in analyses of crabs and shrimps. In their method, the samples were thoroughly scrubbed to remove mud, algae and other residues; they were ground whole and mixed in a Waring blender to make a composite sample. A 20 g subsample of the composite was removed and analysed as follows. The homogenised sample was extracted with a mixture of hexane and isopropanol, and the extract subjected to a concentrated sulfuric acid clean-up. The sulfuric acid destroys dieldrin, endrin, and organophosphorus insecticides. The final extract was cleaned up on a Florisil column and concentrated to the desired level for analysis. If PCBs were suspected in the first analysis, their presence usually being indicated by a series of characteristic peaks, the sample was reprocessed to separate the PCBs from the insecticides as described by Armour and Burke [75], Gaul and Cruz-LaGrange [76] and Markin et al. [77]. After concentrating to the appropriate volume, the extracts from both methods of clean-up were chromatographed on a Hewlett Packard Model 402 dual-column gas chromatograph equipped with dual electron capture detection. Each sample was analysed on two different columns; the first column was a mixture of 1.5% OV-17 and 1.95% QF-1 on Gas Chrom Q. The temperatures of the injector, oven and detector were 250, 200 and 210 °C respectively. The second column was 2% DC-200 on Gas Chrom Q with injector, oven and detector temperatures of 245, 175 and 205 °C respectively. Argon–methane at 80 ml/min was the carrier gas. Limits of detection were 0.001 ppm for DDT and its metabolites, 0.005 ppm for Mirex and 0.01 ppm for Aroclor 1260.

Markin et al. [74] comment that they found Mirex in only a minority of the samples they analysed, contrary to results obtained by earlier workers [78–80].

Tanabe et al. [81] used mussels as bioindicators of PCB pollution. When uncontaminated green-lipped mussels were transplanted in severely contaminated Hong Kong Bay waters, total PCB concentrations increased from 11 μg/kg wet weight to 560 μg/kg wet weight in 17 days. When the remaining mussels were returned to clean waters after 17 days, total PCB concentrations decreased from 630 μg/kg wet weight to 12 μg/kg wet weight within 32 days. Lower chlorinated PCBs (isomer and congeners containing 2 – 4 chlorine atoms) were taken- up and depurated more rapidly than the more lipophilic higher chlorinated PCBs (hexa-, hepta- and octachlorobiphenyls). It was suggested that time-bulking (combining samples collected at frequent intervals from a single location) in PCB monitoring studies involving bivalves would provide a more accurate picture of average contamination conditions.

Ya Ma and Bayne [82] discriminated polychlorobiphenyls in clam tissue using electron capture negative ion chemical ionization mass spectrometry.

Gas chromatography has been applied to the determination of PCBs in mussels [83] and in biota samples [84].

Polychlorinated Terphenyls

The presence of polychlorinated terphenyls has been reported in oyster tissue. To determine polychlorinated terphenyl in oyster, a hexane extract [85] was cleaned on alumina or Florisil, and the analysis performed using a combination of a mass spectrometer and a gas chromatograph used in the mass fragmentography mode. Two m/e values were selected ($m/e = 436$ and $m/e = 470$). Approximately 0.15 μg/kg polychlorinated terphenyl and 0.2 μg/kg PCB were found in oyster tissue.

Polychlorinated Dibenzo-p-Dioxins and Dibenzofurans

Taguchi et al. [86] used high-resolution mass spectrometry to determine these substances in crustacea tissue.

Buser [87] determined polychlorobenzothrophenes, their sulfur analogues or polychlorodibenzofurans in crab, lobster and worm using various gas chromatographic–mass spectrometric techniques.

Chlorinated Paraffins

The procedure described in Sect. 1.2.3 could be applied to the determination of chloroparaffins in invertebrates [134].

2.2.3.1
Organophosphorus Insecticides

Deusch et al. [88] determined Dursban in crustacea. After a preliminary clean-up, the extract is chromatographed on a column packed with 3% Carbowax 20 m on Gas-Chrom (60–80 mesh), which gives excellent separation of Dursban from other organophosphorus insecticides. Both thermionic and flame photometric detectors are satisfactory. Recoveries range from 75 to 105% depending on the nature of the sample. This procedure will detect as little as 0.5 ng of Dursban, corresponding to a level of 0.01 mg/kg in a 10 g sample.

2.2.4
Organosulfur Compounds

Organosulfur compounds are minor components of crude oil and of some fuel oils. Although the quantity of these compounds depends on the source of production, it generally ranges from 0.002 to nearly 30% in crude oil, found as sulfur-containing hydrocarbons (Nakamura and Kashimoto [89] and 1600 ppm I in # 2 fuel oils (Dillon et al. [90]). In a field study, these compounds were found in benthic organisms after an oil spill (Grahl-Nielsen et al. [91]). Researchers have presented several papers on the accumulation of these compounds in eels and short-necked clams (Ogata et al. [92]) and have also identified dibenzothiophene through GC–MS (Ogata and Miyake [93] and Ogata [92]) in biota samples after experimental exposure to crude oil suspension.

Moreover, mussels are a well-known biological monitor of marine pollutants in 'the mussel watch' (Goldberg [94]). Many investigators have reported the susceptibility of this organism to petroleum hydrocarbons (Lee et al. [95]) and polynuclear aromatic hydrocarbons (Dunn et al. [96]), and Kira et al. [97] identified several organosulfur compounds through GC–MS and measured the levels of dibenzothiophene, using a GC–flame photometric detector (GC-FPD), in both mussels and in environment waters. The calculated concentration ratio of dibenzothiophene in mussels to that in water ranged up to 500 in the field sample and 800 or higher after experimental exposure. The estimated biological half-life of dibenzothiophene from field mussel samples was about nine days in clean seawater.

Dibenzothiophene levels were measured by gas chromatography–flame photometric detection. In field samples, the levels of dibenzothiophene ranged from less than 0.1 to over 800 µg/kg. Dibenzothiophene was clearly separated from other organosulfur compounds, even at levels of under a part per billion. The presence of dibenzothiophene was indicated by the simultaneous detection of M+184 and 186 on the GC–MS single ion monitor. Accumulation of the compound in mussel was approximately 600 and 800 times higher than the levels in water after four and eight days' exposure, respectively. The concentration ratio of 800 obtained after eight days' exposure was close to that of petroleum hydrocarbons.

2.2.5
Toxins

Draisci et al. [103] determined diarrheic shellfish toxins in mussels and phytoplankton using mass spectrometry/mass spectrometry with negative and positive ionisation.

Algal and bacterial toxins have been responsible for large fish kills, the poisoning of shellfish and for causing illness in swimmers, resulting in the temporary closure of many beaches used for recreation. Red-tide algae are the best-known of these toxin-producing organisms. Recent publications have focused on identifying microcystins, which are toxins produced by blue-green algae. Ells et al. [104] developed an electrospray ionisation–high-field asymmetric waveform ion mobility spectrometric/mass spectrometric method for measuring the microcystins LR, RR and YR in water, with detection limits of 4.2 and 1 nM, respectively. Field asymmetric waveform spectrometry was shown to reduce the chemical background in the mass spectra of these organisms, and it offered a tenfold improvement in signal/noise ratio over conventional electrospray ionisation–mass spectrometry.

2.2.6
Miscellaneous

Ascorbic Acid-2-Sulfate

Wels et al. [102] determined this substance by solid-phase extraction of the cystis of the brine shrimp *Artemia franciscona*.

Siriraks et al. [52] used chelation ion chromatography to analyse miscellaneous organic compounds in crustacea.

Coprostanol

Matusik et al. [98] used capillary column gas chromatography to separate coprostanol, while mass spectrometry in the electron ionisation mode was used to confirm its identity at the 75 ng level in sewage-contaminated crustacea.

Neutral Priority Pollutants

An extraction gas chromatographic procedure [100] for determining neutral priority pollutants has been used to determine these substances in mussel homogenates. The tissue was macerated with distilled water in a blender, anhydrous sodium sulfate was added, and the mixture ground until dry and powdery. The powder was sonified with acetonitrile and the clear phase recovered. Table 2.16 shows some results obtained by this procedure (final column) with those obtained by other workers on reference ERL-N mussel homogenate (US Mussel Watch Program sample). With the exception of benzo(*a*)anthracene plus chrysene, the mean concentrations obtained by Ozretich et al. [100] were within the range of mean values reported by the laboratories involved in the intercomparison study [101].

2.2.7
Mussel Watch

Sericano [105] has reviewed the mussel watch approach and its application to global chemical contamination monitoring programs.

Table 2.16. ERL-N mussel homogenate I: intercomparison results (from [100])

Compound	Mean concentration[a], mg/kg (dry), X (%RSD)				
	Laboratory A	Laboratory B	Laboratory C	Laboratory D	Cretch
PCB as	0.470 (10)	0.90 (31)[c]	0.412 (6)	0.51 (27)[c]	0.559 (7)
Arochlor 1254	(12)[b]	(10)	(4)	(3)	(3)
Naphthalene	0.005 (40)	0.10 (123)[c]	0.003 (239)		0.036 (21)
Phenanthrene	0.013 (38)	0.032 (125)[c]	0.008 (20)		0.0216 (23)
plus anthracene					
Fluoranthene	0.056 (32)	0.042 (88)[c]	0.080 (15)		0.075 (15)
Pyrene	0.046 (28)[c]	0.034 (91)	0.092 (14)[c]		0.0615 (5)
Benzo[a]	0.029 (21)	0.028 (114)[c]	0.047 (13)		0.059 (24)
anthracene	[6][b]	[10]	[4]		[3]
plus chrysene					

[a] From Galloway et al. [101]
[b] Parentheses denote n for PCB; brackets denote n for PAH
[c] % RSD significantly greater than this study ($P < 0.05$)

2.3
Organometallic Compounds

2.3.1
Organoarsenic Compounds

Francesconi et al. [99] give details of the equipment and a procedure used to identify and determine organoarsenic compounds in species of crab—Alaskan King crab (*Paralithodes camtschatica*), Alaskan snowcrab (*Chinocetee bairdii*), and Dungeness crab (*Cancer magister*)—using HPLC and ICP–AES. The only water-soluble arsenic compound in the crabs was the organic compound, arsenobetaine.

2.3.2
Organolead Compounds

Birnie and Hodges [106] have described the combination of solvent extraction and differential pulse anodic scanning voltammetric techniques for the isolation and determination of trialkyl lead (Et_3Pb^+, Me_3Pb^+) and dialkyl lead ($Et_2Pb^{2+}Me_2Pb^{2+}$) species in oyster and *Macoma*.

In this method, the sample is homogenised in the presence of a mixture of salts (lead nitrate, sodium benzoate, potassium iodide, sodium chloride), which effectively releases the di- and trialkyl lead species present and facilitates their transfer to toluene before back-extraction into dilute nitric acid. The differentiation and determination of the alkyl lead species is achieved by differential pulse anodic stripping voltammetry. The efficiency of the extraction procedure was examined at alkyl lead concentrations of up to 2 mg/kg as lead, and a detection limit of 0.01 mg/kg was established. The recoveries of ionic diethyl lead and the two trialkyl lead species from various

Table 2.17. Analysis of *Macoma* from estuarine locations (from [106])

Sample location	R_4Pb	R_3Pb^+	R_2Pb^{2+}	Total Pb
		mg/kg as Pb on wet basis		
Site A	< 0.2	< 0.01	< 0.01	1.3
Site B	< 0.2	0.03	< 0.01	1.1
Site C	< 0.2	0.05	< 0.01	1.8

Table 2.18. Comparison of organolead levels (mg/kg dry weight) in crustacea and fish (from [107])

	In creatures other than fish	Range in fish
R_4Pb	< 0.2	0.92 – 7.93*
R_3Pb^+	< 0.01 – 0.5	0.54 – 6.16*
R_2Pb^{2+}	< 0.01	0.54 – 0.79*
Pb^{+2}	1.1 – 1.8	0.25 – 4.13

* Ethyl plus methyl compounds

marine vertebrates and molluscs were in the range 80 – 90% whilst those of dimethyl alkyllead was appreciably lower (30 – 40%) using this method. Table 2.17 shows results obtained on *Macoma* samples taken from estuarine locations off the North Wales Coast. Investigations revealed that no interference from phenyl lead, alkyltin, alkylthallium or alkylarsenic compounds was obtained in this procedure.

Langlois et al. [107] has compared organolead levels in fish and crustacea (Table 2.18). Organic lead levels found in fish are appreciably higher than those found in crustacea.

2.3.3
Organomercury Compounds

Uthe et al. [108] have described a rapid semi-micro method for determining methylmercury in crustacea. The procedure involves extracting the methylmercury into toluene as methylmercury(II) bromide, partitioning the bromide into aqueous ethanol as the thiosulfate complex, re-extracting methylmercury(II) iodide into benzene, followed by GC on a glass column (4 ft × 0.25 in.) packed with 7% of Carbowax 20M on Chromosorb W and operated at 170 °C with nitrogen as carrier gas (60 ml/min) and electron capture detection. Down to 0.01 ug/kg of methylmercury in a 2 g sample could be detected. A comparison of the results with those obtained by atomic absorption (total Hg content) indicated that all of the fish samples examined contained more than 41% of the mercury as methylmercury.

Beauchemin et al. [109] determined organomercury compounds in lobster hepatopancreas by ICP–MS using flow injection analysis. Mercury was extracted as chloride with toluene and back-extracted with aqueous cysteine acetate in chloride medium. Comparison of the results with GC suggested that the only significant organic compound-s containing mercury was dimethylmercury, which comprised 39% of the total mercury content of lobster hepatopancreas.

2.3.4
Organotin Compounds

This is the only type of organometallic compound occurring in crustacea which has been studied extensively, and these studies are continuing.

Han and Weber [110] studied the speciation of methyl- and butyltin compounds and inorganic tin in oysters by hydride generation atomic absorption spectrometry. Recoveries from spiked samples of oyster tissue were about 100%, and no organotin decomposition products were observed. Detection limits of inorganic tin were, respectively, 0.023, 0.025 and 0.011 µg/kg oyster sample (wet weight). A comparative study of monomethyltin levels in shellfish from the Great Bay Estuary, NH and the Mediterranean Sea (Turkish coast) suggested that monomethyltin in Great Bay oysters resulted from biological methylation of inorganic tin, whereas in the Mediterranean Sea, mono- and dimethyltin compounds resulted from the degradation of anthropogenic trimethyltin. Comparisons were also made between butyltin levels in oysters from the Great Bay Estuary and English shellfish samples.

Jones (private communication) carried out speciation studies of methyl- and butyltin compounds and inorganic tin using hydride generation atomic absorption spectometry. Down to 0.01 to 0.02 mg/kg organotin compounds could be determined.

High-performance liquid chromatography–hydride generation direct current plasma emission spectrometry has been applied to the analysis of clams and tuna fish [111].

Organotin compounds are considerably more toxic than the corresponding free metals. It is well-known that the trisubstituted species, especially tributyltin (TBT) and triphenyltin (TPT), are the most toxic. Tributyltin has been widely used for many years as an antifouling compound added to paints intended for boats, and TPT is still frequently used as a fungicide in agriculture, mainly to protect against potato blight (*Phytopthora infestans*). Although the use of TBT has now been drastically restricted, it is still allowed on larger boats, and the desorption of organotin compounds from contaminated sediment could be an important source of future organotin pollution [112]. Therefore, it is very important to have a method that allows fast detection in the subnanogram per litre range. In earlier studies, organotins were often extracted using tropolone and *n*-hexane and determined with GC–flame photometric detection after Grignard derivatisation [113,124]. In more recent work, the Grignard derivatisation is replaced by in situ ethylation with sodium tetraethylborate [120–122], and flame photometric detection is sometimes replaced by atomic emission detection [121,123]. Quite recently, solid-phase microextraction was introduced as an elegant and practicable extraction technique for volatile organotins, where it was combined with capillary GC.

Vercauteren et al. [124] have investigated the extraction and preconcentration capabilities of a new extraction technique in which stir bar sorptive extraction was combined with the separation power of capillary gas chromatography and the low limits of detection of ICP–MS for the determination of the organotin compounds TPT and TBT in aqueous standard solutions, harbour water and mussels after digestion with tetramethylammonium hydroxide. Throughout, tripropyltin for TBT and tricyclohexyltin for TPT were used as internal standards to correct for variations in the derivatisation and extraction efficiency. Calibration was accomplished by means of a single standard addition. Derivatisation to transform the trisubstituted compounds into sufficiently volatile compounds was carried out with sodium tetraethylborate. The compounds were extracted from their aqueous matrix using a stir bar of 1 cm length, and coated with 55 µL of poly(dimethylsiloxane) after 15 minutes of extraction; the stir bar was then desorbed in a thermal desorption unit at 290 °C for 15 minutes, during which the compounds were cold-trapped on a precolumn at −40 °C. Flash heating was used to rapidly transfer the compounds to the gas chromatograph, where they were separated on a capillary column with a poly(dimethylsiloxane) coating. After separation, the compounds were transported to the ICP by means of a heated (270 °C) transfer line. Monitoring the ^{120}Sn signal by ICP–MS during the run of the gas chromatograph provided extremely low detection limits for TPT in water: 0.1 pg/l (procedure) and 10 fg/l (instrument) and a repeatability of 12% RSD ($n = 10$). Concentrations of 200 pg/l for the TPT were found in harbour water.

A concentration of 7.2 µg/kg TPT was found in fresh mussels.

Gas chromatography coupled with mass spectrometry has been used to determine organotin compounds in mussels [125].

Triphenyltin has been determined in mussels in amounts down to 0.007 µg/kg by extraction with ethanolic potassium hydroxide, derivatisation with sodium tetraethylboron and solid-phase microextraction/GC/ICP–MS [126].

Jiang et al. [127] determined butyltin compounds in mussels by GC with flame photometric detection using quartz surface-induced luminescence. Down to 3 pg tin absolute could be determined as TPT, or 2 – 3 pg tin as trimethylamyltin, dimethyl diamyltin or methyltriamyltin.

Between 40 and 90 mg/kg tin were found in mussels.

Quevauvilla et al. [128] have carried out a collaborative evaluation of methods of determining TBT in mussel tissue.

2.4
Nonmetallic Elements

2.4.1
Iodine

Fassett and Murphy [129] used isotope dilution resonance ionisation mass spectrometry to determine iodine in oysters at the mg/kg level. Rao and Chatt [130] employed neutron activation of microwave acid digests of oyster samples to determine iodine.

2.4.2
Total Nitrogen and Phosphorus

Collins et al. [137] has described a method for determining both of these elements in oysters. The sample is replaced in borosilicate glass with sulfuric acid and microwave heating is applied. Analysis of the extract is carried out by conventional procedures.

2.5
Detection Limits for the Analysis of Invertebrates

2.5.1
Cations

Available information is tabulated in Table 2.19. It is seen that detection limits of between 0.001 and 0.01 mg/kg (1 – 10 µg/kg) have been achieved for arsenic, antimony, chromium, indium, lead, mercury and selenium, whilst limits in the range 0.01 – 0.1 mg/kg (10 – 100 µg/kg) have been achieved for cadmium, copper, nickel, plutonium, rubidium, uranium, vanadium and zinc. Hydride generation AAS, ICP–MS and neutron activation analysis are particularly sensitive techniques.

On the whole, the detection limits achieved are sufficient to meet most of the needs of environmental analysis for many, but not all, elements.

A comparison of the cation detection limits obtained for fish (Table 1.30) and invertebrates (Table 2.22) shows that, with a few exceptions, greater sensitivity has been achieved for invertebrate samples than for fish samples.

2.5.2
Organic Compounds

Very sensitive methods are available for polyaromatic hydrocarbons, hexachlorobenzene, pentachlorophenol, pentachloroterphenyl, phthalates, chlorinated insecticides and Mirex in invertebrata, and slightly less sensitive methods (10 – 100 µg/kg) are available for polychlorobiphenyls and Dursban. Aliphatic hydrocarbons can be determined in the 100 – 500 µg/kg range (Table 2.20).

Table 2.19. Detection limits for metals in vertebrates (from author's own files)

Element	Type of sample	Method	LD	Reference
Arsenic	Crustacea	Hydride generation atomic absorption spectrometry	11 mg/kg	[2]
Arsenic	Oyster	Hydride generation atomic absorption spectrometry	11 ng absolute	[3]
Arsenic	Oyster	Graphite furnace atomic absorption spectrometry	0.5 mg/kg	[5]
Arsenic	Oyster	Inductively coupled plasma atomic emission spectrometry	0.5 mg/kg	[5]
Arsenic	Oyster	Flow injection/hydride generation inductively coupled plasma atomic emission spectrometry	0.5 mg/kg	[5]
Arsenic	Crustacea	Spectrophotometry	0.05 mg/kg	[1]
Arsenic	Lobster hepato-pancreas	Derivitisation with 2,3-dimercapto propanol–gas chromatography	10 pg absolute	[4]
Arsenic	Oyster	Neutron activation analysis	0.5 mg/kg	[5]
Indium	Biological materials	Ion-pair extraction atomic absorption spectrometry	0.001 mg/kg	[132]
Iodine	Oyster	Neutron activation analysis	5 ng absolute	[130]
Iodine	Oyster	Isotope dilution–laser resonance ionisation mass spectrometry	mg/kg level	[129]
Mercury	Mussel, oyster	Cold vapour atomic absorption spectrometry	0.01 mg/kg	[16–18]
Mercury	Oyster	Neutron activation analysis	0.00015 mg/kg	[12]
Selenium	Biological materials	ICP mass spectrometry	0.001 mg/kg 6.4 ng absolute	[20, 133]
Tin	Oyster	Hydride generation atomic absorption spectrometry	0.011 - 0.025 mg/kg	[25]
Vanadium	Shrimp, crab, oyster	Neutron activation analysis	0.03 mg/kg	[26]
Arsenic, selenium, mercury	Lobster, scallop	Hydride generation atomic absorption spectrometry	Arsenic: 0.3 mg/kg Selenium: 0.2 mg/kg Mercury: 0.05 mg/kg	[30]

Table 2.19. Continued

Element	Type of sample	Method	LD	Reference
Lead, cadmium, mercury, zinc, copper, nickel, plutonium, uranium	Mussel	Atomic absorption spectrometry	All elements 0.05 mg/kg	[47–49]
Cadmium, copper, iron, lead, zinc	Mussel	Cold vapour atomic absorption spectrometry	Cadmium: 0.03 mg/kg Copper: 0.7 mg/kg Lead: 0.7 mg/kg Zinc: 6.0 mg/kg	[31]
Arsenic, antimony, selenium	Oyster	Hydride generation atomic absorption spectrometry	Arsenic: 0.001 mg/kg Antimony: 0.001 mg/kg Selenium: 0.001 mg/kg	[131]
Silver, chromium, cadmium, copper, manganese, nickel, lead, selenium	Oyster	Zeeman atomic absorption spectrometry	Silver, cadmium < 0.001 mg/kg Manganese, chromium, lead: 0.001 mg/kg Copper, nickel: 0.005 mg/kg Selenium: 0.015 mg/kg	[29,32]
Arsenic, bromine, iron, manganese, molybdenum, nickel, rubidium, selenium, strontium, zinc	Lobster	Neutron activation analysis	Arsenic: 0.16 mg/kg Bromine: 0.26 mg/kg Iron: 2.8 mg/kg Manganese: 0.16 mg/kg Molybdenum: 0.16 mg/kg Nickel: 0.1 mg/kg Rubidium: 0.07 mg/kg Selenium: 0.02 mg/kg Strontium: 1.5 mg/kg Zinc: 0.18 mg/kg	[44]
Manganese, nickel, copper, zinc, arsenic, cadmium, lead, sodium, magnesium, chlorine, calcium, strontium	Oyster hepato-pancreas	Photoactivation analysis	Manganese: 4 mg/kg Nickel: 0.6 mg/kg Copper: 3.4 mg/kg Zinc: 2.0 mg/kg Arsenic: 0.3 mg/kg Cadmium: 2 mg/kg Lead: 3 mg/kg Sodium: 200 mg/kg Magnesium: 0.4 mg/kg Chlorine: 100 mg/kg Calcium: 3 mg/kg	

Table 2.20. Detection limit for organic compounds in invertebrates (from author's own files)

Compound	Type of sample	Method	Limit of detection	Reference
Hydrocarbon oils	Marine organisms	Gas chromatography	500 µg/kg	[53]
Benzopyrene, hexachlorobenzene, pentachlorophenol	Marine organisms	Gas chromatography	0.2 µg/kg	[64]
Benzopyrene	Mussel	Spectrofluorimetry	1×10^{-2} µg/kg	[62]
Pentachlorophenol	Oyster	Gas chromatography–mass spectrometry	< 0.1 µg/kg	[87]
Phthalates	Shrimp, crab	Gas chromatography	< 1 µg/kg	[65]
Polychlorobiphenyls	Shrimp, crab	Gas chromatography	10 µg/kg	[74–77]
Polychlorobiphenyls	Oyster, clam	Electron capture ionisation– mass spectrometry	6.5 µg/kg	[73, 82]
Chlorinated insecticides	Oyster	Gas chromatography	10 µg/kg	[68]
Chlorinated insecticides	Oyster	Gas chromatography–mass spectrometry	0.04 µg/kg (lindane)	[73]
DDT	Shrimp, crab	Gas chromatography	1 µg/kg	[74–77]
Mirex	Shrimp, crab	Gas chromatography	5 µg/kg	[7]
Dursban	Crustacea	Gas chromatography	10 µg/kg	[88]
Dibenzothiophen	Mussels	Gas chromatography (flame photometric detection)	< 0.1 µg/kg	[49, 97]
Microcystins, LR, RR and YR	Biological tissue	Mass spectrometry	1 – 4 nM absolute	[104]
Coprostanol	Crustacea	Electron capture ionisation–mass spectrometry	0.75 ng absolute	[98]

2.5.3
Organometallic Compounds

Again, adequate sensitivity can be achieved: 0.01 and 0.02 µg/kg for organomercury and tin compounds, respectively. Methods used for organolead compounds are somewhat less sensitive (10 µg/kg) (Table 2.21).

In general, the comments regarding the greater sensitivities that have been achieved in methods of analysis for organic and organometallic compounds in invertebrates compared to those achieved for fish also apply (Table 2.23).

Table 2.21. Detection limits for organometallic compounds in invertebrates (from author's own files)

Compound	Type of sample	Method	Limit of detection	Reference
Organomercury, methyl mercury	Crustacea	Extraction– gas chromatography	0.01 µg/kg	[108]
Organolead	Oyster	Anodic stripping voltammetry	10 µg/kg	[106]
Organotin, butyltin	Oyster	Hydride generation atomic absorption spectrometry	1 – 2 µg/kg	***
Butyltin, $Me_3SnC_5H_{11}$, $Me_2Sn(C_5H_{11})_2$, $MeSn(C_5H_{11})_3$, PR_4Sn	Mussel	Gas chromatography with flame photometric detection	2 – 3 pg absolute (as Sn) 0.3 pg absolute (as Sn)	[127]
Methyl and butyltin	Oyster	Hydride generation atomic absorption spectrometry	Me Sn: 0.023 µg/kg BuSn: 0.025 µg/kg	[110]
Organotin	Mussel	Inductively coupled plasma mass spectrometry	7 µg/kg	[124]
Butyltin	Mussel	Gas chromatography, flame photometric detection	0.3 pg absolute	[127]

*** Jones (private communication)

Table 2.22. Comparison of best detection limits reported for metals in fish and invertebrates

Element	Fish, g/kg	Invertebrates
Arsenic	0.02 H_2AAS	0.001 AAS
Cadmium	0.02 AAS	0.001 Zeeman AAS
Copper	O_2 AAS	0.05 AAS, 0.005 Zeeman AAS
Lead	0.01 AAS	0.001 Zeeman AAS
Mercury	0.005 H_2AAS	0.0015 NAA
Nickel	0.05 AAS	0.05 AAS, 0.005 Zeeman AAS
Selenium	0.2 H_2AAS	0.01 H_2AAS
Strontium	1 GFAAS	1.5 NAA
Vanadium	0.03 NAA	0.03 NAA
Zinc	0.2 AAS	0.05 NAA

Table 2.23. Comparison of the best detection limits reported for organic and organometallic compounds in fish and crustacea

Compound	Fish, μg/kg	Invertebrates, μg/kg
Polyaromatic hydrocarbons	0.2 – 0.5 capillary GLC	0.2 GLC
		0.01 spectrofluorimetry
Phthalates	5 capillary GLC	1 GLC
Chlorophenols	100 – 1000 methylation GLC	0.2 GLC
Polychlorotriphenols	10 GLC	10 GLC
Chloroinsecticides	1 GLC with electron capture	1 GLC
	detection	0.04 GLC - MS
Dursban	100 thermionic GLC	10 GLC
Mirex	5 GLC	5 GLC
Organolead	25 GLC	10 ASV
	10 ASV	
Organomercury	4 GLC	0.01 GLC
Organotin	0.2 GLC with ethylation	0.02 H_2AAS

GLC: Gas–liquid chromatography
ASV: Anodic stripping voltammetry
MS: Mass spectrometry
H_2AAS: Hydride generation gas chromatography

References

1. Maher WA (1983) *Analyst* **108**:939.
2. Brooke PJ, Evans WH (1981) *Analyst* **106**:514.
3. Uthus, EO, Collings ME, Cornatzer WE, Nielsen FH (1981) *Anal Chem* **53**:2221.
4. Siu KWM, Roberts SY, Berman SS (1984) *Chromatographia* **19**:398.
5. Brzezinska-Paudyn A, Van Loon JC, Hancock RGV (1986) *Atom Spectrosc* **7**:72.
6. Dagnac T, Padro, Rubio R, Rauret G (1999) *Talanta* **48**:763.
7. Gabrielli LF, Marietta GP, Favretto L (1980) *Atom Spectrosc* **1**:35.
8. Ashworth M, Farthing RH (1981) *Int J Anal Chem* **10**:35.
9. Poldoski JE (1980) *Anal Chem* **52**:1147.
10. McLaren JW, Berman SS (1981) *Appl Spectrosc* **35**:403.
11. Mazzucotelli A, Viarengo A, Martino G, Frache R (1988) *Mar Environ Res* **24**:129.
12. Greenberg RR (1980) *Anal Chem* **52**:676.
13. van Raaphorst JG, Van Weers AW, Haremaker HM (1974) *Analyst* **99**:523.
14. Mikac N, Wang Y, Harrison RM (1996) *Anal Chim Acta* **326**:57.
15. Hutton RC, Preston B (1980) *Analyst* **105**:981.
16. Zhe-Ming N, Xioa-Chun L, Heng-Bin H (1986) *Anal Chim Acta* **186**:147.
17. Holak W, Frinitz B, Williams JC (1972) *J AOAC* **55**:741.
18. Kunkel E (1972) *Fresen Z Anal Chem* **258**:337.
19. Louie HW (1983) *Analyst* **108**:1313.
20. Lo JM, Wei JC, Yang MH, Yeh SJ (1982) *J Radioanal Chem* **72**:571.
21. Ahmed RB, Hill JO, Magee RJ (1983) *Analyst* **108**:835.
22. Maher WA (1985) *Mar Poll Bull* **16**:33.
23. Arruda MAZ, Gallego M, Valcarcel M (1996) *J Anal Atom Spectrosc* **11**:169.
24. Florence TM, Farrar YI (1974) *J Electrochem Soc* **51**:191.
25. Brown T (1988) *Anal Chem* **60**:316.

26. Blotcky AJ, Falcone C, Medina VA, Rack EP, Hobson DW (1979) *Anal Chem* **51**:178.
27. Ramelow G, Ozkan MA, Tuncel G, Saydam C, Balkas TI (1978) *Int J Environ Anal Chem* **5**:125.
28. Topping G (1982) *Report of the 6th ICES Trace Metals Intercomparison Exercise for Cadmium and Lead in Biological Tissue*, ICES Coop Research Report No. 111, International Council for the Exploration of the Sea, Copenhagen, Denmark.
29. Schlemmer G, Welz B (1985) *Fresen Z Anal Chem* **320**:648.
30. Welz B, Melcher M (1985) *Anal Chem* **57**:427.
31. Solchaga M, De La Guardia M (1986) *J AOAC* **69**:874.
32. Amiard JC, Pineau A, Boiteau HL, Metayer C, Amiard-Triquet C (1987) *Water Res* **21**:693.
33. Ridout PS, Jones R,Williams JG (1988) *Analyst* **113**:1383.
34. De Oliveira E, McLaren JW, Berman SS (1983) *Anal Chem* **55**:2047.
35. Goulden PD, Anthony DHJ, Austen KD (1981) *Anal Chem* **53**:2027.
36. McLaren JW, Berman SS, Boyko VJ, Russell DS (1986) *Anal Chem* **53**:1802.
37. Richardson CA, Chenery SRN, Cook JM (2001) *Mar Ecol Prog Ser* **211**:157.
38. Raith A. Perkins WT, Pearce NJG, Jefferies TE (1996) *Fresen J Anal Chem* **355**:789
39. Labonne M, Ben Othman D, Luck J-M (2002) *Appl Geochem* **17**:1351.
40. Brewer SW, Sacks RD (1988) *Anal Chem* **60**:1769.
41. Chisela F, Gawlik D, Bratter P (1986) *Analyst* **111**:405.
42. Florence TM (1972) *J Electrochem Soc* **35**:237.
43. Zeisler R, Stone SF, Sanders DW (1988) *Anal Chem* **60**:2760.
44. Landsberger S, Davidson WF (1985) *Anal Chem* **57**:196.
45. Dutton JWR (1969) *Technical Report FRL 4*, Fisheries Radiological Laboratory, Hamilton, Lowestoft, Suffolk, UK.
46. Fourie HO, Peisach M (1977) *Analyst* **102**:193.
47. NAS (1980) *The International Mussel Watch*, US National Academy of Sciences, Washington, DC.
48. Goldberg ED, Bowen VT, Farrington JW, Harvey G, Martin JH, Parker DL, Risebrough RW, Robertson W, Schneider E, Gamble E (1978) *The Mussel Watch Environment Conservation* **5**:101.
49. Patterson D, Settle D, Schaule B, Burnett M (1976) *Transport of Pollutant Lead to the Oceans and Within Ocean Ecosystems*. In: Windem HL, Duce RH (eds) *Marine Pollutant Transfer*, Lexington Books, Lanham, MD, p. 23.
50. Elias R, Hirao Y, Patterson C (1975) *Impact of the Present Levels of Aerosol Lead Concentrations on both Natural Ecosystems and Humans*, International Conference of Heavy Metals in the Environment, 27–31 October 1975, Toronto, Canada, p. 257.
51. Slinkman D, Sacks R (1991) *Anal Chem* **63**:343.
52. Siriraks A, Kingston MH, Riviello JM (1990) *Anal Chem* **62**:1185.
53. Morgan NL (1975) *Bull Environ Contam Toxicol* **14**:309.
54. Meyers PA (1978) *Chemosphere* **7**:385.
55. Chesler SN, Gump BH, Hertz HS, May WE, Wise SA (1978) *Anal Chem* **50**:805.
56. May WE, Chesler SN, Cram SP, Gump BH, Hertz DS, Enagonio DP, Dyszel SM (1975) *J Chromatogr Sci* **13**:535.
57. Chesler SN, Gump BH, Hertz HS, May WE, Dyszel SM, Enagonio DP (1976) *US Technical Note No. 889*, National Bureau of Standards, Washington, DC.
58. Berthou F, Gourmelun Y, Dreano Y, Friocourt MP (1981) *J Chromatogr A* **203**:279.
59. Mason PR (1987) *Mar Pollut Bull* 18:528.
60. Bjorseth A, Knutsen J, Skei J (1979) *Sci Total Environ* **13**:71.
61. Dunn BP, Stich HFJ (1976) *J Fish Res Board Canada* **33**:2040.

62. Kunte H (1967) *Arch Hyg Bakt* **151**:193.
63. Uthe JF, Musal C (1988) *J AOAC* **71**:363.
64. Giam CS, Trujillo DA, Kira S, Hrung Y (1980) *Bull Environ Contam Toxicol* **25**:824.
65. Giam CS, Chan HS, Neff GS (1975) *Anal Chem* **47**:2225.
66. Murray AJ, Riley JP (1973) *Anal Chim Acta* **65**:261.
67. Murray AJ, Riley JP (1973) *Nature* **242**:37.
68. Mills PA, Caley JF, Grithen RA (1963) *J AOAC* **46**:106.
69. Arias C, Vidal A, Vidal C, Maria J (1970) *An Bromat (Spain)* **22**:273.
70. Ernst W, Goerke H, Eder G, Schaefer RC (1976) *Bull Environ Contam Toxicol* **15**:55.
71. Ernst W, Schaefer RC, Goerke H, Eder G, (1974) *Fresen Z Anal Chem* **227**:358.
72. Neudorf S, Khan MAQ (1975) *Bull Environ Contam Toxicol* **13**:443.
73. Teichman J, Bevenue A, Hylin JW (1978) *J Chromatogr* **151**:155.
74. Markin GP, Hawthorne JC, Collins HL, Ford JH (1974) *Pest Monit J* **7**:139.
75. Armour JA, Burke JA (1970) *J AOAC* **53**:761.
76. Gaul J, Cruz-LaGrange P (1971) *Separation of Mirex and PCB in Fish*, Laboratory Information Bulletin, Food and Drug Administration, New Orleans District, LA.
77. Markin GP, Ford JH, Hawthorne JC, Spence JA, Davies J, Loftis CD (1972) *Environmental Monitoring for the Insecticide Mirex*, APHIS (USDA), Riverdale, MD.
78. Butler PA (1969) *Biol Sci* **19**:889.
79. McKenzie MD (1970) *Fluctuations in Abundance of the Blue Crab and Factors Affecting Mortalities*, Technical Report No. 1, South Carolina Wildlife Resources Division, Charleston, SC, USA.
80. Mahood RK, McKenzie MD, Middough DP, Bellar SJ, Davies JR, Spitzbergen D (1970) *A Report on the Co-operative Blue Crab Study in South Atlantic States*, Project Nos. 2-79-R1, 2-81-R-1, 2-82-R-1, Bureau of Commercial Fisheries, US Department of the Interior (now National Marine Fisheries Service, Silver Spring, MD, USA).
81. Tanabe S, Tatsukawa R, Phillips DJH (1987) *Environ Pollut* **47**:41.
82. Ma CY, Bayne CK (1993) *Anal Chem* **65**:772.
83. Alvarez Pineiro E, Simal-Lorenzo J, Lage-Yusty MA (1994) *J AOAC* **77**:985.
84. Pastor MD, Sanchez J, Barcelo D, Albaiges J (1993) *J Chromaogr* **629**:329.
85. Freudenthael J, Grove PA (1973) *Bull Environ Contam Toxicol* **10**:108.
86. Taguchi VY, Reiner EJ, Wang DT, Meresz O, Hallas B (1988) *Anal Chem* **60**:1429.
87. Buser HR, Rappe C (1991) *Anal Chem* **63**:1210.
88. Deusch ME, Westlake WE, Gunther EA (1970) *J Agric Food Chem* **18**:178.
89. Nakamura A, Kashimoto T (1977) *J Food Hyg Soc* **18**:253.
90. Dillon TM, Neff JM, Warner JS (1978) *Bull Environ Contam Toxicol* **20**:320.
91. Grahl-Nielsen O, Steveland JT, Wilhelmsen S (1978) *J Fish Res Board Canada* **35**:615.
92. Ogata M, Miyake Y, Fujisawa K, Kira S, Yoshida Y (1980) *Bull Environ Contam Toxicol* **25**:130.
93. Ogata M, Miyake Y (1980) *J Chromatogr Sci* **18**:594.
94. Goldberg ED (1975) *Mar Pollut Bull* **6**:111.
95. Lee RF, Sauerheber R, Benson AA (1972) *Science* **177**:344.
96. Dunn BP, Young DR (1976) *Mar Pollut Bull* **7**:231.
97. Kira S, Izumi T, Ogata M (1983) *Bull Environ Contam Toxicol* **31**:518.
98. Matusik JE, Hoskin GP, Sphon JA (1988) *J AOAC* **71**:994.
99. Francesconi KA, Hicks P, Stockton RA, Irgolic KJ (1985) *Chemosphere* **14**:1443.
100. Ozretich RJ, Schroeder WP (1986) *Anal Chem* **58**:2041.
101. Galloway WB, Lake JL, Phelps DK, Rogerson PF, Bowen VT, Farrington JW, Goldberg ED, Laseter JL, Lawler GC, Martin JH, Riseborough RW (1983) *Environ Toxicol Chem* **2**:395.

102. Nelis HJ, Merchie G, Lavans P, Sorgeloos P, DE Leenheer AP (1994) *Anal Chem* **66**:1330.
103. Draisci R, Palleschi L, Gianneltti L, Lucentini L, James KJ, Bishop AG, Satake M, Yasumoto T (1999) *J Chromatogr A* **847**:213.
104. Ells B, Froese K, Hrudey SE, Purves RW, Guevremont R, Barnett DA (2000) *Rapid Commun Mass Spec* **14**:1538.
105. Sericano JL (2000) *Int J Environ Pollut* **13**:340.
106. Birnie SE, Hodges DE, (1981) *Environ Tech Lett* **2**:433.
107. Langlois RE, Stemp AR, Liska BJ (1954) *J Milk Food Technol* **27**:202.
108. Uthe JF, Solomon J, Grift B (1972) *J AOAC* **55**:583.
109. Beauchemin D, Siu KWM, Berman SS (1988) *Anal Chem* **60**:2587.
110. Han JS, Weber JH (1988) *Anal Chem* **60**:316.
111. Krull IS, Panaro KW (1985) *Appl Spectrosc* **39**:960.
112. Weidenhaupt A, Arnold C, Muller SR, Haderlein SB, Schwarzenbach RP (1997) *Environ Sci Tech* **31**:2603.
113. Caricchia AM, Chiavarini S, Cremisini C, Morabito R, Scerbo R (1994) *Anal Chim Acta* **286**:329.
114. Fent K, Hunn J (1991) *Environ Sci Tech* **25**:956.
115. Harino H, Fukushima M, Tanaka M (1992) *Anal Chim Acta* **264**:91.
116. Gomez-Ariza JL, Morales E, Ruiz-Benitez M (1992) *Analyst* **117**:641.
117. Nagase M, Hasebe K (1993) *Anal Sci* **9**:517.
118. Muller MD (1987) *Anal Chem* **59**:617.
119. Van den Broek HH, Hermes GBM, Goewie CE (1988) *Analyst* **113**:1237.
120. Folsvik N, Brevik EM (1999) *J High Res Chromatogr* **22**:177.
121. Ceulemans M, Witte C, Lobinski R, Adams FC (1994) *Appl Organomet Chem* **8**:451.
122. Morcillo Y, Porte C (1998) *Trends Anal Chem* **17**:109.
123. Lobinska JS, Ceulemans M, Lobinski R, Adams FC (1993) *Anal Chim Acta* **278**:99.
124. Vercauteren J, Peres C, Devos C, Sandra P, Vanhaecke F, Moens J (2001) *Anal Chem* **73**:1509.
125. Gallina A, Magno T, Tallaudini L, Passaler T, Caravello GH, Pastore P (2000) *Rapid Commun Mass Spec* **14**:373.
126. Vercauteren J, De Meester A, De Smaele T, Vanhaecke F, Moens L, Dams R, Sandra P (2000) *J Anal Atom Spectrom* **15**:651.
127. Jiang G, Maxwell PS, Siu KWM, Luong VT Berman SS (1991) *Anal Chem* **63**:1506.
128. Quevauvilla P, Astruc M, Morabito R, Arieses F, Ebdon L (2000) *Trends Anal Chem* **19**:180.
129. Fassett JD, Murphy TJ (1990) *Anal Chem* **62**:386.
130. Rao RR, Chatt A (1991) *Anal Chem* **63**:1298.
131. De Oliveira E, McLaren JW, Berman SS (1983) *Anal Chem* **55**:2047.
132. Zheng W, Sipes IG, Carter DE (1993) *Anal Chem* **65**:2174.
133. Buckley WT, Budac JJ, Godfrey DV, KoenigKM (1992) *Anal Chem* **64**:724.
134. Gjoes N, Gustavsen KO (1982) *Anal Chem* **54**:1316.
135. Liang L, D'Haese PC, Lamberts LV, Van de Vyver FL, De Broe ME (1991) *Anal Chem* **63**:423.
136. Iosifidou HG, Kilikidis SD, Kamarianos AP (1982) *Bull Environ Contam Toxicol* **28**:535.
137. Collins LW, Chalk SJ, Kingston HMS (1996) *Anal Chem* **68**:2610.

3 Analysis of Water Plant Life

3.1
Cations

3.1.1
Sampling Procedures

Collier and Edmonds (private communication) warn that, when samples of algae are taken from seawater and the samples are shipped to the laboratory in containers in contact with seawater, appreciable elemental transfer occurs from the solid to the liquid phase. Subsequent analysis of the solid could, in these circumstances, lead to low determinations for the solid phase.

One of the first results observed in leaching experiments was the high concentrations of many of the elements in the seawater. Elements initially associated with the algae were released into the seawater in the two hours it took to bring the samples from the tow site into the ship's laboratory for processing. In the case of cadmium, manganese and nickel, the data can be expressed as the percentage of the estimated total particulate element which has been released to the seawater. There was always an initial pulse of copper released, but its concentration then decreased due to some secondary process. The amounts of zinc, barium, iron and aluminium released were always low with respect to their total concentrations.

Suspension of the samples in distilled water accelerated the release of nickel, cadmium, manganese and phosphorus into seawater, reaching a maximum after 72 hours. The implication of this rapid remineralisation process must be considered during the sampling and handling of algal matter by towing, filtration and trapping. The rapid release of elements requires careful containment of the sample, and complete mass balancing from the time of collection, if any systematics of their chemistry are to be understood. Variations in sampling techniques will result in large variations in the collected concentrations of nickel, cadmium and manganese, and perhaps other trace components. Washing the samples with distilled water causes an even more extensive release. This rapid solubilisation, coupled with sample contamination, may account for much of the variability in concentrations seen over the long history of reported plankton elemental analyses.

3.1.2
Americium

Jia et al. [1] has described a simple method for the simultaneous determination of plutonium and americium in lichen and moss samples in which, after leaching with hydrochloric acid, plutonium was isolated using a microwave-TNOA column and americium was isolated using a KL-HDE-HP column followed by purification by kerosene-trioctylphenyl oxide extraction.

3.1.3
Antimony

Abu-Hilal and Riley [2] have described a spectrophotometric procedure for the determination of antimony in algae. After a preliminary oxidative digestion with concentrated nitric acid, then the application of concentrated sulfuric acid, the element is quantitatively coprecipitated at pH 5.0 with hydrous zirconium oxide. The precipitate is dissolved in acid, and, after reduction with titanium(III) chloride, antimony is oxidised to antimony(V) with sodium nitrite. The ion-pair of the $SbCl_6$ ion with crystal violet is extracted with benzene, and its absorbance is measured at 610 nm. The detection limit is 0.005 µg/l. A wide range of anions and cations causes no interference at levels many times those in algae. The relative standard deviation is 1.8% for samples of *Pelvetia canaliculata* (0.19 µg/g Sb). A 98 – 101% recovery of antimony was obtained in spiking experiments.

Kantin [3] gives details of a procedure developed for the determination of antimony in marine algae by atomic absorption spectroscopy with hydride generation, and presents results from the analysis of three species of marine algae from coastal waters of California. Pentavalent antimony was the dominant form found, but *Sargassum sp.* contained up to 30% trivalent antimony.

Dodd et al. [4] has reported a method for determining antimony in freshwater plant extracts using a semi-continuous hydride generator coupled to a gas chromatography–mass spectrometer (GC–MS). These workers reported the detection of organoantimony compounds in freshwater plant extracts for the first time.

3.1.4
Arsenic

Whyte and Englar [5] have described methods for the analysis of inorganic and total arsenic in several species of marine algae. Arsenic trichloride formation and distillation was used to determine inorganic arsenic, and acid-oxidative digestion for total arsenic.

Biodimensional size exclusion anion exchange HPLC with dual ICP–MS and electrospray mass spectrometry/mass spectrometry (ES MS/MS) detection has been used to speciate arsenic in edible algae [6].

Maher [7] has described a procedure for the determination of total arsenic in algae. The sample is first digested with a mixture of nitric, sulfuric and perchloric acids. Then arsenic is converted into arsine using a zinc reductor column, the evolved arsine is trapped in a potassium iodide–iodine solution, and the arsenic determined spectrophotometrically at 866 nm as the arseno-molybdenum blue complex. The detection limit is 0.05 mg/kg dry sediment and the coefficient of variation 5.1% at this level. The method is free from interferences by other elements at levels normally found in algae. Values of 9.7 ± 0.3 and 13.2 ± 0.4 mg/kg obtained for NBS reference waters SRM 1S71 and SRM 1566 were in good agreement, respectively, with the nominal values of 10.2 ± and 13.4 ± mg/kg.

A recovery of 96 – 99% of added arsenic was obtained in spiking recovery experiments carried out on the microalgae *Echlonia radiata*.

3.1.5
Bismuth

Lee [8] has used flameless AAS with hydride generation to determine down to 3 pg/kg of bismuth in marine algae. Precision is 6.7% at the 14 μg/kg bismuth level. The sediment (0.5 g) is completely digested with nitric acid, perchloric acid, and hydrochloric acid–hydrofluoric acid on a hot plate. The residue is dissolved in 50 ml 1 N hydrochloric acid. The bismuth is reduced in solution by sodium borohydride to bismuthine, stripped with helium, and collected in situ in a modified carbon rod atomiser. The bismuth collected is subsequently atomised by increasing the atomiser temperature, and detected by an atomic absorption spectrometer. High concentrations of cobalt, copper, gold, molybdenum, nickel, palladium, platinum, selenium, silver and tellurium interfere in this procedure.

Using this method, 0.005 mg/kg bismuth was found in kelp and 0.009 mg/kg bismuth in *Macrocystis* taken in San Onofre, California.

3.1.6
Cobalt

It has been reported [9] that no loss of cobalt occurs during the dry ashing of seaweed in porcelain crucibles at 450 – 550 °C. The cobalt was removed from the crucible with hydrochloric acid.

3.1.7
Copper

Shengjun and Holcombe [10] extracted copper from algae and, after preconcentration, determined it by slurry graphite furnace AA spectroscopy. One hundred-fold concentration enabled 300 μg/l of copper to be determined.

3.1.8
Mercury

Svasankara-Pillay et al. [11] has applied neutron activation analysis to the determination of mercury in algae.

Kuldvere and Andreassen [12] determined mercury in seawater weed by cold vapour AAS.

Some 50 – 60% of mercury was lost from algae as volatile organic mercury compounds upon attempting to dry algae samples by freeze-drying or upon drying at 60 °C prior to neutron activation analysis. To avoid these errors, wet algae samples were centrifuged to remove excess moisture, and portions of this material were put in polyethylene bags for neutron activation analysis and subjected to weight loss determination at 60 °C to ascertain moisture content, so that neutron activation results could be calculated on a dry weight basis.

Mercury determinations on plankton algae from samples collected in Lake Erie, carried out at various dates between 1970 and 1971, were in the range 31–81 mg/kg.

Mitchell et al. [13] evaluated the use of electrothermal vaporisation–direct-current argon plasma emission spectrometry for the direct determination of mercury compounds in *Chlorella vulgaris*. Here, 5-ml volumes of 20 µg/ml mercury solution were equilibrated with 0.6 to 9-mg algal masses. The amount of mercury taken up (60 – 900 µg) was linearly related to algal mass. Acceptable calibration curves were obtained for up to 20 mg of mercury absorbed into 5-mg algal masses. The effects of mercury(II) chloride, mercury(I) chloride and mercury(II) acetate on the mercury emission signal were examined. The addition of sulfur-containing algae or cysteine made the mercury signal the same regardless of the mercury compound originally present.

3.1.9
Molybdenum

Colborn [14] determined low levels of down to 0.05 mg/kg molybdenum in insects by AAS of 10 N hydrochloric acid digests of the samples.

3.1.10
Plutonium

The method described in Sect. 3.1.2 has been applied to the determination of plutonium in lichens and mosses.

3.1.11
Tin

Hodge et al. [15] determined nanogram quantities of tin(IV) as well as the halides of methyltin, dimethyltin, trimethyltin, *n*-butyltin, di-*n*-butyltin, tri-n-butyltin and phenyltin in microalgae samples using a procedure involving reaction with sodium borohydride to convert to the tin hydrides, which are then detected by AAS. The compounds are separated based on of their differing boiling points, which range from $1.4\,°C$ (CH_3SnH_3) to $280\,°C$ (n-$C_4H_4)_3SnH$). Detection limits range from $0.4\,\mu g/kg$ (CH_3SnH) to $2\,mg/kg$ (tri-n-butyl tin). To digest the sample, $1\,g$ of oven-dried material was digested with nitric acid, perchloric acid and hydrofluoric–hydrochloric acid and the digest made up to $50\,ml$. This solution was injected directly into the hydride generator. Stannane and the organotin hydrides evolve from the hydride trap in such a way that they can be identified by a 'retention time'.

Tin concentrations found in algae samples collected from Narragansett Bay, California were in the range $0.3\,mg/kg$ (inner tissue of algae) to $0.83\,mg/kg$ (algae blade). In these cases the samples were destroyed by acids, and it was assumed that all tin forms end up as Sn(IV) and that no losses occurred during the wet ashing procedure.

Dogan and Haerdi [16] also applied graphite furnace AAS to the determination of tin in algae in amounts down to $0.5\,\mu g/kg$. They digested the sample with Lumatom at $50\,°C$, injected it into a graphite furnace, and ashed at $800\,°C$ for $40\,s$. The material was then atomised at $2860\,°C$ for $5\,s$ before the analysis for tin.

3.1.12
Zinc

Van Raaphorst [9] reported that no loss of zinc occurs during dry ashing of seaweed in porcelain crucibles at $450 – 550\,°C$.

3.1.13
Multi-cation Analysis

Atomic Absorption Spectrometry (AAS)

Bando et al. [17] studied analytical errors associated with trace element determination in algae by AAS. In this method, algae were filtered from the water sample on $47\,mm$ Nucleopane $0.4\,\mu m$ polycarbonate filters, which were then weighed and dried overnight at $65\,°C$. After drying, the filters were reweighed and transferred into the PTFE vessel of a Perkin-Elmer Autoclave-3 with $35\,ml$ of acetone. The tip of the sonifier disruptor was then immersed in the solution for $3\,min$ of pulsed sonification (a cycle of $0.8\,s$ ultrasonic exposure and $0.2\,s$ rest to avoid excessive heating), and the filter was removed whilst rinsing with acetone. The solution was evaporated at $55 \pm 1\,°C$, and the vessel placed in the Autoclave-3, and $5\,ml$ of a mixture of 70% nitric acid and 30% hydrofluoric acid was added. The particulate matter was rendered completely soluble by heating the bomb at $160 \pm 5\,°C$ for 20 minutes.

Table 3.1. Instrumental errors (from [17])

| Element | Sample concentration, ppb | | Relative standard deviation of absorbance | | | |
| | Minimum | Maximum | A_1 | | A_2 | |
			Minimum	Maximum	Minimum	Maximum
Fe	1200	2100	0.00	0.08	4.8	5.3
Mn	40	110	0.65	0.73	0.74	0.78
Cu	10	60	3.0	3.2	3.3	3.4
Cr	10	80	1.6	1.7	1.0	1.1
Zn	500	1200	0.00	0.07	0.58	0.62

Results are the reproducibilities of the measurements ($n = 10$) during an analytical run (A_1) for samples with different concentrations, and variabilities of these mean values on five consecutive days after separate calibration of the atomic absorption instrument (A_2).

Another series of filters was destroyed by ashing overnight at 500 ± 30 °C (the polycarbonate filter was not soluble in the acid mixture), and then the ash was introduced into the PTFE vessel of the bomb for the same treatment in order to render it soluble.

For both of the methods (the sonifier method and the ashing method), the final solutions for the atomic absorption spectroscopic determinations were obtained by carefully washing out the PTFE vessel with a volume of deionised water of up to 15 ml.

Each atomic absorption measurement is biased by two instrumental errors: the first (A_1) refers to the variability of consecutive measurements of the same sample, whilst the second (A_2) is an estimate of the calibration imprecision introduced by the operators setting-up the instrument for the analysis on different days.

Table 3.1 shows the ranges of the A_1 and A_2 errors thus obtained in the range of concentrations considered. As expected, A_2 is usually higher than A_1, but the imprecision of each measurement seems to be small enough to permit the comparison of data collected on different days.

Other sources of error include blank errors (the amount of metal that filters and reagents add to the blank could be between 20 and 90% of the sample signal, RSD 8 – 132%), the variability of the measurement of the final volume (RSD 1%), the errors associated with the sample mass (RSD 0.25%) and sample volume measurements (RSD 0.15%), and the dissolution efficiency and the filtration reproducibility. Considering dissolution efficiency to be a source of error, if the subsamples are taken from a homogeneous sample of candidate algae reference materials and they are analysed independently, both as oven-dried-only samples (sonifier method) and after ashing them (ashing method), then the results reported in Table 3.2 are obtained.

Fleckenstein [18] presents results obtained from the determination of cadmium, lead, copper and zinc in samples of aquatic fungi and algae using direct solid-sampling Zeeman AAS. The results confirmed the accuracy of

Table 3.2. Reproducibility of sample dissolution for different materials and different sample sizes (from [17])

	Fe	Mn	Cu	Cr	Zn
A Ashing method					
Platyphypnidium ripariodes					
2000 mg × ppm	9480	3900	660	550	640
RSD, %	5	1	11	4	1
10 mg × ppm	7080	3470	640	520	700
RSD, %	7	4	2	9	16
Olea europea					
2000 mg × ppm	340	57	50	–	20
RSD, %	·1	1	1	–	16
Lagarosiphon major					
200 mg × ppm	2300	1800	50	40	410
RSD, %	3	3	2	2	2
Freshwater plankton					
10 mg × ppm	3800	250	40	70	–
RSD, %	9	6	5	7	–
B Sonifier method					
Platyphypnidium ripariodes					
200 mg × ppm	9720	4170	690	630	640
RSD, %	3	2	1	1	1
Olea europea					
200 mg × ppm	340	60	50	–	20
RSD, %	2	3	3	–	7
Lagarosiphon major					
200 mg × ppm	2430	1880	50	40	360
RSD, %	1	1	1	1	3
Freshwater plankton					
10 mg × ppm	3700	230	40	60	–
RSD, %	5	16	19	25	–

×: mean
RSD: Relative standard deviation

this method and its usefulness in studies of heavy metal contamination in the environment.

Table 3.3 presents results obtained in the AA spectrometric analysis of sea plants and copepod carried out during 1978 in an international round-robin experiment organised by the International Commission for Exploration of the Seas (ICES) and the International Atomic Energy Authority (IAEA) [19]. It is clear that interlaboratory agreement was very poor at the time.

Yang et al. [20] studied the effect of wet decomposition methods on the electrophoretic determination of cobalt, copper, selenium and zinc in biological materials.

Table 3.3. Comparision of results from intercomparison exercises using marine reference materials (from [19])

Element	Marine reference materials	No. of participants	Range of values, μmol/kg	Mean, μmol/kg	SD	CV
Copper	Seaplant	67	51.8 – 675	198	26	
	Copepod	56	70.7 – 256	121	19	
Zinc	Seaplant	75	23 – 3443	979	27	
	Copepod	66	64.3 – 3779	367	15	
Mercury	Seaplant	75	0.25 – 28	1.7	43	
	Copepod	43	0.20 – 18	1.4	48	
Cadmium	Seaplant	46	1.0 – 214	6.2	96	
	Copepod	43	3.0 – 26.7	6.7	26	

Armannsson [21] used a method based on dithizone extraction and AAS to determine down to 0.03 μg/kg of cadmium, zinc, lead, copper, nickel, cobalt and silver in kelp tissue. The sample is first evaporated with concentrated nitric acid and then nitric–perchloric acid. Following adjustment to pH 8, the solution is extracted with chloroformic dithizone and an aqueous acid extract of the organic phase is analysed by AAS.

Inductively Coupled Plasma Mass Spectrometry (ICP–MS)

Bettinelli et al. [22] determined various trace elements in lichens by ICP–MS. The samples were oxidised in a microwave oven beforehand. Only treatment with hydrofluoric acid ensured complete recovery of many trace elements.

Pietilainen et al. [23] used tube excited energy dispersive X-ray fluorescence to carry out multi-element analysis for 20 elements in plankton algae. The samples were taken from an estuary in the Gulf of Finland and an archipelago outside the estuary. Both molybdenum and titanium–potassium secondary target radiation were used. Detection limits ranged from 0.3 μg/l for bromine to 12 μg/l for aluminium. Most of the environmentally important heavy metals were determined in all samples, and a considerable degree of interrelation was found between several of them.

Bistricki and Munewar [24] found that a combination of scanning electron microscopy and energy dispersive X-ray spectroscopy was an effective tool for characterising heavy metals in green algae, phytoflagellates and diatoms.

Aliquots of the samples were filtered through Nuclepore membranes of 0.45 μm porosity, washed with purified 0.05 mol/l S-collidine buffer which contained no traces of heavy metals (Polysciences, Warrington, PA, USA), and fixed with a solution of osmium tetroxide prepared in the above buffer. After being washed and dehydrated in alcohol, the filtered specimens were dried by the critical point drying technique.

The membrane filters were mounted on a specimen support stub and transferred into a vacuum evaporator. A 20-nm layer of carbon was evap-

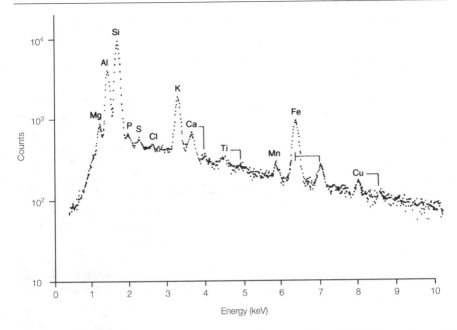

Figure 3.1. X-ray spectrum showing the elemental composition of a dried specimen of *Staurastrum paradoxin*. From [24]

orated on the surface of the specimens to facilitate thermal and electrical conductivity.

Figure 3.1 shows a spectrum obtained for *Staurastrum paradoxin*, a green alga that is commonly found in the eutrophic waters of western Lake Erie. The elemental spectrum demonstrates the presence of metals in a cellular structure. Besides manganese and copper, elevated levels of iron and aluminium were detected. Other elements indicated in the spectrum are inherent structural components found in biota under normal conditions.

These results indicate that the algae, because of their sensitivity and short generation time, can be used as an indicator of heavy metal pollution, thus providing early warning of pollution.

Gamma-Ray Spectrometry

Dutton [25] used gamma-ray spectrometry to measure radioactive elements in bladderwrack. The dry material was packed into a polyethylene tube, which was placed in an aluminium copper or perspex 'hat' that fitted over the detector crystal. Counting of γ-emitting nuclides was carried out with a NaI(Tl) crystal coupled to 200 channels of a pulse height analyser. Methods of spectrum stripping were discussed; a least-squares fitting procedure or a matrix-inversion procedure was preferred, with calculations being performed by computer. The γ-ray spectrometer was calibrated through the use of standardised radioactive solutions added to materials with similar X-ray scattering properties to the samples.

Alpha Spectrometry

Alpha spectrometry has been applied to the determination of alpha emitters in seaweed [26]. After the extraction of actinoid elements from a solution in nitric acid–sodium nitrate with trioctylphosphine oxide–heptane, and back-extraction into ammonium carbonate solution, the actinoids are deposited electrolytically after acidification of the solution and addition of ammonium formate to destroy any excess nitric acid. The separated elements are subjected to β-spectrometry with a stable ($\pm$ 3.5 keV in the 5 MeV region for counting times of 10 000 min) spectrometer that has silicon surface-barrier detectors in conjunction with a 400- or 256-channel pulse height analyser, and spectra are processed by the simultaneous matrix method. The method was applied to the separation of plutonium, americium and other actinoids, and the separation of plutonium and other tetra- and hexavalent actinoids, together with the elimination of americium and other tervalent actinoids.

Miscellaneous

Szefer [27] determined uranium and thorium in samples of seaweeds and plants collected from the Gdansk Bay coastal region of the open Baltic and from Lake Zarnowieckie. Average concentrations of uranium and thorium, respectively, were 0.07 – 0.41 and 0.01 – 0.60 mg/kg dry weight in seaweeds, and 0.09 – 0.22 mg/kg dry weight in lake plants. Calculated concentration factors for uranium and thorium in Baltic seaweeds were 40 – 50 and 280 – 320, respectively. Discrimination factors with respect to calcium were 2.2 – 4.7 for uranium and 16 – 28 for thorium, with the affinities of the seaweeds for uranium and thorium being inversely related to their degree of calcification. Calculated enrichment factors for uranium, thorium, and calcium (sea salt-corrected) with respect to aluminium as a normaliser were 5.8, 1.4 and 24, respectively, indicating that the biological affinities of thorium and, to a lesser extent, uranium were weaker than that of calcium.

Ward [28] studied temporal variations of metals in the seagrass *Posidonia australis*, and its potential as a sentinel accumulator near a lead smelter. *Posidonia australis* from three sites in the hypersaline Spencer Gulf were sampled in October 1980 and February, May and September 1981. Site A was closest to the Port Pirie lead smelter, with sites B and C being 8 and 16 km, respectively, from site A. Site A had the lowest density and smallest standing crop of *P. australis*, and the highest leaf concentrations of cadmium (198 – 541 mg/kg dry weight), manganese (112 – 537 mg/kg), lead (116 – 379 mg/kg) and zinc (728 – 4241 mg/kg). At all sites, concentrations of cadmium, copper, lead and zinc in plant leaves were generally lowest in summer and autumn, and highest at the end of winter. The reverse was true for manganese. Compared with plants from sites B and

C, site A plants also had significantly higher concentrations of cadmium (5.63 mg/kg), lead (167 mg/kg) and zinc (379 mg/kg) in the epibiota. Nickel was highest (3.65 mg/kg) in the epibiota of site B plants, but there were no site-related differences in epibiotic copper. Concentrations of cadmium, copper and zinc in the epibiota were generally higher than those in the leaves, whereas the reverse was true of manganese and nickel. Epibiotic metal concentrations did not vary significantly with time.

3.2
Organic Compounds

3.2.1
Hydrocarbons

Aliphatic Hydrocarbons

Smith [29] classified large sets of hydrocarbon oil spectral data by computer into 'correlation sets' for individual classes of compounds. The correlation sets were then used to determine the class to which an unknown compound belongs according to its mass spectral parameters. A correlation set is constructed by using an ion series representing the contribution to the total ionisation of each of 14 ions. The technique is particularly valuable when examining results from coupled gas chromatography–mass spectrometry of complex organic mixtures. For example, an alkane fraction of lichen extract gave a spectrogram with 24 peaks (molecular weight range 212 – 464), each of which was rapidly classified, generally unambiguously.

Dynamic headspace analysis of an aqueous sodium hydroxide homogenate has been used to determine traces of hydrocarbons in marine algae [32]. A combination of gas chromatography and mass spectrometry was used to identify and determine 'volatiles'.

Law et al. [31] has described procedures for the determination of hydrocarbons (especially petroleum hydrocarbon residues from oil spills) in marine biota and related environmental samples. Basic procedures for sample collection, extraction and clean-up are outlined, together with the methodology for fluorescence analysis.

Methods have been described for the determination of (benzo(a)pyrene and perylene) in aquatic fauna [32,33].

Hydrocarbons are separated from lipids by column chromatography on alumina. The oil fractions were identified and determined by fluorimetry using pyrene as a fluorescence standard. The detection limits of crude, Bunker C and creosote oil were 100, 50 and 100 µg/g of lipid, respectively.

3.2.2
Phenols

Dallakyan et al. [34] has described a method for the determination of low concentrations of phenols and substances containing sulfhydryl groups in microalgae secretions. The method is based on the electrochemiluminescent oxidation of luminol (3-aminophthalic hydrazide) at 14 – 16 °C, pH 6.5, with a potassium iodide electrolyte and a platinum electrode, in order to determine phenols and thiols. The inhibition of the chemiluminescence, specific amongst the substances studied to phenols and thiols, was used as a means of measurement.

Of the substances examined, phenolic compounds possessed the strongest inhibitory properties; among them, monophenols—phenol and tyrosine—exhibited the lowest inhibiting effects. The introduction of a second hydroxyl group on the benzene ring (hydroquinone, pyrocatechol) increased the ability of the phenols to inhibit luminescence. The inhibitory activity varied depending on the position of the hydroxyl groups. Phenols with *ortho-* and *para-*arrangements of the hydroxyl groups (hydroquinone, pyrocatechol, chlorogenic acid) inhibited luminescence more strongly than metaphenols (resorcinol). With some phenols such as phenol, hydroquinone, pyrocatechol and resorcinol, a direct connection was discovered between the redox potential and the ability of luminol to inhibit chemiluminescence. Phenols which contain three hydroxyl groups (propylgallate, gallic acid, pyrogallol) inhibit luminescence more weakly than diphenols. Benzoic acid, which lacks the hydroxyl group on the ring, did not influence luminescence. Out of the complex phenols of plant origin that were studied, an inhibitory effect was exhibited by tannin with a mean molecular weight of 1700. Gossypol inhibited luminescence relatively weakly. This method has a high sensitivity, with the relative error varying from 1.5 to 6%.

3.2.3
Acrylic Acid

Capillary gas chromatography with electron capture detection has been used [35] for the trace level determination of acrylic acid in the algal cultures (*Hymenomonas carterae* and *Skeletonema costatum*). Acrylic acid was extracted using tri-*n*-octylphosphine oxide dissolved in methyl tertiary butyl ether. The extracted acids were analysed following crown ether (18-crown-6)-catalysed derivatisation with pentafluorobenzyl bromide.

3.2.4
Carbohydrates

Cowie and Hedges [36] have described a technique for the extraction and quantitative analysis of neutral monosaccharides from plankton algae, requiring as little as 10 mg of total organic matter. Acid hydrolysis yields monomeric sugars which may exist in up to five isomeric forms when in solution. Lithium perchlorate is used to equilibrate sugar isomer mixtures in pyridine catalytically prior to conversion to their trimethylsilyl ether derivatives. Analysis is carried out using gas-liquid chromatography on fused-silica

Table 3.4. Reproducibilities of the carbohydrate contents of algae (from [36])

| | mg of sugar/100 mg of organic carbon | | | |
	Arabinose	Rhamnose	Ribose	Xylose
SMD	0.07 ± 0.01	0.08 ± 0.000	0.30 ± 0.02	0.08 ± 0.00
% RSD	14.3	0.00	6.67	0.00
	Mannose	Galactose	Glucose	TCH$_2$O
SMD	0.60 ± 0.02	0.32 ± 0.02	2.06 ± 0.05	3.75 ± 0.12
% RSD	3.33	6.25	2.43	3.20

capillary columns. Quantification on the basis of a single clearly resolved peak for each sugar is made possible by the equilibration step.

The freeze-dried sample of algae is ground and homogenised, and a weighed portion is treated with 70% sulfuric acid at 20 °C for two hours, and is then heated to 100 °C for three hours. Adonitol and sorbitol are added as internal standards, and the mixture homogenised prior to neutralisation with anhydrous barium hydroxide and centrifugation to remove barium sulfate. The neutralised solution is passed down a column of 1:1 cation : anion exchange resins, and the eluate evaporated to dryness at 60 °C. The dried residue is dissolved in pyridine and treated with bis(trimethylsilyl) fluoracetamide and trimethylchlorosilane to form the methyl derivatives of the carbohydrates which are then gas-chromatographed. Recoveries of carbohydrates from algae put through the entire analytical procedure ranged from 64% (ribose) and 98% (fucose). Recovery of glucose from cellulose was 81%. Some typical values obtained for carbohydrates in algae are listed in Table 3.4.

3.2.5
Chlorinated Insecticides

Sodergren [37] used digestion with fuming sulfuric acid to clean up samples prior to the determination of chlorinated insecticides in algae by gas chromatography. A hexane extract of the sample is concentrated to 350 μl and 50 μl is sealed in a glass tube with 50 μl of fuming sulfuric acid (10% SO$_3$). After mixing, the phases are then separated and the hexane layer is subjected to gas chromatography. A second portion of the original extract is mixed with an equal volume of 5% propanolic potassium hydroxide in a special pipette, which is sealed and heated in a water bath for ten minutes. After heating, 5 ml of water is added and, after further mixing, the hexane fraction is allowed to separate for analysis. The remainder of the original extract is evaporated, and the residue of extractable lipids is weighed. Sample recoveries are 78 – 94%, with losses occurring mainly at the extraction stage.

Sodergren [38] investigated the simultaneous detection of PCBs, chlorinated insecticides, and other compounds by electron capture and flame ionisation detectors combined in series using an open tube capillary column.

He combined the electron capture detector and flame ionisation detection in series to obtain a dual detection system capable of simultaneous detection of environmental pollutants of different character, e.g. organochlorine residues and the oil and lipid constituents in samples from aquatic environments. In order to avoid the limit of detection being adversely affected when capillary columns were employed, a splitless system without a scavenging gas was used. Since electron capture is a nondestructive process, the effluent from the column passes undisturbed through the electron capture detector. The effluent was then directed to the jet-tip of the flame ionisation detector by means of a glass capillary tube. With capillary columns, a minimum flow rate of 1.9 ml/min was required to operate the electron capture detector. The flows of hydrogen and air to the flame ionisation detector were around 25 and 250 ml/min, respectively.

Organochlorine insecticides and methyl esters of fatty acids were detected simultaneously using this system. Sodergren [38] used his detection system to study the degradation and fate of persistent pollutants in aquatic model ecosystems. Usually these pollutants are closely associated with lipids. Therefore, it is an advantage to be able to study the occurrences and amounts of both lipids and, for example, organochlorine residues. A cell extract from a continuous flow culture of the green alga *Chlorella pyrenoidosa*, to which polychlorinated biphenyls had been added, was hydrolysed by treatment with a solution of acetyl chloride in methanol and then presented to the detection system. The PCBs added to the culture were efficiently taken up by the algal cells: the lipids detected in the extract were palmitic acid and stearic acid. Lipids and substances of lipophilic character tend to accumulate in aquatic environments at the interface between water and air. To assess the ability of the electron capture flame ionisation detector system to detect mineral oil and PCBs simultaneously, a mixture of these substances was injected into the water of an aquarium below the surface. The surface film thus created was sampled [39], extracted, and an aliquot of the extract injected into the gas chromatograph equipped with a capillary column and a low-volume electron capture detector and flame ionisation detector. The mineral oil was eluted before the main PCB components appeared. Due to the high sensitivity of the electron capture detector to changes in temperature, programming the column resulted in severe baseline drift at the beginning of the run. However, neither class of components affected the detection of the other. Thus, both the mineral oil and the halogenated compounds can be conveniently analysed and quantified simultaneously after single injection.

Liquid scintillation counting of [^{14}C] DDT has been used to study the pick-up and metabolism of DDT by freshwater algae [40]. Neudorf and Khan [41] investigated the uptake of ^{14}C-labelled DDT, dieldrin, and photodieldrin by *Ankistrodesmus amalloides*.

The results from their liquid scintillation spectrometric analyses show that the total pick-up of DDT during a one- to three-hour period was 2–5 times higher than that of dieldrin, and ten times higher than that of photodieldrin. The algae metabolised 3 – 5% of DDT to DDE and 0.8% to DDD. The metabolism of DDT by *Daphnia pulex* was also monitored by exposing 100 organisms to 0.31 ppm of the labelled pesticide for 24 hours without feeding. The metabolites were then extracted and separated by thin-layer chromatography, and the R_f values of radioactive spots were compared to R_f values for nonradioactive DDD and radioactive DDE. The results show a conversion of DDT to DDE of about 13.6%.

Sackmauerova et al. [43] has described a method, given below, for the determination of chlorinated insecticides (BHC isomers, DDE, DDT and hexachlorobenzene) in water weeds. In this method, a weighed portion of sample is dried at room temperature and homogenised. From the pulverised sample, 20 g are taken and extracted with petroleum ether in a Soxhlet apparatus for 12 hours. The concentrated extract is purified from coextracts on a Florisil-filled column activated by heating at 120 °C for 48 hours with 5% water added to the cooled column. Insecticides are eluted from the column with 15% dichloromethane in petroleum ether. The eluate is concentrated to a volume of 1 ml and analysed by gas chromatography. The following chlorinated insecticides can be determined in amounts down to the concentrations stipulated below.

α-BHC	0.15 mg/kg
β-BHC	0.75 mg/kg
γ-BHC	0.2 mg/kg
δ-BHC	0.15 mg/kg
p,p'-DDE	0.45 mg/kg
o,p'-DDT	0.5 mg/kg
p,p'-DDD	1.5 mg/kg
p,p'-DDT	1.5 mg/kg

Sackmauerova et al. [42] used thin-layer chromatography on silica plates to confirm the identities of chlorinated insecticides in water plants previously identified by gas chromatography. The compounds can be separated by single or repeated one-dimensional development in *n*-heptane or in *n*-heptane containing 0.3% ethanol. The plate is dried at 65 °C for ten minutes and detected by spraying with a solution of silver nitrate plus 2-phenoxyethanol. Thereafter, the plate was dried at 65 °C for ten minutes and illuminated with ultraviolet light ($\lambda = 254$ nm) until spots representing the smallest amounts of standards were visible (10 – 15 min).

Using gas chromatography methods, Sackmauerova [42] obtained the four BHC isomers at recoveries of 93 – 103.5% from spiked samples. Yields of the four bioisomers were 85.6 – 94%, 90 – 93.2%, 90 – 102.4% and 92 – 105.8%. Purification on a Florisil column was used when determining chlorinated insecticides unstable at low pH (aldrin, dieldrin). The type and activity of Florisil influence the yield and accuracy of the method. Therefore, the activity of this adsorbent had to be verified and adjusted. The average content of the γ isomer of BHC found in water plants was 0.026 mg/kg, while the β isomer was not present at 0.00 mg/kg, $\gamma + \delta$ at 0.032 mg/kg, DDE at 0.003 mg/kg and DDT at 0.002 mg/kg. These results suggest that chlorinated insecticides, due to their physical and chemical properties, can accumulate and adsorb onto solid particles.

3.2.6
Polychlorobiphenyls

Photoactivated luminescence spectroscopy has been used as a method for rapid screening polychlorobiphenyls in biota [44].

3.2.7
Organophosphorus Insecticides

Szeto et al. [45] have described a gas chromatographic method for the determination of acephate and methamidophos residues in plant tissue.

3.2.8
Polychorodibenzo-p-Dioxins and Dibenzofurans

Hashimoto and Morita [46] studied two different methods for determining dioxins in seaweed. Soxhlet extraction from dried sample gave higher recoveries than extraction from alkali-digested samples.

3.2.9
Chlorophyllous Pigments

Youngman [47] of the Water Research Centre (UK) has studied methods of extracting these pigments from algae and the spectrophotometric determination of chlorophylls in extracts.

Jensen [48] showed that there was no difference between the chlorophyll extracted by either methanol or acetone from fresh diatom material, but methanol gave 30% higher results than acetone with green algae. Low levels were found in stored filters, and this was most pronounced with methanol. Immediate analysis of the sample is recommended; even storage by freezing should be avoided.

Wun et al. [49] have described a method for the simultaneous extraction of the water quality marker algal chlorophyll a and the faecal sterol coprostanol from water. This method utilises a column of Amberlite XAD-1 for the simultaneous extraction of both markers. Chlorophyll content was determined by the trichromatic method [52] using a double beam spectrophotometer.

Wun et al. [49], using ^{14}C-labelled cholesterol c, showed that 100% of this substance is adsorbed from a 500 to 1000 ml water sample by XAD-1 resin. When the column was subsequently eluted with 50 ml of basic methanol and 30 ml of benzene, up to 97% of the added sterol was recovered in the benzene eluate.

To ascertain the suitability of the column method for the simultaneous extraction of coprostanol and chlorophyll a in actual field testing situations, various unialgal cultures were mixed with sewage samples. The effectiveness of the neutral resin column extraction for these markers was evaluated with that of the conventional procedure.

The efficiency of the resin column used for the simultaneous extraction of coprostanol and phytoplankton chlorophyll a is comparable to or better than the conventional extraction procedures for the respective compounds. Dilution of the samples and the presence of extraneous materials did not affect the recovery efficiency significantly. The column technique was effective at isolating chlorophyll a from various algae. The superiority of the column method was more pronounced when the small green alga (*Oocystis sp.*) and the blue-green alga (*Oscillatoria sp.*) were used as test organisms. The coprostanol extraction efficiency was again shown to be comparable to that of the hexane liquid–liquid partitioning process.

It was apparent that a more complete extraction of phytoplankton chlorophyll a from water samples could be obtained by the column method. Furthermore, the column procedure is much faster; a 1–1 sample required a processing time of approximately 1–2 hours, compared to 24 hours for the conventional aqueous acetone method. Coprostanol contents of these samples were too low (< 0.2 ppb) for meaningful comparisons to be made and are not presented.

Garside and Riley [51] have used thin-layer chromatography to achieve a preliminary separation of chlorophylls on solvent extracts of water and algae prior to a final determination by spectrophotometry or fluorimetry. Garside and Riley [51] filtered seawater samples (0.5–5 l) through Whatman GF/C glass fibre coated with a layer, 1–2 mm thick, of light magnesium carbonate. This retains the smallest particles of organic matter and it is easy to extract the pigment from it. The filter is extracted with acetone and then with methanol using ultrasonic vibration. The solution is passed through anhydrous sodium sulfate to remove water and then evaporated in vacuo at less than 50 °C. The residue is dissolved in ethyl ether–dimethylamine (99 : 1, 1–2 ml) and this is applied as a spot to a plate coated with silica gel PF_{254}. The chromatogram is developed with light petroleum (60–80)–ethyl acetate–dimethylamine (55 : 32 : 13) and the plate scanned by reflecting the light passing through an Ilford 601 filter (603 nm). The integration reading for each peak is measured and the R_f values noted relative to chlorophyll a. Xanthophylls are identified by scraping off the spots and measuring the absorption spectrum of an extract of the scrapings. Chlorophyll c remains at the origin and can be developed in light petroleum–ethyl acetate–dimethylformamide (1 : 2 : 2) and scanned as before. Chlorophylls a, b and c, carotene, xanthophylls, and certain degradation products can be determined. The sensitivity for chlorophyll is about 0.12 µg and the precision for most pigments is ± 5% or better at the 0.5 µg level.

Shoaf and Lium [52] used thin-layer chromatography to separate algal chlorophylls from their degradation products. Chlorophyll is extracted from the algae with dimethyl sulfide and chromatographed on thin-layer cellulose sheets, using 2% methanol and 98% petroleum ether as solvents, before determination by either spectrophotometry or fluorimetry.

Reference values and columns obtained are reported below:

Compound	R_f	Colour
Phaeophytin a	0.89	Grey
Chlorophyll a	0.76	Blue Green
Phaeophytin b	0.61	Greenish Yellow
Chlorophyll b	0.34	Yellowish Green
Phaeophytin c	0	Yellowish Green
Chlorophyll c	0	Yellowish Green

Recoveries of pure chlorophylls a and b were 98% and 96%, respectively. Thus chlorophylls a and b and their phaeophytins may be readily separated and determined by this method. It is not possible to determine chlorophyll c accurately by applying this method to natural samples for two reasons. In some samples, other degradation products of chlorophyll a and b and phaeophytin a and b (apparently chlorophylides and phaeophorbides, i.e. chlorophylls or phaeophytins missing part or all of the phytol tail) do not migrate, but remain at the origin with chlorophyll c and phaeophytin c, each of which also lacks a phytol tail. Chlorophyll c is only sparingly soluble in diethylether or acetone, so that total recovery is not possible. Chlorophyll a and b are accurately determined by this method without interference from degradation products.

The applicability of HPLC to the determination of various chlorophylls has been examined.

Chlorophyll in natural waters is frequently estimated by trichromatic spectrophotometry of algal extracts [53–67]. When no interfering compounds are present, these trichromatic equations are good estimates. The major criticism is that, in natural plankton extracts, spectrally similar chlorophyll breakdown products are frequently present. Thus, chlorophyll cannot be accurately determined by this method. One way to avoid the problem is to chromatographically separate the breakdown products from the chlorophylls before measurement. Separation of the spectrally similar chlorophylls a and b (as well as degradation products) will result in a more accurate determination of the chlorophylls.

Tests carried out with the HPL chromatographic technique endorsed the claim that it causes negligible degradation of both the chlorophylls and the xanthophylls [51]. A more efficient separation of plant pigments could be achieved on a silica stationary phase than on a C_{18} reversed-phase medium. A 30 cm column packed with Partisil 10 gave an efficient separation of the individual carotenoids, chlorophylls a and b, and many of the degradation products of the latter pair. The solvent consists of light petroleum (bp 60 – 80 °C), acetone, dimethyl sulfoxide and diethylamine in the ratio 75 : 23.25 : 1.5 : 0.25 by volume. Unfortunately, this solvent is not sufficiently polar to elute phaeophorbide and chlorophyll c. With samples containing these pigments, it is necessary to carry out an additional, stepwise, elution with a more polar solvent. Further tests showed that excellent resolution of these compounds could be achieved via a mixture containing light petroleum (bp 60 – 80 °C), acetone, methanol and dimethyl sulfoxide in the ratio 30 : 40 : 27 : 3 by volume, respectively.

In order to identify the various peaks on the chromatograms, extracts of a range of algae from various classes were injected repeatedly onto the Partisil 10 column operated at a flow rate of 2 ml/min. The eluates corresponding to individual peaks of known retention times were collected and identified from their absorption spectra. Chlorophylls and their degradation products were characterised by their spectra in ether and in (1 + 9) water–acetone, based on data from Strickland [60]. Carotenoids were identified by comparison of their wavelengths of maximum absorption in hexane, ethanol and carbon disulfide with the tabulated values published by Davies [61]. Identifications were confirmed by thin-layer chromatography on silica gel G [62]. The retention times of various pigments are shown in Table 3.5. The retention times are extremely reproducible.

Since the HPL chromatographic method provides pigments with a high degree of purity, it is relatively easy to standardise the technique.

Table 3.5. Retention times and corresponding coefficients of variation for various phytoplankton pigments on a 30 cm Partisil 10 column with a solvent flow rate of 2 ml/min (from author's own files)

Pigment	Retention time(s) First mobile phase	Coefficient of variation, %
β-Carotene	107 ± 1	0.9
Echinenone	134 ± 1	0.7
Phaeophytin b	141 ± 1	0.7
Phaeophytin a	183 ± 1	0.5
Chlorophyllide a	237 ± 0.3	0.1
Chlorophyll a	293 ± 2	0.6
Chlorophyll b	424 ± 2	0.6
Diatoxanthin	478 ± 2	0.4
Myxoxanthophyll	488 ± 1	0.6
Lutein	533 ± 1	0.3
Diatinoxanthin	580 ± 3	0.5
Violaxanthin	664 ± 1	0.2
Fucoxanthin	809 ± 5	0.6
Neoanthin	1773 ± 15	0.9
	Second mobile phase	
Phaeophorbide a	2395 ± 19 (622 in second solvent)	0.8
Chlorophyll c	2497 ± 21 (724 in second solvent)	0.8

The sensitivity of the method varies considerably from one pigment to another, varying from about 5 ng for β-carotene to about 80 ng for chlorophyll a.

Evans et al. [57] separated phaeophytins a and b on Corasil II with a mobile phase consisting of a 1:5 (v/v) mixture of ethyl acetate and light petroleum. Eskins et al. [58] have employed two 0.62 m columns of C_{18}-Porasil B for the preparative separation of plant pigments by means of programmed stepwise elution with methanol–water–ether. However, the method is of little value for routine application because of the time required, and also because the chlorophyll degradation products, other than phaeophytin, are not separated. Shoaf [59] has used HPLC to separate the chlorophylls a and b of a pigment extract from which the carotenoids had been previously removed. Good resolution of the two pigments and several of their unspecified degradation products was achieved on a 25 cm column of Partisil PXS 1025 by elution with aqueous 95% methanol; however, chlorophylls were not determined quantitatively.

Abayachi and Riley [56] compared results obtained by the HPL chromatographic method with those obtained by a reflectometric thin-layer chromatographic method and the SCOR/UNESCO polychromatic procedure for the determination of chlorophylls a, b, and c, β-carotene, fucoxanthin, diatinoxanthin, lutein, violaxanthin, neoxanthin, echinenone, and myxoxanthophyll. The results obtained from the latter were evaluated by the SCOR/UNESCO equations and also by the more recent ones of Jeffrey and Humphrey [66]. The carotenoids were determined collectively from the absorbance of the 90% acetone extract at 480 nm by means of the equations of Strickland and Parsons [67]. The results of these comparative studies show that there is satisfactory agreement for all pigments between the two chromatographic methods. However, although the results for chlorophyll a obtained by the polychromatic method were in reasonable accord with those derived chromatographically, many of those for the other chlorophylls showed a higher discrepancy. Obviously, the polychromatic method is particularly unsatisfactory with respect to the interference of chlorophyll degradation products, as these are nearly always present in environmental samples.

Liebezeit [63] has described a HPL chromatographic method for determining chlorophyll a in marine phytoplankton.

Sartory [64] used a combination of HPLC and spectrometry to determine algal pigments. He discussed sample clean-up procedures which allowed the determination of chlorophylls free from carotenoids, and a HPLC procedure with fluorescence detection which allowed the separation of all chlorophyll pigments within 40 min, using a simple solvent programme. The method had detection limits of 10 pg for chlorophyll a and phaeophytin b, 15 pg for chlorophyll b, and 20 pg for phaeophytin a. Comparative analyses of carotenoid-free extracts by HPLC and several spectrophotometric procedures tended to overestimate chlorophyll a and phaeophytin a, and to underestimate chlorophyll b.

3.2.10
Organosulfur Compounds

Bechard and Rayburn [68] determined volatile organic sulfides in freshwater algae.

Andreae [69] has described a gas chromatographic method for the determination of nanogram quantities of dimethyl sulfoxide in phytoplankton. The method involves chemical reduction to dimethyl sulfide with chromium(II) on sodium borohydride, which is then determined gas chromatographically using a flame photometric detector. Andreae [69] investigated two different apparatus configurations. One consisted of a reaction/trapping apparatus connected by a six-way valve to a gas chromatograph equipped with a flame ionisation detector, and the other combined the trapping and separation functions in one column, which was attached to a flame photometric detector. The gas chromatographic flame ionisation detector system was identical to that described by Andreae [70] for the analysis of methylarsenicals, with the exception that a reaction vessel which allowed the injection of solid sodium borohydride pellets was used.

Andreae tested a large number of sulfur compounds in order to investigate potential positive interferences due to the formation of dimethylsulfide from the reaction of sulfur compounds, other than dimethyl sulfoxide, to form dimethyl sulfide. The only compound other than dimethyl sulfoxide that gave a positive reaction with sodium borohydride was dimethylpropiothetin (($CH_3)_2S^+CH_2CH_2COO^-$), an organosulfur compound occurring in some algae.

Andreae [70] showed that dimethyl sulfoxide is a common constituent in natural waters. Its occurrence in seawater is restricted to the zone of light penetration. This fact, and the abundance of dimethyl sulfoxide in the medium after the growth of phytoplankton, suggest that it occurs as an end-product of algal metabolism.

3.2.11
Miscellaneous

Humic Substances

Gadel and Bruchet [71] applied pyrolysis–gas chromatography–mass spectrometry to the characterisation of humic substances resulting from the decay of aquatic algae and macrophytes. The compositions of humic substances from decaying algae and aquatic macrophytes in a coastal Mediterranean lagoon and from sediments from different sources, including a lake on the Greenland icecap were investigated. The material was also analysed by gas elemental analysis and by infrared spectroscopy. The humic matter included varying proportions of carbohydrates, n-acetylamino sugars, proteinaceous material and lignin derivatives: lesser amounts of phthalates and aliphatic compounds were found in some of the samples. There were marked differences between humic and fulvic fractions, with most of the carbohydrates being included in the fulvic acids.

Adenosine Triphosphate

Shoaf and Lium [72] compared various extraction methods involving dimethyl sulfoxide, boiling tris buffer and butanol–octanol for the isolation of adenosine triphosphate from algae prior to its luminometric determination using luciferin–luciferase assay. All were equally effective on algae *Chlorella vulgaris*, and measurement of the activity by either peak height or integration of the area under the peak was equally sensitive and reproducible. Determination of adenosine triphosphate was inhibited by mercuric chloride, cadmium chloride, calcium chloride, potassium or sodium phosphates, and high concentrations of the extracted dimethyl sulfoxide (Table 3.6). Of the methods investigated for preserving samples for analysis, field extraction of the adenosine triphosphate followed by quick freezing in an acetone–dry ice bath is recommended.

Table 3.6. Inhibition of the ATP assay (from [72])

Compound	Final concentration in assay cuvette	Activity
Distilled water	–	100
Mercuric chloride, μmol/l	8.3	28
Cadmium chloride, mmol/l	0.42	38
Calcium chloride, mmol/l	4.2	63
Potassium phosphate, pH 4.7, mmol/l	21	33
Sodium phosphate, pH 7.4, mmol/l	21	41

Note: The standard volume and concentration of luciferin–luciferase was used. Then 10 μl of low-response water or the appropriate inhibitor was added and mixed. Samples were then immediately assayed by addition of 10 μl of 0.1 μg ATP per ml.

Martin [73] describes the chemistry and biology of adenosine triphosphate. The extraction of adenosine necessitates rupturing cell envelopes and inhibiting adenosine triphosphatase and other intracellular enzymes. Adenosine triphosphate has up to now been determined by bioluminescence techniques, but these have been subject to interference problems. High-performance liquid chromatography and nuclear magnetic resonance using phosphorus-31 have shown promise in overcoming these problems. The adenosine triphosphate content in algae cells is discussed, and the relationship between quantity of adenosine triphosphate and other parameters used for the evaluation of biomass is reviewed.

Anatoxin *a*

High-performance liquid chromatography combined with ultraviolet spectroscopy has been used to determine this toxin in algae [74]. A temperature of 55 °C is recommended, in either a normal phase silica-A column, using

80% isopropane in hexane or a reverse phase C_{18} column, using an aqueous solution of acetonitrile. The toxin was separated from other components in aqueous resuspensions of dried extracts of algal cultures.

Uronic Acids and Aldose

Walters and Hedges [75] carried out a simultaneous determination of uronic and aldoses in plankton and plant tissues by capillary gas chromatography, following conversion to N-hexylaldonamide and alditol acetates. The sample was first hydrolysed with hydrofluoric acid at 135 °C to produce the N-alkylaldonamide acetates.

Oligosaccharides

Franco and Garrido [76] developed a method for the determination of oligosaccharides in complex samples with high salt contents. Reversed phase HPLC with water used as eluent and refractive index detection was employed. A diagrammatic scheme of the proposed treatment for sample purification was included together with chromatograms of the enzymic assays of the amylolytic activity of *Aspergillus oryzae* and *Endomyces fibuliger* grown in mussel processing wastes. Samples were treated with ethanol to precipitate macromolecular organic matter and desalted by a modified mixed bed resin before chromatographic analysis was carried out. Retention times of sodium chloride, glucose, maltose and maltotriose on three columns with different lengths (15, 25 and 40 cm) were determined. The method allows for the easy removal of substances interfering with the identification of sugars, namely glucose and maltose.

Lewis and Wang [77] have reviewed the use of biomonitoring using aquatic vegetation.

Weiss and Grauk [78] developed a bioassay method based on the production of carbon dioxide, as measured by infrared gas analysis, by submerged macrophytes. Results obtained with the water moss *Fontinalis antipyretica* show that the carbon dioxide balance becomes unstable at the same concentration of copper that affects the metabolism. The method can therefore be used to determine the threshold concentration of toxic substances for aquatic plants.

3.2.12
Domoic Acid (DA)

Amnesic shellfish poisoning, a new malady [79,80], entered the public health lexicon in 1987. In the years following this first incident, many hundreds of people were made ill and a number of deaths were attributed to amnesic shellfish poisoning [81, 82, 84]. The causative agent was shown to be DA, a neuroexcitatory amino acid. The DA is produced by plankton diatoms [85–92] from the genus *Pseudonitzschia*. Humans eating shellfish that fed upon *Pseudonitzschia* fall victim [93] to amnesic shellfish poisoning when DA in the shellfish approaches or exceeds 20 ppm. Concentrations as high as 100 ppm [94] have been observed in shellfish, and 20 pg/cell in the diatom cells [94] themselves. Also, a number of ecologically and commercially valuable animals are affected adversely by DA as it is vectored [94–96] through the food web. In recent years, the occurrence of DA in *Pseudonitzschia* has been demonstrated [79,83,85–92] in many US coastal locations and worldwide.

There are a number of ways to detect DA in algae and in food. The low levels, which must be detectable, mandate prior separation of DA from many interfering substances in most currently used chemical methods [80,97–100]. Such procedures are time-consuming. One such procedure is that described by Wu et al. [101] for the determination of DA in plankton.

These workers harvested cultures of the phytoplankton diatom *Pseudonitzschia multiseries* under controlled growth conditions ranging from late logarithmic to late stationary phase (17 – 58 days). The amount of DA present in the growth media and in the homogenised cells were determined by HPLC. Defined samples of media, homogenised cells, whole cells and whole cells in media were laser-excited at 251 nm for the purpose of selectively exciting intense UV resonance Raman spectra from DA in the samples. Neither media nor cell component spectra from algae seriously interfere with DA spectra. The spectral cross-sections for the dominant 1652 cm^{-1} mode of DA have been determined for 242, 251, and 257 nm excitation. Maximum sensitivities are achieved with 251 nm excitation because cross-sections for DA are a maximum and interference from other algal components becomes very small. Domoic acid concentrations that have been determined with 251 nm excitation by resonance Raman methods correlate closely with values determined independently with HPLC, especially at higher DA concentrations. The UV resonance Raman analysis of DA in phytoplankton algae is shown to be very sensitive and quantitative as well as rapid and nonintrusive.

3.3
Organometallic Compounds

3.3.1
Organoarsenic Compounds

Maher [102] has described ion-exchange chromatographic–hydride generation atomic absorption and spectrometric methods for the determination of inorganic arsenic, monomethyl arsenic and dimethylarsenic acid in the analysis of algae. The organoarsenic compound was extracted from the algae by digestion with 0.1 mol/l sodium hydroxide followed by filtration, concentration to dryness and dissolution of the residue in 8.5 mol/l hydrochloric acid. A toluene extract of the acidic phase was prepared, and arsenic back-extracted from this phase with hydrochloric acid–potassium dichromate reagent. Arsenic species were separated on an ion-exchange column. Arsenic in each of the fractions was reduced to arsine on a zinc reductor column. Arsine was evaluated in the extract by carbon tube AAS at 193.7 nm.

Typical results obtained in a study of the forms of arsenic in several species of macro algae, tissues of *Mercenaria mercenaria*, and estuarine sediments collected from the southern coast of England were discussed.

Inorganic arsenic in algae was in the range 0.1 – 3.2 mg/kg, whilst monomethyl arsenic acid and dimethyl arsinic acids, respectively, were in the ranges 0.2 – 0.6 and 7.6 – 15.6 mg/kg, giving total arsenic contents in the range 20 to 49 mg/kg.

White and Englar [103] have described a procedure for the determination of inorganic and organically bound arsenic in marine brown algae. In this method, inorganic arsenic was removed from the brown algae by distillation as the corresponding trichloride, and assessed by absorption of the arsine–silver diethyldithiocarbamate complex. Severe digestion conditions for organic-bound arsenic were required to determine total arsenic analysed subsequently by the silver diethyldithiocarbamate–pyridine reagent. Arsenic in the inorganic and organic forms ranged from 0.5 to 2.7 and 40.3 to 89.7 µg/g dry weight, respectively, in seven species of the *Laminariaceae* and three species of the *Alariaceae* and *Lessoniaceae* collected in British Columbia, providing levels of inclusion similar to commercially available brown seaweed products. By contrast, *Sargassum muticum* contained 20.8 µg/g dry weight of inorganic arsenic, about 38% of the total concentration, and a commercial specimen of *Hizikia fusiforme* contained 71.8 µg/g inorganic arsenic, some 58% of the total concentration, suggesting the propensity of members of the *Sargassaceae* family to accumulate the inorganic form of arsenic.

3.3.2
Organolead Compounds

Chau et al. [104] has described a procedure involving gas chromatography with an atomic absorption detector for the determination of organolead compounds (Me_3EtPb, Me_2Et_2Pb, $MeEt_3Pb$, Et_4Pb, Et_3Pb^+ and Et_2Pb^{2+}) in macrophytes taken in the St. Lawrence River, Ontario, Canada.

In this method, the sample is extracted with benzene in the presence of added sodium chloride, potassium iodide, sodium benzoate and sodium diethyldithiocarbomate. After centifugation, a measured volume of the benzene extract is butylated using *n*-butyl magnesium chloride to convert ethylmethyl lead compounds to their corresponding tetraalkyl lead forms R_nPbBu_{4-n} and Bu_4Pb (R = Et, Me)[4], respectively, all of which can be determined by gas chromatography using an atomic absorption detector.

Low concentrations of alkyl lead compounds found in macrophytes taken 4 m deep in the water were as follows; Me_3EtPb 0.038 mg/kg, Me_2Et_2Pb 1.50 mg/kg, $MeEt_3Pb$ 3.61 mg/kg, Et_4Pb 16.5 mg/kg, Et_3Pb^+ 0.59 mg/kg,

Me_3Pb^+ 0.11 mg/kg, and total lead 59.2 mg/kg. Surface macrophyte samples had lower alkyl lead contents, e.g. Et_4Pb 0.07 mg/kg, Et_3Pb^+ 0.13 mg/kg and total lead 4.3 mg/kg.

Wong et al. [105] observed that various species of algae were capable of converting trimethyl lead acetate, but not inorganic lead salts, to tetramethyl lead, and that the tetramethyl lead appreciably decreased the rate of algal cell growth.

3.3.3
Organomercury Compounds

Houpt and Campaan [106] used emission spectrographic analysis to determine traces of organomercury compounds isolated from plant matter by gas chromatography. The method permits the determination of 5 pg of methylmercury.

3.3.4
Organothallium Compounds

Schedlbauer and Heumann [107] using preconcentrated extracts of 500 ml seawater and positive thermal ionisation isotope dilution mass spectrometry, were able to determine down to less than 0.4 µg/l of dimethyl thallium in Atlantic seawaters. Levels found were in the range 0.4 – 3.0 ng/l and are believed to be of biogenic origin, such as from biomethylation of algae.

3.3.5
Organotin Compounds

Francois and Weber [108] and Garside and Riley [51] used hydride generation AAS to speciate and determine methyltin and butyltin compounds in eelgrass (*Zostera marina L.*) leaf tissue from Great Bay estuaries, New Hampshire.

Extraction with hydrochloric acid yielded 63 – 87% recoveries of methyltin, mono- and dibutyltin. Hydride formation with sodium borohydride was effective. Tributyltin was extracted with dichloromethane-methanol, giving 77% recovery, and detection was improved by using tetrabutylammonium borohydride for hydride formation. Analysis of eelgrass samples indicated minimal pollution by methyl- and butyltin compounds, and results from an experimental mesocosm suggested that determination of butyltin concentrations in seagrass tissue could provide a sensitive method for detecting low levels of butyltin contamination in the aquatic environment.

Hodge et al. [15] determined nanogram quantities of the halides of methyltin, dimethyltin, trimethyltin, diethyltin, triethyltin, *n*-butyltin, di-*n*-butyltin, tri-*n*-butyltin, phenyltin and inorganic tin(IV) in algae by a procedure involving reacting them with sodium borohydride to convert them to tin hydrides, which are then detected by AAS. The compounds are separated on the basis of their different boiling points, which range from 1.4 °C (CH_3SnH_3) to 280 °C (n-$C_4H_4)_3SnH$). Detection limits range from 0.4 µg/kg (Sn(IV)) to 2 µg/kg (tri-n-butyltin chloride). To digest the sediment sample, 1 g of oven-dried material was digested with nitric acid, perchloric acid and hydrofluoric–hydrochloric acid and the digest made up to 50 ml. This solution was injected directly into the hydride generator. Stannane and the organotin hydrides evolve from the hydride trap in such a manner that they can be identified by a 'retention time'.

3.4
Nonmetallic Elements

3.4.1
Iodine

Kuldvere [109] used cold vapour AAS to determine iodine in seaweed. The method is based on the interfering effect of iodine on the determination of mercury.

3.4.2
Halogens

Ion-selective electrodes have been used to determine halogens in marine algae [106].

3.4.3
Phosphorus

Harwood et al. [114] have reviewed methods for the determination of phosphate in algae.

Phosphate levels of 1 – 10 µg/l in water can have a significant influence on algal productivity. Methods for determining these levels have been described [112, 113].

Darich et al. [111] determined the effectiveness of chemical extraction for predicting the biological availability of phosphorus in suspended solids, as determined by two-day and 14-day incubations with the algae *Selenastrum capricornutum* collected from several sites in an agricultural catchment in Indiana. The extraction procedures involved the sequential use of sodium hydroxide and hydrochloric acid, ammonium fluoride, sodium hydroxide and hydrochloric acid, nitrilotriacetic acid, and hydroxyaluminium resin. Some 21 – 25% of total sediment phosphorus was biologically available.

Inorganic phosphorus extracted with sodium hydroxide was significantly correlated with both two-day and 14-day available phosphorus, although extraction removed amounts of inorganic phosphorus that exceeded available sediment phosphorus by a relatively constant amount. A single extraction with 0.1 M sodium hydroxide could be used to estimate both short- and longer-term algal-available phosphorus in stream sediments derived from noncalcareous soils where amorphous iron and aluminium were principally responsible for phosphorus retention.

3.4.4
Halogens, Phosphorus and Sulfur

Houpt and Compaan [106] used emission spectrographic analysis to identify traces of organic matter containing halogens, phosphorus, sulfur and isolated them from plant matter by gas chromatography. They transferred the gas chromatographic fractions sequentially through a heated stainless-steel capillary tube to a silica tube (3 mm id) in which they were submitted to a plasma discharge (2.45 MHz) in helium at 10 Torr. The emission spectrum arising from the fragmentation, ionisation and excitation of the organic molecule was then analysed with the aid of two monochromators, the intensities of the required analytical lines being measured photoelectrically. One monochromator was focused on a characteristic line (e.g. the 247.86 nm carbon line) as a chromatographic detector and, when the intensity of this line was a maximum for any one fraction detected in the discharge tube, a 10 s sweep over the range 200 – 600 mm was made by the other monochromator. Examination of the resulting complete spectrograms revealed the presence or absence of phosphorus, sulfur, chlorine, bromine and iodine in samples.

3.4.5
Organic Carbon Compounds

Weiss and Grauk [78] developed a bioassay method based on the production of carbon dioxide, as measured by infrared gas analysis, by submerged macrophytes.

Results obtained with the water moss *Fominalis antipyretica* show that the carbon dioxide balance becomes unstable at the same concentration of copper that affects the metabolism. The method can therefore be used to determine the threshold concentration of toxic substances for aquatic plants.

3.5
Anions

3.5.1
Iodide

Chakrabarty and Das [115] used cold vapour AAS to carry out indirect determinations of iodide in seaweed.

3.6
Detection Limits

Only limited information is available on detection limits that have been achieved in the analysis of water plant and algal matter.

3.6.1
Inorganic Elements

Detection limits achieved by AAS and X-ray fluorescence spectroscopy range from 0.03 mg/kg (cadmium, zinc, lead, copper, nickel, cobalt and silver [21]) to 0.001–0.01 mg/kg (bromine aluminium [23]).

3.6.2
Organometallic Compounds

The only reported value is 0.0004 – 0.002 mg/kg for organotin compounds in algae [116].

References

1. Jia G, Desideri D, Guerra F, Meli MA, Testa C (1997) *J Radioanal Nucl Chem* **220**:15.
2. Abu-Hilal AH, Riley JP (1981) *Anal Chim Acta* **131**:175.
3. Kantin R (1983) *Limnol Oceanogr* **28**:165.
4. Dodd M, Pergantis SA, Cullen WR, Li H, Eigendorf GK, Reimer KJ (1996) *Analyst* **121**:223.
5. Whyte JNC, Englar JR (1983) *Botanica Marina* **26**:159.
6. McSheehy S, Szpunar J (2000) *J Anal Atomic Spectrom* **15**:79.
7. Maher WA (1983) *Analyst* **108**:939.
8. Lee DS (1982) *Anal Chem* **54**:1682.
9. Van Raaphorst JG, Van Weers AW, Haremaker HM (1974) *Analyst* **99**:523.
10. Shengjun M, Holcombe JA (1990) *Anal Chem* **62**:1994.
11. Sivasankara Pillay K, Thomas, CC Jr, Sondel JA, Hyche CM (1971) *Anal Chem* **43**:1419.
12. Kuldvere A, Andreassen BT (1979) *Atomic Spectrosc* **18**:106.
13. Mitchell PG, Greene B, Sneddon J (1986) *Microchimica Acta* **34**:249.
14. Colborn T (1982) *Bull Environ Contam Toxicol* **29**:422.
15. Hodge VF, Seidel SL, Goldberg ED (1979) *Anal Chem* **51**:1256.
16. Dogan S, Haerdi W (1980) *Int J Environ Anal Chem* **8**:249.
17. Bando R, Galanti G, Varini PG (1983) *Analyst* **108**:722.
18. Fleckenstein J (1987) *Fresen Z Anal Chem* **328**:396.
19. IAEA (1978) *Intercalibration of Analytical Methods on Marine Environmental Samples*, Progress Report No.19, International Atomic Energy Agency, Monaco.
20. Yang JY, Yang MH, Lin SM (1990) *Anal Chem* **62**:146.
21. Armannsson H (1979 *Anal Chim Acta* **110**:21.
22. Bettinelli M, Spezia S, Bizzarri G (1996) *Atom Spectrosc* **17**:133.
23. Pietilainen K, Adams F, Nullens H, Van Espen P (1981) *X-Ray Spectrom* **10**:31.

24. Bistricki T, Munewar M (1982) *Can J Fish Aq Sci* **39**:506.
25. Dutton JWR (1969) *Gamma Spectrometric Analysis of Environmental Materials, Technical Report FR14*, Fisheries Radiobiological Laboratory, Lowestoft, UK.
26. Hampson BL, Tennant D (1973) *Analyst* **98**:873.
27. Szefer P (1987) *Mar Pollut Bull* **18**:439.
28. Ward TJ (1987) *Mar Biol* **95**:315.
29. Smith DH (1972) *Anal Chem* **44**:536.
30. Chesler GM, Gump BH, Hertz HS, May WE, Wise SA (1978) *Anal Chem* **50**:805.
31. Law RJ, Fileman TW, Partmann JE (1988) *Methods for Analysis of Hydrocarbons in Marine and Other Samples*, Report No.2, Ministry of Agriculture, Fisheries and Food, Lowestoft, UK.
32. Maher WA, Bagg J, Smith JD (1979) *Int J Environ Anal Chem* **7**:1.
33. Zitro V (1975) *Bull Environ Contam Toxicol* **14**:621.
34. Dallakyan GA, Veselovski VA, Tarusov BN, Peagasyan SI (1978) *Hydrobiol* **14**:90.
35. Vairavamurthy A, Andreae MO, Brooks JM (1986) *Anal Chem* **58**:2684.
36. Cowie GL, Hedges JI (1984) *Anal Chem* **56**:497.
37. Sodergren A (1973) *Bull Environ Contam Toxicol* **10**:116.
38. Sodergren A (1978) *J Chromatogr A* **160**:271.
39. Sperling ER (1977) *Fresen Z Anal Chemie* **287**:23.
40. Picer N, Picer M, Strohal P (1975) *Bull Environ Contam Toxicol* **14**:565.
41. Neudorf S, Khan MAQ (1975) *Bull Environ Contam Toxicol* **13**:443.
42. Sackmauerova M, Pal'usova O, Szokolay A (1977) *Water Res* **11**:537.
43. Sackmauerova M, Pal'usova O, Hluchan E (1972) *Vodni Hospod* **10**:267.
44. Vo-Dihn T, Pal A, Pal T (1994) *Anal Chem* **66**:1264.
45. Szeto SY, Brown MJ, Vlotts PC (1982) *J Chromatogr* **240**:526.
46. Hashimoto S, Morita M (1995) *Chemosphere* **31**:3887.
47. Youngman RE (1978) *Report TR 82*, Water Research Centre, Medmenham, UK
48. Jensen KS (1976) *Vatten* **32**:337.
49. Wun CK, Rho J, Walker RW, Litski W (1979) *Water Air Soil Pollut* **11**:173.
50. Ackermann F (1977) *Deutsche Gewasserkundliche Mittelungen* **21**:53.
51. Garside C, Riley JP (1969) *Anal Chim Acta* **46**:179.
52. Shoef, WI, Lium BW (1977) *J Res US Geol Survey* **5**:263.
53. Jeffrey NW (1968) *Biochim Biophys Acta* **162**:271.
54. Daley RJ, Gray CBJ, Brown SR (1973) *J Chromatogr A* **76**:175.
55. Daley RJ (1973) *J Arch Hydrol* **72**:400.
56. Abayachi JK, Riley JP (1969) *Anal Chim Acta* **107**:1.
57. Evans N, Games DE, Jackson AH, Matlin SA (1975) *J Chromatogr A* **115**:325.
58. Eskins K, Scholfield CR, Dutton HH (1977) *J Chromatogr A* **135**:217.
59. Shoaf WT (1978) *J Chromatogr A* **152**:247.
60. Strickland DH (1965) In: Riley JP, Skirrow G (eds) *Chemical Oceanography*, Vol. 1, Academic, London, UK, p. 494.
61. Davies BH (1976) In: Goodwin TW (eds) *Chemistry and Biochemistry of Plant Pigments*, Vol. 2, Academic, London, UK, p.108.
62. Wiley JP, Wilson TR (1965) *J Mar Biol Assoc* **45**:583.
63. Liebezeit G (1980) *J High Res Chromatogr* **3**:531.
64. Sartory DP (1985) *Water Res* **19**:605.
65. UNESCO (1966) *UNESCO Monographs on Oceanographic Methodology*, No. 1, United Nations Educational, Scientific and Cultural Organization, Paris, France.
66. Jeffrey SW, Humphrey GH (1975) *Biochem Physiol Pflanz* **167**:191.

67. Strickland JDH, Parsons TF (1968) In: *A Practical Handbook of Seawater Analysis*, Fisheries Research Board of Canada, Ottawa, Canada.
68. Bechard MJ, Rayburn WR (1979) *J Phycol* **15**:379.
69. Andreae MO (1980) *Anal Chem* **52**:150.
70. Andreae MO (1977) *Anal Chem* **49**:820.
71. Gadel F, Bruchet A (1987) *Water Res* **21**:1195.
72. Shoaf WT, Lium BW (1976) *J Res US Geol Survey* **4**:241.
73. Martin G (1983) *Rev Francais Sci Eau* **2**:407.
74. Wong SH, Hindin E (1982) *J Am Water Works Assoc* **74**:528.
75. Walters JS, Hedges JI (1988) *Anal Chem* **60**:988.
76. Franco JM, Garrido L (1987) *Chromatographia* **23**:557.
77. Lewis MA, Wang W (2000) *Environ Res Forum* **9**:273.
78. Weiss G, Grauk AH (1975) *Environ Protect Eng* **1**:137.
79. Bates SS, Bird CJ, De Freitas AS, Foxall R, Gilgan M, Hanic LA, Johnson GR, McCulloch AW, Odense P, Pocklington R, Quilliam MA, Sim PG, Smith JC, Subba Rao DV, Todd ECD, Walter JA, Wright JLC (1989) *Can J Fish Aquat Sci* **46**:1203.
80. Wright JLC, Boyd RK, De Freitas ASW, Falk M, Foxall RA, Jamieson WD, Laycock MV, McCulloch AW, McInnes AG, Odense P, Pathak VP, Quilliam MA, Ragan MA, Sim PG, Thibault P, Walter JA (1989) *Can J Chem* **67**:481.
81. Noguchi T, Arakawa O (1996) *Adv Exp Med Biol* **391**:521.
82. Wekell JC, Gauglitz EJ Jr, Barnett HJ, Hatfield CL, Eklund M (1994) *J Shellfish Res* **13**:587.
83. Lundholm N, Skov L, Pocklington R, Moestrup O (1994) *Phycologia* **33**:475.
84. Altwein DM, Foster K, Doose G, Newton RT (1995) *J Shellfish Res* **14**:217.
85. Hasle GR, Lange CB, Syvertsen EE (1996) *Helgol Meeresunters* **50**:131.
86. Fryxell GA, Reap ME, Valencic DL (1990) *Nova Hedwigia Beiheft* **100**:171.
87. Hillebrand H, Sommer U (1996) *J Plankton Res* **18**:295.
88. Buck KR, Uttal-Cooke L, Pilskaln CH, Roelke DL, Villac MC, Fryxell GA, Cifuentes L, Chavez FP (1992) *Mar Ecol Prog Ser* **84**:293.
89. Garrison DL, Conrad SM, Eilers PP, Waldron EM (1992) *J Phycol* **28**:604.
90. Villac MC, Roelke DL, Chavez FP, Cifuentes LA, Fryxell GA (1993) *J Shellfish Res* **12**:457.
91. Martin JL, Haya K, Burridge LE, Wildish DJ (1990) *Mar Ecol Prog Ser* **67**:177.
92. Worms J, Bates SS, Smith JC, Cormier P, Legar C, Pauley K (1991) *Proceedings of the Symposium on Marine Biotoxins*, Centre National d'Etudes Veterinaires et Alimentaires, Paris, p. 35.
93. Douglas DJ, Kenchington ER, Bird CJ, Pocklington R, Bradford B, Silvert W (1997) *Can J Fish Aquat Sci* **54**:907.
94. Wright JLC, Quilliam MA (1995) *Intergovernmental Oceanographic Commission of UNESCO Manuals and Guides*, Vol. 33, IOC, Paris, p. 113.
95. Fritz L, Quilliam MA, Wright JLC, Beale AM, Work TM (1992) *J Phycol* **28**:439.
96. Work TM, Barr B, Beale AM, Fritz L, Quilliam MA, Wright JLC (1993) *J Zoo Wildl Med* **24**:54.
97. Quilliam MA, ie M, Hardstaff WR (1995) *J AOAC Int* **78**:543.
98. Subba Rao DV, Quilliam MA, Pocklington R (1988) *Can J Fish Aquat Sci* **45**:2076.
99. Hummert C, Reichelt M, Luckas B (1997) *Chromatographia* **45**:284.
100. Lawrence JF, Charbonneau CF, Menard C, Quilliam MA, Sim PG (1989) *J Chromatogr* **462**:349.
101. Wu Q, Nelson WH, Treubig, J., JM, Brown PR, Hargreaves P, Kirs M, Feld M, Desari R, Manoharan R, Hanlon EB (2000) *Anal Chem* **72**:1666.

102. Maher WA (1981) *Anal Chim Acta* **126**:157.
103. White JNC, Englar JR (1983) *Botanica Marina* **26**:159.
104. Chau YK, Wong PTS, Bengert GA, Dunn JL (1984) *Anal Chem* **56**:271.
105. Wong PTS, Chau YK, Luxon PL (1975) *Nature* **253**:263.
106. Houpt DM, Campaan H (1972) *Analusis* **1**:27.
107. Schedlbauer OF, Heumann KG (1999) *Anal Chem* **71**:5459.
108. Francois R, Weber JH (1988) *Mar Chem* **25**:279.
109. Kuldvere A (1982) *Analyst* **107**:1343.
110. Whyte JNC, Englar JR (1976) *Analyst* **101**:815.
111. Darich RA, Nelson DW, Sommers LE (1985) *J Environ Qual* **14**:400.
112. Mackereth FJH (1963) *Some Methods of Water Analysis for Limnologists*, Sci. Publ. No. 21, Freshwater Biological Association, Windermere, UK.
113. *Methods for the Determination of Low Level Phosphate in Water* (1973) Water Research Centre Report TIR 272, Medmenham, UK.
114. Harwood JE, Van Steenderen RA, Kuhn AL (1969) *Water Res* **3**:425.
115. Chakrabarty D, Das AK (1988) *Atom Spectrosc* **9**:189.
116. Kersten M, Förstner U (1986) *Water Sci Technol* **18**:121.

4 Pollution Levels in the Aqueous Environment

4.1
Fish

The health of aquatic animals is very dependent on the concentrations of various pollutants in their flesh and organs. It is the purpose of this chapter to summarise available information on pollutant levels in fresh and seawater aquatic animals (Sects. 4.1 and 4.2), aquatic plants (Sect. 4.3) and sediments (Sect. 4.4).

4.1.1
Cations

Results obtained for the determination of cations are summarised in Table 4.1. More detailed information is given in Appendix 4.1.

It will become apparentthat a wide range of concentrations occur in fish or in organs and that, in general, the highest concentrations of metals are found in fish organs rather than whole fish tissue, and this is particularly so for cadmium and lead [1].

The US Safe Water Drinking Act points out that concentrations exceeding 0.05 µg/l of mercury and 0.4 µg/l of cadmium in water will cause levels of these elements in fish and other creatures that may be harmful to aquatic life and human consumers.

4.1.2
Organic Compounds

Hydrocarbons

The occurrence of organic compounds in whole fish tissue is summarised in Tables 4.2 to 4.4. The main classes of compounds investigated so far are alicyclic and aromatic hydrocarbons, polyaromatic hydrocarbons (Table 4.2), various chlorinated compounds, including chlorinated aliphatics and aromatics, polychlorinated biphenyls and chlorinated insecticides (Tables 4.3 and 4.4), and a compound that is currently causing great environmental concern, 2,3,7,8-tetrachlorodibenzodioxin (Table 4.4).

Table 4.1. Metal content of whole fish and fish organs (from author's own files)

Element	Reported values, mg/kg, dry weight								Maximum value in organs / Maximum value in whole creature
	(A) Whole fish tissue				(B) Fish organs				
	Minimum reported value	Fish Type	Maximum reported value	Fish type	Minimum reported value	Organ	Maximum reported value	Organ	
As	Inorganic 0.02	Herring and Haddock	0.44	Smelt					
	Total 1.1	Herring	2.9	Tuna					
Cd	0.02	Sardine	0.17	Horse mackerel	0.038	Gill	9.5 10.9 9.0 7.1	Opercle Skin Liver Kidney	64
Cr	0.10	Grey mullet	2.2	Rainbow trout	0.8	Muscle	23.7 26.0 18.4	Kidney Opercle Liver	12.6
Cu	0.39	Flathead	3.46	Crayfish	0.6	Gill	48 62	Perch Liver White fish Liver	13.9 – 17.9
Cu	0.53	Rainbow trout	2.18	Sardine					
Pb	0.12	Striped mullet	1.36	Grey mullet	0.12	Muscle	36	Kidney	26.5
Mn	0.22	Striped mullet	1.63	Sardine					
Hg	0.09	Chub, crappie	2.4 7.23	Carp Unidentified					
Mo	73.6	Carp	3.6						
Ni	0.15	Rainbow trout	0.2	Rainbow trout	0.34	Trout	1.9	Kidney	9.5
Se	0.19	Shark	0.55	Coho salmon					
Ag	0.02	Whale meat	0.04	Trout					
Ag	0.04	Trout							
Zn	6.3	Sardine Striped mullet	39	Trout	12.6	Liver	150 120 57	Liver Opercle Kidney	2.5 – 3.8

Polyaromatic hydrocarbons are present in the exhaust gases of most vehicles fuelled by heavy hydrocarbon fuels. The total concentrations of these in fish are in the range 0.1 – 5 mg/kg (Table 4.2), and certainly the concentrations at the higher end of this range give cause for environmental concern, not only due to the effect on the fish but also due to the effect on consumers of that fish. The maximum permitted WHO level of polyaromatic hydrocarbons in drinking water is, for example 0.2 µg/l (six compounds[1] fluoranthene, benzo(d)fluoranthene, benzo(k)fluoranthene, benzo(a)pyrene, benzo(ghi)perylene and indeno(1,2,3,ed)-pyrene).

---

[1]   specified by WHO

Table 4.2. Concentrations of hydrocarbons occurring in environmental fish samples, mg/kg dry weight (from author's own files)

Compound	Tuna	Trout	White fish
Pristane	2.4		
Methylcyclohexane		0.002 [2]	
Ethylcyclohexane		0.001 [2]	
Propylcyclohexane		0.002 [2]	
Benzene		0.008 [2]	
Toluene		0.008 [2]	
m/p-xylene		0.005 [2]	
Methyl-3-methyl benzene		0.04 [2]	
1, 3, 5-Trimethylbenzene		0.05 [2]	
1-Methyl-1,4-propylbenzene		0.01 [2]	
2-Ethyl-1,4-dimethylbenzene		0.01 [2]	
2-Methyl-1,4-cyclopentanol		0.090 [2]	
4-Ethyl-1, 2-dimethylbenzene		0.001 [2]	
4-Methylindan		0.002 [2]	
Naphthalene		0.001 [2]	
2-Methylnaphthalene			0.001 – 0.006 [3]
Biphenyl			0.001 – 0.014 [3]
C_2 naphthalenes			0.017 – 0.18 [3]
Acetnaphthalene			0.043 – 0.27 [3]
Acenaphthalene			0.007 – 0.039 [3]
Dibenzothiophene			0.021 – 0.27 [3]
Phenanthracene			0.002 – 2.7 [3]
Methyldibenzothiophenes			0.17 [3]
Fluoranthene			0.004 – 1.8 [3]
Phenanthro(4,5,b,c,d)-thiophene			0.016 – 0.078
Pyrene			0.004 – 1.5 [3]
Benzo(b)naphtha(2,1-d)thiophene			0.006 [3]
Benz(a)anthracene			0.004 – 0.022 [3]
Chrysene			0.003 – 0.061 [3]
Benzo(e)pyrene and benzo(a)pyrene			0.014 [3]
Perylene			0.001 – 0.007 [3]
Phenanthrene plus anthracene			0.008 [3]

4.1.3
Organic Chlorine Compounds

Concentrations of polychlorinated biphenyls in fish can be up to 2.2 mg/kg (Table 4.3), and at this level adverse effects would be expected in birds, fish and humans who consume the fish. Much the same can be said for chlorinated insecticides, which have been detected at 0.05 and 20 mg/kg levels in fish.

2,3,7,8-Tetrachlorodibenzodioxin is one of the more toxic substances produced in the combustion of higher boiling point organic chlorine compounds. As was demonstrated in the Savisesio incident, even minute traces

Table 4.3. Concentration of organic chlorine compounds occurring in environmental fish, mg/kg dry weight (from author's own files)

	Perch	Trout	Pike	Cod	Salmon	White bass	Whale	Herring
Bromochloromethane		0.008 [2]						
Pentachlorobenzene				0.001 – 0.002 [6]				
Hexachlorobenzene				0.08 – 0.17 [6]				
Octachlorostyrene				1.3 – 4.2 [6]				
Polychlorinated biphenyl 1242			0.89 [4]					
Polychlorinated biphenyl 1254			1.01 [4]					
Polychlorinated biphenyl 1260			0.48 [4]					
Polychlorinated biphenyl				2.2 [8]			0.69 – 5.0 [11]	
DDE				2.2 – 20 [8]				
DDD				0.59 – 8.0 [8]				
DDT				0.47 – 7.5 [8]				
Mirex		0.05 – 0.36 [7]	0.05 [4]		0.09 – 0.33 [10]	0.06 – 0.43 [10]	1.25 – 7.4 [11]	
Polychloro-2-(chloromethyl sulfonamide) diphenylether (Eulan WA)	0.3 – 0.33 [5]							
Dieldrin							0.007 – 0.04 [11]	
Toxaphane				1.1 [9] liver				0.4 – 1.0 [9]

Table 4.4. Concentrations of 2,3,7,8-tetrachlorodibenzo-p-dioxin occurring in environmental fish, mg/kg dry weight (from author's own files)

Lake trout	Carp	Ocean herring
< 0.004 – 0.014 [12]	0.001 – 0.094 [13]	< 0.001 – < 0.01 [13]
0.054 – 0.058		

Rainbow trout	Edible fish	Catfish
0.031 – 0.038 [13]	0.48 [14]	0.04 – 0.05 [14]

Buffalo fish	Predator fish	Bottom-feeding fish
< 0.01 [14]	0.015 – 0.23 [14]	0.077 [14]

of this substance in the soil for many miles surrounding its release point had severe health effects. The results in (Table 4.4) indicate that up to 0.48 mg/kg of this substance has been detected in edible fish, and is of concern.

As in the case of metals, certain organic substances tend to concentrate in the organs of fish. For example, the concentration of polychlorinated biphenyl found in the liver of long nose gar (1.11 – 3.7 mg/kg) is appreciably higher than that found in whole fish tissue (0.5 – 1.0 mg/kg) [14].

Rogers and Hall [15] determined PCBs in starry flounder (*Platichthys stellatus*) in muscle, bone and liver in polluted sites. Polychlorobiphenyls have been found in the flesh of starry flounder (*Platichthys flesus*) caught in the Elbe estuary, the German Bight [16] and San Francisco Bay [17].

O'Connor and Pizza [18] studied the pick-up by tissues and the elimination routes of PCBs in striped bass (*Morone saxatilu*) from the Hudson River. In a single close study, measurable quantities of polychlorobiphenyls were detected in tissues six hours after dosing, and peaked 1 – 2 days after dosing. Approximately 53% of the administered dose was eliminated by the fish within 120 hours. Polychlorobiphenyl burdens of the fish increased with successive doses of PCBs.

Concentrations of α-BHC, β-BHC, aldrin, heptachlor, heptachlor expoxide, α-endosulfan, β-endosulfan, α-chlordane and γ-chlordane have been determined in samples of 13 commercially significant fish species caught in the North West American Gulf [19]. Concentrations in all of these in fish tissues were below the analytical detection limit of 1 μg/kg wet weight. DDT was the most prevalent organochlorine pesticide, with average concentrations in fish ranging from 1 to 28 μg/kg. Dieldrin was detected in about 25% of fish species examined at 1 – 4 μg/kg wet weight. Total DDT and endrin residues in fish caught in an insecticide-sprayed lake were 5 – 72 μg/kg and 3 – 67 μg/kg, respectively.

Levels of DDT and PCB have been determined in liver from immature cod species *(Gadus morhua)* and in herring *(Clupea larengus)* muscle [20]. DDT levels in herring muscle between 1979 and 1986 were 0.3 – 2.2 mg/kg, but dropped to 0.010 – 0.017 mg/kg in 1988. Polychlorobiphenyl levels in cod liver between 1979 and 1986 were 0.3 to 3.7 mg/kg, but dropped to 0.013 – 0.19 mg/kg in 1988. Data have been obtained on the concentrations of aldrin, endrin, endosulfan, heptachlor, heptachlor epoxide, lindane and the DDT group in black bullhead, bleak, chub, common carp, eel and tench collected in Italian rivers in 1986 [21]. Aldrin, lindane, heptachlor and endosulfan were detected in less than 20% of the fish examined, whereas dieldrin was found in almost all the fish studied. Total DDT group residue concentrations in fish were between 17 and 153 µg/kg, depending on the river. Other pesticides were at lower concentrations (up to 39 µg/kg).

Fingerling rainbow trout *(Salmo gairdner)* exposed for up to four days to 10 µg/kg of Aminocarb in water at pH 4.6 – 8.2 picked up 9.1 mg/kg Aminocarb in fish tissue in the first six hours of exposure [22]. At pH 8.2, whole body Aminocarb increased to 12 mg/kg in one hour and remained elevated until the fish died in 72 hours.

Various other workers have reported on the concentrations of chlorinated insecticides found in fish and fish tissues, including benzene hexachloride, hexachlorocyclohexane, heptachlor, aldrin, DDT and polychlorobiphenyl, toxaphene [23,24], PCBs, DDT and hydrocarbons [25], polychlorobiphenyls and *p,p'*-DDE [26], and lindane [27].

Table 4.5. Organic compounds in fish (from author's own files)

Compounds	Reference
Aliphatic hydrocarbons	[29,30]
Unsaturated fatty acids	
Phthalate esters	[31]
Volatile chloroaliphatics	[32–39]
Hexachlorobenzene	[40]
Chlorophenols	[42–46]
α,α,α-Trifluoro-4-nitro-*m*-cresol	[47]
Polychlorostyrenes	[48,49]
Polychloronitrobenzene	[50]
Chlorinated insecticides	[6,51–64]
PCB	[64–69]
Mirex	[70,71]
Toxaphene	[72,170]
Nitrogen bases	[73]
Trialkyl and triaryl phosphates	[74]
Organophosphorus insecticides	[75]
Organosulfur compounds	[76]
Dioxins	[77,78]
Squoxin	[79]
Geosmin	*** [79,80]
Fluridone	[81]
Priority pollutants (EPA)	[8]

*** Uthe J, *J Fish Res Board, Ottawa, Canada*, private communication

Concentrations of Eulan WA (polychloro-2-(chloromethyl sulfon-amido)diphenyl ethers) in perch livers (4.5 – 5.5 mg/l) are appreciably higher than those found in the whole fish tissue (0.30 – 0.33 mg/kg) [29].

Diethylhexyl Phthalate

Between 0.002 and 0.02 mg/kg diethylhexyl phthalate, a PVC, plasticiser, has been found in fish [28].

Other Organic Chlorine Compounds

Wong et al. [169] reported on enantioselective bioaccumulation measure-, ments of α-hexachlorocyclohexane, *trans*-chlordane and PCBs in rainbow trout.

Sources of further information on the concentrations of various other organic compounds that have been found in fish are summarised in (Table 4.5).

4.1.4
Organometallic Compounds

Organic compounds of both lead and mercury have been found in fish (Tables 4.6 and 4.7). These originate predominantly from the use of alkyl lead compounds in petroleum and the methylation of inorganic mercury released into the ecosystem as effluents in the chloralkali process.

Organomercury Compounds

Richman et al. [82] have discussed the factors that might govern the uptake of mercury by fish in acid-stressed lakes. It was concluded that mercury cycling and uptake in aquatic systems were governed by a variety of interconnecting and sometimes covarying factors, the relative importance of which could differ from lake to lake.

4.2
Invertebrates

4.2.1
Cations

Available information on the concentrations of metals found in these creatures is given in Table 4.8 (more detailed information and references appear in Appendix 4.2). Again, as in the case of fish, metal concentrations vary over a wide range and certainly cover the region where adverse effects or

Table 4.6. Concentrations of organolead compounds found in environmental fish, mg/kg dry weight (from [86])

	Miscellaneous fish	Carp	Bass	Small moult bass	Pike	White sucker	Range
Me_4Pb	0.43 [218]	0.14					0.14 – 0.43
Me_3EtPb		< 0.001	< 0.001				< 1.001
Me_2Et_2Pb		1.43		0.057	0.10		0.057 – 1.43
$MeEt_3Pb$		0.14		0.19 – 0.25	0.15 – 0.17	0.29	0.14 – 0.29
Et_4Pb		0.78 – 7.5		1.20 – 1.83	1.02 – 1.12	2.95 – 4.38	0.78 – 7.5
Me_3Pb^+		0.16 – 2.73		< 0.01	0.20 – 0.21	0.09 – 0.20	< 0.01 – 2.73
Me_2Pb^{2+}		0.36					0.36
Et_3Pb^+		0.09 – 1.21		0.22 – 0.86	0.053	2.17 – 2.43	0.53 – 3.43
Et_2Pb^{2+}		0.71 – 1.31		0.09 – 2.75		2.2 – 4.3	0.09 – 4.3
Pb^{2+}		1.28 – 4.13		0.25 – 0.30	1.04 – 1.19	3.61 – 3.48	0.25 – 4.13
Total excluding Pb^{2+}		5.09 – 18.94	< 0.01	1.76 – 5.55	1.52 – 1.65	7.70 – 12.60	

Table 4.7. Concentrations of organomercuy compounds found in environmental fish, mg/kg dry weight (from author's own files)

	Whiting	Sardine	Turbot	Halibut	Coho salmon	Salmon	Red tuna	White tuna
$MeHg^+$								0.93 [86]
$EtHg^+$								< 0.01 [86]
$PhHg^+$								< 0.01 [86]
CH_3HgCl	0.08 [84]	0.03 [84]	< 0.01 [84]	5.65 [85]	0.18 – 0.20 [92]	0.06 – 0.11 [170]	1.89 – 8.3 [87]	0.54 [84] 0.61 [85] 0.33 [87] 0.08 [88] 0.36 [89] 0.35 – 0.74 [90] 0.15 [93] 0.3 – 11.5 [92]

	Pike	Trout	Rainbow trout	Whale	Shark	Swordfish	Octopus	Squid
CH_3HgCl	0.11 – 0.88 [90,170]	0.06 – 1.46 [85,88,92]	0.05 – 1.97 [88,92]	8.41 [88,92]	0.177 [86,91]	0.57 – 1.01 [84,87]	< 0.01 [84,87]	< 0.01 [84]

Table 4.8. Metal contents of sea creatures other than fish, mg/kg (from author's own files)

Element	Oyster	Lobster	Crab	Whelk	Canned lobster	Prawn	Scallop	Mollusc	Mussel	Clam	Shrimp	Range
Sb	0.4	0.071–0.089										0.071–0.4
As	13.4	11.9–25.5 (13.4 hepatopancreas)	1.5 (total) 0.06–0.1 (inorganic)	3.2 (total) 0.06–0.18 (inorganic) 26 (total) 0.10–0.18 (inorganic)	3.6 (total) 0.06–0.08 (inorganic)	14 (total) 0.02–0.04 (inorganic)	7.0–7.7	2–23.2				0.02–26
Bi	0.0042								0.0007–0.0025			0.0007–0.0042
Br		50.6–51.7										50.6–51.7
Cd	0.0025–2.56	0.5–1.1 3.5 (hepatopancreas)	0.07–7.0						0.02–20.2	1.3	0.07–0.24	0.0025–20.2
Cr		0.75										0.75
Co	0.34–0.44	0.34–0.44										0.34–0.44
Cu		0.75–2.65 63 (hepatopancreas)	0.75–2.65						0.75–2.65		0.75–2.65	0.75–63
Fe		212–219										212–219
Pb	0.48–0.61	0.11–3.2 2.5–12.4 (hepatopancreas)	0.48–2.8						0.48–0.61	0.83	0.48–0.61	0.48–12.4
Mn		16.57 17.5 (hepatopancreas)										16.6–57
Hg	0.14–0.16	0.16–0.31 (hepatopancreas)	0.02–0.05					0.1	0.02–0.05		0.02–0.05	0.02–0.16
Ni		0.98 19.4 (hepatopancreas)										0.98–19.4
Pu										0.3–13.9		0.3–13.9
Se	1.7–2.26	2.0–6.7				4.01	0.71–1.24					0.71–6.7
Sc		0.015										0.015
Ag	0.86–0.93	0.86–0.93										0.86–0.93
Sr		11.0 84.9 (hepatopancreas)										11.0–84.9
V	0.53–1.42		1.90–1.84									
Zn		548–888 852 (hepatopancreas)										

mortalities in the creatures would occur and where the suitability of the creature for human consumption would be queried.

High and variable concentrations of cadmium have been reported in the tissues of the mollusc *Murex trunlus* taken in Calvi Bay, Corsica, during the tourist season [94]. Doherty et al. [95] have reported the occurrence of a metallothionein-like metal-binding protein in the soft tissues of Asiatic clams following exposure to dissolved cadmium and zinc. It was found that clams exposed to dissolved cadmium had higher concentrations of dissolved cadmium and metal binding protein in the gill, mantle and aductor muscle. Tissue concentrations increased with time of exposure.

Drabaek et al. [96] has reported concentrations of rare earth elements in the tissues of mussels (*Cyprina islandica, Mytilus edulis*) and flounder fish caught in waste waters discharging into the sea from a fertiliser production plant located at Lillebaelt, Denmark. Samples of the barnacle *Balanus amphitrite* collected in the Zuan Estuary, India, had zinc contents in tissue of 203.6 – 1937.5 mg/g. The zinc content of the overlying waters was 13 – 46 µg/l [97]. Gil et al. [98] has reported iron, zinc, manganese, copper, cadmium, lead and nickel concentrations in scallops (*Chalamys tehuelcha*) and mussels (*Aulacomya ater* and *Mylitus platensis*) from a rural uncontaminated site in San Jose Gulf, and from an urban industrialised site in the Nuevo Gulf, both in Argentina. In scallops taken in both areas, iron, manganese, copper and cadmium were primarily concentrated in the liver and kidney. Zinc was concentrated in the mantle and gills. Nickel and lead were below detection limits.

Lyngby and Brix [99] determined the heavy metals in mercury, cadmium, zinc, lead and copper in mussel tissues (*Mytilus edulis*), and compared values obtained with those obtained in eelgrass (*Zostera marina*). The object of this study was to compare mussels and eelgrass as biological indicators of water pollution. Metal contents of oysters taken from Darwin Harbour have been determined, including lead, nickel, copper, cadmium and iron [100].

4.2.2
Organic Compounds

A variety of organic substances can occur in sea creatures other than fish, as they do in fish (Table 4.9). It is seen that in many instances the concentrations in fish and in creatures other than fish are of a similar order of magnitude.

Rice and White [108] determined the concentrations of polychlorobiphenyls (PCBs) in fathead minnows (*Pimephales promelas*) and fingernail clams (*Sphaerium striatinum*) before, during and up to six months after the completion of the dredging of polychlorobiphenyl-contaminated sediments in the Shiawasse River, USA, 1 km downstream of the pollution outfall. The concentrations of PCBs found in fathead minnow tissue were 32.1 – 61.1 mg/kg dry weight and the concentrations in clams were

Table 4.9. Occurrences of organic substances in creatures other than fish (from author's own files)

Substance determined	Type of creature	Concentration found, mg/kg	Reference	See Tables 4.4 and 4.8 (concentration found in fish for comparison), mg/kg	Reference
Polyaromatic hydrocarbons					
Naphthalene	Mussel	0.003 – 0.1	[8]	0.001	[2]
Phenanthrene and anthracene		0.008 – 0.032	[8]	0.008	[15]
Fluoranthrene		0.042 – 0.080	[8]	0.004 – 1.8	[3]
Pyrene		0.034 – 0.092	[8]	0.004 – 1.5	[3]
Benz(a)anthracene and chrysene		0.029 – 0.059	[8]	0.004 – 0.022	[3]
Benz(a)pyrene		0.55	[102]	0.007 – 0.083	[3]
PCB		0.41 – 0.9	[8]	0.48 – 5.0	[8, 101]
		0.011 – 0.56	[103]		
	Oyster	0.0002	[104]		
Polychloroterphenyl		0.00015	[104]		
Dibenzothiophane	Mussel	0.0001 – 0.8	[105]	–	
Diethylhexylphthalate	Shrimp, crab	0.003 – 0.02	[31]	–	
Aliphatic hydrocarbons	Mussels	0.54	[106]		
	Oyster	0.65	[106]		
	Clam	0.49 – 1.41	[106]		

13.2 – 15.3 mg/kg dry weight. It was concluded that dredging had increased the bioavailability of PCBs to these organisms.

Organochlorine Insecticides

Exposure of (*Asellus aquaticus*) crustacea to water containing 5 µg/l of lindane for five days led to a pick-up by the organism of 0.2 mg/kg lindane in the tissue [164]. Depuration was rapid, with over 40% of accumulated lindane being eliminated within one day.

The bioconcentration factors of short-chain PCBs (polychlorinated decane, 69% chlorine) reached a value of 1.4×10^5 in mussels [107]. Other information on organic compounds is summarised in Appendix 4.3.

4.2.3
Organometallic Compounds

Organotin Compounds

Bailey and Davies [109] determined tributyl tin concentrations in dogwhelk samples taken at various locations in Sullom Voe, Shetland. These ranged from values of 0.1 mg/kg inside Sullom Voe down to less than 0.03 mg/kg in Yell Sound. Concentrations of 0.02–0.03 mg/kg were found in edible tissue of queen scallops inside the Voe, but tin was rarely detected in commercial shellfish outside the Voe. Only very low concentrations of tributyl tin ($\sim$2 ng/l) were found in a small proportion of seawater samples taken in the area.

4.3
Water Plants

4.3.1
Cations

The limited information available on the occurrence of metals in phytoplankton, algae and weed is reviewed in Table 4.10. Significant correlations have been found between the concentrations of mercury, lead, cadmium

Table 4.10. Metal and nonmetal contents of algae, phytoplankton and weeds (from author's own files)

Determined	Type of sample	Concentration, mg/kg unless otherwise stated)	Reference
		(a) Algae	
Aluminium	Plankton, algae	12 µg/l	[110]
Arsenic	Algae	20.0 – 56.1	[111]
Bismuth	Marine algae	3 pg absolute	[111]
Bismuth	Macrocystis	0.089	[112]
Bromine	Plankton, algae	0.3 µg/l	[110]
Chromium	Algae	40 – 630	[110]
Copper	Algae	50 – 660	[110]
Iron	Algae	340 – 9720	[110]
Mercury	Plankton, algae (Lake Eyrie)	31 – 81	[112]
Manganese	Algae	230 – 4170	[114]
Tin	Algae	0.03 – 1.06	[113]
Tin	Macroalgae (Narragansett Bay, CA, USA)	0.03 (inner tissue)	[113]
		0.83 (algal blade)	[113]
Zinc	Algae	20 – 700	[120]
Iron	Freshwater	340 – 9720	[114]
Manganese	Plankton	87 – 4170	[114]

Table 4.10. Continued

Determined	Type of sample	Concentration, mg/kg unless otherwise stated)	Reference
Copper	*Platihypnidium riparoides*	40- 690	[114]
Chromium	*Olea europa*	40 – 630	[114]
Zinc	*Lagarosiphon major*	20- 700	[114]
		(b) Plankton	
Chromium	Plankton	60 – 70	[112]
Copper		40	[112]
Iron		3700 – 3800	[114]
Manganese		230 – 250	[114]
Mercury		31.2 – 81.0	[112]
Rare earths			
Lanthanum		0.15	[112]
Cerium		0.24	[119]
Neodymium		0.03	[119]
Samarium		0.006	[119]
Gadolinium		0.019	[119]
Europium		0.003	[119]
Dysprosium		0.008	[119]
Ytterbium		0.001	[119]
		(c) Weeds	
Bismuth	Kelp	0.005	[111]
	Macrocystis	0.009	[111]
Thorium	Seaweed (Baltic)	0.01 – 0.06	[115]
Uranium		0.07 – 0.41	[115]
Copper	Sea plant, copod	198 μmol	[116]
		121 μmol/kg	
Zinc		979 μ/kg	
		367 μmol/kg	
Mercury		1.7 μ/kg	[116]
		1.4 μmol/kg	
Cadmium		6.2 μ/kg	[116]
		6.7 μmol/kg	
Cadmium	Seagrass, *Posidonia australis* (near lead smelter)	198 – 541	[117]
Manganese		112 – 537	
Lead		116 – 379	
Zinc		728 – 4241	
Potassium	Macrophyte, *Juncus bulbosus*	0.83 – 26.8 mg/g dry weight	[118]
Sodium		0.39 – 11.7	
Magnesium		1.5 – 2.85	
Nitrogen		19.4 – 25.5	
Phosphorus		0.14 – 0.29	
Manganese		3.40 – 17.17	
Iron		4.47 – 35.81	
Zinc		0.12 – 0.42	

and zinc in Limfjord, Denmark, and the concentrations of these elements found in eelgrass leaves and root rhizomes (*Zostem marino*) present in the fjord [99].

Enhanced levels of metals can occur in algae and weeds in areas of high pollution, making them useful indicators of such pollution.

4.3.2
Organic Compounds

Miscellaneous Organic Compounds

Limited information is available on the concentrations of various organic compounds that occur in plankton, namely aliphatic hydrocarbons [121], phenols [122], unsaturated fatty acids [123], carbohydrates [133], chlorinated insecticides [124–128], adenosine triphosphate [129,130], organophosphorus insecticides [129, 130], humic and fulvic acids [131], and anatoxin A [132].

4.3.3
Organometallic Compounds

Organoarsenic and Organolead Compounds

Organoarsenic and organolead compounds have been found in algae and plankton (Table 4.11).

Table 4.11. Concentrations of organometallic compounds found in algae and plankton (from author's own files)

Compound	Type of sample	Concentration, mg/kg unless otherwise stated		Reference
Organoarsenic compounds	*(a) Algae* Macroalgae	0.2 – 0.6 monomethyl arsenic and 7.6 – 15.6 dimethyl arsenic acid		[120]
	Marine brown algae (*Haminariaceae*)	40.3 – 89.7		[133]
Organolead compounds	*(b) Plankton* Macrophytes	4 m-deep samples	Surface water samples	[171]
		Me_3EtPb – 0.038	–	
		Me_2Et_2Pb – 1.5	–	
		$MeEt_3Pb$ – 3.61	–	
		Et_4Pb – 16.5	0.07	
		Et_3Pb^+ – 0.59	0.13	
		Et_2Pb^{2+} – 0.11	–	
		Total Pb – 59.2	4.3	

4.4
Sediment

Pollutants entering freshwater and oceans remain partly in solution and are partly absorbed onto the surface of sedimentary matter. Both sources of pollution, i.e. dissolved and sedimentary, are capable of entering living creatures with possible adverse effects. The concentration of a toxicant in sediment is a measure of its concentration in the water over a period of time and is, therefore, a measure of its risk to creatures. In the case of bottom-feeding creatures, there is the additional risk of direct ingestion of sediments via the gills and mouth, with consequent adverse effects. The concentrations of particulate pollutants are considered below.

Because of the tendency of pollutants to concentrate in sediments, their concentrations in the latter can be appreciably higher than in an equivalent volume of water. For these reasons, much work has been carried out on the determination of toxicants in sediments.

4.4.1
Cations

Cherry et al. [134] carried out a study to identify the toxic sediments of fly ash and bottom ash obtained from the Glen Lyn Power Station Plant, Virginia, in acute laboratory bioassays using a warm-water rainbow trout (*Lepomis macrochirus*) and a cold-water bluegill sunfish species to evaluate the surface availability of trace elements at various pH values. Rainbow trout (*Salmo gairdneri*) were highly sensitive to fly ash when dissolved metal availability was high, but not to high particulate concentrations (up to 2350 mg/l dissolved solids) when metals were removed. Bluegill were much less sensitive to cadmium, chromium, copper, nickel, lead and zinc. Both species were acutely sensitive below pH 4.0 and above pH 9.1.

Hammer et al. [135] have studied the effect of low dissolved oxygen concentrations in eutrophic lakes on the release of mercury from sediments and subsequent bioaccumulation by aquatic plants (*Ceratophyllum demersum*) and clams (*Anodonto grandis*). The mercury concentration in plants and clams present in water which a reduced dissolved oxygen content (1.8 mg/l) was considerably higher than in water has a higher dissolved oxygen content (6.7 – 7.2 mg/l).

Tables 4.12 and 4.13 summarise the concentrations of metallic elements that have been found in freshwater and seawater sediments. A more detailed breakdown of the results obtained for freshwater sediments can be found in Appendix 4.4.

List 1 in Table 4.12 shows the results obtained for toxic elements that have been discussed in various EU directives. List 2 covers the major naturally occurring elements, and List 3 the minor elements, many of which occur naturally and most of which are of little toxicological concern.

Table 4.12. Elements in freshwater sediments (from author's own files)

Element	Concentration, mg/kg	
	River sediments	Lake water sediments
(1) *Elements covered in EU directives*		
Al	9,890 – 46200	26,200 – 63,000
As	0.22 – 7.1	1.9 – 26
Sb		0.01 – 2.9
Ba		163 – 2,700
Cd	0.06 – 27.5	3.5 – 40
Cr	0.48 – 1143	16 – 110
Co	1.8 – 53	3.9 – 200
Cu	0.07 – 244	50
Pb	0.11 – 5,060	20 – 180
Hg	0.91 – 46.8	1.95 – 6.8
Ni	1.4 – 238	1 – 218
Se	0.09 – 0.93	0.03 – 1.0
Ag	1.5.53	0.1 – 8.05
Ti		800 – 3,800
U		0.78 – 4.3
V		28 – 68
Zn	0.31 – 9040	10 – 450
(2) *Naturally occurring elements*		
Br		23 – 96
Ca		12,300 – 40,000
Cl		20 – 609
Fe	16.9 – 31,000	14,700 – 30,600
Li		50
Mg		5,900 – 16,800
Mn	0.34 – 9640	214 – 4,500
P	675 – 1,870	
Na		3000 – 9200
Sr		10 – 242
(3) Minor elements (*few or no toxicity data*)		
Ru		19 – 49
Cs		0.5 – 14
Au		0.25 – 19
Th		4.0 – 9.4
Hf		1.7 – 12
Zr		55 – 488
In		5.3 – 19
Ru		45 – 500
Sc		3.3 – 9.2
Ta		0.4 – 1.4
Tm		0.19 – 7.4
Ce		53 – 160

Table 4.12. Continued

Element	Concentration, mg/kg	
	River sediments	Lake water sediments
Yb		2.3 – 9.3
Dy		5.3 – 15
Gd		6.4 – 22
La		28 – 73
Tb		0.95 – 2.4
Nd		15 – 137
Sm		7.9 – 28
Ir		0.5 – 48
Os		1 – 4.5
Pt		0.3 – 8.1

Table 4.13. Metals in marine sediments (from author's own files)

Element	Location	Concentration, mg/kg		Reference
Bismuth	Narragonsett Bay,	Surface	0.40	[112]
	USA	49 – 54 mm core	0.27	[112]
	Pacific		0.1	[136]
Mercury		Sand	< 0.1 – 1.4	
		Clay	< 0.1 – 0.8	
	River Loire estuary, salinity 20 – 35%.		13.2	[137]
	River Loire 0 – 10 km, upstream of estuary.		28.0	[137]
	River Loire 10 – 15 km upstream of estuary.		22.9	[137]
	River Loire 15 – 30 km upstream of estuary.		46.8	[113]
Tin	Narragonsett Bay,	1 cm core	20	[113]
	USA	80 cm core	1	[113]
Lanthanum	Deep-sea sediments		65.1	[119]
Cerium			91.0	
Neodymium			92.5	
Samerium			22.9	
Europium			5.7	
Gadolinium			25.2	
Dysprosium			23.0	
Erbium			13.4	
Ytterbium			13.1	

Table 4.14. Accumulation of metals from Severn and Humber estuary waters into sediments (from author's own files)

	Accumulation factor = $\dfrac{\mu g/kg \text{ weight of toxicant sediment}}{\mu g/l \text{ of toxicant in water}}$		
Metal	Copper	Lead	Nickel
Severn estuary	15,710 – 16,300	26,830 – 67,330	15,280 – 22,600
Humber estuary	57,350 – 430,000	68,000 – 136,000	2,130 – 3,200
Metal	Zinc	Arsenic	Cadmium
Severn estuary	13,090 – 25,640	–	1,280 – 3,230
Humber estuary	4,060 – 102,500	3,700 – 37,000	800 – 4,000

During the accumulation of cations in sediments as mentioned above, there is a very strong tendency for metals and organic compounds present in waters to concentrate in sediments in those waters. This process is known as bioaccumulation. Two competing factors operate in bioaccumulation: the rate of uptake of metals or organics by sediment, and the rate of loss. These will govern whether there is a net decrease or increase in the toxicant content in the sediment with the passage of time.

Bioaccumulation may be expressed as an accumulation factor as follows:

$$\text{Accumulation factor} = \frac{\mu g/kg \text{ weight of toxicant sediment}}{\mu g/l \text{ of toxicant in water}}$$

Table 4.14 presents some accumulation factors for metals from Humber and Severn estuary waters into sediments.

Let us take the case of cadmium in the Humber estuary. It is seen in Table 4.4 that for this element a range of values of 800 – 4000 $\mu g/kg$ has been obtained for the concentration ratio:

$$\frac{\mu g/kg \text{ (in sediments)}}{\mu g/l \text{ (in water)}} = \frac{\mu g/kg}{\mu g/l}.$$

The range of cadmium content found in river sediment (Table 4.12) is 60 – 27,500 $\mu g/kg$. When the river is relatively unpolluted, i.e. the cadmium content of the sediment is 60 $\mu g/kg$, then:

$$\frac{\mu g/kg \text{ (in sediment)}}{\mu g/l \text{ (in water)}} = 800 \text{ to } 4,000.$$

Concentration of cadmium in water (60 $\mu g/kg$ Cd in sediment):

$$60 = V_o = 60/4,000 = 0.075 \text{ to } 0.015 \text{ mg/l.}$$

When the river water is more polluted, i.e. there is a higher cadmium content of sediment and concentration of cadmium in the water (25.5 mg/kg Cd in sediment):

$$27,500/800 \text{ to } 27,500/4,000 = 34.3 \text{ to } 6.87 \, \mu g/l.$$

The S_x value (i.e. the maximum safe concentration of cadmium in water for survival of nonsalmonids for periods exceeding one year (see Table 10.5, Chap. 10) is $4 \, \mu g/l$.

Thus it is apparent that waters over sediments containing $60 \, \mu g/kg$ cadmium would enable fish to survive long-term, as the cadmium content of the water would be $0.075 - 0.015 \, \mu g/l$, i.e. below the safe limit of $4 \, \mu g/l$, while waters over sediments containing $27,500 \, \mu g/kg$ cadmium would not enable fish to survive long-term, as the cadmium content of the water would be $34.3 - 6.87 \, \mu g/l$, i.e. above the safe limit of $4 \, \mu g/l$.

Similar calculations (Table 4.15) show that for the higher concentrations of metals in sediments, reflecting as they do higher average of metal concentrations in water over a period of time, toxic effects towards fish would also be expected for lead at the higher end of the range, while toxic effects might not be expected for these elements at lower levels or for arsenic/nickel at any of the concentrations studied.

Table 4.15. Survivals of nonsalmonid fish in Humber and Severn estuaries (from author's own files)

Metal		Copper	Lead	Nickel	Zinc	Arsenic
Accumulation factor	Min.	15,710	26,830	2,130	4,060	3,700
(Table 4.14). $\dfrac{\mu g/kg \text{ in sediment}}{\mu g/l \text{ in water}}$	Max.	430,000	136,000	32,000	102,500	37,000
Concentration in	Min.	70	110	140	310	22
sediment (Table 4.12), $\mu g/kg$	Max.	244,000	5,060,000	238,000	904,000	7,100
Calculated concen-	Min.	0.0002*	0.00081*	0.044*	0.0030*	0.0059*
tration in water. $\dfrac{\left(\dfrac{\mu g/kg \text{ in sediment}}{\text{accumulation factor}}\right)}{}$	Max.	15.53	188.5	111.7*	222.7	0.19*
$5 \times 5 \times \mu g/l$		4	20	220	23	80

* Satisfactory

4.4.2
Organic Compounds

Following an aviation kerosine spill, hydrocarbons were detected in trout stream sediments and fish up to 14 months after the spill [138]. After a fire at a weed treatment plant in 1970, a large area of mixed forested ecosystem became contaminated with polycyclic aromatic hydrocarbons and creosote [139]. High polyaromatic concentrations in stream sediments adversely affected micro- and meiobenthic communities at all trophic levels. Stein et al. [140] has studied the uptake by benthic fish (English sole, *Parophrys vetulus*) of benzopyrene and polychlorinated biphenyls from sediments. Accumulation of contaminants from sediments was a significant route of uptake by English sole. It has been shown that microorganisms in river sediments rapidly dechlorinated PCBs [148].

Table 4.16. Organic compounds in river, lake and marine sediments (from author's own files)

Compound	Type of sediment	Concentration, mg/kg	Reference
Aromatic hydrocarbons	Freshwaters	0.001 – 3	[143, 145]
1,3-dihexachlorobutadiene	River and lake	0.05	[143, 145]
1,3,5-trihexachlorobutadiene		0.25	[143, 145]
1,2,4-trihexachlorobutadiene		0.07	[143, 145]
1,2,3-trihexachlorobutadiene		0.10	[143, 145]
1,2,3,5-tetrahexachlorobutadiene		0.01	[143, 145]
1,2,3,4-tetrahexachlorobutadiene		0.27	[143, 145]
Pentachlorobutadiene		0.15	[143, 145]
Hexachlorobutadiene		1.2	[144]
1,3-Dichlorobenzene	Marine	0.031	[144]
1,4-Dichlorobenzene		0.081	[144]
1,3,5-Trichlorobenzene		0.004	[144]
1,2,4-Trichlorobenzene		0.020	[144]
1,2,3,5-Tetrachlorobenzene		0.004	[144]
1,2,4,5-Tetrachlorobenzene		< 0.001	[144]
Chlorobenzenes	Estuary	0.003 – 0.07	[152]
Perchlorobenzene		0.004	[144]
Hexachlorobenzene		0.007	[144]
Diethylhexylphthalate		0.1 – 70.5	[146]
Dibutyl phthalate		< 0.1 – 15.5	[146]
Alkyl benzene sulfonates		16.9 – 96.3	[147]
Methylene blue active substances		107 – 288	[147]
Nitrogen-containing aromatics	Marine	200 – 1200	[148]
Fluorescent whitening agents		0.25 – 1.35	[147, 150]
Total organic carbon		2.4 – 65.6	[149]
		1.4 – 6.2	[151]
Total phosphorus		610 – 1870	[161]
Total sulfur		229	[162]

Sediments containing 50 – 1600 mg/kg of triphenyl phosphate altered the drift dynamics of benthic invertebrates. Invertebrates exposed to contaminated sediments drifted almost immediately when threshold toxicity was reached [142].

Some data on the occurrence of organic compounds in river, lake and marine sediments are given in Table 4.16. References to further information are available in Appendix 4.5. Chlorinated aliphatic and aromatic compounds as well as phthalate esters are among the compounds that have been detected in sedimentary matter.

4.4.3
Organometallic Compounds

Butyl and cyclohexyl tin compounds have been found in river and lake sediments (Table 4.17). These probably originate from the use of organotin antifoulants on boats and pier works [154]. Further information on the occurrence of organometallic compounds is given in Table 4.18.

Table 4.17. Organotin compounds in river and lake sediments (from author's own files)

Compound	Concentration, mg/kg
$BuSn^{3+}$	0.055
$BuSn^{2+}$	0.14
$BuSn^{+}$	0.28
$Cyclohexyl_2Sn^{2+}$	0.01
$Cyclohexyl_2Sn^{+}$	0.075

Table 4.18. References for organometallic compounds in sediments (from author's own files)

	River	Lake	Marine
Arsenic			[120]
Lead	[73]	[155]	[25]
Mercury	[156–159]	[160]	
Silicon	[161]		
Tin	[161–166]	–	[161, 168]

References

1. Behar JY, Schuck A, Stanley RE, Morgan GB (1979) *Environ Sci Technol* **13**:34.
2. Hiatt MH (1983) *Anal Chem* **55**:506.
3. Vassilaros DL, Stoker PW, Booth GM, Lee ML (1982) *Anal Chem* **54**:106.
4. Kaiser KLE (1974) *Science* **185**:523.
5. Ichinose N, Addachi K, Schwedt G (1985) *Analyst* **110**:1505.
6. Norheim G, Okland EM (1980) *Analyst* **105**:990.
7. Sperling KR (1982) *Fresen Z Anal Chem* **58**:2041.
8. Ozretich RJ, Schroeder WP (1986) *Anal Chem* **58**:2041.
9. Musial CJ, Uthe JF (1983) *Int J Environ Anal Chem* **14**:117.
10. Laseter JL, DeLeon R, Remele PC (1978) *Anal Chem* **50**:1169.
11. Gaskin DE, Smith GJD, Arnold PW, Louisy MV, Frank R, Moldrinet M, McWade JW (1974) *J Fish Res Board Canada* **31**:1235.
12. Lamparski LL, Nestrick TJ, Stehl RH (1979) *Anal Chem* **51**:1453.
13. Smith LM, Stalling DL, Johnson JL (1984) *Anal Chem* **56**:1830.
14. Mitchum RK, Moier GF, Korfmacker WA (1980) *Anal Chem* **52**:2278.
15. Rogers IH, Hall KJ (1987) *Water Pollut Res J Canada* **22**:197.
16. Luckas B, Harms U (1987) *Int J Environ Anal Chem* **29**:215.
17. Spies RB, Rice DW (1988) *Mar Biol* **98**:191.
18. O'Connor JM, Pizza JC (1987) *Estuaries* **10**:68.
19. Douabul AAZ, Al-Saad HT, Al-Obaidy SZ, Al-Rekabi HN (1987) *Water Air Soil Pollut* **35**:187.
20. Haahti H, Perttila M (1988) *Mar Pollut Bull* **19**:29.
21. Amodio-Cocchiero R, Arnese A (1988) *Bull Environ Contam Toxicol* **40**:233.
22. Doe KG, Ernst WR, Parker WR, Julien GRJ, Hennigar PA (1988) *Can J Fish Aquatic Sci* **45**:287.
23. Venant A, Cumont G (1987) *Environ Pollut* **43**:163.
24. Gooch JW, Matsumura F (1987) *Arch Environ Contam Toxciol* **16**:349.
25. Albaiges J, Farran A, Soler M, Gallifer A, Martin P (1987) *Mar Environ Res* **22**:1.
26. Devaux A, Monod G (1987) *Environ Monit Assess* **9**:105.
27. Cossarini-Dunier M. Monod G, Demael A, Lepot D (1987) *Ecotoxicol Environ Safety* **13**:339.
28. Giam CS, Chau HS, Neff GS (1975) *Anal Chem* **47**:2225.
29. Smith DH (1972) *Anal Chem* **44**:536.
30. Chesler SN, Gump BH, Hertz HS, May WE, Wise SA (1978) *Anal Chem* **50**:805.
31. Giam CS, Chau HS, Neff GS (1975) *Anal Chem* **47**:2225.
32. Itoh K, Chikuma M, Tanaka H (1988) *Fresen Z Anal Chem* **330**:600.
33. Sinex SA, Cantillo AY, Helz GR (1980) *Anal Chem* **52**:2342.
34. Cantillo AY, Sinex SA, Helz GR (1984) *Anal Chem* **56**:33.
35. Sturgeon RE, De Sauliniers JAH, Berman SS, Russell DS (1982) *Anal Chim Acta* **134**:283.
36. McQuaker NR, Kluckner PD, Chang GN (1979) *Anal Chem* **51**:888.
37. Walsh JN, Howie RA (1980) *Mineral Mag* **43**:967.
38. Parejko RW, Johnston R, Keller R (1975) *Bull Environ Contam Toxicol* **4**:480.
39. Solomon J (1979) *Anal Chem* **51**:1861.
40. De Leon IR, Maberry MA, Overton EB, Roschke CK, Remele PC, Steele CF, Warren VL, Leister JL (1980) *J Chromatogr Sci* **18**:85.
41. Johnson JL, Stalling DL, Hogen JW (1974) *Bull Environ Contam Toxicol* **11**:393.
42. Stark A (1969) *J Agric Food Chem* **17**:871.

43. Rudling L (1970) *Water Res* **4**:533.
44. Renberg L (1974) *Anal Chem* **46**:459.
45. Hoben HJ, Ching SA, Casarett LJ, Young RA (1976) *Bull Environ Contam Toxicol* **15**:78.
46. Sackmauerova M, Vennigerova M, Uhnak J (1981) *Vodni Hospodarstvi Series B* **31**:133.
47. Allen JL, Sills JB (1974) *J AOAC* **57**:387.
48. Kuehl DW, Kopperman HL, Veith GD, Glass GE (1976) *Bull Environ Contam Toxicol* **16**:127.
49. Steinwandter H, Zimmer L (1983) *Fresen Z Anal Chem* **316**:705.
50. Steinwandter H (1987) *Fresen Z Anal Chem* **326**:139.
51. Markin GP, Hawthone JC, Collins HL, Ford JH (1974) *Pest Monit J* **7**:139.
52. Frank R, Armstrong AF, Boelus RG, Braun HH, Douglas CN (1974) *Pest Monit J* **7**:165.
53. Hesselberg RJ, Johnson JL (1972) *Bull Environ Contam Toxicol* **7**:115.
54. Simal J, Crous Vidal J, Maria-Chareo Arias A, Boado MA, Diaz R, Vilas D (1971) *An Bromat (Spain)* **23**:1.
55. Chau ASY (1972) *J AOAC* **55**:519.
56. Kuehl DW (1977) *Anal Chem* **49**:521.
57. Langlois RE, Stamp AR, Liska BJ (1954) *Milk Food Technol* **27**:202.
58. Luckas B, Pscheidl H, Haberland P (1978) *J Chromatogr A* **147**:41.
59. Luckas B, Pscheidl H, Haberland P (1976) *Nahrung* **20**:K1-K2.
60. Neeley WB (1977) *Sci Total Eviron* **7**:117.
61. Frederick LL (1975) *J Fish Res Board Canada* **32**:1705.
62. Jan J, Malnersic S (1978) *Bull Environ Contam Toxicol* **19**:772.
63. Olsson M, Jenson S, Reutergard L (1978) *Ambio* **7**:66.
64. Szelewski MJ, Hill DR, Speigal SJ, Tifft EC (1979) *Anal Chem* **51**:2405.
65. Ludke JL, Schmitt CJ (1980) In: *Proceedings of the 3rd USA-USSR Symposium on the Effect of Pollutants upon Aquatic Ecosystems, Theoretical Aspects of Aquatic Toxicology*, US Environmental Protection Agency, Duluth, MN, USA, p. 97-100.
66. Sackmauerova M, Pal'Usova O, Szokolay A (1977) *Water Res* **11**:551.
67. Tausch H, Stehlik G, Wihlidal H (1981) *Chromatographia* **41**:403.
68. Bush B, Barnard EL (1982) Anal Lett **15**:1643.
69. Tuinstra LGMT, Dreissen JJM, Keukens HJ, Van Munsteren TJ, Roos AH, TRaag WA (1983) *Int J Environ Anal Chem* **14**:147.
70. Armour JA, Burke JA (1970) *J AOAC* **53**:761.
71. Gaul J, Cruze-LaGrange P (1971) *Separation of Mirex and PCBs in Fish*, Laboratory Information Bulletin, Food and Drug Administration, New Orleans, LA, USA.
72. Hughes RA, Lee GF (1973) *Environ Sci Tech* **7**:934.
73. Chau YK, Wong PTS, Bengert GA, Kramar O (1979) *Anal Chem* **51**:186.
74. Murray DAJ (1975) *J Fish Res Board, Canada* **32**:457.
75. Szeto SY, Yee J, Brown MJ, Oloffs PC (1982) *J Chromatogr A* **240**:526.
76. Lindström K, Schubert R (1984) *J High Res Chromatogr* **7**:68.
77. Chisela F, Gawlik D Brätter P (1986) *Analyst* **111**:465.
78. Kiigemagi U, Burnard J, Terriere LC (1975) *J Agric Food Chem* **23**:717.
79. Martin JF, McCoy CP, Greenleaf W, Bennett L (1987) *Can J Fish Aqua Sci* **44**:909.
80. Persson P-E (1980) *Water Res* **14**:1113.
81. West SD, Day W (1986) *J AOAC* **69**:856.
82. Richman LA, Wren CD, Stokes PM (1988) *Water Air Soil Pollut* **37**:465.
83. Chau YK, Wong PTS, Bengent GA, Dunn JL (1984) *Anal Chem* **56**:271.

84. Holak W (1982) *Analyst* **107**:1457.
85. Shum GTC, Freeman HC, Uthe JF (1979) *Anal Chem* **51**:414.
86. MacCrehan WA, Durst RA (1978) *Anal Chem* **50**:2108.
87. Capelli R, Fezia C, Franchi A, Zanicchi G (1979) *Analyst* **104**:1197.
88. Westoo F (1967) *Acta Chem Scand* **21**:1790.
89. Callum GI, Ferguson MM, Lenihan JMA *Analyst* **106**:1009.
90. Analytical Methods Committee, Society for Analytical Chemistry, London (1977) *Analyst* **102**:769.
91. Matsunaga K, Takahashi S (1976) *Anal Chim Acta* **87**:487.
92. Uthe JF, Solomon J, Griff B (1972) *J AOAC* **55**:583.
93. Jones P, Nickless G (1978) *Analyst* **103**:1121.
94. Bouquegneau JM, Martoja M (1987) *Bull Environ Contam Toxicol* **39**:69.
95. Doherty FG, Failla ML, Cherry DS (1988) *Water Res* **22**:927.
96. Drabaek I, Eichner P, Rasmussen L (1987) *J Radioanal Nucl Chem Articles* **114**:29.
97. Anil AC, Wagh AB (1988) *Mar Pollut Bull* **19**:177.
98. Gil MN, Harvey MA, ESteves JL (1981) *Mar Pollut Bull* **19**:181.
99. Lyngby JE, Brix H (1987) *Sci Total Environ* **64**:239.
100. Peerzada N, Dickinson C (1988) *Mar Pollut Bull* **19**:182.
101. Micallef S, Tyler PA (1987) *Mar Pollut Bull* **18**:180.
102. Dunn Bp, Stich HFJ (1976) *J Fish Res Board Canada* **33**:2040.
103. Tanabe S, Tatsukawa R, Phillips DJH (1987) *Environ Pollut* **47**:41.
104. Freudenthal J, Greve PA (1973) *Bull Environ Contam Toxicol* **10**:108.
105. Kira S, Izumi T, Ogata M (1983) *Bull Environ Contam Toxicol* **31**:518.
106. Chesler SN, Gump BH, Hertz HS, May WE, Wise SE (1978) *Anal Chem* **50**:805.
107. Thybaud E, Le Bras S (1988) *Bull Environ Contam Toxicol* **40**:731.
108. Rice CP, White DS (1987) *Environ Toxicol Chem* (1987) **6**:259.
109. Bailey SK, Davies IM (1988) *Environ Pollut* **55**:161.
110. Pietilainen K, Adams F, Nullens H, Van Espen P (1981) *X-Ray Spectrom* **10**:31.
111. Lee DS (1982) *Anal Chem* **54**:1682.
112. Sivasankara-Pillay KK, Thomas Jr CC Jr, Sondel JA, Hyche CM (1971) *Anal Chem* **43**:1419.
113. Hodge VF, Seider SL, Goldberg ED (1979) *Anal Chem* **51**:1256.
114. Bando R, Galanti G, Varini PG (1983) *Analyst* **108**:722.
115. Szefer P (1987) *Mar Pollut Bull* **18**:439.
116. *Intercalibration of Analytical Methods in Marine Environment Samples* (1978) Progress Report No. 19, IAEA International Atomic Energy Agency, Monaco.
117. Ward TJ (1987) *Mar Biol* **95**:315.
118. Aulio K (1987) *Environ Pollut* **44**:1.
119. Edenfield H, Greaves HJ (1981) In: Wong CS et al. (eds) *Trace Metals in Seawater*, Proceedings of a NATO Advanced Research Institute on Trace Metals in Seawater, Sicily, Italy, Plenum, New York, USA.
120. Maher WA (1981) *Anal Chim Acta* **126**:157.
121. Law RJ, Fileman TW, Portman J (1988) *Methods of Analysis of Hydrocarbons in Marine and Other Samples, Report No. 2*, Ministry of Agriculture and Fisheries and Food, Lowestoft, UK.
122. Dallakyan GA, Veselvski VA, Tarusov BN, Peogosyan SI (1978) *Hydro J* **14**:90.
123. Vairavamurthy A, Andreae MO, Brooks JM (1986) *Anal Chem* **58**:2684.
124. Cowie G, Hedges JI (1984) *Anal Chem* **56**:497.
125. Sodergren A (1973) *Bull Environ Contam Toxicol* **10**:116.
126. Sodergren A (1978) *J Chromatogr A* **160**:271.

127. Picer N, Picer M, Strohal P (1975) *Bull Environ Contam Toxicol* **14**:565.
128. Sakmauerova M, Pal'Usova O, Szokolay A (1977) *Water Res* **11**:537.
129. Shoaf WT, Lium BW (1976) *J Res US Geol Survey* **4**:241.
130. Martin G (1983) *Rev Francais Sci Eau* **2**:407.
131. Gadel F, Bruchet A (1987) *Water Res* **21**:1195.
132. Wong SH, Hindin E (1982) *J Am Water Works Assoc* **74**:528.
133. Whyte JNC, Englar JR (1983) *Bot Marina* **26**:159.
134. Cherry DS, Van Hassel JH, Ribbe PH, Cairns J Jr (1987) *Water Resour Bull* **23**:293.
135. Hammer UT, Merkowsky AT, Huang PM (1988) *Arch Environ Contam Toxicol* **17**:257.
136. Jirka AM, Carter MJ (1978) *Anal Chem* **50**:91.
137. Frenet-Rabin M, Ottman F (1978) *Est Mar Coast Sci* **7**:425.
138. Guiney PD, Sykora JL, Kelett G (1987) *Environ Toxicol Chem* **6**:105.
139. Catello III WJ, Gambrell RP (1987) *Chemosphere* **16**: 1053.
140. Stein JE, Hom T, Casillas E, Friedman A, Varanasi U (1987) *Mar Environ Res* **22**:123.
141. Quensen III JF, Tiedje JM, Boyd SA (1988) *Science* **242**:752.
142. Fairchild JF, Boyle T, English WR, Rabeni C (1987) *Water Air Soil Pollut* **36**:271.
143. Hargrave BT, Phillips GA (1975) *Environ Pollut* **8**:193.
144. Onuska FI, Terry KA (1985) *Anal Chem* **57**:801.
145. Lee H-B, Hong You RL, Chau ASY (1986) *Analyst* **111**:81.
146. Schwartz HW, Anzion GSM, Von Vleit HPM, Peereboom JWL, Brinkman UAT (1979) *Int J Environ Anal Chem* **6**:133.
147. Uchiyama M (979) *Water Res* **13**:847.
148. Krone CA, Burrows DG, Brown DW, Robisch PA, Friedman AJ, Halins DC (1986) *Environ Sci Tech* **20**:1144.
149. Mills GL, Quinn JG (1979) *Chem Geol* **25**:155.
150. McQuaker NR, Fung T (1975) *Anal Chem* **47**:1435.
151. Kerr RA (1977) *The Isolation and Partial Characterisation of Dissolved Organic Matter in Seawater*, PhD Thesis, University of Rhode Island, Kingston.
152. Aspila KI, Agemian H, Chau ASY (1976) *Analyst* **101**:187.
153. Landers DH, David MP, Mitrchell MJ (1983) *Int J Environ Anal Chem* **14**:245.
154. Müller MD (1987) *Anal Chem* **59**:617.
155. Wong PTS, Chau YK, Luxon PL (1975) *Nature* **253**:263.
156. Leong PC, Ong HP (1971) *Anal Chem* **43**:940.
157. Anderson DH, Evans JH, Murphy JJ, White WW (1971) *Anal Chem* **43**:1511.
158. Brettaur EW, Moghissi AA, Snydr SS, Mathews NW (1974) *Anal Chem* **46**:445.
159. Feldman C (1974) *Anal Chem* **46**:1606.
160. Jensen S, Jernelov MP (1969) *Nature* **223**:753.
161. Pellenbarg R (1979) *Mar Pollut Bull* **10**:267.
162. Hattori Y, Kobayashi A, Takemoto S, Takami K, Kuge Y, Sugimae A, Nakemoto N (1984) *J Chromatogr* **315**:341.
163. Müller MD (1984) *Fresen Z Anal Chem* **37**:32.
164. Gilmour CC, Tuttle JH, Means JC (1986) *Anal Chem* **58**:1848.
165. Rapsomankis S, Donard OFX, Weber JH (1987) *Appl Organomet Chem* **1**:115.
166. Randall L, Han JS, Weber JH (1986) *Environ Tech Lett* **7**:571.
167. Unger MA, MacIntyre WG, Huggett RJ (1988) *Environ Tech Chem* **7**:907.
168. Renberg L, Tarkpea M, Sundstrom G (1987) *Ecotoxicol Environ Safety* **11**:361.
169. Wong CS, Lau E, Clark M, Mabury SA, Muir DCG (2002) *Environ Sci Tech* **36**:1257.
170. Farrington JW, Teal JM, Quinn JG, Wade T, Burns KA (1973) *Bull Environ Contam Toxicol* **10**:129.
171. Chau YK, Wong PTS, Bengert GA, Dunn IL (1984) *Anal Chem* **56**:271.

5 Quantitative Toxicity Data for Cations in Fish and Invertebrates

The concentrations of substances picked up from water by sea creatures such as fish and invertebrates and by algal and plant matter are dependent upon the concentrations of the substances in the water and, to some extent, upon their concentrations in sedimentary matter. Many creatures bioaccumulate toxicants from the water, and as a consequence their concentration in the organism is many times higher than that present in the water. Once the concentration of toxicant in the organism exceeds a certain level then harmful effects or mortalities occur.

Determinations of the concentrations of pollutants in creatures therefore provide a very useful way to ascertain the causes of adverse effects or deaths in creatures, and analysis of plant and algal material is a valuable way to obtain an early warning that excessive levels of toxicants may be present. Much work has been carried out on the determination of toxicants in creatures and plant life, and this is discussed below.

The concentrations of metals in fish organs are a useful indicator of the cause of mortalities, while the metal contents of gill, muscle and skin do not provide any such indicator. Thus, the data in Table 5.1 shows the maximum concentrations of chromium, zinc, copper and cadmium found in opercle, liver and kidney organs taken from environmental fish samples that would lead to fish mortalities. Fish would survive at the lowest concentrations encountered in environmental fish samples.

5.1
Cations in Fresh (Nonsaline) Water

5.1.1
Fish

Aluminium (and Manganese)

Reader et al. [1] studied growth, mineral uptake and skeletal calcium deposition in brown trout (*Salmo trutta*) exposed to aluminium and manganese in soft acid water.

Two 30-day experiments were conducted with brown trout alevins. Experiment 1 investigated the effects of manganese (0. 0.7, 2.2, 6.6, 20.0 µmol/l),

Table 5.1. Effect of metal contents of fish organs on mortality (from author's own files)

Exposure time	Laboratory tests on Rudd fish				Maximum concentrations found in organs of a wide variety of fish		Minimum concentrations found in organs of a wide variety of fish	
	10 weeks	3 weeks	12h	<12h	Dry weight, mg/kg (See Appendix 6.1)	Comments	Dry weight, mg/kg (See Appendix 6.1)	Comments
Element: chromium								
Concentration of metal in water, µg/l	3	16	20	80–45				
Condition of animal (Rudd fish)	Good	Good	Good	100% mortality				
Concentration of metal in organ, mg/kg dry weight								
Opercle	<0.2	<2	8.3	20–26	26	Mortalities during 12 h exposure	8.3	No mortalities during 12 h exposure
Liver	<0.2	<2	5.6	15–18	18.4	Mortalities during 12 h exposure	5.6	No mortalities during 12 h exposure
Kidney	<0.2	<2	10.3	24–27	23.7	Mortalities during 12 h exposure	10.3	No mortalities during 12 h exposure
Element: zinc								
Concentration of metal in water, µg/l	180	800	1600	7,500–18,000				
Condition of animal (Rudd fish)	Good	Good	Good	100% mortality				
Concentration of metal in organ, mg/kg dry weight								
Opercle	48	20	115.3	91–174	120	Mortalities during 12 h exposure	No data	
Liver	120	15	42.5	34–63	150		13	No mortalities
Kidney	29	28	154.6	92–216	57		No data	

Table 5.1. Continued

Exposure time	Laboratory tests on Rudd fish				Maximum concentrations found in organs of a wide variety of fish		Minimum concentrations found in organs of a wide variety of fish	
	10 weeks	3 weeks	12 h	< 12 h	Dry weight, mg/kg (See Appendix 6.1)	Comments	Dry weight, mg/kg (See Appendix 6.1)	Comments
Element: copper								
Concentration of metal in water, µg/l	11	50	–	250 – 1,600				
Condition of animal (Rudd fish)	Good	Good	–	100% mortality				
Concentration of metal in organ, mg/kg dry weight								
Opercle	12	31	–	52-104	12.4		No data	
Liver	7	20	–	22–40	62	Mortalities during 12 h exposure	1.7	No mortalities
Kidney	6	28	–	30 – 100	6		0.7	No mortalities
Element: cadmium								
Concentration of metal in water, µg/l	3	250	–	1,100 – 11,000				
Condition of animal (Rudd fish)	Good	Good	–	100% mortality				
Concentration of metal in organ, mg/kg dry weight								
Opercle	9.5	9	–	6- 29	9.5	Mortalities during 12 h exposure	No data	
Liver	5	10	–	4 – 12	9		0.2	No mortalities
Kidney	4	14	–	14 – 28	7.1		3	No mortalities

calcium (25, 500 µmol/l) and pH (4.8, 6.5) on growth, mineral uptake and calcium deposition in the developing skeleton. Experiment 2 investigated the effects of aluminium (0, 2, 4, 6, 8 µmol/l), calcium (10, 50 µmol/l) and pH (4.5, 5.4) on these parameters. Trout survival was adversely affected only by aluminium (6, 8 µmol/l) in low calcium water (10 µmol/l), pH 4.5 or 5.4. Irrespective of pH, manganese (6.6, 20.0 µmol/l) in low-calcium water (25 µmol/l) impaired net calcium uptake and/calcium deposition. Aluminium (2 – 8 µmol/l) impaired gross development, net calcium, potassium and sodium uptake, and skeletal calcification, and slightly increased the net loss of magnesium. Aluminium toxicity was ameliorated at higher calcium concentrations.

Sadler and Turnpenny [2] showed that in the absence of aluminium, brown trout growth and survival were reduced at pH 4.3 but independent of pH in the range 4.4 – 5.2. At pH 5.2 and below, the threshold aluminium concentration for growth rate suppression was approximately 20 µg/l. Aluminium toxicity was reduced at higher pH (5.9, 6.3). Field surveys in 61 upland streams (pH 4.45 – 7.30) showed that pH, labile monomeric aluminium and heavy metal concentrations were the factors most highly correlated with the biomass of the 1-plus age group of trout. A mobile bioassay laboratory, which enabled controlled toxicity studies to be carried out in the field, was being used in Fleet loch (an acid lake devoid of brown trout) to assess treatment options for restoring trout fisheries.

Muniz et al. [3] exposed sexually mature brown trout (*Salmo trutta*) to acidic stream water (pH 4.83, 240 µg/l aluminium) and to a limed control (pH 5.70, 55 µg/l aluminium) for 28 days. Neither pronounced stress (as assessed from plasma chloride, osmolality and haematocrit) nor mortality was observed at the control site, but at the exposure site there were significant but moderate stress responses and 15% mortality. The relatively high haematocrit measured at the exposure site is explained in terms of a reduction in plasma volume.

Sadler and Lynam [4] exposed yearling brown trout (*Salmo trutta*) to various pHs (4.3 – 6.5) and inorganic aluminium concentrations (0.55 – 3.7 µM/l) for six weeks. In the absence of aluminium, only pH 4.3 adversely affected trout growth. Total aluminium concentrations of 1 µM/l (27 µg/l) and above markedly reduced trout growth rates at pH 5.5 and below. At higher pH, the effects of the given aluminium concentrations were inversely related to pH. Variations in aluminium toxicity were attributed to pH-related differences in aluminium speciation. Multiple regression analysis involving growth rates and the different aluminium species (calculated using literature values for thermodynamic constants) indicated that the aluminium hydroxide species accounted for most of the toxicity, with a small contribution from polymeric aluminium complexes.

Five days of exposure to aluminium concentrations of up to 200 µg/l at pH 5.0 and recovery periods of 21 and 32 days produced no mortality in lake trout (*Salvelinus namaycush*), although alevins from the exposure period were smaller, had less calcified skeletons and lower concentrations of calcium and potassium, and were less effective predators on *Daphnia magna*. The latent effects of pulse exposure might help to explain the disappearance of lake trout populations from lakes with fairly high ambient pH.

Segner et al. [6] exposed recently hatched brown trout to pH 5.0 for five days in high-calcium water with and without aluminium (230 µg/l) at 4 or 12 °C. Acid stress had no effect on fish behaviour, growth or mucous cell concentration and volume. Acid plus aluminium stress inhibited growth and increased whole-body aluminium concentrations, but had no effect on mucous cell morphometrics. Only temperature had a significant influence on mucous cell numbers, which were lower in experiments conducted at 4 °C. Field exposure (eight days) of newly hatched fish to pH 5.1, under conditions of low calcium and 60 µg aluminium per litre, caused high mortality (57%) but had no effect on mucous cells. Brown trout juveniles (three months old) exposed to pH 5.0 in the laboratory for eight days at 12 °C exhibited behavioural responses to acid stress during the first 24 hours, mucous cell hyperplasia (without hypertrophy) within three hours, and sloughing of the integument after 120 hours of exposure. Epithelial structure was not fully restored during the four-day recovery period.

Klauda et al. [7] exposed blueback herring (*Alsoa aestivalis*) embryos and yolk-sac larvae to a range of pH (5.0, 5.7, 6.5, 7.6 – 7.8) and aluminium concentrations (nominally 0, 0.05, 0.1, 0.2, 0.4 mg/l) in soft freshwater (hardness 23 – 25 mg/l as calcium carbonate). MIBK-extracted total monomeric aluminium concentrations were generally lower than total aluminium concentrations, the ratio between the two varying with pH. In the absence of aluminium, significant embryo mortality (69%) was only observed at pH 5.0. In aluminium-exposed embryos, mortality ranged from 39 – 100% at pH 5.0, and from 27 – 81 at pH 5.7. In the absence of aluminium, larval mortality at pH 5.0, 5.7, 6.5 and 7.8 was 99, 89, 35 and 16%, respectively. Total aluminium concentrations of up to 0.34 mg/l (monomeric aluminium up to 0.21 mg/l) had a negligible effect on larval survival at near-neutral pH values, but larvae exposed to pH values of 5.0 and 5.7 died faster in the presence of aluminium. All larvae exposed to a predicted total monomeric aluminium concentration of only 0.03 mg/l at pH 5.0 died within 24 hours. Toxicity appeared to be due to physiological effects rather than cellular damage.

Ammonium

Ram and Sathyanesan [8] exposed adult *C. punctatus* to safe (100 mg/l) and sublethal (500 mg/l) concentrations of ammonium sulfate for six months. Liver histology revealed exhaustion and degeneration, probably after prolonged hyperactivity. Hepatocytes exhibited degranulation, nuclear pyknosis and focal necrosis. Additional changes included atrophy, cellular infiltration and degeneration of the pancreatic islet cells. Histological changes observed in the thyroid included hypertrophy, hyperplasia, hyperaemia and reduced colloid content. The severity of the changes in both liver and thyroid was dose-related.

Unionised Ammonia

De et al. [9] showed that 70% mortality occurred, particularly among eggs of the rainbow trout (*Salmo gairdneri*), at concentrations of unionised ammonia as low as 0.027 mg/l when exposure began within 24 hours of fertilisation and proceeded for 73 days in hard fresh water. When exposure did not begin until the eyed-egg stage (about 24 days), only 40% of eggs, yolk-sac fry, and fry (but especially the fry) died at an ammonia concentration of 0.27 mg/l.

Cadmium (and Lead)

The cadmium content of muscle taken from juvenile (*Tilapia aurea*) fish exposed to water containing 6.8 – 522 µg/l cadmium ranged from 0.12 mg/kg at the 6.8 µg/l level in water to 0.92 mg/kg at the 52 µg/l level in water. Few national or international authorities have set limits for cadmium in foodstuffs; Norway and the Netherlands are reported to have set a limit of 0.5 mg/kg in fish [10].

Carrier and Beitinger [11] noted the reduction in thermal tolerance of *Nitropis lutensis* and *Pimephales promelas* exposed to cadmium.

For both of these species, the 96-hour LC_{50} value for cadmium was less than 10 mg/l. Exposure to sublethal concentrations reduced the ability of the fish to tolerate heat stress; this reduction in tolerance increased with increasing cadmium concentration and period of exposure.

Gill et al. [12] studied the bronchial pathogenesis of the freshwater fish *Puntius conchonius* chronically exposed to sublethal concentrations of cadmium.

Puntius conchonius were exposed to 630 or 840 µg/l cadmium (96 h LC_{50} 12.6 mg/l) for 4, 8 or 12 weeks. Branchial lesions were observed in all treated fish. Severity appeared to be influenced more by duration of exposure than by exposure concentration. Observed changes included epithelial disruption, necrosis, accumulation of cell debris, capillary congestion, wilting of the pillar cell system, hypertrophy and hyperplasia of chloride cells, and fusion of the secondary lamellae. During cadmium exposure, all fish secreted copious mucus from all over the body surface and, at sacrifice, a film of coagulated mucus was invariably present on the gills. The observed branchial lesions and coagulation film anoxia were likely to result in respiratory disorders and associated tissue hypoxia.

Sehgal and Saxena [13] determined the acute toxicity of cadmium and lead to male and female *Lebistes reticulatus* over 96 hours. Static LC_{01}, LC_{16}, LC_{50}, LC_{84} and LC_{100} values were calculated. Visible signs of toxicity were a decrease in swimming activity, rapid opercular movement and gulping of air. The 96 h LC_{50} values for cadmium were 350 and 371 mg/l for male and female fish, respectively. The corresponding 96 h LC_{50} values for lead were

1620 and 1630. Safe concentrations of cadmium for males and females were estimated as 112.9 and 116.5 mg/l, respectively, and those for lead as 492 and 487 mg/l.

Calcium (and Aluminium)

Thomsen et al. [92] exposed fertilised rainbow trout eggs to pH 3.7, 4.6, 5.7, 6.6 or 7.6 in hard water (150 mg/l calcium). Mortality peaked during days 1–3, 6–9, 14–16 and 24–27. Eggs exposed to pH 3.7 and 4.6 did not survive beyond days five and nine, respectively. Hatching was delayed and hatching success reduced by exposure to pH 5.7 and 6.6. In larval (25 days post-hatch) tests at pH 7, aluminium had an LC_{50} of 3.8 mg/l in soft water (1 mg/l calcium) and 71 mg/l in hard water. Experiments conducted at pH 5 and 7 in hard and soft water with and without aluminium (0.5 mg/l) determined effects on egg and larval survival, and on larval (14 days post-hatch) physiology. All three variables reduced survival (low calcium was more lethal than high aluminium), larval dry weight and larval respiration rates. Cardiac rates were decreased by low pH and low calcium but not by high aluminium. Soft water exposure reduced larval body calcium concentrations by about 47%, but high aluminium had little effect. Previous exposure to low pH, low calcium and high aluminium had little effect on the growth of larvae subsequently transferred to control conditions.

Cobalt

Nath and Kumar [93] noted cobalt-induced alterations in the carbohydrate metabolism of a freshwater tropical perch, *Colisa fasciatus*.

The effect of cobalt intoxication on the muscle glycogen and blood lactate levels of the freshwater tropical teleost *C. fasciatus* was investigated. A sublethal dose (232.8 ppm) was administered, and the effects monitored for 3–96 hours. At all time intervals (except at three hours) there was a significant decrease in muscle glycogen content, and hyperlacticaemic response was observed at all time intervals except three and six hours. Blood lactic acid level was at its maximum of 78.07% at 96 hours.

Chromium

Kranz and Gereken [14] studied the effects of sublethal concentrations of potassium dichromate on the occurrence of splenic-macrophage cations on the juvenile plaice *Pleuronectes platessa* L.

The effects of two sublethal concentrations of potassium dichromate on macrophages (cells of the immune system responsible for removing foreign particles and damaged cells from an organism by phagocytosis) in the spleen of plaice (*Pleuronectes platessa*) were studied. Uptake of chromium by the fish resulted in an increase in splenic melanomacrophage centres, but a reduction in their average size; the percentage area of melanomacrophage centres in the spleen tissue remained unchanged. Other histological changes were also observed.

Greene et al. [15] used *Selenastrum capricornutum* to assess the toxicity potential of surface and groundwater contamination caused by chromium waste.

The toxic potential of ground and surface water samples from a site used for the disposal of chrome-plating wastewaters was assessed using 96-h *Selenastrum capricornutum* bioassays. EC_{50} values calculated on the basis of bioassays of one set of samples for different chromium concentrations gave an excellent correlation with values from a later set for groundwater and drainage ditch samples, but no correlation was obtained with off-site surface water samples. Even in simple systems, toxicity was difficult to predict solely on the basis of chemical analysis.

Boge et al. [16] studied the effects of hexavalent chromium on enzyme activities and transport processes of the intestinal brush border membrane in rainbow trout (*Salmo gairdneri*).

Chromium inhibited alkaline phosphatase activity, the degree of inhibition increasing with increasing chromium concentration and contact period. No effect of chromium on maltase activity was observed. High concentrations of chromium also inhibited glycine absorption.

Copper

Starved or fed yearling roach (*Kutilus rutilus*) exposed to sublethal copper contamination (80 µg/l copper) for seven days accumulated 19 µg copper/kg (dry weight in gill tissue), but only starved fish accumulated significant quantities of copper in water (95 mg/kg copper, dry weight). Refeeding after cessation of copper exposure resulted in a significant loss of copper from the liver, which fell to 70 mg/kg copper dry weight [17].

Exposure of juvenile rainbow trout (*Salmo gairdneri*) to 55 µg/l copper in water for 28 days led to whole-body copper uptakes of 1.2 µg/g copper on day 1 to 6.6 mg/kg copper on day 28. Liver copper increased from 25 µg copper dry weight on day 0 to 69 mg/kg on day 2 and 113 mg/kg copper on day 28, both dry weight [18].

Lauren and McDonald [19] studied the accumulation of copper by rainbow trout (*Salmo gairdneri*).

Sodium uptake kinetics and whole-body sodium concentrations were monitored in juvenile rainbow trout exposed to 55 µg copper per litre for 28 days, followed by seven days in uncontaminated water. The maximal sodium uptake rate was reduced by 55, 41 and 23% after 1, 7 and 14 days of copper exposure, but was not significantly different from the control by day 28. The apparent affinity for sodium was reduced by 42–49%.

Nemcsok and Hughes [20] showed that exposure of rainbow trout (*Salmo gairdneri*) 250–500 g) to 2.0 ppm copper sulfate for 24 or 48 hours increased plasma aspartate aminotransferase (ASAT) activity, plasma alanine aminotransferase (ALAT) activity and blood glucose, and decreased acetylcholinesterase activity. Exposure to 0.2 ppm copper sulfate caused qualitatively similar changes, but only after 48 hours. Alterations in plasma transaminase enzyme activities were indicative of serious tissue necrosis. Increases in blood glucose were reflective of stress. Inhibition of acetylcholinesterase seriously interfered with nervous system functioning.

Ellgaard and Guillot [21] measured the behaviour of bluegill sunfish, *Hepomis marrochirus rafinesque* exposed to sublethal concentrations of copper (less than 4 ppm) using a kinetic method in which the locomotor rate constant provided an index of activity. After exposure to 0.04, 0.08 and 0.4 ppm copper for eight, eight and six days, fish were 67, 61 and 44% as active, respectively, as they were during the pretreatment period. At all test concentrations, the locomotor activity decreased sharply during the first four days of treatment and then remained approximately constant. Sublethal concentrations of copper might decrease the metabolic rate of bluegills.

Cyanide

McGeachy and Leduc et al. [22] studied the influence of season and exercise on the lethal toxicity of cyanide to the rainbow trout (*Salmo gairdneri*).

Results are reported from bioassays performed to compare the response of continuously exercised and unexercised juvenile rainbow trout (*Salmo gairdneri*) to lethal concentrations of cyanide at different seasons of the year. These showed that the 96-h LC_{50} value of cyanide varied seasonally and with exercise (swimming at a rate of one body length per second). In summer there was no significant difference in sensitivity between exercised and unexercised fish, but in winter the exercised fish survived twice as long as the unexercised fish. However, the resistance of unexercised fish in winter could be increased by increasing the acclimation period from four to ten weeks.

Ruby et al. [23] studied changes in plasma, liver and ovary vitellogenin in land-locked Atlantic salmon, following exposure to sublethal concentrations of cyanide.

It has been stated that cyanide concentrations of 0.005 – 0.03 mg/l are safe levels for aquatic organisms. Based on these reports, the lower concentration of 0.005 mg/l was selected to test its effect on the mechanism of yolk synthesis in *Salmo salar*. Females were maintained in experimental flow-through tanks supplied with hydrogen cyanide for 12 days at about 7 °C. The fish were then killed, and the mechanism of vitellogenesis examined through direct measurement of vitellogenin, utilising a homologous radioimmunoassay specific for *S. salar*. Plasma vitellogenin increased in cyanide-exposed fish. However, gonad vitellogenin levels declined by day 12. The experiments were carried out in October during late vitellogenesis, and the results indicated that exposure of female salmon to sublethal cyanide at this stage inhibited vitellogenin uptake at the ovarian level. This effect on yolk synthesis differed seasonally; the recommended safe concentration for cyanide (0.005 mg/l) altered patterns of plasma vitellogenin just prior to the spawning period.

Manganese

See under Sect. 'Aluminium (and Manganese)' above.

Mercury

Ram and Joy [24] studied the mercury-induced changes in the hypothalamo–neurohypophysial complex in relation to reproduction in the teleostean fish *Channa punctatus* (Block).

Adult *Channa punctatus* were exposed to sublethal concentrations of inorganic mercuric chloride (0.01 ppm) and the organic fungicide methoxyethyl mercuric chloride (Emisan) for six months, starting in January when the gonads were in the immature stage I condition. At termination, control gonads were fully mature, neurons in the nucleus preopticus (thought to be responsible for secretion of gonadotrophin-releasing hormone) were large, actively secreting and contained an adequate quantity of neurosecretory material, and the proximal pars distalis was dominated by large, actively secreting, hypertrophied vacuolated gonadotrophs. By contrast, mercury-treated fish were gonadally immature with ovaries in the stage I condition, testes devoid of sperm, and Leydig cells inactive and atrophied. Nucleus preopticus neurons were small, inactive, contained little secretory material, and exhibited varying degrees of degeneration (pyknotic and necrotic changes were more pronounced in Emisan-treated fish). These changes were accompanied by significant ($P < 0.01$) inhibition of brain monoamine oxidase activity. In addition, gonadotrophs in the pituitaries of treated fish were small, inactive, involuted, and fewer in number compared with controls. Mercury-induced inhibition of gonadal growth in *C. punctatus* would appear to have been mediated via impairment of the hypothalamoneurohypophysial–gonadal axis.

Selenium

Mosquito fish *(Gambusia affinis)* in a reservoir at San Joaquin Valley, California, were found to contain 30 µg/kg selenium (as selenate) originating in drainage waters. All other species of fish had died [25].

Woock et al. [26] noted decreased survival and teratogenesis during laboratory exposure of bluegill (*Lepomis macrochirus*) to selenium.

The effects of dietary selenite (13, 30 µg/g selenium), dietary selenomethionine (3, 13, 30 µg selenium per g), and dietary selenomethionine plus waterborne selenite (13 µg per g plus 10 µg/l) on bluegill sunfish were compared. Treatment begun with juveniles, and continued for 323 days. Spawning experiments were started after 260 days exposure. Cumulative mortality (day 260) was increased ($P < 0.05$) in both the 30 µg selenium per g groups.

Day 260 body weights and lengths were significantly lower in bluegills exposed to 30 μg selenite per g; 37% of fish exposed to 30 μg organoselenium per g had developed lens cataracts. Parental exposure to selenite (30 μg per g), selenomethionine (13, 30 μg per g), and organoselenium plus waterborne selenite reduced larval survival but not percentage hatched. Abnormal larvae represented 21% in the controls, 2 and 15% in the 13 and 30 μg/g inorganic selenium groups, 3, 10 and 100% in the 3, 13 and 30 μg/g organoselenium groups, and 50% in the group exposed to dietary plus waterborne selenium.

Zinc

Concentrations of zinc found in *Tilapia zilli* gills, liver and muscle after four days of exposure to zinc were, respectively, 38,000, 23,000 and 2000 mg/kg zinc dry weight. The corresponding figures for *Clarius lazera* were 49,000, 34,000 and 5,000 mg/kg dry weight [27].

Meisner and Hum [28] studied the acute toxicity of zinc to juvenile and subadult rainbow trout, *Salmo gairdneri*.

Previous workers had indicated that sensitivity to zinc increased as fish progressed from the embryo stage to just after transition to exogenous food, and then decreased through the juvenile and adult stages. However, in flow-through bioassays, 96 h LC_{50} for zinc in juvenile (15.7 cm fork length, 25 – 70 g) and sub-adult (25.8 cm fork length, 160 – 290 g) rainbow trout were 26.0 and 24.0 mg/l, respectively, suggesting that both of these life stages were equally tolerant of the metal. Further data on the acute toxicity of zinc to adult salmonids is required before relationships between tolerance and body size can be formulated with confidence.

Hardness and pH

Lemly and Smith [29] studied the mechanisms and ecological implications of the effects of chronic exposure to acidified water on the chemoreception of feeding stimuli in flathead minnows (*Pimephales promelas*).

Fathead minnows were exposed to hard or soft water (160 – 200, 5 – 10 mg/l as calcium carbonate) at pH 8.0 (control), 7.0, 6.5, 6.0 or 5.5 for 72 hours or 30 days. Results obtained using an automated behavioural assay showed that the response to a single-pulse dose of a chemical feeding stimulus was eliminated in fish exposed to pH 6.0 or below. The response was restored by transferring affected fish to control water for 24 hours. It was probable that impairment of chemoreception involved mechanical and chemical inhibition of receptor cells in the olfactory and gustatory epithelia, as effects were reversible and scanning electron microscopy revealed no pathological or gross morphological changes in chemosensory tissue. Impairment of feeding behaviour at low pH would reduce the long-term survival of fathead minnow populations, and could account for the observed elimination of this species from natural waters at pH 5.8 – 6.0 and below.

Aluminium, Copper and Zinc

Hutchinson and Sprague [30] noted a reduced lethality of aluminium, zinc and copper mixtures to American flagfish upon complexation with humic substances in acidified soft waters.

American flagfish *(Jordanella floridae)* fry were acutely exposed to a 12.5:2:1 alumimum:zinc:copper mixture in synthetic soft water and in soft water from four Ontario lakes. All experiments were conducted at pH 5.8. LC_{50} values for the metal mixture in waters from the Blue Chalk, Walker, Chub and Brandy lakes were 2.1, 4.2, 7.4 and 14.5 times higher, respectively, than the corresponding value obtained in synthetic soft water. Reductions in trace metal lethality were directly related to the humic substance content of the lake water, as estimated by measurements of total carbon, total organic carbon, ultraviolet absorbance, apparent colour and Secchi depth. Dialysis data indicated that the toxicity of aluminium and copper was reduced by complexation with high molecular weight humic substances in the coloured water. Zinc was mostly present in low molecular weight forms, and the effects of complexation on zinc lethality remained unclear. Acid lakes in which the TOC was below 2.2 mg/l and the absorbance at 310 nm was less than 0.016 would be more susceptible to fish loss due to trace metal toxicity.

Bleached Kraft Pulp Mill Effluent

Gouillard et al. [31] presented results from studies on histopathological changes in rainbow trout *(Salmo gairdneri)* exposed to either high or sublethal concentrations of untreated bleached kraft pulp mill effluent. Extensive fusion of gill lamellae was found in fish exposed to lethal concentrations. No such lesions were observed in fish exposed to sublethal concentrations of effluent for up to 60 days, but the exposed fish had a higher incidence of fin necrosis and damaged gills than the control fish. Loss of resistance to bacterial pathogens was a significant stress-related effect in fish exposed to low concentrations of the effluent for long periods.

5.1.2
Invertebrates

Ammonia

Zischke and Arthur [32] studied the effects of elevated ammonia levels in the fingernail clam *Musculium transversum* in outdoor experimental streams.

Caged fingernail clams were exposed to ammonia in experimental streams for 2 – 8 weeks during the summers of 1983 and 1984. Weekly mean unionised ammonia concentrations were 0.02 – 0.08, 0.04 – 0.25 and 0.14 – 0.56 mg/l in 1983, and 0.04 – 0.20, 0.07 – 0.38 and 0.48 – 1.17 mg/l in 1984. In 1983, clam survival was 50 – 85% in control and low-treatment streams, 30 – 55% in the medium-treatment stream, and zero in the high-treatment stream. Older clams appeared more sensitive to ammonia than younger clams. Growth was adversely affected by medium ammonia concentrations. In the 1984 experiments, total yields of clams in control, low, medium and high ammonia streams were 12.2, 4.7, 1.1 and 0.03 times the original stock, respectively. Reproduction, in terms of numbers of newborn clams recovered from the cages after four weeks of exposure, was 90% lower in the low-treatment stream than in the control stream. Based on the numbers of clams recovered from low and medium ammonia streams in 1984, the lowest mean concentration affecting survival was between 0.09 and 0.19 mg/l ammonia.

Arthur et al. [33] carried out seasonal toxicity studies of ammonia on five fish and nine invertebrate species.

These flow-through laboratory studies (48- and 96-hour acute toxicity tests) were conducted in winter, spring, summer and autumn river water at ambient seasonal temperatures (3.4 – 26.1 °C). The five fish and nine invertebrate species were collected from outdoor experimental streams. Ammonia was supplied as ammonium chloride. In fish, ammonia LC_{50} values were lowest in rainbow trout (geometric mean 0.53 mg/l) and highest in fathead minnow (geometric mean 2.17 mg/l). The order of decreasing sensitivity to ammonia was rainbow trout, walleye, channel catfish, white sucker, fathead minnow. LC_{50}s in invertebrates ranged from 1.10 mg/l in the fingernail clam *Musculium transversum* to 18.3 mg/l in the crayfish *Orconectes immunis*. Within a species, LC_{50} values at different test temperatures varied by factors of less than 1 – 4. With the exception of channel catfish, none of the tests showed a progressive increase in LC_{50} values with increasing temperature. Colder temperatures were associated with an increase in dissolved oxygen, and a slight decrease in pH. It might therefore be more appropriate to attribute the temperature/ammonia toxicity relationships found to seasonal variation in water quality. Results did not indicate that fish size influenced the ammonia LC_{50} value.

Cadmium and Selenium

Bouquegneau and Martoja [34] have studied the seasonal variation of the cadmium content of *Murex trunculus* in an environment not polluted with cadmium.

Previous studies had indicated that Calvi Bay, Corsica, was unpolluted by cadmium but was subject to seasonal organic pollution from the tourist trade. However, high concentrations of cadmium had been reported in the mollusc *Murex trunculus*, collected from the bay, and a two-year study was carried out on this phenomenon.

Guidici et al. [35] has conducted acute and long-term studies on the toxicity of cadmium to *Idothea baltica* (crustacea, Ispoda).

Male, female and juvenile *Idothea baltica* were acutely exposed to cadmium concentrations ranging from 0.01 to 15 mg/l. Comparison of LC_{50} values indicated that males were more sensitive than females over the range

0.01 – 0.1 mg/l cadmium, but that both sexes were equally sensitive to the higher cadmium concentrations. Juveniles were more sensitive than adults to concentrations of 0.01 – 10 mg/l cadmium, but adults and juveniles were equally sensitive to 15 mg/l cadmium. In chronic experiments, exposure of I, baltica to 0.5 mg/l cadmium during embryonic development, juvenile development or both reduced the LC_{50} from approximately 63 days (controls) to 31, 28.5 or 12 days, respectively. Animals exposed to cadmium during embryonic development were larger than controls for the first 60 days after hatching, but were smaller than controls by days 90 and 120. Animals exposed during juvenile development were similar in size to controls on day 30, but smaller on days 60, 90 and 120. Those exposed during embryonic and juvenile development showed the same growth rate as controls up to day 30, but did not survive to day 60. Results suggested that testing heavy metal pollutants at different stages of the I. baltica life cycle, and over the long term, would provide more reliable criteria for water standard evaluation.

Hong and Reish [36] have studied the long-term toxicity of cadmium to eight species of marine amphipod and isopod crustaceans from Southern California; 96-hour and seven-day LC_{50} values were reported by these workers.

Knowles and McKee [37] determined the protein and nucleic acid contents of Daphnia magna during chronic exposure to cadmium.

Survival, reproduction, protein, RNA, DNA, glycogen and lipid contents were determined in Daphnia magna exposed to cadmium (0, 0.4, 0.8, 2.1, 4.3, 7.2 µg/l) for 4, 7 or 21 days. The 21-day no observable effect concentrations (NOEC) for survival, reproduction and protein growth were 2.1, 0.8 and 0.8 µg cadmium per litre, respectively. Effects of cadmium on protein growth were most pronounced on day 7, immediately after the rapid growth phase, whereas effects on RNA:protein and protein:RNA:DNA ratios were most appropriately measured on day 4, before the rapid growth phase. The NOEC based on the day 4 protein:RNA:DNA ratio was 0.8 µg/l cadmium, but the most sensitive parameter was protein growth at day 7, which was significantly reduced by all cadmium exposure concentrations.

Winner and Whitford [38] have studied the interactive effects of cadmium stress, selenium deficiency and water temperature on the survival and reproduction of Daphnia magna straus.

In this work, neonate D. magna were exposed for 25 days to 0, 1.5 or 3.0 µg/l cadmium at 20 or 25 °C. Half of the animals in each group were fed a selenium-enriched diet (Chlamydomonas reinhardtii cultured in medium supplemented with 24 µg/l selenium). At 20 °C, cadmium exposure increased mortality among selenium-deprived daphnids but not among selenium-enriched animals. At 25 °C, cadmium had no effect on daphnid survival, regardless of dietary selenium levels. Interactive effects on daphnid reproduction were complex. Abortion ratios were markedly increased in all selenium-deficient animals at 25 °C, regardless of cadmium exposure.

Mean brood size was unaffected by selenium in the absence of cadmium stress, but was decreased in selenium-deprived animals exposed to 1.5 µg/l cadmium at 20 °C (the 3.0 µg cadmium per litre group died before reproducing) or 3.0 µg cadmium per litre at 25 °C. Reductions in total young per female were related to selenium deficiency and to cadmium exposure concentrations. Selenium-deprived animals were more sensitive to cadmium stress at 20 °C, whereas selenium-enriched daphnids were more sensitive at 25 °C. At both temperatures, selenium-deficient animals were more sensitive to cadmium stress than were selenium-enriched animals.

Bodar et al. [39] studied the effect of cadmium on the reproductive strategy of *Daphnia magna*.

Young *Daphnia magna* (12 h old) were exposed to cadmium (0 – 50 ppb) for 25 days. The 25-day LC_{50} was approximately 10 ppb. Reproduction (number of broods per female and neonates per brood) was stimulated by exposure to 0.5, 1.0 and 5.0 ppb cadmium, but inhibited by 10 ppb. (Data on animals exposed to 20 and 50 ppb were not analysed). Onset of reproduction was delayed by exposure to 5 and 10 ppb cadmium. Data on the fifth and sixth breedings showed a treatment-related decrease in neonatal length. It would appear that, at low cadmium concentrations, daphnids produced larger broods but smaller neonates. They discussed whether this was a cadmium-induced change in reproduction strategy or a nonspecific effect on the growth control mechanism (hormesis).

Doherty et al. [40] has shown that metallothionein-like heavy metal binding protein levels in Asiatic clams are dependent on the duration and mode of exposure to cadmium.

These metallothionein-like heavy metal binding proteins have previously been identified in clams, (*Corbicula fluminea*). The concentrations of this compound increased following exposure to other cadmium and zinc, and to a lesser extent following exposure to other stresses (residual chlorine and extremes of temperature). Further experiments were carried out to examine the effects of period of exposure and mode of exposure on tissue concentrations of cadmium and of the metal-binding protein. It was found that clams exposed to dissolved cadmium had higher concentrations of cadmium and metal-binding protein associated with gill, mantle, and adductor muscle, while animals ingesting cadmium-contaminated algae had higher proportions associated with the visceral mass. Tissue concentrations of both metal and protein increased with increasing period of exposure.

De Lisle and Roberts [41] has studied the effects of salinity on the toxicity of cadmium to the estuarine mysid *Mysidopsis bahia*.

Mysidopsis bahia were exposed for 96 hours to cadmium chloride at salinities of 6, 14, 22, 30 and 38 per thousand g/l. LC_{50} values, expressed in terms of total cadmium, CdCl ion and cadmium chloride increased with increasing salinity.

Calcium

Shumway et al. [42] has studied the effects of calcium oxide (quicklime) on non-target organisms on mussel beds.

Three experimental tanks (area 1.2×0.9 m), set up to simulate conditions in a natural mussel bed, were stocked with mussels (*Mytilus edulis*), bloodworms (*Glycera dibranchiate*), sandworms (*Nereis virens*), periwinkles (*Littorina littorea*), juvenile lobsters (*Homarus americanus*), and starfish (*Asterias vulgaris*). To avoid predation, the worms were confined to mud-filled basins within each tank. After one week of acclimation, quicklime (1.5 kg per tank) was applied by hand to the surface waters of each tank. All starfish died within 12 hours, but no detrimental effects were detected in the other species. A histological examination conducted one week after quicklime application revealed no abnormalities in lobster gills, worm parapodia or mussel gills, and all animals except starfish were still alive six months after the experiment ended. These results indicated that the use of quicklime to control starfish predation in natural mussel beds would not adversely affect other species likely to be present.

Chromium

Dorn et al. [43] has studied the use of hexavalent chromium as a reference toxicant in effluent toxicity tests.

These studies, conducted concurrently at different laboratories, compared accuracy in preparing chromium solutions and the acute toxicity (EC_{50}, LC_{50}) results obtained with four invertebrates (*Mysidopsis bahia*, *M. almyra*, *Daphnia pulex*, *Ceriodaphnia* sp.) and three vertebrates (*Cyprinodon variegatus*, *Lepomis macrochirus*, *Pimephales promelas*) exposed to chromium. The coefficients of variation associated with the preparation of chromium solutions in fresh and saltwater, respectively, were 51 and 63% at one laboratory, and 136 and 14.8% at the other. Hexavalent chromium remained relatively stable during the 48-hour tests, with recoveries ranging from 77 to 114%. Precision in measuring chromium-spiked fresh- and saltwaters ranged from 0.2 to 9.1%. The use of measured, rather than nominal, chromium concentrations had little effect on the calculated EC_{50} and LC_{50}. Acute toxicity values ranged from 0.031 (*Ceriodaphnia* sp.) to 182.9 (*L. macrochirus*) mg/l chromium(VI). Over a three-week period, 48-hour LC_{50} values in *M. bahia* did not differ significantly within or between laboratories. Inter-laboratory differences (almost one order of magnitude) in *D. pulex* results were attributable to differences in feeding. Reference tests, run concurrently with effluent toxicity tests, were a valuable method of monitoring the health and baseline responses of test organisms, laboratory precision, and overall toxicity test quality. Chromium was a suitable reference toxicant.

Van der Meer et al. [44] has studied the toxicity of sodium chromate to crustaceans. These workers reported on the survival and development of young adults and larval crustaceans at different salinities (3.3, 23, 33 per thousand g). Test species were *Palaemonetes varians, Palaemon eleganes, Neomysis integer, Praunus flexosus* and *Daphnia magna*. Exposure concentrations ranged from 0.001 to 2.0 µ/mol sodium chromate per litre. Exposure durations ranged from 8 hours to 40 days. Calculated values included minimal concentrations effecting adult mortalities, larval mortalities and larval development. No observed effects concentrations, four- and ten-day LC_{50}, ET_{50} (number of days until 50% of larvae had reached the first post larval stage), ET_{20} and ET_{80} values were measured. Based on its comparative sensitivity, ease of culture and the fact that it has been widely used for toxicity experiments, *Daphnia magna* was the organism of choice.

Copper

Guidici et al. [45] has examined the sensitivity of *Asellus aquaticus (L)* and *Proasellus coxalis Dollf* (crustaceans, isopoda).

Samples of natural populations of *Asellus aquaticus* and *Proasellus coxalis* were taken from the River Sarno, Naples, and bred under controlled laboratory conditions. In all, 1032 adults and 1942 juveniles were tested in triplicate. Copper sulfate (0.01 – 15 mg/l) was added to the experimental cultures. ST_{50} (the time elapsed from the beginning of the experiment to the death of half of the individuals) was recorded. Death was determined by checking for respiratory movements. Both species proved highly sensitive to copper, and their ST_{50} values decreased with increasing metal concentration. Tolerance to copper was the same in both species, and no statistically significant differences between males and females were found, but the tolerance of juveniles was significantly lower than that of adults. Very low copper concentrations (0.005 mg/l) did not affect adult survival.

Samples of the barnacle (*Balanus amphitrite*) collected in the Zuan estuary, India, had tissue copper concentrations of 39.7 – 864.8 mg/kg. The copper contents of the waters in the region were 1 – 11 µg/l [46].

Iron

Maltby et al. [47] carried out acute toxicity tests on the freshwater isopod *Asellus aquaticus* using iron sulfate heptahydrate as toxicant.

Hydrous ferrous sulfate was added to artificial pondwater to give final iron(II) concentrations ranging from 50 to 1000 mg/l. Solution pH was adjusted to pH 4.5 using 1 M hydrochloric acid, or buffered to approximately 6.5 (0.0025 M sodium potassium tartrate was the most appropriate buffer). *Asellus aquaticus* were obtained from two different sites, one downstream from a disused coal mine (2.48 mg total dissolved iron per litre, pH 6 – 6.7) and the other upstream (0.88 mg total dissolved iron per litre, pH 7 – 7.5). LC_{50} values (calculated by probit analysis) obtained in the low pH experiments suggested that *A. asellus* exposed to mine-waste pollution were more sensitive to iron(II) ($LC_{50} = 256$ mg/l) than were isopods from the unpolluted site ($LC_{50} = 383$ mg/l). This difference in sensitivity was masked in the tartrate buffer experiments (LC_{50} = 431 and 467 mg/l iron(II))in polluted and unpolluted mine waste water, respectively, possibly due to the formation of less toxic buffer–iron(II) complexes.

Selenium

Johnson [48] determined the acute toxicity of *Daphnia magna* (Straus) to inorganic selenium and the effect of the element upon growth and reproduction.

In adult *Daphnia magna* exposed to sodium selenite or sodium selenate, the 48-hour LC_{50} values for selenium were 0.68 ppm and 0.75 ppm, respectively. Exposure of juvenile *D. magna* to sodium selenate yielded a 48-hour LC_{50} of 0.55 ppm selenium, while the 72-hour LC_{50} for selenium in selenate-exposed eggs and embryos was 1.4 ppm. Subacute exposure of *D. magna* to selenium (0.025 – 0.5 ppm as selenate) adversely affected growth (measured over instars 1 – 5) and reproduction (measured over instars 6 – 9). Effects were dose-related and, although they occurred at selenium concentrations which were unlikely to be attained in natural waters as a result of sediment sequestration, the use of selenium as an ameliorating agent for mercury toxicity in freshwaters (proposed dose range 10 – 100 μg/l) could significantly affect daphnid population dynamics.

Draback et al. [49] showed that selenium is toxic to the mussels *Mytilus edulis*, and concentrations of this element appear in the tissues.

Micallef and Tyler [50] examined whether selenium modified the acute lethal toxicity of mercury in *M. edulis*, and measured mercury (and selenium) accumulation under different treatments. The best protection was obtained by simultaneous addition of selenium on an equimolar basis with mercury. The simultaneous addition of a high concentration of selenium was more toxic than mercury alone. Analyses in the soft tissues showed that mercury and selenium did interact, but simultaneous addition of a low concentration of selenium on an equimolar basis with mercury did not produce the expected redistribution of mercury in the tissues. The mechanism for the antagonistic effect between selenium and mercury was consequently still unclear. Although the concentrations used in these experiments were higher than in real environmental conditions, such studies might aid the design of more detailed chronic toxicity tests where life cycles may be taken into account. See also under 'Cadmium' alone [38].

Mercury

Harrison et al. [51] exposed slipper limpet (*Crepidula fornicata*) to mercury.

Slipper limpets were exposed to 5, 25 or 50 μg soluble mercury per litre for 16 weeks. Total body mercury concentrations increased rapidly during the first 60 days, then tended to plateau at concentrations of around 28, 55 and 75 μg/g wet weight (5, 25, 50 μg/l exposures, respectively). Mercury associated with low molecular weight proteins increased with exposure time, although absolute amounts (25 µg/g wet weight on day 112) were similar in all treatment groups. Acute (48-hour) mercury toxicity experiments were conducted on days 0, 13, 47 and 114, when water temperatures were 13.5, 9, 6 and 3 °C, respectively. Mercury toxicity was temperature-dependent (LC_{50} of 1100 µg/l mercury at 13.5 °C, no mortality among limpets exposed to 1600 µg/l at 3 °C). The sensitivity of limpets to acute mercury exposure was unaffected by pre-exposure to 5 µg/l mercury, but was increased by pre-exposure to 25 or 50 µg/l mercury.

Slipper limpets (*Crepidula fornicata*) exposed to 5 – 50 µg/l mercury for 16 weeks contained 28 – 75 mg/kg wet weight of mercury in their tissues [50]. The mortality of limpets was water temperature-dependent. Thus, at 13.5 °C a 114-day LC_{50} value of 1100 µg/l was obtained while at 3 °C no mortalities of limpet occurred upon exposure to 1600 µg/l mercury in water for up to 114 days.

Micallef and Tyler [50] also measured the toxicity to *Medulis* (see above).

Zinc

Hall et al. [52] studied the effects of suspended solids on the acute toxicity of zinc to *Daphnia magna* and *Pimephales promelas*. It was found that the suspended solids concentration was a significant factor in reducing toxicity due to the sorption of zinc to the suspended solids or other changes in water chemistry. In the case of *P. promelas*, only suspended solids values of 483 – 734 mg/l and of such a type that total hardness, total alkalinity and total dissolved carbon were increased were effective at reducing toxicity. The forms of zinc toxic to these organisms were apparently restricted to the aqueous phase. Dose–response curves obtained in the studies were useful for assessing the ability of an organism to respond to aqueous phase zinc concentrations.

Gil et al. [53] determined concentrations of iron, zinc, manganese, copper, cadmium, lead and nickel in scallops (*Chlamys tehuelcha*) and mussels (*Aulacomya ater, Mytilus platensis*) from a rural, uncontaminated site in the San Jose Gulf and from an urban, industrialised site (Puerto Madryn City) in the Nuevo Gulf. Samples collected in winter showed that, although iron, zinc and copper concentrations were higher in Nuevo gulf molluscs, values were of the same order of magnitude as those recorded in the San Jose Gulf. In scallops, iron, manganese, copper and cadmium were primarily concentrated in the liver and kidney, though iron was also found in gonad and foot. Zinc was concentrated in mantle and gills. Lead and nickel were below detection limits. Metal concentrations in the Nuevo Gulf molluscs were in the baseline range: the Gulf was not contaminated by urban–industrial effluents.

Abbasi et al. [54] calculated LC_{50} and safe concentration (SC) values for mercury, chromium and zinc in the copepod *Cyclops* sp. and mosquito larvae *Aedes aegypti*. Static bioassays and computer-aided analysis of mortality data were used. The LC_{50} and SC values were compared with minimal levels permitted in drinking water and irrigation water.

Willis [55] carried out experimental studies of the effect of zinc on *Ancylus fluviatilis* (Muller) *(Mollusca: Gastropoda)* from the River Afon Crafnant in North Wales.

Groups of *Ancylus fluviatilis* in three size ranges were exposed to concentrations of 1.0 to 18.0 mg/l zinc (in a logarithmic series) for 96 hours and 30 – 100 days (in steps of ten days). In a second experiment, concentrations of 100–1000 µg/l zinc were used. LC_{50} values showed that *A. fluviatilis* was relatively tolerant of zinc in solution in the short term, with larger snails being 2–4 times more tolerant than smaller ones, although the difference between sizes decreased with increasing exposure time. At 60 days, LC_{50} values were 600 and 200 µg/l zinc, respectively. Adults exposed to levels of up to 320 µg/l zinc would live long enough to breed, after which they would normally die, but levels of 180 µg/l zinc and above reduced reproductive capacity, and the effects increased with increasing zinc concentrations. No effects on growth nor any avoidance or attraction to zinc were noted.

Heavy Metals

Khangarot and Ray [56] used bioassays to determine the sensitivity of toad tadpoles, *(Bufo melanostictus, Schneider)* to heavy metals. The 12-, 24-, 48-, and 96-hour LC_{50} values and 95% confidence limits for silver, mercury, copper, cadmium, zinc, nickel, and chromium were tabulated. Silver was the most toxic metal, and chromium the least toxic.

Sarkar and Juna [57] determined the effects of combinations of heavy metals on the Hill activity of *Azolla pinnata*.

Azolla pinnata (7 – 10 days old) were exposed for 14 or 28 days to six heavy metals, both singly and combined, at pH 8.5. Each metal was used at a concentration of 1 mg/l in both the single and combined tests. All treatments decreased the Hill activity of the *A. pinnata* chloroplasts, relative to controls. Effects were most pronounced with the (mercury plus arsenic plus lead plus copper plus cadmium plus chromium) and (mercury plus arsenic plus lead plus copper plus cadmium) treatments. Comparison with single metal inhibitory effects indicated that the two- and three-metal combinations were antagonistic, while the four- and five-metal combinations were not.

De Zwart and Slooff [58] determined the toxicities of mixtures of heavy metals and petrochemicals to *Xenopus laevis*.

The test substances comprised three metals (mercury, cadmium, copper, alcohols, six amines, six hydrocarbons and six halogenated hydrocarbons). The individual 48-hour LC_{50} values of these compounds were determined in the clawed toad, *Xenopus laevis*. The 31 mixtures tested in subsequent LC_{50} experiments contained between 3 and 33 of the test compounds at equitoxic concentrations (i.e. at a given fraction of their individual 48-hour LC_{50} values). The toxicity of a mixture was calculated in the same way as the individual LC_{50}s, but concentration was expressed as the sum of the fractions of the LC_{50} values of the individual constituents.

Results were used to assess whether mixture constituents were acting additively or antagonistically with respect to acute toxicity.

Von Leeuwen et al. [59] studied the effects of chemical stress on the population dynamics of *Daphnia magna*. They compared two methods used in chronic toxicity studies with *Daphnia magna*. In semistatic experiments with cohorts (life table studies), survival appeared to be the dominant factor in exponential population growth. With cadmium, dichromate, metavanadate and bromide, individual growth (carapace length) was a sensitive parameter.

Roesijadi and Fellingham [60] studied the influence of pre-exposure to copper, zinc and cadmium on the effect of mercury in the mussel *Mytilus edulis*.

Mussels were pre-exposed to copper (5 ug/l), cadmium (1, 10, 50 µg/l), zinc (10, 50, 250 µg/l) or a mixture of all three metals (5 µg copper, 1 µg cadmium, 10 µg/l zinc) for 28 days prior to challenge with 75 µg/l mercury. Survival data indicated that tolerance to mercury toxicity was significantly increased by all pre-exposures except 1 µg/l cadmium, 10 µg/l zinc and 250 µg/l zinc. Analysis of gill tissue indicated that increased tolerance was associated with bioaccumulation of the metal and, in the case of copper and cadmium, with metallothionein induction. Cadmium and zinc did not bioaccumulate at their lowest exposure concentrations. In the case of 250 µg/l zinc, zinc toxicity superseded its protective effect.

Once the concentration of toxic metal has reached a certain level in water, the exposure of creatures to that concentration for a certain period of time will produce adverse effects or mortality in them. Adverse effects are commonly identifiable with the occurrence of disease, reduced growth or impaired reproducibility, although other adverse effects exist.

Regarding mortalities, it is possible to draw up tables correlating, for each toxicant and type of creature, the concentration and exposure time to that toxicant above which mortalities occur (i.e. the concentration above which the species is at risk), based on LC_{50} values. Such correlations are not always rigidly correct, as creatures can develop a reversible immunity to toxic substances when exposed for a period of time. Nevertheless, such correlations do provide a useful indicator for the species at risk.

It is, of course, true to say that even though the concentration of a toxicant is not sufficiently high to cause mortalities in a particular type of fish during exposure for a stipulated period of time, colonies of that creature may still suffer from ill health, resulting in a reduction in colony size due to impaired reproducibility or illnesses in the young, or because species involved in their food chain are at risk and diminish in numbers.

The toxicity of an element to a particular type of creature can differ appreciably between freshwater (e.g. rivers, ponds, streams), estuarine and bay waters, and open seawater. For this reason, toxicities in freshwater and seawater are discussed separately below.

Table 5.2. Effect of trace concentration (µg/l) of metals on the wellbeing of freshwater nonsalmonid and salmonid fish (from author's own files)

Element	Lethal dose (LC_{50}) Days	µg/l	Reduced growth (LC_{50}) Days	µg/l	Impaired repro-ducibility (LC_{50}) Days	µg/l	
Arsenic	100	1400	100	800	–	–	Nonsalmonid
	23	5000	23	1000	–	–	Salmonid
Zinc	30	2000	30	210	–	–	Nonsalmonid
	140	600	140	210	–	–	Nonsalmonid
Cadmium	10	4000	–	–	10	240	Nonsalmonid
	100	180	–	–	100	15	Nonsalmonid
Chromium	60	2000	60	720	–	–	Nonsalmonid
	10	18300	10	2300	–	–	Salmonid
	100	1150	100	100	100	100	Salmonid
Copper	30	200	30	100	30	–	Salmonid
	40	150	40	–	40	140	Salmonid
	72	80	72	30	72	30	Salmonid
Lead	40	900	–	–	40	70	Nonsalmonid
	90	550	–	–	90	400	Nonsalmonid
Nickel	4	35000	–	–	4	100	Nonsalmonid
	15	8000	–	–	15	50	Nonsalmonid
	100	2200	–	–	100	100	Nonsalmonid

5.1.3
Examples of Toxic Effects, Fish and Invertebrates

As discussed above, substances can cause either mortalities (characterised by LC_{50}) or adverse effects (characterised by LE 50) in exposed creatures. Table 5.2 shows LC–exposure time relationships for salmonid and nonsalmonid fish for various toxicants. As would be expected, the toxicant concentrations for a particular exposure time that causes adverse effects are lower than those that cause mortalities. Thus in the case of chromium, 100 days' exposure to 1150 µg/l chromium causes 50% mortality of salmonid fish, while exposure to 100 µg/l chromium for the same time causes reduced growth and impaired reproducibility, but no or few mortalities.

Further information on the effects of metals on freshwater fish and invertebrates is summarised in Table 5.3. The durations of the toxicity tests are not included in this table. They are, however, generally short-term tests, 4 – 14 days' exposure, and the concentrations listed are toxic effect concentrations. This information should be treated with some caution, as factors such as water hardness pH, salinity, temperature and acclimatisation of species to metals can influence toxicity (see Chap. 10). Table 5.3 is, nevertheless, a useful guide in that it highlights those species that are at risk due to adverse effects or mortality when the concentrations of the element exceed the levels quoted. Thus, it can be seen at a glance that when concentrations of mercury, copper and cadmium exceed 1 µg/l then crustaceans are

Table 5.3. Toxicity of metals to freshwater fish and invertebrate species at risk during 4 – 14 days of exposure to stated concentrations ($\longrightarrow$ indicating toxic range; from author's own files)

Concentration above which mortalities can occur		Annelids	Bivalve molluscs	Crustacea
µg/l	mg/l	Hg Cu Cd Zn Pb Cr As Ni	Hg Cu Cd Zn Pb Cr As Ni	Hg Cu Cd Zn Pb Cr As Ni
0.1 – 1	0.0001 – 0.001			
1 – 10	0.001 – 0.01			
10 – 100	0.01 – 0.1			
100 – 1000	0.1 – 1			
	1 – 10			
	10 – 100			

Concentration above which mortalities can occur		Gastropods	Rotifers	Insects
µg/l	mg/l	Hg Cu Cd Zn Pb Cr As Ni	Hg Cu Cd Zn Pb Cr As Ni	Hg Cu Cd Zn Pb Cr As Ni
0.1 – 1	0.0001 – 0.001			
1 – 10	0.001 – 0.01			
10 – 100	0.01 – 0.1			
100 – 1000	0.1 – 1			
	1 – 10			
	10 – 100			

Concentration above which mortalities can occur		Non-salmonid fish
µg/l	mg/l	Hg Cu Cd Zn Pb Cr As Ni
0.1 – 1	0.0001 – 0.001	
1 – 10	0.001 – 0.01	
10 – 100	0.01 – 0.1	
100 – 1000	0.1 – 1	
	1 – 10	
	10 – 100	

at risk, when concentrations of copper exceed 10 µg/l bivalve molluscs are at risk, when concentrations of zinc and nickel exceed 10 µg/l crustaceans are at risk, and when concentrations of lead and cadmium exceed 10 µg/l then gastropods and nonsalmonid fish are at risk. These observations regarding mercury, copper, zinc, lead and cadmium, in general, confirm those reported in Table 9.6.

There have been some general comparisons of observed water quality and biological status for a wide range of UK rivers for chromium, lead, zinc, nickel and copper [61–65]. These provide information on the ranges and annual average concentrations of metals, both dissolved and insoluble, and provide information on fishery status, e.g. none, poor, fair, good, and whether salmonid or nonsalmonid fish are found in the rivers. Some of these data are tabulated in Fig. 5.1, and they show the effect of rises in average metal levels on both yield and type of fish obtained. Increased zinc levels

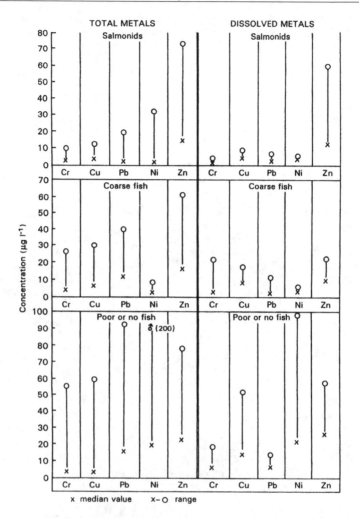

Figure 5.1. Fishery status versus element concentrations. Toxicants: chromium, copper, lead, nickel and zinc. From author's own files

does not seem to have an adverse effect, whereas increased levels of total and dissolved lead, copper, nickel and total chromium do.

It is interesting to compare the data given in Fig. 5.1 with the toxicity data for long-term (365 days) maximum safe concentrations (S_x) and the 95th percentile data (S_{95}) derived by UK authorities (as discussed in Table 10.5). This information confirms that a high mortality of fish is generally observed when the yearly, average concentration and the maximum metal concentration (Column E, Table 5.4) exceed the maximum safe concentration S_x (Column A, Table 5.4) for copper and zinc. It also confirms the adverse effect of increased metal content on fish yield and type.

Table 5.4. Comparison of data in Fig. 5.1 with S_x and S_{95} toxicity standard data taken from Table 10.5 (from author's own files)

Toxicity standards (from Table 10.5)		Concentrations (µg/l; yearly average of dissolved and total metals found in UK rivers). Supported by Table 10.5 on page 374:					
Maximum safe concentration S_x (µg/l) (365 days)	Percentile maximum metal concentration S_{95} (µg/l) for 17 days out of 365 days	Salmonids		Coarse Fish		No Fish	
A	B	Maximum	Median	Max.	Median	Max.	Median
		C		D		E	
Cr 100	800 (s)	d 2	1	22	2	18	4
100	1000 – 3000 (ns)	t 10	2	27	3	55	2
Cu 4	17 (s)	d 9	5	20	8	54	14
		t 12	3	30	5	60	2
Pb 20	100 (ns)	d 5	2	15	1	12	6
		t 20	2	40	11	95	15
Ni 220	900 (ns)	d 6	3	5	2	100	20
		t 30	1	5	1	200	18
Zn 23	200 (ns)	d 60	14	22	10	60	24
		t 75	20	60	18	80	21

5.1.3.1
Example of River Draining Rural and Urban Catchments

The results in panel A of Table 5.5 show the effect on animal life of the mean toxic metal concentrations found in UK rivers draining rural and urban catchments.

Panel A in Table 5.5 and Table 5.3 show that in rivers draining rural catchments, concentrations of cadmium, copper, lead and zinc will put crustacea at risk. In addition, high lead concentrations ($10 – 100$ µg/l) might lead to mortalities in fish and gastropods. Rivers draining urban catchments (panel B in Table 5.5) have—as would be expected—higher trace metal concentrations, and in addition to the above adverse effects, concentrations of mercury and nickel might reach levels which have adverse effect on crustaceans. In addition, molluscs might be affected by copper, gastropods by copper and lead, and fish by lead. Thus, the animal life in many rivers are at risk, and it is only in the upper reaches of rivers (where industrial activity is small or nonexistent) or in clean streams that the risk to such species is consistently low.

River Gwyddan

Table 5.6 shows analysis and fish mortality profiles of the River Gwyddan, Wales, both before and after installation of an effluent treatment plant at a steelworks which discharges into the river [66, 67]. The beneficial effect of the installation of the effluent treatment on the metal content of the river and on the survival of fish is immediately apparent.

Table 5.5. Adverse effects of metals on animal life in river waters (from author's own files)

	As	Cd	Cr	Cu	Pb	Hg	Ni	Zn
(A) Rivers draining rural catchments								
Composition of water, µg/l	1.0	0.19 – 0.33	1.4 – 6.0	3.2 – 8.8	2.4 – 13.0	0.03 – 0.09	7.2 – 9.3	10.0 – 26.0

Adverse effects expected: Crustaceans: cadmium, copper, lead and zinc; Fish and gastropods: lead

	As	Cd	Cr	Cu	Pb	Hg	Ni	Zn
(B) Rivers draining urban catchments								
Composition of water, µg/l	3.0	1.22 – 1.56	7.4 – 11.0	11.4 – 13.2	11.0 – 15.6	0.24 – 0.7	11.0 – 45.0	68.0 – 69.0

Adverse effects expected: Crustaceans: cadmium, copper, lead, zinc, mercury and nickel; Molluscs: copper; Gastropods: copper and lead; Fish: lead

	As	Cd	Cr	Cu	Pb	Hg	Ni	Zn
(C) River Carnon								
Composition of water, µg/l	60	8	–	300	–	–	–	8000

Adverse effects expected: Crustaceans: cadmium, copper and zinc; Annelids: copper and zinc; Molluscs, gastropods, rotifers, insects and nonsalmonid fish: copper

Table 5.6. Adverse effect of metals on animal life in River Gwyddan, Wales, before and after installation of effluent treatment plant (µg/l). From [66]

	Zn	Pb	Ni	Cr	Cu	Fe	Total metals	Fish found
(A) 1969 – 1973: Prior to installation of effluent treatment plant								
Upstream steelworks	0.3	0.1	< 0.1	< 0.1	< 0.05	0.63	1.06	All types, no mortalities
Downstream steelworks	1.26	1.32	0.3	1.0	0.27	325	3.29	No fish survive
Further downstream	0.29	< 0.1	0.02	0.34	0.06	50	57	No fish survive
Upstream of tidal limit	0.7	0.3	< 0.1	0.8	0.09	289	291	No fish survive
(A) 1974 – 1976: After installation of effluent treatment plant								
Upstream steelworks	0.1	0.01	< 0.01	< 0.1	0.01	3.4	3.5	All types, no mortalities
Downstream steelworks	0.12	0.01	0.02	1.4	0.01	3.0	4.5	No fish survive
Further downstream	–	0.01	–	0.02	0.01	2.5	2.5	*Salmo trutta, Anguilla, Gasterosteus aculeatus, Noemach- heilus barbatulus* all found
Upstream of tidal limit	0.08	0.01	0.01	0.01	0.01	1.2	1.3	Above plus flounder all found

5.1.3.2
Example of a Polluted River

River Carnon

The River Carnon, UK, drains an area of natural mineral enrichment which has a long history of mining for tin and other metals [68–70]. The high concentrations of metals in this river (Table 5.5) decimated most forms of animal life; the only forms that have survived are those that have developed some degree of tolerance to these metals.

Tables 5.7 and 5.8, respectively, show information on the toxic effects (LC_{50} values) and adverse effects of metals in fresh (nonsaline) waters on fish (Table 5.7) and also invertebrates (Table 5.8). See also Table 9.1.

Table 5.9 summarises the concentrations found in various environmental freshwaters (see Appendix 5.1). It is clear that concentrations vary over a very wide range, the lowest of which would produce no adverse effects on freshwater creatures upon short-term exposure, and the highest of which would have severe toxic effects, as illustrated in Table 5.10. Thus, in the case of copper at the lower end of the quoted concentration range (0.11 µg/l), no creatures are at risk during 4–14 days of exposure, while all creatures are at risk at the higher end of the concentration range quoted (200 µg/l).

It must be emphasised here that the data quoted here refer to the well-being of freshwater creatures. A further consideration is the well-being of the humans that eat these fish or crustacea. Organisms which survive might well be inedible to humans due to the presence of high levels of metals in their tissues. Unfortunately, no data are available on the long-term toxic effects of metals on creatures other than nonsalmonid fish.

Table 5.7. Concentration of metals in freshwaters and effects on fish (from author's own files)

Element	Fish type	Water type	pH	Exposure concentration	Toxicity index	Adverse effects	Reference
Acidity	Fathead minnows (*Pimephales promelas*)	Natural	5.5 – 7.0	Hard water: 160 – 200 µg/l as CaCO₃. Soft water: 5 – 10 mg/l as CaCO₃.	–	Impairment of feeding behaviour at low pH. This species virtually eliminated at pH 5.8 – 6.0.	[29]
pH	Brown trout (*Salmo trutta*)	Loch	4.3	–	–	Reduced growth and survival.	[71]
Aluminium	Brown trout (*Salmo trutta*)	Loch	≤ 5.2	–	–	Growth suppression.	[71]
pH and aluminium	Lake trout (*Salvelinus namaycush*)	Lake	5-0 pulse exposure	200 µg/l AC pulse exposure	–	No mortality in 21 – 32 days exposure.	[5]
Aluminium and calcium	Rainbow trout (*Salmo gairdneri*)	River	3.7 – 7.6	150 mg/l Ca	Larvae pH, aluminium had LC₅₀ of 3.8 mg/l in softwater (1 mg/l Ca)	pH 3.7: no survival after 5 days. pH 4.6: no survival after 9 days. pH: 5.7 – 6.6: delayed hatching.	[12]
Aluminium acidity	Rainbow trout (*Salmo gairdneri*)	Stream	4.83	240 µg/l (Al)		15% mortality and stress response.	[3]

Table 5.7. Continued

Element	Fish type	Water type	pH	Exposure concentration	Toxicity index	Adverse effects	Reference
Aluminium acidity	Blackback herring (*Alosa pseudoharengus*)	Soft fresh-water	5.0 – 7.8	0 – 400 μg/l		At hardness 23 – 25 mg/l Ca (as $CaCO_3$), in absence of aluminium, 6.9% embryo mortality at pH 5.0. In aluminium-exposed embryos 39 – 100% mortality at pH 5.0. In absence of aluminium, larval mortality ranged from 99% (pH 5.0) to 16% (pH 7.8). 100% mortality within 24 h if larvae exposed to 0.03 mg/l aluminium at pH 5.0.	[72]
Aluminium acidity	Brown trout (*Salmo trutta*)	Natural	4.3 – 6.5	14.8 – 100 μg/l		In 6 weeks' exposure, only pH 4.3 adversely affected trout growth at aluminium concentrations of > 27 μg/l, and pH 5.5 reduced growth.	[4]
Ammonium	Teleost fish (*Channa punctatus*)	River	–	–		Exposure to 100 mg/l NH_4, the safe concentration (6 months). Exposure to 500 mg/l, sublethal concentration (6 months) revealed liver and thyroid degeneration and signs of hyperactivity.	[8]

Table 5.7. Continued

Element	Fish type	Water type	pH	Exposure concentration	Toxicity index	Adverse effects	Reference
Ammonia	Rainbow trout (*Salmo gairdneri*)	Natural	–	0.027 – 0.27 mg/l		0.027 mg/l ammonium for up to 73 days gave 70% mortality when exposure began within 2 h of fertilisation of eggs. When exposure did not begin until eggs were 24 days old, 40% of eggs, yolk-sac fry and fry died at an ammonia concentration of 0.27 mg/l.	[9]
Aluminium and acidity	Brown trout (*Salmo trutta*)	Natural	5.0	60 – 230 mg/l		5 days' exposure to acid plus 230 μg/l aluminium inhibited growth. 8 days' exposure of newly hatched fish to 60 mg/l at pH 5.1 caused 57% mortality.	[16]
Ammonia (as NH$_4$Cl)	Rainbow trout (*Salmo gairdneri*), Fathead minnows (*Pimephales promelas*)	River	–	–	LC$_{50}$ 0.53 mg/l LC$_{50}$: 2.17 mg/l	– –	[33]
Cadmium	Red shiner (*Notropsis lutrensis*), Fathead minnows (*Pimephales promelas*)	Natural	–	–	96 h LC$_{50}$: < 10 mg/l	Exposure to sublethal concentration reduced ability to withstand heat stress.	[29]

Table 5.7. Continued

Element	Fish type	Water type	pH	Exposure concentration	Toxicity index	Adverse effects	Reference
Cadmium	Guppy (*Lebistes reticulates*)	Natural	–	–	96 h LC$_{50}$: 350 mg/l (male fish), 371 mg/l (female fish)	Safe concentration of cadmium: 112.9 mg/l (male), 116.5 (female).	[11]
Cadmium	Rosy barb (*Puntius conchonis*)	Freshwater	–	630 – 840 μg/l	96 h LC$_{50}$: 12.6 mg/l	Branched lesions observed. Copious mucus secretion from all over the body surface.	[12]
Cadmium	Blue tilapia (*Tilapea aurea*)	Natural	–	6.8 – 52 mg/l	–	In 16 weeks no adverse effects on survival or growth observed.	[15]
Chromium	Green alga (*Selenastrum capricornutum*)	Ground and surface water	–	–	96 h EC$_{50}$ reported	Inhibition of alkaline phosphatase activity	[16]
Copper	Midge (*Polypedilum nubifer*)	Natural	–	10 – 40 μg/l	–	Emergence success of eggs decreased from 74% to 2% of the control as the copper content of the water increased from 10 to 40 mg/l.	[74]
Copper	Bluegill sunfish (*Lepomis macrochirus*)	Natural	–	40 – 400 μg/l	–	40 μg/l copper for 8 days: fish 67% as active as control. 80 μg/l copper for 8 days: fish 61% as active as control. 400 μg/l copper for 6 days: fish 44% as active as control.	[21]

Table 5.7. Continued

Element	Fish type	Water type	pH	Exposure concentration	Toxicity index	Adverse effects	Reference
Copper	Rainbow trout (*Salmo gairdneri*)	Natural	–	200 μg/l	–	24 – 48 h exposure: tissue necrosis and stress, interference in nervous system functioning	[20]
Cyanide	Rainbow trout (*Salmo gairdneri*)	Natural	–	–	96 h LC$_{50}$	96 h LC$_{50}$ varied seasonally and with exercise.	[22]
Lead	(*Hebistes reticulates*)	Natural	–	–	96 h LC$_{50}$: 1620 mg/l (male), 1630 mg/l (female)	Safe concentrations of lead: 492 mg/l (male), 487 mg/l (female)	[73]
Mercury (as HgCl$_2$)	Teleost fish (*Channa punctatus*)	Natural	–	10 μg/l	–	6 months' exposure: mercury-induced inhibition of gonadal growth.	[24]
Selenium as selenite. selenome-thionine	Bluegill sunfish (*Lepomis macrochirus*)	Natural	–	3 – 30 μg/g Se	–	In spawning experiments after 260 days' exposure: cumulative mortality increased, cataract development occurred and larval survival reduced.	[26]
Zinc	Rainbow trout (*Salmo gairdneri*)	Natural	–	–	96 h LC$_{50}$: 26 mg/l (juveniles), 24.0 mg/l (adults)	–	[28]

Table 5.7. Continued

Element	Fish type	Water type	pH	Exposure concentration	Toxicity index	Adverse effects	Reference
Zinc	Redbelly tilapia (*Tilapia zilla*), Walking catfish (*Clarias lazera*)	Natural	–	–	96 h LC_{50}: 13–33 mg/l (*Tilapia zilla*), 26–52 mg/l (*Clarias lazera*) depending on season	–	[27]

Table 5.8. Concentration of metals in freshwaters and effects on organisms other than fish (from author's own files)

Element	Organism	Water type	pH	Exposure	Toxicity index	Adverse effects	Reference
Ammonia	Long fingernail clam (*Musculium transversum*) Crayfish (*Oronectes immunis*)	Natural	–	–	LC_{50} values increased with temperature	–	[33]
Ammonia	Long fingernail clam (*Musculium transversum*)	Stream	–	0.002–0.08 mg/l 0.04–0.25 mg/l 0.14–0.56 mg/l	– – –	Survival rate: 50–85% 30–55% Nil (growth adversely affected)	[32]
Cadmium	Crusteacean isopo (*Idothea baltica*)	Natural	–	0.01–15 mg/l	–	In chronic exposure to 0.5 mg/l Cd, did not survive 60 d	[35]
Cadmium	Water flea (*Daphnia magna*)	Natural	–	0–50 µg/l	25 d $LC_{50} = 10$ µg/l	–	[39]

Table 5.8. Continued

Element	Organism	Water type	pH	Exposure	Toxicity index	Adverse effects	Reference
Cadmium	Amphipod (Allorchestes compressa)	Natural	–	–	96 h LC_{50} = 0.78 mg/l	–	[75]
Cadmium	Water flea (Daphnia magna)	Natural	–	0–72 µg/l	No observed effect concentrations (NOEC) for survival and reproduction: 4 d = 2.1 µg/l 7 d = 0.8 µg/l 21 d = 0.8 µg/l	–	[76]
Chromium	Ceriodaphnia Sunfish (Lepomis macrochirus)	Natural	–	–	Acute toxicity value: 0.031 mg/l Cr(VI) (Ceriodaphnia) 182.9 mg/l Cr(VI) (Lepomis macrochirus)	–	[43]
Copper	Freshwater isopod (Ascellus aquaticus), Crustacean isopod (Proasellus coxalis Dollf)	River	–	0.01–0.15 mg/l	ST_{50} values recorded	0.005 mg/l Cu did not affect adult survival, juveniles had lower survival at any particular Cu level.	[45]
Copper	Amphipod (Allorchestes compressa)	Natural	–	–	96 h LC_{50} = 0.48 mg/l	–	[75]
Cobalt	Freshwater tropical perch (Colisa fasciatus)	Freshwaters	–	–	–	In presence of 232.8 mg/l Co, a decrease in muscle glycogen content observed.	[13]

Table 5.8. Continued

Element	Organism	Water type	pH	Exposure	Toxicity index	Adverse effects	Reference
Iron (ferrous)	Freshwater isopod (*Ascellus aquaticus*)	Pondwater	4.5 or 6.5	50 – 1000 mg/l	4 d LC$_{50}$ at pH 4.5 in the presence of 0.88 mg/l iron = 256 mg/l 4 d LC$_{50}$ at pH 6.5 in the presence of 0.88 mg/l iron = 431 mg/l	–	[47]
Mercury	Slipper limpet (*Crepidula fornicate*)	Natural	–	5 – 50 µg/l	2 d LC$_{50}$ was 1100 µg/l at 13.5 °C. No mortality when exposed to 1600 µg/l at 3 °C for up to 114 d.	–	[51]
Selenium (as sodium and selenite) Selenite	Water flea (*Daphnia magna*)	Fresh-waters	–	–	48 h LC$_{50}$ (for adults = 0.68 mg/l (sodium selenite) = 0.75 mg/l (sodium selenate). For juveniles = 0.55 mg/l (sodium selenate)	0.025 – 0.5 mg/l Se (as sodium selenate) adversely affected growth and reproduction	[48]
Zinc	Snail (*Ancylus fluviatilis*)	Natural	–	1 – 18 µg/l 100 – 1000 µg/l	60 d LC$_{50}$ 600 µg/l (large snails) 200 µg/l (small snails)	Adults exposed to 320 µg/l Zn survived long enough to breed, but levels above 180 µg/l Zn reduced reproductive capacity	[55]
Zinc	Amphipod (*Allorchestes compressa*)	Natural	–	–	96 h LC$_{50}$ = 2 mg/l	–	[77]

Table 5.9. Metal concentrations found in environmental freshwaters (µg/l; from author's own files)

Element	River and lake upstream	Surface water	Groundwater	All types
Aluminium (total)	73 – 3600	20 – 1430	–	73 – 3600
Aluminium (labile)	14 – 520	–	–	14 – 520
Antimony	0.08 – 0.42	–	0.77	0.08 – 0.77
Arsenic	0.42 – 4.90	–	2.3	0.42 – 4.90
Barium	10 – 23	100 – 103	4.1	10 – 103
Beryllium	0.4	< 0.01 – 1	–	< 0.01 – 1
Bismuth	0.005	–	< 0.00015 – 0.006	0.005 – 0.006
Cadmium	0.03 – 5	4 – 130	100 – 2600	0.03 – 2600
Chromium	0.05 – 23	0.2 – 180	1.0	0.05 – 180
Cobalt	0.2 – 10	–	0.11	0.11 – 10
Copper	0.11 – 200	14 – 110	3.7	0.11 – 200
Europium	0.00008 – 0.018	–	–	0.00008 – 0.018
Gold	< 0.001 – 0.036	–	–	< 0.001 – 0.036
Iron	1 – 3925	150 – 5000	0.15	0.15 – 5000
Lead	0.13 – 60	17 – 42	–	0.13 – 60
Manganese	0.97 – 1835	70 – 500	3.2	0.97 – 1835
Mercury	0.009 – 1.3	–	–	0.009 – 1.3
Molybdenum	0.74 – 4.08	–	–	0.74 – 4.08
Nickel	1.5 – 40	8 – 40	–	1.5 – 40
Scandium	–	–	0.009	0.009
Selenium	< 0.002 – 750	–	0.002 – 0.7	< 0.0002 – 750
Silver	0.3 – 32	–	–	0.3 – 32
Titanium	–	3 – 31	–	3 – 31
Uranium	0.37 – 1.36	–	–	0.37 – 1.36
Vanadium	0.1 – 24	3.9 – 24	0.63	0.1 – 24
Zinc	0.86 – 630	2.5 – 250	8.9	0.86 – 630
Bromine	0.7 – 7500	40 – 140	200 – 28000	0.7 – 28000
Iodine	–	–	10	10
Nitrogen	190 – 2940	1500 – 91000	–	1060 – 9100
Phosphorus	20 – 800	–	–	20 – 800
Silicon	3000 – 5800	–	–	3000 – 5800
Sulfur	20	–	–	20
Borate	0.12 – 0.25	–	44	0.12 – 44
Bromide	–	–	7.8	7.8
Fluoride	100 – 600	–	–	100 – 600
Phosphate	160 – 550	–	–	160 – 550

Table 5.10. Creatures at risk in typical environmental freshwaters: short-term and long-term exposure to metallic contaminants (from Tables 5.3 and 5.9). From author's own files

Creatures	Short-term (4-14 days') exposure leading to mortalities. Element concentration (µg/l) at which creatures do not survive								Long-term (365 days') exposure leading to mortalities. Element concentration (µg/l) at which salmonids and coarse fish do not survive				
	Hg	Cu	Cd	Zn	Pb	Cr	As	Ni	Cu	Zn	Pb	Cr	Ni
Annelids	>100	>100	>100	>1000	>10000	-	-	>1000	-	-	-	-	-
Molluscs	-	>10	-	-	-	-	-	-	-	-	-	-	-
Crustaceans	>0.1	>1	>0.1	>10	>10	-	>100	>10	-	-	-	-	-
Gastropods	-	>10	>10	-	>10	-	>10000	>10000	-	-	-	-	-
Rotifers	-	>100	-	-	>10000	-	>1000	-	-	-	-	-	-
Insects	>1000	>100	>100	>10000	>10000	-	>10000	>1000	-	-	-	-	-
Nonsalmonid fish	>100	>100	>10	>1000	>10	>10	>100	>100	14	24	6	4	20
Range of metal concentrations found in environmental freshwaters (Table 5.9)	0.009–1.3	0.11–200	0.03–2600	0.86–630	0.13–60	0.05–180	0.42–490	1.5–40	0.11–200	0.86–630	0.13–60	0.05–180	1.5–40
Creatures at risk during short-term exposure in environmental waters	Crustaceans	All types	All types	Crustaceans	Crustaceans, gastropods, fish	Fish	Crustaceans non-salmonid fish	Crustaceans					
Creatures at risk during long-term exposure in environmental waters									Nonsalmonid fish				

Table 5.11. Metal determinations (µg/l) in seawater (1974 – 1980; from author's own files)

Element	1974	1977	1978/1980
Copper	0.5 – 6 [77,78]	0.1 [79]	0.1 – 0.2 [80,81]
Cadmium	–	0.17 [82]	15 – 17 [81,83]
Zinc	–	15 [82]	0.03 – 0.35 [80,81]
Nickel	–	2.0 [82]	0.25 – 0.39 [80,81,83]
Lead	–	0.4 [82]	100***
Cobalt	–	0.2 [84]	0.02 – 0.03 [83] ***
		0.2 [84]	0.02 – 0.03 [83] ***

*** Sturgeon RE, Berman SJ, Desauliniers A, Mytytiuk A, McClaren JW, Russell J, private communication

5.2
Cations in Saline, Sea and Coastal Waters

The natural concentrations of trace metals in relatively unpolluted open seawater (where the effect of coastal discharges is minimal) are very low, and the accurate determination of these concentrations has presented a great challenge to the analytical chemist. As such, low-level contamination of the sample by sampling equipment and neighbouring ships are important factors that affect the accuracy of results, and it is only in recent years, in fact, that reliable techniques have evolved. For this reason, in the discussion that follows, only results obtained since 1975 are quoted, as these represent the most accurate available. In general, lower values for metals in seawater have been obtained in recent years compared with in earlier years due to the control of contamination (Table 5.11).

5.2.1
Fish

Aluminium

Atlantic salmon smolts were exposed to acidic, aluminium-rich salt water, pH 5.1, (calcium 1 mg/l, labile aluminium (60 µg/l). Mortalities occurred when the pH was 6.05 or below and did not occur at pH 6.45 or above [87].

Striped bass (*Morone saxatilis*) aged between 4 and 195 days undergo mortalities at lower values upon exposure to 25 – 400 µg/l aluminium at a pH of between 5.0 and 7.2.

Cyanide

Cyanide concentrations of 0.005 – 0.03 mg/l have been reported to be safe levels for aquatic organisms. Ruby et al. [23] have shown, however, that a level of 0.005 mg/l has adverse effects on yolk synthesis in Atlantic salmon (*Salmo salar*).

Heavy Metals and Silver

Hardy et al. [85] studied the relationship between the concentrations of these contaminants and their toxicity to fish in the sea.

This study was conducted in three polluted embayments (Elliott Bay, Commencement Bay, Port Angeles Harbor) and two reference sites (Sequim Bay, Central Sound). The sea-surface microlayers of the urban bays are toxic to fish eggs and larvae, with toxicity being associated with the presence of surface slicks. Concentrations of metals (cadmium, lead, zinc, silver, copper) in the sea-surface microlayer showed temporal and spatial variations, though levels were generally highest in Elliott and Commencement Bays and lowest in Sequim Bay. Total contaminant concentrations were inversely correlated with the percentage of fish eggs producing normal live larvae, but principal component analysis failed to identify any single inorganic or organic component of the complex contaminant mixture as being responsible for the observed toxicity.

5.2.2
Invertebrates

Cadmium

Exposure of the estuarine mysid (*Mysidopsis bahia*) for 96 hours to cadmium chloride at salinities of between 6 and 38 g per thousand g show increased toxic effects, as indicated by their 96-hour LC_{50} values, with increasing salinity [41].

Calcium Oxide

Calcium oxide is used to treat mussel (*Mytilus edulis*) beds. It has been found that whereas quicklime has no adverse effect on mussels, it does adversely affect starfish (*Asterias vulgaris*) in coastal waters. It has no adverse effect on other creatures likely to be present in mussel beds, i.e. bloodworms (*Glycra dibranchiata*), sandworms (*Littorina littorea*) and juvenile lobsters (*Homarus americanus*) [42].

Chromium

The effect of chromium exposure on the survival and development of young adult and larval crustaceans (*Palaeomonetes varians, Palaemon elegans, Neomsis integer, Praunus flexosus* and *Daphnia magna*) has been studied by exposing them to sodium chromate (0.001 – 2 mol/l) for 40 days. Minimal concentrations affect adult and larval mortality and larval development (MEC). No observed effect concentration (NOEC) and four- and ten-day LC_{50} values were obtained [44]. Uptake of chromium by juvenile plaice (*Pleuronectes platessal*) has been shown to reduce average fish size and to produce other histological changes [14].

Lanthanides

Drabaek et al. [49] determined concentrations of lanthanum, cerium, neodymium, samarium, europium, terbium, ytterbium and lutetium in wastewater from a fertiliser production plant, and in sediments, mussels (*Cyprina islandica, Mytilus edulis*) and fish (flounder) sampled at various sites in the contaminated Danish marine environment at Lillebaelt. To improve upon the detection limits offered by instrumental neutron activation analysis (INAA), an attempt was made to combine INAA with a simple destruction (using a mixture of nitric, sulfuric and perchloric acid) and preconcentration (Drabaek [49]) method. The data obtained are reported, and the experience gained using these techniques is discussed. Only in the case of fish were the authors unable to produce any of the results.

Lead

It has been shown that there is significant relationship between lead concentrations in mussel (*Mytilus edulis*) and lead concentrations in seawater. The lead level in seawater should not exceed 1.27 µg/l in order to avoid adverse effects on the mussels and on the humans that eat them.

Heavy Metals

The acute toxicities of copper, mercury, cadmium and zinc in fiddler crabs (*Uca annuliges* and *Uca triangularis*) collected in Visakhapatan Harbour have been determined. The 96-hour LC_{50} values for the two creatures were 2.75 and 43.23 mg/l Cd and 76.95 and 66.42 mg Zn/l, respectively [87]. Verriopoulos et al. [88] determined LC_{50} values for copper and chromium in *Artemia salina*.

Ahsanullah et al. [75] studied the individual and combined effects of zinc, cadmium and copper on the Marine amphipod *Allorchestes compressa*.

The 96-hour LC_{50} values of copper, cadmium and zinc for *A. compressa* were 0.48, 0.78, and 2.00 mg/l, respectively. The potency ratios were cadmium:zinc 2:57, copper:zinc 4:13 and copper:cadmium 1:61. The models used to predict the mortality of *A. compressa* (independent dissimilar and simple similar action) were noninteractive. In the combinations zinc/cadmium and cadmium/copper, the expected mortalities were significantly different from the observed values. In the zinc/copper mixture the two metals acted antagonistically. For the combination of three metals, the mortalities were predicted by the simple similar action model. Except for the zinc/cadmium combination, the toxic unit concept underestimated the toxicities of combinations of two and three metals.

Peersada and Dickinson [89] reported on the levels of lead, nickel, zinc, copper, cadmium and iron in the oysters *Saccostrea cucullata* and *Saccostrea echinata* in Darwen Harbour. Values ranged from 0 to 100 µg/g wet weight of oyster.

5.2.3
Examples of Toxic Effects

Information on the effects of metals on marine creatures is summarised in Table 5.12. The durations of the toxicity tests are not included in this table, as discussed in Sect. 5.1.3. They are, however, generally short-term tests and the concentrations quoted are toxic effect concentrations, i.e. concentrations above which mortalities can occur. Table 5.12 is a useful guide in that it highlights those species at risk when concentrations of the stated elements exceed the levels quoted in short-term exposures. If the analytical composition of a marine water is known (Table 5.13), then reference to Table 5.12 shows the adult and larval species at risk. It can be seen, for example, that when concentrations of copper or mercury exceed 1 µg/l then adult and larval bivalve molluscs are at risk, when concentrations of mercury exceed 1 µg/l then crustacea larval are at risk, and when concentrations of cadmium exceed 1 µg/l then adult bivalve molluscs are at risk. When concentrations of mercury exceed 10 µg/l, the following species are at risk: molluscs and crustacea (adult and larval) and adult fish. When concentrations of copper exceed 10 µg/l then adult larval annelids, bivalve molluscs, crustacea and fish are at risk, as are adult echinoderms and hydrozoans. With cadmium concentrations above 10 µg/l, adult bivalve molluscs and crustacea as well as hydrozoans are at risk. This concentration of zinc causes mortalities in adult and larval bivalve molluscs, and nickel similarly affects larval echinoderms and fish.

From the reported metal concentrations in open seawaters and coastal waters (Table 5.13 and 5.14), respectively (for more details see Appendix 5.2), it is seen that in each case the observed metal concentrations can vary over a wide range. In the case of open seawater (Table 5.12), the only creatures at risk are adult and larval bivalve molluscs due to short-term exposure to copper at the higher end of the concentration range found as quoted (i.e. > 1 µg/l), and the same creatures when exposed to zinc at the higher end of the concentration range found as quoted in Table 5.13 (i.e., 10.9 µg/l) for a short exposure period.

A much more serious situation exists in the case of coastal, bay and estuary water. While the maximum concentrations of lead, chromium, arsenic and nickel found in these waters do not present any risk to creatures during short-term exposures, the same cannot be said for mercury, copper, cadmium or zinc (Table 5.12). Some of the creatures at risk in short-term exposures include juvenile and adult bivalve molluscs (mercury, copper, zinc and cadmium), juvenile and adult crustacea (mercury, copper and zinc), juvenile and adult annelids (copper), adult annelids (zinc), juvenile and adult fish (mercury and cadmium), hydrozoans (copper and cadmium), echinoderms and gastropods (copper).

It is interesting at this point to compare the relative toxicities of different elements to freshwater and marine adult creatures when subjected to short-term exposure (4 – 14 days) to these elements. The metal toxicity data

Table 5.12. Toxicity of metals to marine fish and invertebrates: species at risk during 4 – 14 day exposure to stated concentrations (lines indicate the toxic range; from author's own files)

Concentration above which mortalities can occur		Adult species		
		Annelids	Bivalve molluscs	Crustaceans
μg/l	mg/l	Hg Cu Cd Zn Pb Cr As Ni	Hg Cu Cd Zn Pb Cr As Ni	Hg Cu Cd Zn Pb Cr As Ni
0.1 – 1	0.0001 – 0.001			
1 – 10	0.001 – 0.01			
10 – 100	0.01 – 0.1			
100 – 1000	0.1 – 1			
	1 – 10			
	10 – 100			

Concentration above which mortalities can occur		Adult species		
		Echnioderms	Fish	Gastropods
μg/l	mg/l	Hg Cu Cd Zn Pb Cr As Ni	Hg Cu Cd Zn Pb Cr As Ni	Hg Cu Cd Zn Pb Cr As Ni
0.1 – 1	0.0001 – 0.001			
1 – 10	0.001 – 0.01			
10 – 100	0.01 – 0.1			
100 – 1000	0.1 – 1			
	1 – 10			
	10 – 100			

Concentration above which mortalities can occur		Adult species	Larval species	
		Hydrozoans	Larval annelids	Larval bivalves
μg/l	mg/l	Hg Cu Cd Zn Pb Cr As Ni	Hg Cu Cd Zn Pb Cr As Ni	Hg Cu Cd Zn Pb Cr As Ni
0.1 – 1	0.0001 – 0.001			
1 – 10	0.001 – 0.01			
10 – 100	0.01 – 0.1			
100 – 1000	0.1 – 1			
	1 – 10			
	10 – 100			

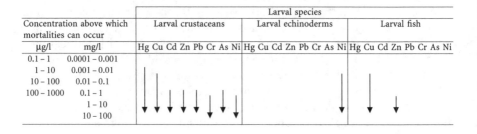

Concentration above which mortalities can occur		Larval species		
		Larval crustaceans	Larval echinoderms	Larval fish
μg/l	mg/l	Hg Cu Cd Zn Pb Cr As Ni	Hg Cu Cd Zn Pb Cr As Ni	Hg Cu Cd Zn Pb Cr As Ni
0.1 – 1	0.0001 – 0.001			
1 – 10	0.001 – 0.01			
10 – 100	0.01 – 0.1			
100 – 1000	0.1 – 1			
	1 – 10			
	10 – 100			

Table 5.13. Ranges of metal concentrations found in open seawaters (post 1975; from author's own files)

Element	Concentration range found in open surface seawater (µg/l)	Consensus value (µg/l)
Aluminium	0.1 – 0.6	–
Bismuth	< 0.000003 – < 0.000005	–
Cadmium	0.01 – 0.126	0.03
Chromium (total)	0.005 – 1.26	–
Cobalt	0.003 – 0.16	0.005
Copper	0.0063 – 2.8	0.05
Iron	0.2 – 320	0.2
Lead	0.000041 – 9.0	–
Manganese	0.018	0.02
Mercury	0.002 – 0.078[a]	< 0.2
Molybdenum	3.2 – 12.0	–
Nickel	0.15 – 0.93	0.17
Rare earths	61.7 (nmole/kg)	–
Rhenium	6 – 8	–
Selenium	0.00095 – 0.029	–
Silver	0.08	–
Thorium	≤ 0.0002	–
Tin	0.02 – 0.05	–
Uranium	1.9 – 2.6	–
Vanadium	0.45 – 2.0	2.5
Zinc	0.05 – 10.9	0.49

[a] Generally < 0.2 µg/l except in parts of the Mediterranean, where additional contributions due to man-made pollution are found [41,81,86,93].

in Table 5.10 and 5.17 are compared in Table 5.15 for those cases where comparable data exists for both types of water. It is seen that in some cases (where

$$\frac{\text{The concentration of metal (µg/l) producing mortalities in freshwater fish}}{\text{The concentration of metal (µg/l) producing mortalities in seawater fish}}$$

$= (a/b) = $ unity),

creatures are equally sensitive to metals

e.g., annelids to mercury and zinc,
gastropods to copper,
fish to zinc,
crustacea to arsenic, and
fish and gastropods to nickel.

In other cases, where $a/b < 1$, the freshwater creatures are more sensitive than the marine creatures:

Table 5.14. Ranges of metal concentrations (µg/l) found in coastal waters compared to concentrations in open seawater (from author's own files)

Element	Concentration range found in surface estuary, bay and coastal waters (µg/l)		Concentration range found in open surface seawater (µg/l)		C_{Wmax}/S_{Wmin}
	C_{Wmin}	C_{Wmax}	S_{Wmin}	S_{Wmax}	
Aluminium	6.4	63	0.1	0.6	630
Antimony	0.3	0.82	–	–	–
Arsenic	1.0	1.04	–	–	–
Barium	4.8		–	–	–
Bismuth	0.00005	0.68	< 0.000003	0.000005	> 226,000
Cadmium	0.015	5.0	0.01	0.126	500
Cerium	1.6	16.7	–	–	–
Chromium (total)	0.095	3.3	0.005	1.26	600
Cobalt	< 0.01	0.25	0.003	0.16	83
Copper	0.069	20.0	0.0063	2.8	3,200
Iron	1	250	0.2	322	1,250
Lanthanum	0.17	0.72	–	–	–
Lead	0.038	7.44	0.000041	9.0	181,500
Manganese	0.35	250	0.018	–	13,900
Mercury	≤ 0.00002	15.1	0.002	0.078	7,550
Molybdenum	2.1	200	3.2	12.0	63
Nickel	0.2	5.33	0.15	0.93	36
Rare earths	–	–	61.7 nmol/kg	–	–
Rhenium	–	–	6	8	–
Scandium	0.00095	0.098	–	–	–
Selenium	0.4		0.0095	0.029	–
Thorium	0.0002			≤ 0.0002	
Tin	–	–	1.9	2.6	–
Uranium	1.36	1.9	–	–	1
Vanadium	0.01	5.1	0.45	2.0	11.3
Zinc	0.007	200	0.05	10.9	5000

e.g., crustacea to mercury and copper,
annelids, crustacea and fish to cadmium,
crustacea to zinc,
crustacea, fish and gastropods to lead, and
fish to chromium.

In yet other cases, where $a/b > 1$, the seawater creatures are more sensitive than the freshwater creatures:

e.g., fish to mercury,
annelids, bivalve molluscs and fish to copper, and
annelids to lead.

Table 5.15. Comparison of short-term (4 – 14 days) concentrations of metals (µg/l) producing mortalities in adult creatures in (a) freshwater and (b) seawater (from author's own files)

Creature	Hg			Cu			Cd			Zn		
	a	b	a/b	a	b	a/b	a	b	a/b	a	b	a/b
Annelid	> 100	> 100	1	> 100	> 10	10	> 100	> 1000	0.1	> 1000	> 1000	1
Bivalve mollusc	–	–	–	> 10	> 1	10	–	–	–	–	–	–
Crustacea	> 0.1	> 10	0.01	> 1	> 10	0.1	> 0.1	> 10	0.01	> 10	> 100	0.1
Fish	–	> 10	10	> 100	> 10	10	> 10	> 100	0.1	> 1000	> 1000	1
Gastropod	–	–	–	> 10	> 10	1	–	–	–	–	–	–

Creature	Pb			Cr			As			Ni		
	a	b	a/b	a	b	a/b	a	b	a/b	a	b	a/b
Annelid	> 10000	> 100	100	–	–	–	–	–	–	> 1000	> 10000	0.1
Bivalve mollusc	–	–	–	–	–	–	–	–	–	–	–	–
Crustacea	> 10	> 1000	0.001	–	–	–	> 100	> 100	1	> 10	> 100	0.1
Fish	> 10	> 1000	0.01	> 1000	> 10000	0.1	> 100	> 1000	0.1	> 100	> 100	1
Gastropod	> 10	> 100	0.1	–	–	–	–	–	–	> 10000	> 10000	1

Thus it is dangerous to conclude what the toxic effect of an element on creatures will be in seawater from measurements made on freshwater creatures and vice versa.

While the concentrations of toxic metals in open seawaters might be sufficiently low to cause no adverse effects on sea creatures (see below), the same cannot be said for estuary and coastal waters, as these might be contaminated by metals originating from industrial discharges (Table 5.16, panels A and B) or coastal sewage discharges (panel C). Reference to Table 5.16 shows that the estuary waters discussed (Severn and Humber, UK) seem to be quite clean and both support fisheries. Only occasional adverse effects on bivalve molluscs, hydrozoans and echinoderms are to be expected, while fish, crustacea, annelids and gastropods survive.

Examples of Contaminated Estuaries

The higher concentrations of metals present in a coastal water adjacent to a sewage outlet (Table 5.17) have more severe adverse effects on a wide range of creatures, including adult and larval molluscs, hydrozoans, annelids, echinoderms, gastropods, crustacea and fish. Similar comments can be made for estuary samples taken at the river outfall from a mining area (Table 5.17), and it is not surprising that the waters in both of these areas do not support any animal life.

It must also be recalled that increasing the water temperature, say from 10 °C to 30 °C, causes up to a hundredfold increase in the toxicity of cadmium, chromium, copper, lead, mercury, nickel and zinc. Therefore, more severe adverse effects would be expected in the summer months than in winter.

Table 5.16. Adverse effects of metals on marine life in estuary, coastal and seawaters (from author's own files)

Composition of water (µg/l)	As	Cd	Cr	Cu	Pb	Hg	Ni	Zn
(A) Severn Estuary water (1975 – 1980)								
	nd	0.31 – 1.48	nd	2.2 – 4.2	1.5 – 4.1	nd	1.9 – 3.6	11 – 22

Adverse effects reported: bivalve molluscs and hydrozoan adults (cadmium), bivalve molluscs, adults and larvae (copper and zinc). The water quality of this estuary is having no adverse effects on harvested creatures and is in fact sustaining salmon, eel and shellfisheries.

(B) Humber Estuary (1980 – 1982)								
	–	0.2 – 0.7	–	0.1 – 8.0	0.5 – 1.0	1.0 – 10	1.0 – 15	2.0 – 50

Adverse effects expected: bivalve molluscs and their larvae (zinc and copper), bivalve molluscs (mercury) and echinoderm larvae (nickel). The water in this estuary at times contains sufficiently high concentrations of nickel to adversely affect echinoderm larvae and of zinc and copper to affect adult bivalve molluscs. Nevertheless, these waters support fisheries of salmon, eels, sole, flounders, sprat, shrimp, cockles, whelks, crabs and lobster, bass, whiting, pouting, weever, coley, mullet, mackerel, dab and plaice.

(C) Coastal water adjacent to sewage discharge								
	1.0	1.5 – 2.5	13.0 – 16.5	48.6 – 49.5	30 – 31	0.03 – 0.09	17 – 18	113 – 115

Adverse effects expected: Adult bivalve molluscs (cadmium, copper, zinc), mollusc larvae (copper, zinc), adult hydrozoans (cadmium, copper), adult annelids (copper), annelid larvae (copper, zinc), adult echinoderms (copper), echinoderm larvae (nickel), adult gastropods (copper), adult crustacea and crustacea larvae (copper and zinc), and adult fish and fish larvae (copper). Little or no fishing would be expected in this area.

(D) Restronguet Creek (estuary of River Carnon, UK)								
	4- 8	–	2 – 8	–	2 – 30	–	–	500 – 700

Adverse effects expected: Adult bivalve molluscs (cadmium, copper, zinc) and their larvae (copper, zinc), hydrozoans (cadmium, copper), adult annelids (copper), and their larvae (copper, zinc), adult echinoderms (copper), adult gastropods (copper), adult and larvae crustacae (copper, zinc), and adult and larvae fish (copper). The high concentration of metals has decimated most forms of adult and larval life in these waters, the only forms surviving being those which have developed some tolerance to metals.

Table 5.13 presents a summary of the best available values for trace metals in open surface seawater (see Appendix 5.2 for more details). With the exception of iron, manganese, zinc and aluminium, metal concentrations are usually below 1 µg/l, and except for molybdenum, uranium, arsenic and barium this applies to all of the toxic metals. In general, minimum metal concentrations reported in numerous surveys agree with the consensus values reported in 1986 by Paulson [90].

It would be expected, and it is indeed found, that the concentrations of metals in coastal waters and estuaries are higher than those in open seawater due to pollution from rivers and coastal discharges. That this is so is shown in Table 5.14, which compares the metal contents in coastal waters with those in open seawater. Some idea of the relative concentrations of metals in coastal waters compared to those in open seawater can be

Table 5.17. Creatures at risk in typical environmental open seawaters and coastal waters adjacent to a sewage outlet: short-term (4 – 14 days) exposure to metallic contaminants (from Tables 5.12 to 5.14). From author's own files

Creature	Element concentration (μg/l) at which creatures do not survive						
	Hg	Cu	Cd	Zn	Pb	Cr	Ni
Annelids (adult)	>100	>10	>1000	>1000	>100	>100	>10000
Annelids (larval)		>10		>100	>1000	>10000	>10000
Bivalve mollusc (adult)	>0.1	>1	>1	>10	>100	>1000	>100
Bivalve mollusc (larval)	>1	>1	>100	>10	>1000	>1000	>100
Crustacea (adult)	>10	>10	>10	>100	>100	>1000	>100
Crustacea (larval)	>1	>10	>100	>100	>1000	>1000	>1000
Echinoderm (adult)		>10		>1000	>1000	>10000	>10
Gastropods		>10	>1		>100		>10000
Hydrozoans		>10	>100	>1000			
Fish (adult)	>10	>10		>1000	>1000	>10000	>100
Fish (larval)		>10					
Range of metal concentrations found in open seawaters (μg/l) (Table 5.13)	0.002 – 0.078	0.0063 – 2.8	0.01 – 0.126	0.05 – 10.9	0.000041 – 9.0	0.005 – 1.26	0.15 – 0.93
Creatures at risk during short-term exposure in open seawater	None	Bivalve molluscs (adult and larval) at high end of concentration range	None	Bivalve molluscs (adult and larval) at high end of concentration range	None	None	None
Range of concentrations found in bay, coastal and estuary waters adjacent to sewage outlet (μg/l) (Table 5.14)	0.00002 – 15.1	0.069 – 20.0	0.015 – 5.0	0.007 – 200	0.038 – 7.44	0.095 – 3.3	0.2 – 5.33
Creatures at risk during short-term exposure in above waters whn metal concentrations are at the higher end of the range quoted	Bivalve molluscs (adult and larval) Crustacea (adult and larval) Fish	Annelids (adult and larval) Bivalve molluscs (adult and larval) Crustacea (adult and larval) Echinoderm gastropods Hydrozoans Fish (adult and larval)	Bivalve molluscs (adult) Hydrozoans	Bivalve molluscs (adult and larval) Annelids (larval) Crustacea (adult and larval)	None	None	None

Table 5.18. Comparison of metal contents (µg/l) of coastal waters and rivers (from author's own files)

	Coastal water		River water		
	C_{Wmin}	C_{Wmax}	R_{Wmin}	R_{Wmax}	C_{Wmax}/R_{Wmax}
Aluminium (total)	6.4	63	73	6300	0.01
Antimony	0.3	0.82	0.08	0.42	2
Arsenic	1.0	1.04	0.42	490	0.002
Barium	0.48		10	23.0	0.021
Bismuth	0.00005	0.68	0.005		1.43
Cadmium	0.015	5.0	0.03	5.0	1
Chromium (total)	0.095	3.3	0.05	23.0	1.14
Cobalt	< 0.01	0.25	0.2	10.0	0.025
Copper	0.069	20.0	0.11	200	0.1
Iron	1	250	1	3925	0.062
Lead	0.038	7.44	0.13	60	0.12
Manganese	0.35	250	0.97	1835	0.14
Mercury	0.00002	15.1	0.009	1.3	11
Molybdenum	2.1	200	0.74	4.1	50
Nickel	0.2	5.3	1.5	4.40	0.13
Selenium	0.4		< 0.0002	> 50	0.008
Uranium	1.3	1.9	0.37	1.36	0.14
Vanadium	0.1	5.1	0.1	24	0.20
Zinc	0.007	200	0.86	630	0.33

Maximum concentration in coastal water less than maximum concentration in rivers, i.e., $C_{Wmax}/R_{Wmax} \leq 1$: Al, As, Ba, Cd, Co, Fe, Pb, Mn, Ni, Se, V, U, Zn
Maximum concentration in coastal water up to ten times greater than maximum concentration in river water, i.e., $C_{Wmax}/R_{Wmax} > 1$ to < 10: Sb, Hg, Cr
Maximum concentration in coastal water more than ten times greater than maximum concentration in river water, i.e., $C_{Wmax}/R_{Wmax} > 10$, Bi, Mo

obtained by dividing the maximum concentration found in coastal water (i.e. C_{Wmax}) by the minimum concentration found in open seawater (i.e. S_{Wmin}). It is seen in Table 5.14 that values of C_{Wmax}/S_{Wmin} range from about 200,000 (bismuth, lead) through intermediate values of 1000 to 8000 (iron, copper, mercury, zinc) to relatively low values of below 1000 (aluminium, cadmium, chromium, cobalt, molybdenum, nickel, uranium and vanadium). For relatively unpolluted coastal waters, the quotient C_{Wmin}/S_{Wmin} is, as would be expected, close to unity.

Table 5.18 compares metal concentrations in coastal waters with metal concentrations of rivers discharging into the sea. It is seen that, with the possible exception of antimony, mercury, bismuth and molybdenum, the maximum metal contents of coastal waters are considerably lower than those in river waters, as would be expected from the diluting effect of seawater. Consequently, some creatures that do not survive in rivers may do so in coastal waters.

Metal Load on the North Sea

In Table 5.19, the inputs of metals to the North Sea [91] from all land-based European sources are compared with the median estimates for the direct atmosphere fallout of metal from European sources onto the surface of the North Sea. It is seen that atmosphere fallout is a major proportion of total metal contamination from all sources, ranging from less than 20% (chromium and zinc) to more than 30% (copper, arsenic, lead, zinc, mercury, nickel and cadmium). This fallout will be due to smokestack emissions from metal industries, and a large proportion of these metals are swept from the atmosphere into the oceans by rain (Table 5.20).

When total annual land-based plus atmospheric emissions of metals are considered as percentages of the estimated quantities of metals in the North Sea water column, it is seen that the input of arsenic is small (annual input is 24% of column load), copper and nickel are intermediate (about 50% of column load) while cadmium, chromium, mercury and zinc are major contributors, being comparable with those present in the water column. The annual and land-based plus atmospheric input of lead is four times the column load, with this mainly being contributed by land-based sources.

Also included in Table 5.19 are the average soluble metal contents found in open waters in the North Sea. As the data show, the average annual concentrations of metals from land-based and atmospheric sources range from 24% (arsenic) to 439% (lead) of the weight of metals in the water column. If all of these annual additions of metals were to distribute themselves evenly through the water column and remain in solution, then the soluble metal contents of the North Sea would be expected to undergo quite dramatic annual increases, ranging from 24% to 439% of the original metal content of the water.

Thus, a lead content of $0.02\,\mu g/l$ lead at the start of 1990 would, by the end of the year 1990, have increased to

$$0.02 + \frac{439}{100} \times 0.02 = 0.108,$$

and by the end of 2001 it would have increased to

$$0.02 + \frac{11 \times 439 \times 0.02}{100} = 0.99\,mg/l$$

in other words, a fifty fold increase in eleven years.

Careful monitoring of the metal levels in the North Sea has shown that such increases in dissolved or suspended metal contents do not occur with time. This is presumably because metals absorbed onto bottom sediments are accumulated by animal life, and are converted into chemically insoluble forms which settle on the sea bed (in other words they are transferred into

Table 5.19. Total quantities of metals entering the North Sea from all land-based sources and from direct atmospheric deposition compared with the estimated mass of metals in the water column of the North Sea (from author's own files)

	Total land-based, tonnes/y A	Total atmospheric, tonnes/y B	Grand total land-based plus atmospheric, tonnes/y C = A + B	North Sea water column, tonnes D	Total land-based as % of mass in water column E = A × 100/D	Total atmospheric as % of mass in water column F = B × 100/D	Grand total land-based plus atmospheric as % of mass in water column E + F	Concentration of soluble metal in North Sea, µg/l
As	800	230	1030	4300	18.6	5.3	24.0	1.0
Cd	245	569	814	860	28.4	66.1	94.6	0.02
Cr	666	667	7328	8600	77.4	7.75	85.2	0.4
Cu	4705	3942	8647	17200	27.3	22.9	50.3	0.2
Pb	6521	2920	9441	2150	0.303	135.8	439	0.05
Hg	66.5	51	117.5	86	77.3	59.3	136.6	0.002
Ni	4052	1569	5621	10750	37.7	14.6	52.3	0.25
Zn	31508	7008	38516	43000	73.3	16.3	89.5	1.0

Table 5.20. Metal contents (µg/l) of aqueous precipitation (rain and snow) into oceans (from author's own files)

Element	Sample	Concentration (µg/l)	Range of metal content in open ocean (Table 5.13) (µg/l)	Atmospheric fallout of metals as % of metal contamination of North Sea from all sources (Table 5.19)
Mercury	Rain	0.0017 – 0.0023	0.002 – 0.078	43.6
Mercury inorganic	Rain	0.014	0.002 – 0.078	43.6
Mercury total	Rain	0.015	0.002 – 0.078	43.6
Cadmium	Snow	0.034	0.010 – 0.126	69.9
Cadmium	Snow	0.005	0.010 – 0.126	69.9
Copper	Snow	0.097	0.0063 – 2.8	45.5
Copper	Snow	0.02	0.0063 – 2.8	45.5
Lead	Snow	2.48	0.000041 – 9.0	30.9
Lead	Snow	0.05	0.000041 – 9.0	30.9
Lead	Rain	2 – 40	0.000041 – 9.0	30.9
Lead	Rain	4.7	0.000041 – 9.0	30.9
Nickel	Rain	5.0	0.15 – 0.93	27.9
Selenium	Snow	0.005 – 0.025	0.00095 – 0.029	–
Silver	Snow	3 – 300	0.08	–
Tin	Rain	0.025	0.02 – 0.05	–
Bismuth dissolved	Rain	0.0006	–	–
Bismuth total	Rain	0.003	< 0.000003 – < 0.000005	–
Antimony	Rain	0.002 – 0.089	–	–

Table 5.21. Total quantities of metals entering the North Sea from all land-based sources and from direct atmospheric deposition compared with estimated masses in the water column of the North Sea (from author's own files)

	Land-based, tonnes/y				Atmospheric deposition, tonnes/y	Grand total (land-based plus atmospheric, tonnes/y)	As percentage of grand total				
	River	Direct discharges (incl. input from industrial discharges)	Sea dumping	Dredging spoils	Total			River	Direct discharges	Sea dumping	Dredging spoils	Atmospheric
As	584	206	5	5	800	230	1030	56.6	20.0	0.48	0.48	22.3
Cd	157	22	6	60	245	569	814	19.3	2.7	0.7	7.4	69.9
Cr	1761	355	596	3949	6661	667	7328	24.0	4.8	8.1	53.9	9.1
Cu	2600	276	360	1469	4705	3942	8647	30.0	3.2	4.2	17.0	45.5
Pb	2554	150	377	3440	6521	2920	9441	27.0	1.6	4.0	36.4	30.9
Hg	27	7	2.5	30	66.5	51	117.5	22.9	6.0	2.1	25.6	43.6
Ni	2466	500	97	989	4052	1569	5621	43.9	8.9	1.7	17.6	27.9
Zn	14017	1160	950	15381	31508	7008	38516	36.4	3.0	2.5	40.0	18.2

appreciable increases in the metal contents of oceanic sediments and sea creatures).

Of course, soluble and total metals will also be swept to areas beyond the North Sea water column. The fact that the soluble metal contents in the North Sea water column are not increasing perceptibly with time is thus not as reassuring as it may seem at first: the total metal load on the North Sea and surrounding areas will be increasing each year.

The results in Table 5.21 present a more detailed breakdown of the sources of pollution for the North Sea water column. The data shows that rivers and atmospheric pollution are major contributors to metal pollution in the North Sea, and that sewage and sea dumping are the lowest contributors, with dredging spoils occupying an intermediate position:

Metal load as % of total metal load

Rivers   19.3% (cadmium) to 56.6% (arsenic)
Atmospheric   9.1% (chromium) to 69.9% (cadmium)
Dredging spoil 0.5% (arsenic) to 53.9% (chromium)
Sewage   1.6% (lead) to 20% (arsenic)
Sea dumping   0.5% (arsenic) to 8.1% (chromium)

Sea dumping of sewage and industrial waste (direct discharge) and also sea dumping from ships are clearly minor contributors to the metal pollution of the North Sea, amounting in total to no more than 3.4% (cadmium) to 20.5% (arsenic) of the pollution from all sources. Having said this, a framework for regulating these sources of pollution is essential. However, in order to stop direct discharge it would be necessary to find alternative routes for disposing of some 10.7 million tonnes per year of sewage sludge and industrial waste by methods such as landfill, incineration and farm application, which would themselves pose threats to the environment.

By far the largest reductions in pollution load on the North Sea would be achieved by controlling pollution from rivers, atmospheric pollution and dredging spoils, which together contribute to between 70% (arsenic) and 96.6% (cadmium) of the metals entering the North Sea. Controlling these sources will be essential to the future well-being of the North Sea and its surrounding oceans.

References

1. Reader JP, Dalziel TRK, Morris R (1988) *J Fish Biol* **32**:607.
2. Sadler K, Turnpenny AWH (1986) *Water Air Soil Pollut* **30**:593.
3. Muniz IP, Andersen R, Sullivan TJ (1987) *Water Air Soil Pollut* **36**:371.
4. Sadler K, Lynam S (1987) *J Fish Biol* **31**:209.
5. Gunn JM, Noakes DLG (1987) *Can J Fish Aquat Sci* **44**:418.
6. Segner H, Marthaler R, Linnenbach M (1988) *Environ Biol Fish* (1988) **21**:153.
7. Klauda RJ, Palmer RE, Lenkevich MJ (1987) *Estuaries* **10**:44.
8. Ram RN, Sathyanesan AG (1987) *Ecotoxicol Environ Safety* **13**:185.
9. Solbe JF, De LG, Shurben DG (1989) *Water Res* **23**:127.
10. Papoutsoglou SE, Abel PD (1988) *Bull Environ Contam Toxicol* **41**:404.
11. Carrier R, Beitinger TL (1988) *Water Res* **22**:511.
12. Gill TS, Pant JC, Tewari H (1988) *Ecotoxicol Environ Safety* **15**:153.
13. Sehgal R, Saxena AB (1987) *Int J Environ Studies* **29**:157.
14. Kranz H, Gereken J (1987) *J Fish Biol Supplement A*, **31**:75.
15. Greene JC, Miller WE, Debacon M, Long MA, Bartels CL (1988) *Environ Toxicol Chem* **7**:35.
16. Boge G, Bussierre D, Peres G (1988) *Water Res* **22**:441.
17. Segner H (1987) *J Fish Biol* **30**:423.
18. Lauren DJ, McDonald DG (1987) *Can J Fish Aquat Sci* **44**:105.
19. Lauren DJ, McDonald DG (1987) *Can J Fish Aquat Sci* **44**:99.
20. Nemcsok JG, Hughes GM (1988) *Environ Pollut* **49**:77.
21. Ellgaard EG, Guillot JL (1988) *J Fish Biol* **33**:601.
22. McGeachy SM, Leduc G (1984) *Arch Environ Contam Toxicol* **17**:313.
23. Ruby SM, Idler DR, So YP (1987) *Arch Environ Contam Toxicol* **16**:507.
24. Ram RN, Joy KP (1988) *Bull Environ Contam Toxicol* **41**:329.
25. Presser TS, Ohlendorf HM (1987) *Environ Management* **11**:805.
26. Woock SE, Garrett WR, Partin WE, Bryson WT (1987) *Bull Environ Contam Toxicol* **39**:998.
27. Hilmy DL, El-Domiaty NA, Daabees AY, Latife HAA (1987) *Comp Biochem Physiol* **86C**:263.
28. Meisner JD, Hum WQ (1987) *Bull Environ Contam Toxicol* **39**:898.
29. Lemly AD, Smith RJF (1987) *Environ Toxicol Chem* **6**:225.
30. Hutchinson NJ, Sprague JB (1987) *Environ Toxicol Chem* **6**:755.
31. Couillard CM, Berman RA, Panisset JC (1988) *Arch Environ Contam Toxicol* **17**:319.
32. Zischke JA, Arthur JW (1987) *Arch Environ Contam Toxicol* **16**:225.
33. Arther JW, West CW, Allen KN, Hedtke SF (1987) *Bull Environ Contam Toxicol* **38**:324.
34. Bouquegneau JM, Martoja M (1987) *Bull Environ Contam Toxicol* **39**:69.
35. Guidici MDN, Migliore L, Guarino SM, Gamardella C (1987) *Mar Pollut Bull* **18**:454.

36. Hong J-S, Reish DJ (1987) *Bull Environ Contam Toxicol* **39**:884.
37. Knowles CO, McKee MJ (1987) *Ecotoxicol Environ Safety* **13**:290.
38. Winner RW, Whitford TC (1987) *Aqua Toxicol* **10**:217.
39. Bodar CMW, Van Leeuwen CJ, Voogt PA, Zandee DI (1988) *Aqua Toxicol* **12**:301.
40. Doherty FG, Failla ML, Cherry DS (1988) *Water Res* **22**:927.
41. De Lisle PF, Roberts MH Jr (1988) *Aqua Toxicol* **12**:357.
42. Shumway SE, Card D, Getchell R, Newell C (1988) *Bull Environ Contam Toxicol* **40**:503.
43. Dorn PB, Rodgers JH Jr, Jop JC, Raia JC, Dickson KL (1987) *Environ Toxicol Chem* **6**:435.
44. Van der Meer C, Teunissen C, Boog TFM (1988) *Bull Environ Contam Toxicol* **40**:204.
45. de Nicola Giudici M, Migliore L, Guariano AM (1987) *Hydrobiol* **146**:63.
46. Anil AC, Wagh AB (1988) *Mar Pollut Bull* **19**:177.
47. Maltby L, Snart JOH, Calow P (1987) *Environ Pollut* **43**:271.
48. Johnson PA (1987) *Aqua Toxicol* **10**:335.
49. Drabaek I, Eichner P, Rasmussen L (1987) *J Radioanal Nuc Chem Articles* **114**:29.
50. Micallef S, Tyler PA (1987) *Mar Pollut Bull* **18**:180.
51. Harrison FL, Watness K, Nelson DA, Miller JE, Calabreses A (1987) *Estuaries* **10**:78.
52. Hall WS, Dickson KL, Saleh FY, Rodgers JH Jr, Wilcox D, Entazami A (1986) *Water Res Bull* **22**:913.
53. Gil MN, Harvey MA, Esteves JL (1988) *Mar Pollut Bull* **19**:181.
54. Abbasi SA, Nipaney PC, Soni R (1988) *Int J Environ Stud* **32**:181.
55. Willis M (1988) *Archiv Hydrobiol* (1988) **112**:299.
56. Khangarot BS, Ray PK (1987) *Bull Environ Contam Toxicol* **38**:523.
57. Sarkar A, Jana S (1987) *Water Air Soil Pollut* **35**:141.
58. De Zwart D, Slooff W (1987) *Bull Environ Contam Toxicol* **38**:345.
59. Van Leeuwen K, Niebeck G, Luttmer W (1987) H_2O **20**:170.
60. Roesijadi G, Fellingham GW (1987) *Can J Fish Aquat Sci* **44**:680.
61. Mance G, Brown VM, Yates J (1984) *Proposed Environmental Quality Standards for List III Substances in Water: Copper*, Water Research Centre, Marlow, Buckinghamshire, UK.
62. Mance G, Brown VM, Gardiner J, Yates J (1984) *Proposed Environmental Quality Standards for List III Substances in Water: Chromium*, Technical Report TR207, Water Research Centre, Marlow, Buckinghamshire, UK.
63. Mance G, Brown VM, Gardiner J, Yates J (1984) *Proposed Environmental Quality Standards for List II Substances in Water: Inorganic Lead*, Technical Report TR208, Water Research Centre, Marlow, Buckinghamshire, UK.
64. Mance G, Yates J (1984) *Proposed Environmental Quality Standards for List II Substances in Water: Zinc*, Technical Report TR209, Water Research Centre, Marlow, Buckinghamshire, UK.
65. Mance G, Yates J (1984) *Proposed Environmental Quality Standards for List II Substances in Water: Nickel*, Water Research Centre, Marlow, Buckinghamshire, UK.
66. Williams R, Williams PE, Benson-Evans K, Hunter MD, Harcup MF (1976) *Water Pollut Control* **76**:428.
67. Turnpenny AWH, Williams R (1981) *Environ Pollut Series A* **26**:39.
68. Klumpp PW, Peterson PJ (1979) *Environ Pollut* **19**:11.
69. Luoma Sn, Bryan GW (1982) *Est Coastal Shelf Sci* **15**:95.
70. Bryan GW, Gibbs PE (1983) *Heavy Metals in the Fal Estuary, Cornwall: A Study of Long-Term Contamination by Mining Water and its Effects on Estuarine Organisms*, Occasional Publication No.2, Marine Biology Association, Plymouth, UK.

71. Sadler K, Turnpenny AWH (1986) *Air Water Soil Pollut* **30**:593.
72. Klauder RJ, Palmer RE, Lenkevich MJ (1987) *Estuaries* **10**:44.
73. Sehgal R, Saxena AB (1987) *Int J Environ Stud* **29**:157.
74. Hatakeyama S (1988) *Ecotoxicol Environ Safety* **16**:1.
75. Ahsanullah M, Mobley MC, Rankin P (1988) *Aust J Mar Freshwater Res* **39**:33.
76. Skogheim OK, Rosseland BO, Hoell E, Kroglund F (1986) *Water Air Soil Pollut* **30**:587.
77. Muzzareli RAA, Rocchetti R (1974) *Anal Chim Acta* **69**:35.
78. Ediger RD, Peterson GE, Kerber JD (1974) *Atomic Absorption Newslett* **13**:61.
79. Sperling KRZ (1977) *Fresen Z Anal Chem* **287**:23.
80. Bruland KW, Franks RP, Knauer GA, Martin JH (1979) *Anal Chim Acta* **105**:233.
81. Smith RG Jr, Windom HL (1980) *Anal Chim Acta* **113**:39.
82. Campbell WC, Ottaway JM (1977) *Analyst* **102**:495.
83. Yeats PA, Bewers JM, Walton A (1978) *Mar Pollut Bull* **9**:264.
84. Batley GE, Matousek JP (1977) *Anal Chem* **49**:2031.
85. Hardy JT, Crecelius EA, Antrim LD, Broadhurst VL, Apts CW, Gurtisen JM, Fortman TJ (1987) *Mar Environ Res* **23**:251.
86. Talbot V (1987) *Mar Biol* **94**:557.
87. Devi VU (1987) *Bull Environ Contam Toxicol* **39**:1020.
88. Verriopoulos G, Moraitou-Apostolopoulou M, Milliou E (1987) *Bull Environ Contam Toxicol* **38**:483.
89. Peerzada N, Dickinson C (1988) *Mar Pollut Bull* **19**:182.
90. Paulson AJ (1986) *Anal Chem* **58**:183.
91. Hill JM, O'Donnell AR, Mance G (1984) *The Quantity of Some Heavy Metals Entering the Sea*, Technical Report TR205, Water Research Centre, Stevenage, Hertfordshire, UK.
92. Thomsen A, Korsgaard B, Joensen J (1988) *Aqua Toxicol* **12**:291.
93. Nath K, Kumar N (1988) *Chemosphere* **17**:465.

6 Qualitative Toxicity Data for Organic Compounds in Fish and Invertebrates

A wide variety of organic compounds can occur in fresh and marine waters. Also, naturally occurring organic compounds such as amino acids and fatty acids that are involved in food chains are present. The majority of organic compounds found result from human activities. Possible causes of water, land or atmospheric pollution from organics are industrial and other discharges, whether accidental or deliberate, land use of chemicals, substances produced due to fires and industrial smoke emissions, and domestic waste and discharges.

6.1
Fresh Waters

6.1.1
Fish

LC_{50} values and data on the adverse effects of organics on fish and creatures other than fish for a range of organic compounds in freshwaters are reviewed in Tables 6.1 and 6.2 (see also Sect. 9.2 and Tables 9.9 and 9.11).

Information on the concentrations of organic compounds encountered in nonsaline (fresh) water is given in Table 6.3 (further information is given in Appendix 6.1).

The effects of some particular organic compounds on freshwater fish and creatures other than fish are discussed next.

Polyaromatic Hydrocarbons

Polycyclic aromatic hydrocarbons constitute an important class of ubiquitous environmental pollutants [109, 110]. Because of their generally high carcinogenicities, mutagenicities and toxicities [111–113], they are considered to be priority pollutants by both the European Environmental Agency and the Environmental Protection Agency [114, 115]. The anthropogenic contribution to their presence in the environment can be essentially related to pyrolytic and petrogenic factors. Polycyclic aromatic hydrocarbons may enter the aquatic compartment through leaching of contaminated soils. In the marine environment, tank washing and accidental oil spillage represent

Table 6.1. LC$_{50}$ values for substances in freshwaters and their effects on fish (from author's own files)

Substance	Fish type	Water type	Exposure concentration	Toxicity index	Adverse effects	Reference
1	2	3	4	5	6	7
Methylene dichloride	Juvenile fathead minnow (*Pimphales promelas*)	Natural		48 h LC$_{50}$ = 502 mg/kg 192 h LC$_{50}$ = 471 mg/kg		[1]
Methyl bromide and sodium bromide	Guppy (*Poecilia reticulata*)	Natural	Methyl bromide 0.032 – 3.2 mg/l Sodium bromide 0 – 32000 mg/l		NaBr concentrations of ≥ 100 mg/l caused paralysis in 1 – 3 months. No observed lethal concentration in 1 month at 10 mg/l. Exposure to 1.8 mg/l methyl bromide for 4 days caused degenerative changes in gills and oral mucosa.	[2]
1,3-dichlorobenzene 1,4-dichlorobenzene 1,2,3,4-tetrachlorobenzene Pentachlorobenzene Hexachlorobenzene	Fathead minnows (*Pimphales promelas*)	Natural		96 h LC$_{50}$ 1,3-dichlorobenzene = 7800 µg/l 1,2,3,4-tetrachlorobenzene = 1100 µg/l 1,4-dichlorobenzene = 4200 µg/l No effect concentrations (NOEC) (highest) 1,3-dichlorobenzene = 1000 µg/l 1,4-dichlorobenzene = 570 µg/l		

Table 6.1. Continued

Substance	Fish type	Water type	Exposure concentration	Toxicity index	Adverse effects	Reference
1	2	3	4	5	6	7
				1,2,3,4-tetrachlorobenzene = 250 µg/l lowest effect concentrations (LOEC) 1,3-dichlorobenzene = 2300 µg/l 1,4-dichlorobenzene = 1000 µg/l 1,2,3,4-tetrachlorobenzene = 400 µg/l Pentachlorobenzene and hexachlorobenzene are nontoxic at 5.5 and 4.8 µg/l, respectively.		[4]
Diethylhexyl phthalate	Daphnia and fish	Natural	–	Oral 4-day LC_{50} 10 – 100 kg/g	Chronic sublethal exposure causes deterioration of reproductive capacity and immune system and carcinogenic activity	[3]
Acrylates and methacrylates	Juvenile fathead minnows (*Pimphales promelas*)	Natural	–	6 h LC_{50} reported to be lower than predicted values	Toxicity ratio 4 – 6; respiratory and metabolic inhibition toxicity ratio: 42 – 56; neurotoxicity (toxicity ratio = predicted LC_{50} divided by observed LC_{50})	[4]

Table 6.1. Continued

Substance	Fish type	Water type	Exposure concentration	Toxicity index	Adverse effects	Reference
1	2	3	4	5	6	7
1,2,4-Trichloro-benzene	Fathead minnows (*Pimphales promelas*)	Natural	–	96 h LC_{50} 2.76 mg/l	–	[5]
Pentachlorophenol	*Selenastrum capricornutium*	Natural	–	96 h LC_{50} 0.11 – 0.15 mg/l (soft water), 0.76 mg/l (hard water), (pH 7.6 – 8.4)	–	[6]
Bleached Kraft mill effluent containing 2,4,6-trichlorophenol, pentachlorophenol, 2,3,4,6-tetrachloro-phenol, and resin acid	Roach (*Rutilus rutilus*)	Kraft mill effluent diluted × 2000	–	2,4,6-Trichlorophenol, 0.05 mg/l, 2,3,4,6-trichlorophenol, 0.071 mg/l, pentachlorophenol, 0.028 mg/l	Exposure of roach to 0.035 LC_{50} for 38 days, then to 0.07 LC_{50} for 14 days, produced no evident adverse effects	[7]
Pentachlorophenol 2,4-dichlorophenol Tricaine methane-sulfonate, 1-octanol	Rainbow trout (*Salmo gairdneri*)	Natural		96 h LC_{50} (1) Pentachlorophenol, 0.09 mg/l (2) 2,4-Dichlorophenol, 4.64 mg/l (3) Tricaine methane sulfonate, 50.2 mg/l (4) 1-Octanol, 15. 8 mg/l	Survival time (h) (1) 31.6 (2) 15.2 (3) 47 (4) 5.6	[8]
Aroclor 1254	Rainbow trout (*Salmo gairdneri*)	Natural	3 – 300 mg/l	–	Severe weight reduction in 12 months exposure (also liver weight reduction)	[9]

Table 6.1. Continued

Substance	Fish type	Water type	Exposure concentration	Toxicity index	Adverse effects	Reference
1	2	3	4	5	6	7
Aroclor 1254 plus Mirex	Rainbow trout	Natural	30 mg/l Aroclor plus 5 mg/l Mirex	–	Severe weight reduction in 12 months' exposure (also liver weight reduction)	[9]
3,4-Dichloroaniline	Fathead minnows (*Pimephales promelas*)	Natural		24 h LC_{50} = 9.03 – 12 mg/l 48 h LC_{50} = 8.8 – 10.0 mg/l 96 h LC_{50} = 6.99 – 8.06 mg/l	Exposure to 5.1 – 15.7 µg/l 3,4-dichloroaniline had no effect on egg hatchability or egg survival. In 5 days fry survival, reduced on exposure to 23 µg/l 3,4-dichloroaniline for 28 days	[10]
Insecticides						
Carbaryl	Catfish (*Clarias batrachus*)	Freshwater		24 h LC_{50} = 6.1 – 16.1 mg/l 48 h LC_{50} = 53.7 – 134 mg/l 72 h LC_{50} = 48.6 – 123 mg/l 96 h LC_{50} = 46.9 – 107.7 mg/l	–	[11]
Carbaryl	Rainbow trout (*Salmo gairdneri*)	Natural	5.17 mg/l	–	Survival time 13 hours	[12]
Malathion	Rainbow trout (*Salmo gairdneri*)	Natural	0.3 mg/l	–	Survival time 40.1 hours	[12]

Table 6.1. Continued

Substance	Fish type	Water type	Exposure concentration	Toxicity index	Adverse effects	Reference
1	2	3	4	5	6	7
Malathion	Freshwater teleosts (*Channa punctatus*)	Freshwater	–	96 h LC$_{50}$ = 1.73 mg/l	–	[13]
Trichlorofon	*Cichalsoma urophthalmus* fry	Natural	0 – 80 mg/l	100% mortality in 24 hours in the presence of 60 – 80 mg/l trichorofon	Erratic swimming and hyperventilation prior to death seen at 30 – 80 mg/l. At lower concentrations fish showed loss of reflexes but survived	[14]
Roundup and Rodeo herbicides	Rainbow trout, chinock, Coho Salmon	Lake and stream	–	Roundup 96h LC$_{50}$ = 7.2 – 12 mg/l Rodeo-x-77 surfactant 96 h LC$_{50}$ = 120 – 290 mg/l Rodeo 96 h LC$_{50}$ = 580 mg/l	–	[15]
Bromacil and Diuron	Fathead minnows (*Pimphales promelas*)	Natural	–	Bromacil 24 h LC$_{50}$ = 185 mg/l 48 h LC$_{50}$ = 183 mg/l 96 h LC$_{50}$ = 182 mg/l 168 h LC$_{50}$ = 167 mg/l	1.0 – 29.0 mg/l Bromacil had no effect on fish hatch % mortality or juvenile fish survival. These concentrations did affect	[16]

Table 6.1. Continued

Substance	Fish type	Water type	Exposure concentration	Toxicity index	Adverse effects	Reference
1	2	3	4	5	6	7
				Diuron 24 h $LC_{50} = 23.3$ mg/l 48 h $LC_{50} = 19.9$ mg/l 96 h $LC_{50} = 14.2$ mg/l 168 h $LC_{50} = 7.7$ mg/l	fish growth. 2.6 – 78 mg/l Diuron did not affect fish hatch or growth. 78 µg/l Diuron decreased fish survival and increased numbers of dead fish and deformed fry	
Lindane	Carp (*Cyprinus carpio*)	Natural	0 – 1000 mg/l in food pellets	–	No adverse effects	[17]
Lindane	Teleost fish (*Anguilla anguilla*)	Lake	–	96 h LC_{50} = 0.32 – 0.54 mg/l (15 °C) 0.67 – 0.68 mg/l (22 °C) 0.5 – 0.55 mg/l (29 °C)		[18]
Methoxychlor	Rainbow trout (*Salmo gairdneri*)	Natural	0 – 580 µg/l for 2 days and 0 – 30 µg/l for 68 days	–	Methoxychlor had no adverse effect on survival, growth and development parameters. Long-term effects on reproduction not clarified.	[19]
Benomyl, Corbendazim	Rainbow trout, Channel catfish, bluegills	Natural	–	96 h LC_{50} = values reported	–	[20]

Table 6.2. LC$_{50}$ values of organic substances in fresh waters and effect on organisms other than fish

Substance	Fish type	Water type	Exposure concentration	Toxicity index	Adverse effects	Reference
1	2	3	4	5	6	7
Polycyclic aromatic hydrocarbons	Daphnid (*Daphnia magna*)	Natural	–	LC$_{50}$ values reported for various PAH	–	[21]
Sodium dodecyl sulfate, Triton-X-100 1 Sodium dodecylbenzene sulfonate	Lugworm (*Arenicola marina*)	Natural	–	48 h LC$_{50}$ sodium dodecyl sulfate 15.2 mg/l Triton X-100, 15.2 mg/l sodium dodecylbenzene sulfonate 12.5 mg/l	Lugworm gills and epidermic receptors are the most sensitive to detergents	[22]
Di-2-ethyl hexylphthalate	Daphnid (*Daphnia magna*)	Natural	0 – 811 µg/l	21-day maximum allowable toxicant concentration (MATC) between 158 and 811 µg/l	Surfacing behaviour increased on day 1 of test on dapnids exposed to 158 or 811 µg/l di-2-ethylhexylphthalate	[23]
Phenol	*Ascellus aquaticus*	Natural	–	–	Immobilisation, paralysis and mortality reported	[24]
Phenol, *o/m/p* Trimethylphenol Cresol, Xylenols	Daphnid (*Daphnia magna*)	Natural	–	–	24 LC$_{50}$ reported	[25]
2,4-Dichlorophenol and aniline	Daphnid (*Daphnia magna*)	Natural	–	MATC: 24.6 – 46.7 µg/l (aniline) 0.7 – 1.48 mg/l (2,4-dichlorophenol)	–	[26]

Table 6.2. Continued

Substance	Fish type	Water type	Exposure concentration	Toxicity index	Adverse effects	Reference
1	2	3	4	5	6	7
Ethylene dibromide	*Hydra oligactis*	Natural	–	72 h LC_{50} = 50 mg/l	–	[27]
2,2-Dichloro-biphenyl	*Daphnia pulicarria*	Lake	50 ng/l to 10 µg/l	–	Significant mortality in inhibition of reproduction at concentrations down to 50 – 100 ng/l	[28]
Insecticides Carbaryl	Lugworm (*Arenicola marina*)	Natural	–	48 h LC_{50} = 7.2 mg/l	–	[22]
Parathion ethyl	Lugworm (*Arenicola marina*)	Natural	–	48 h LC_{50} = 2.7 mg/l	–	[22]
Malathion	Toad embryos (*Bufo arenarum*)	Natural	0 – 70 mg/l	–	0 – 30 mg/l embryonic development normal. 44 mg/l, 67% mortality in 5 days	[29]
Methyl parathion	Penaeid prawn (*Metapenaeus monoceros*)	Freshwaters	0.04 – 1.2 mg/l	Sublethal concentration = 0.04 mg/l Lethal concentration = 1.2 mg/l	–	–

Table 6.2. Continued

Substance	Fish type	Water type	Exposure concentration	Toxicity index	Adverse effects	Reference
1	2	3	4	5	6	7
Fenitrothion Carbofuran	Freshwater mullet (*Channa punctatus*)	Freshwaters	1.5 mg/l (fenitrothion) 5 mg/l (carbofuran)	–	Growth abnormalities of follicle and epithelium reported	[30]
Permethrin	Snail (*Lymnaea acuminata*)	Natural	–	–	At 40 and 80% of 24 h LC_{50} dose, evidence for nerve poisoning blocking aerobic and anaerobic metabolism of snail	[31]
Lindane, endosulfan	European eel (*Anguilla anguilla*)	Natural	–	LC_{50} values reported	Toxicity at 15 °C and 29 °C greater than at 22 °C	[32]
Phosphamidon	Freshwater prawn (*Macrobrachum lamarrei*)	Freshwaters	–	Extremely toxic to *Macrobrachum lamarrei*. LC_{50} caused glycogen depletion in muscle	–	[33]
Phosphamidon	Penaeid prawn (*Metapenaeus monoceros*)	Freshwaters	0.04 – 1.2 mg/l	Sublethal concentration = 0.04 mg/l Lethal concentration = 1.2 mg/l	–	[34]
2,4-Dinitrophenoxyl acetic acid	Macro invertebrates	Pond	–	–	No adverse short-term effect on macro invertebrate communities	[35]

Table 6.2. Continued

Substance	Fish type	Water type	Exposure concentration	Toxicity index	Adverse effects	Reference
1	2	3	4	5	6	7
p-Chloroaniline	South African clawed toad (Xenopus laevis)	Natural			Death of macrophytes over one year. p-Chloroaniline at 100 mg/l kills embryos	[36]
Aniline					Aniline at ≥ 1 mg inhibited embryo development	–
Sodium dodecyl benzene sulfonate					Sodium dodecyl benzene sulfonate at 50 mg/l kills embryos	–
Methylene bisthiocyanate	(1) Nitromonas nitrobacter (2) Photobacterium phosphoreum (3) Chlorella pyrenoidosa (4) Daphnia magna (5) Poecilia reticulata (6) Salmo gairdneri	Freshwater		(1) 3 h MIC = 3200 µg/l at pH 6–8 (2) 15 min EC_{50} = 54 µg/l at pH 6–8 (3) 96h EC_{50} = 42 µg/l at pH 6–8 (4) 48h EC_{50} = 25 µg/l at pH 6–8 48h EC_{50} = 73 µg/l at pH 6–8 (5) 96h LC_{50} = 390 µg/l (6) 14 day LC_{50} = 84 µg/l 60 day LC_{50} = 65 µg/l	Phenol at 50 mg/l was lethal to tadpoles at early stages of larval development	[37]

Table 6.3. Summary of organics in natural waters (rivers, lakes and surface waters; concentration µg/l). From author's own files

Haloform	Rivers	Lakes	Surface waters
$CHCl_3$	0.02 – 0.75	54.6 – 59.1	–
$BrCl_2CH$	< 0.1 – 7.6	–	–
Br_2ClCH	< 0.1 – 4.66	–	–
Br_3CH	< 0.1 – 0.51	–	–
CCl_4	0.02 – 0.12	11.8 – 14.3	–
CH_2CHCH_2Cl	0.05 – 0.09	7.8 – 11.4	–
Cl_2CHCH_2Cl	–	8 – 20	–
$Cl_2CHCHCl_2$	–	2 – 5	–
Total haloforms	0.92 – 13.4	62.4 – 70.5	–
Total polyaromatic hydrocarbons	< 0.1 – 4.3	–	–
Chlorinated insecticides			
α-BHC	0.002 – 0.003	–	–
β-BHC	0.0004 – 0.023	–	0.006 – 0.078
γ-BHC	0.006 – 0.69	–	0.004 – 0.02
δ-BHC	0.16	–	–
DDT	0.042	–	–
p,p'DDT	0.051	–	0.009–0.037
o,p'DDT	–	–	0.005 – 0.025
DDE	0.022	–	–
p,p'DDE	–	–	0.002 – 0.010
Lindane	0.001 – 0.01	–	–
Dieldrin	0.031	–	–
Aldrin	0.02	–	–
Endrin	0.035	–	–
γ-Chlordane	0.03	–	–
Heptachlor	–	–	0.001 – 0.007
Methoxychlor	0.12	–	–
Endosulfan	0.028 – 0.28	–	–
Hexachlorobenzene	–	–	0.002 – 0.008
Total chlorinated insecticides	0.003 – 0.76	–	0.029 – 0.185
Other insecticides			
Ronnel	0.002	–	–
Dursban	0.030 – 0.043	–	–
Diazinon	0.020 – 0.037	–	–
Malathion	0.027 – 0.032	–	–
Parathion	0.037 – 0.039	–	–
Parathion methyl	0.021 – 0.038	–	–
Total organophosphorus insecticides	0.14 – 0.8	–	–
Polychlorinated biphenyls	0.0001 – 0.002 (as Aroclor 1016)	–	–
Pentachlorophenol	0.1 – 250	–	–

Table 6.3. Continued

Haloform	Rivers	Lakes	Surface waters
Dibutylphosphate	< 0.1 – 45	–	–
Di-2-ethylhexylphthalate	0.1 – 4.2	–	–
Nonionic detergents	8 – 70	–	–
Alkyl benzene sulfonates	10 – 600	–	–
Fatty acids	4.1 – 527	–	–
Nitriloacetic acid	0.4	–	–
Dissolved organic carbon	1,500 – 10,000	1,500 – 3,080	300 – 6,300
Dissolved inorganic carbon	–	1,060 – 6,190	1000
Gaseous organic carbon	–	1,900 – 2,310	–
Particulate organic carbon	–	100 – 300	–

relevant sources of polycyclic aromatic hydrocarbons. Because of their high hydrophobicities, polycyclic aromatic hydrocarbons are normally present at very low concentration levels in water, but sediments and sea crustaceans are effective polycyclic aromatic hydrocarbon collectors; therefore, monitoring of the sediment contamination level is of primary importance.

Polycyclic Aromatic Hydrocarbons

It has been postulated that polycyclic aromatic hydrocarbons cause liver, lip and skin tumours in brown bullhead trout (*Ictalurus nebulosus*) [38]. The concentrations of polycyclic aromatic hydrocarbons including benzo(*a*)-anthracene and benzo(*a*)pyrene found in organisms so affected were high (up to 16 µg/kg wet weight benzo(*a*)anthracene and up to 6.4 µg/kg wet weight benzo(*a*)pyrene).

Reference to Table 6.3 shows that up to 4.3 µg/l of polycyclic aromatic hydrocarbons have been found in river water. At these concentrations there is cause for ecological concern.

Thus, the World Health Organisation quotes a maximum permitted level of 0.2 µg/l for six carcinogenic polyaromatic hydrocarbons (fluoranthrene, benzo(*d*)-fluoranthene, benzo(*k*)-fluoranthene, benzo(*a*)pyrene, benzo(*ghi*)perylene and indeno-1,2,3-(*ed*)-pyrene), while Germany's specification for total polyaromatic hydrocarbons is 0.25 µg/l.

Chlorobenzenes

Guppies (*Poecilia reticulata*) have been exposed to 1,2,3-trichlorobenzenes (1.92, 3.78, 55.9 µmol/l), 1,2,3,4-tetrachlorobenzene (1.13, 1.69 µmol/l) or pentachlorobenzene (0.40, 0.54 µmol/l) in acute flow-through tests. In each experiment, the time of death was inversely related to toxicant concentration. Irrespective of the test compound or exposure concentration, death occurred when the internal toxicant concentration reached 2.0 – 26 µmol/g fish [39].

Carlson and Kosian [40] studied the toxicity of chlorinated benzenes to fathead minnows (*Pimephales promelas*). Compounds studied were 1,3-dichlorobenzene, 1,4-dichlorobenzene, 1,2,3,4-tetrachlorobenzene, pentachlorobenzene and hexachlorobenzene. The mean tissue residue concentrations were:

	No effect concentration (NOEC) mg/kg	Lowest effect concentration (LOEC) mg/kg
1,3-dichlorobenzene	120	160
1,4-dichlorobenzene	70	103
1,2,3,4-tetrachlorobenzene	640	1100

Tissue residue concentrations in fish chronically exposed to maximal test concentrations of pentachlorobenzene and hexachlorobenzene were 380 and 97 mg/kg, respectively.

Chlorophenols

As assessment of the sublethal effects on rainbow trout (*Salmo gairdneri*) of 2,4-dichlorophenol and 2,4,6-trichlorophenol has been carried out [41]. Both compounds were accumulated in fish even at the lowest concentration in water tested (5 µg/l), the greatest amount of chlorophenol being accumulated in the liver, adversely affecting liver enzyme activity. Rogers and Hall [42] determined three tetrachlorophenol isomers in starry flounder (*Platychthys stellatus*) muscle, bone and liver in polluted sites.

McKim et al. [8] used respiratory–cardiovascular responses of rainbow trout (*Salmo gairdneri*) to identify acute toxicity syndromes in fish. Pentachlorophenols, 2,4-dichlorophenol, tricaine, methyl sulfonate and 1-octanol were included in their studies. Decreased heart rates were observed.

Oikara and Kukkonen [7] studied the acclimatisation of roach (*Rutilus rutulus L*) to toxic components of kraft pulp mill effluents. These effluents contained 2,4,6-trichlorophenol, 2,3,4,6-tetrachlorophenol and pentachlorophenol.

Average weight gains during the period were 16.2%, but no differences were noted between the groups and no assimilatory or energetic changes were seen. Pre-exposed fish showed a significantly decreased accumulation of pentachlorophenol, but this was not associated with enhanced tolerance to the effluent itself. The unchanged growth rate in polluted waters was due to compensatory acclimatisation under potentially toxic environmental conditions.

Polychlorobiphenyls

Reijnders [43] has reported that seals that feed on polychlorobiphenyl contaminated fish undergo reproductive failure.

Cleland et al. [9] studied the effect of dietary exposure to Aroclor 1254 and Mirex on humoral immune expression of rainbow trout (*Salmo gairdneri*). No treatment-related effects were observed.

Chlorinated Insecticides

Kawano et al. [44] reported on the concentrations of chlordane compounds present in fish, seabirds, invertebrates and mammals. The metabolite oxychlordane, which is much more toxic than the parent compounds and very persistent, was found in higher concentrations in seabirds than in marine mammals.

Allyl Formate

Rainbow trout (*Salmo gairdneri*) which picked up a body burden of 100 µg/kg of allyl formate developed severe liver damage [45].

Acrolein and Benzaldehyde

McKim et al. [12] used respiratory–cardiovascualr responses of rainbow trout (*Salmo gairdneri*) to identify acute toxicity syndromes in fish.

In trout exposed to mucous membrane irritants (acrolein, benzaldehyde), an initial rapid increase in cough rate was accompanied by moderate-low increases in ventilation volume (Vg) and oxygen consumption (VO$_2$), followed by a rapid decline in Vg and VO$_2$ from midway through the survival period. Ventilation rate, oxygen utilisation and heart rate declined throughout survival time. Arterial pH, total arterial oxygen and carbon dioxide decreased in the latter half of the survival period, while haemoglobin steadily increased. These results were analysed by principal components analysis, and used to characterise fish acute toxicity syndromes for acetylcholinesterase inhibitors and respiratory irritants.

Fungicides and Weed Killers

Matthiessen et al. [46] have tabulated data on the toxicity (96 h LC$_{50}$) of mixtures of fungicides and weedkillers on rainbow trout. No evidence of synergism was found.

Hydrothiol-191 (Alkylamine Salt of Endothal) Weed Killer

Keller et al. [47] studied the effect of temperature on the chronic toxicity of hydrothiol-191 to the fathead minnow (*Pimephales promelas*). Chronic toxicity values were two (at 15 °C) to six (at 25 °C) times lower than acute toxicity values.

Phenolic Wastes

Ward et al. [48] studied the effect of phenolic silt wastes in Lake Washi, New Zealand, on the common smelt (*Retropinnia retropinna*). A reduction in numbers of fish caught was ascribed to increased loadings of silt over a period of time.

Cross-checking of the toxicity data (Tables 6.1 and 6.2) and actual concentrations occurring in water samples (Table 6.3 and Appendix 6.1) makes it possible to evaluate creatures that will be subject to adverse effects or mortalities for any particular water.

Thus, in the case of pentachlorophenol, up to 250 µg/l of this compound has been found in river waters (Table 6.3). The 96 h EC_{50} value of this compound lies in the range 90 – 760 µg/l (Table 6.1). Thus, concentrations of 250 µg/l could cause adverse effects or even fatalities, water hardness being an important parameter in this respect. Di-2-ethyl hexylphthalate can occur in rivers at concentrations of up to 4.2 µg/l (Table 6.3). As the reported 21-day maximum allowable concentration (MATC) for this compound lies between 158 and 811 µg/l (Table 6.2), no abnormal effects such as abnormal surfacing behaviour would be expected at the maximum concentration of this compound likely to occur in rivers.

Many other examples of the correlation can be obtained by comparing concentration data (see Table 6.3) and toxicity data (see Tables 6.1 and 6.2). Thus, 96 hours' exposure of *Selanastrum capricornutum* to amounts of pentachlorophenol (see Table 6.1) above 150 µg/l will kill 50% of creatures after 96 hours of exposure in soft waters (96 h LC_{50} = 110 – 150 µg/l), but not in hard waters (96 h LC_{50} = 760 µg/l).

6.1.2
Invertebrates

Endosulfan

Rajeswari et al. [49] showed that endosulfan had an acute toxicity to freshwater crabs due to its effect on the hydromineral balance of these creatures.

Lindane and Endosulfan

Thybaud and Le Bras [50] studied the adsorption and elimination of lindane by a crustacean isopod (*Asellus aquaticus*).

In 48-h experiments, bioaccumulation of lindane by *Asellus aquaticus* increased linearly with the aqueous concentration of lindane (1 – 10 µg/l). *A. aquaticus* were exposed to 2 µg/l lindane for five days, followed by three days' depuration. Uptake was rapid, with concentrations of lindane in the organisms reaching a plateau (approximately 200 ng/g) after three days. Depuration was also rapid, with over 40% of the accumulated lindane being eliminated within 24 hours. Aqueous concentrations of lindane decreased during uptake by *A. aquaticus* and vice versa.

Fenitrothion and Carbaryl

Saxena and Mani [30] exposed the freshwater mussel to supposedly safe concentrations of fenitrothion and carbaryl and histopathological changes in the thyroid gland were noted. The diameters of the follicle and the colloid of the thyroid declined significantly and the height of the epithelium increased significantly. Fenithrothion exposure also caused invasion by blood corpuscles into the follicular lumen following breakdown of the epithelium.

Takimoto et al. [51] studied the comparative metabolism of fenitrothion in the freshwater snails (*Cipangopaludina japonica* and *Physa acuta*).

In both species, fenitrothion was metabolised primarily through demethylation, hydrolysis and reduction, with the liberated phenol being conjugated with sulfate in *C. japonica* and with glucose in *P. acuta*. Whole body autoradiography of fenitrothion-exposed *P. acuta* showed that almost all of the [14]carbon was located in the liver, with small amounts in the mantle but none in the intestine.

Takimoto et al. [52] carried out comparative metabolic studies of fenitrothion in the crustaceans *Daphnia pulex* and *Palaemon paucidens*. They were exposed to 1.0 ppm radiolabelled fenitrothion in a flow-through system for one or three days, respectively, followed by 1 – 2 days' depuration. In *Daphnia pulex* and *Palaemon paucidens*, maximal bioaccumulation ratios of fenitrothion were 71 and 6, respectively, and the parent compound had a biological half-life of 5 and 1.5 hours. In both species, fenitrothion was metabolised primarily through oxidation, hydrolysis and demethylation, with the liberated phenol being conjugated with sulfate in *Daphnia pulex*.

Clark et al. [54] compared toxicity test results obtained in the laboratory with field results on estuarine animals—mysids (*Mysodopsis bahia*), grass shrimp (*Palaemonestes pugio*) and sheepshead minnow (*Papeus duorum*).

Results indicated that laboratory-derived LC_{50} values provided a reasonable basis for predicting acute mortality in field situations where fenthion persisted in the water for over 24 hours, but overestimated toxicity in habitats where fenthion concentrations decreased rapidly because of dilution and flushing. Laboratory pulse-exposure tests with rapidly changing concentrations for 12 hours were predictive of the nonlethal and lethal effects observed during these short-term (less than 24 h) field exposures.

3,4-Dichloroaniline

Van der Meer et al. [53] studied the toxicity of 3,4-dichloroaniline to various crustaceans. The effects of 3,4-dichloroaniline on the survival and development of young adult and larval crustaceans at different salinities (3.3, 23, 33 per thousand) are reported. Test species were *Palaemonetes varians*, *Palaemon elegans*, *Neomysis integer*, *Praunus flexosus* and *Daphnia magna*. Exposure concentrations ranged from 0.03 to 100 μmol 3,4-dichloroaniline per litre. Exposure duration ranged from 8 h to 40 days. Calculated values included MECs (minimum effective concentrations affecting adult mortality, larval mortality and larval development), no observed effect concentrations, four- and ten-day LC_{50} values, ET_{50} (number of days until 50% of larvae had reached the first postlarval stage), ET_{20} and ET_{80}. On the basis of comparative sensitivity (results obtained in this study), ease of culture, and the fact that it has been widely used for toxicity testing in freshwater toxicity experiments, *Daphnia magna* was the organism of choice.

Methoxychlor

Henning et al. [19] studied the effects of pulsed and spiked exposure to methoxychlor on the early life stages of rainbow trout (*Salmo gairdneri*). No adverse effects were observed in short-term tests.

2,4-Dinitrophenoxyacetic Acid (2,4-D)

Stephenson and Mackie [35] studied the effects of 2,4-D on benthic macroinvertebrate communities in artificial ponds. No primary effects were observed. However, secondary effects caused by the death and decay of the macrophytes appeared over several months, and after 338 days the diversity in the treated ponds was significantly lower than in control ponds. The treated ponds were dominated by tubificids.

Detergents

Commercial detergents may cause behavioral changes in aquatic organisms, e.g. *Brachiodontes solisianus* [55], even at low concentrations. Thus, sodium lauryl sulfate has an acute toxicity to snails (*Limnaea peregrina*) [56]. As calcium carbonate is the primary inorganic constituent of mollusc shells, this detergent might have a chelating effect on calcium or may alter epithelial permeability, thus decreasing the ability of *L. peregrina* to maintain a calcium shell.

Polychlorobiphenyls

Bridgham [28] demonstrated a chronic effect of 2,2'-dichlorobiphenyl on the reproduction, mortality, growth and respiration of *Daphnia pulicaria*. Rice and White [57] monitored polychlorobiphenyl levels in water, caged fathead minnows (*Pimephales promelas*) and caged fingernail clams (*Sphaerium striatinum*) in river water over a period of six months.

By the end of the study, polychlorobiphenyl concentrations in water and clams had declined to around pre-dredge values, but remained above control values. At the site 11 km downstream, pre- and post-dredge polychlorobiphenyl concentrations in fish were 32.1 and 61.1 µg/g dry weight, whereas corresponding concentrations in clams were 13.2 and 15.3 µg/g. Additional in situ experiments were conducted to determine uptake rates and bioconcentration factors for Aroclor 1242 and Aroclor 1254 in fathead minnows and clams. Results were similar to corresponding laboratory-derived constants.

Phenols

Devilliers [25] measured 24-hour LC_{50} values for the effects of phenol, o-, m- and p-cresol, 6-xylenol and 3-methylphenols on *Daphnia magna Straus 1820*. Cresols were found to be more toxic than phenols.

Xylenols were not significantly less toxic than cresols. Trimethylphenols were significantly less toxic than the cresols. There was no direct relationship between the number and position of methyl groups on the phenol nucleus and their toxicity to *Daphnia magna*.

Chlorophenols, Chlorobenzenes, Chloroanilines

Van Leeuwen et al. [58] studied the effects of chemical stress on the population of *Daphnia magna* caused by pentachlorophenol, pentachlorobenzene, and 2,4-dichloroaniline. 2,4-Dichloroaniline caused inhibition of reproduction. Dumpert [36] showed that 2,4-dichloroaniline inhibited embryo development in South African clawed toad (*Xenopus laevis*) and decreased survival. Le Blanc et al. [59] studied the relationship between structures of chlorinated phenols, their toxicity and their ability to induce glutathione S-transferase activity in *Daphnia magna*. There seemed to be no evidence of a relationship between induction potency and compound structure in the case of pentachlorophenol, 2,4,5-trichlorophenol, 2,4- and 2,6-dichlorophenol and 2- and 4-chlorophenol.

Di(2-Ethylhexyl) Phthalate

Woin and Larsson [60] showed that phthalate esters reduce the predation efficiency of dragonfly larvae (*Odonata aeshna*). A sevenfold increase in the phthalate content of body tissue was observed in a 40-day experiment.

6.2
Estuary and Coastal Waters

6.2.1
Fish

Malathion, Endosulfan and Fenvalerate

Trim [61] has discussed the results obtained in static 96-h toxicity tests with malathion, endosulfan and fenvalerate in estuarine waters on the Mummichog (*Fundulas heteroclitus*). All three insecticides were highly toxic to estuarine and coastal water fish.

Invertebrates

Diflubenzuron

Weis and Ma [62] studied the effects of the pesticide diflubenzuron on larval horseshoe crabs (*Limulus polyphemus*).

Endosulfan

The effect of endosulfan on the transport properties of haemocyanin in crab has been investigated [63].

At lethal concentrations, a decrease in haemocyanin synthesis coupled with a decrease in the affinity of oxygen for the pigment reduces oxygen supply to the tissues and contributes to the onset of terminal conditions.

Fenthion

Ram and Sathanesan [64] reported the LC_{50} values obtained when mysid shrimps (*Mysidopsis bahia*), grass shrimps (*Palaemonetes pugio*), pink shrimps (*Penaeus duorarum*) and sheepshead minnow (*Cyprinodon variegatus*) were exposed to spray applications (336 g per hectare) of fenthion to water on an estuarine shoreline. Mortalities and nonlethal effects occurred in these species.

Aromatic Amines

Knezovitch et al. [65] exposed bay mussels (*Mytilus edulis*) to labelled *p*-toluidine, 2-aminofluorene or 2-acetylaminofluorine, and observed high losses of tissue residues within four hours.

Phosphamidon and Methylparathion

Reddy and Rae [34] exposed intermoult penaeid prawns to sublethal and lethal concentrations of phosphamidon (0.4, 1.2 ppm) for 48 hours; this reduced acetylcholinesterase activity in the nervous tissue by 28 and 54%, respectively, relative to controls. After four and seven days' of depuration, respectively, acetylcholinesterase activity was reduced by only 9 and 2% in sublethally exposed prawns, and by only 22 and 6% in lethally exposed prawns. In similar experiments with sublethal (0.04 ppm) and lethal (0.12 ppm) methylparathion, acetylcholinesterase activity in prawn nervous tissue was reduced by 35 and 64% after 48 hours' exposure, but recovered to 20 and 32% below control values after five days of depuration, and to 7 and 15% below controls after seven days of depuration.

Dehydroabietic Acid and Benzopyrene

Kukkonen and Oikari [66] noted the effect of humic acid in water on the uptake by *Daphnia magna* and the toxicities of various organic pollutants. Accumulations in *Daphnia magna* were 50% less from the humic water in the case of dehydroabietic acid and benzopyrene. Consequently, the toxic effects of these to *Daphnia magna* was reduced. Humic acid did not affect pentachlorophenol uptake.

6.3
Seawater

Available toxicity data for organisms in seawater are reviewed in Table 6.4 (for fish) and Table 6.5 (for creatures other than fish). See also Section 9.2.

Due to the diluting effect, much lower concentrations of organics are to be expected in seawater. This is borne out by comparing the total haloform content of river water (up to 13.4 µg/l; Table 6.3) with that in seawater (0.119 µg/l; Table 6.6). An exception is, of course, the naturally occurring amino acids found in seawater where concentrations for total combined amino acids of up to 1350 µg/l have been found in the North Sea and up to 120 µg/l in the open ocean.

Total organic carbon levels in seawater are a reflection of the total amount of carbon present originating from natural and polluting sources. Levels range from about 500 to 3000 µg/l, of which only a negligible proportion is particulate or volatile (Table 6.6).

The effects of some particular organic compounds in seawater on fish and invertebrates are now discussed.

6.3.1
Fish

Molinate

Tjeerdema and Crosby [106] studied the biotransfornmation of molinate (ordram) in the striped bass (*Morone saxatilis*).

Bioconcentration, depuration and metabolism of the thiocarbamate herbicide molinate in the striped bass (*Morone saxatilis*) were investigated in a flow-through metabolism chamber. Fish were exposed to molinate for 50 hours. During the first two hours (acclimation), the fish were observed for signs of stress. During the next 24 hours (absorption), 7 ml of molinate in methanol (0.29 mg/ml) was introduced, providing a water concentration of 5 ug/l. During the final 24 hours (depuration), only molinate-free water flowed through the system. Fish were homogenised in acetonitrile and fish tissue filtered and dried before oxidising. The acetonitrile filtrate was diluted with aqueous sodium chloride before passing through an XAD-4 resin to collect metabolites. Metabolites were identified by high-performance liquid chromatography against reference standards. They included molinate sulfoxide, carboxymolinate, 4-hydroxymolinate, molinate mercapturic acid, 4-ketomolinate and hexahydroazepine. Juvenile fish were exposed to 100 ug/l of ^{14}C ring-labelled molinate for 24 hours to give a bioconcentration factor of 25.3 and a ^{14}C total concentration factor of 30.9 (in molar equivalents of molinate). After 24 hours of depuration, 90.5% of the absorbed ^{14}C had been excreted. The metabolites accounted for 19.2% percent of the depurated and 7.92% of the retained ^{14}C label. The lower toxicity of molinate to striped bass observed gave a 96 h LC_{50} of 12.1 mg/l.

Table 6.4. LC$_{50}$ values for organic substances in seawater and effect on fish (from author's own files)

Substance	Fish type	Water type	Exposure pH	Toxicity concentration	Toxicity index	Adverse effects	Reference
Hydrocarbon oils	*Myagropsis myagroides* Fensholt	Sea	–	10–10,000 mg/l	–	Allometric growth inhibited by 10,000 mg/l oil or 10–10,000 mg/l dispersant and 10–10,000 mg/l oil plus dispersant	[67]
	Pink salmon (*Oncorhynchus gorbuscha*)	Sea	–	0.7–2.4 mg/l	–	Exposure to 0.7–2.4 mg/l oil for 30 days reduced yolk reserves	[68]
	Plaice	Sea in vicinity of *Amoco Cadiz* spill	–	–	–	Polyaromatics in sediments 10 mg/kg; adverse effect on reproduction	[69]
	Baltic herring (*Clupea horengu*)	Sea	–	3–200 µg/l	–	180–200 µg/l hydrocarbon toxic to developing fish	[70]
Phthalate esters	Atlantic cod (*Gadus morhua*)	Sea	–	7–25 µg/l	–	7–25 µg/l phthalate esters toxic to developing fish	[70]
Hexazinone	Juvenile Pacific salmonids	Sea	–	–	96 h LC$_{50}$ hexazinone = 276 mg/l Pronone 109 = 904 mg/l Velpar L = 1686 mg/l	–	[71]

Table 6.4. Continued

Substance	Fish type	Water type	Exposure pH	Toxicity concentration	Toxicity index	Adverse effects	Reference
					Pronone carrier = 4330 mg/l Velpar carrier = 20,000 mg/l		
3-Trifluoromethyl-4-nitrophenol lampricide	Walleye (*Stizostedium vitreum*) Larval sea lampreys (*Petromyzon marinus*)	Sea	–	–	12 h LC_{25} 4.1 mg/l (gametes) 2.6 mg/l (eggs) 8 h $LC_{99,9}$ 1 mg/l	Adverse effects in eggs and fry	[72]
Roundup herbicide	Coho salmon smolts (*Oncorhynchus kisutch*)	Sea	–	0.03 – 2.78 mg/l	–	10 days' exposure to 0.03 – 2.78 mg/l. Roundup had no effect on survival or growth	[73]
Garlon 3A, Garlon 4A, Trichloropyrester, 3,5,6-Trichloro-2-pyridinol, 2-Methoxy-3,5,6-trichloropyridine	Juvenile Pacific salmonids	Sea	–	–	24 h to 96 h LC_{50} reported	Garlon 4 and 3,5,6-trichloro-2-pyridinol were highly toxic	[74]

Table 6.5. LC$_{50}$ values for organic substances in seawater and effect on organisms other than fish (from author's own files)

Substance	Organism	Water type	pH	Exposure, concentration	Toxicity index	Adverse effects	Reference
Hydrocarbon oil	Mussel (Mytilus edulis) Periwinkle (Littorina littorea)	Sea	–	30–129 µg/l	–	Mytilis edulis, 4–5 months exposure, investigation of mixed function oxidase system and its recovery during subsequent 2–9 months in absence of oil	[75]
Diesel oil	Mussel (Mytilus edulis)	Sea	–	27.4–127.7 µg/l	–	Storage for 6- and 9-month, comparison of survival rates in high and low nutrient seawater	[76]
Oil dispersants	Mussel (Mytilus edulis)	Sea	–	2–2.5 µg/l oil ctg 5% dispersants	–	Growth rate reduced 80–90% in 170h exposure	[77]
Hydrocarbon oil	Marine bivalve (Venus verrucosa)	Sea	7.2–7.8	–	–	Reduced pumping activity of lateral cilia. Mucus production increased	[78]
Phenol Fuel oil Kerosene Solar oil	Crustaceans (1) Shrimps (Palaemon elegans) (Palaemon adsperus) (2) Crab (Rhithropano-peusharrisi tridentatus)	Sea	–	–	–	Mean critical concentrations Phenol: 0.001 mg/l; Fuel oil: 0.4 mg/l; Kerosene: 0.01 mg/l; Solar oil: 0.01 mg/l	[79]

Table 6.5. Continued

Substance	Organism	Water type	pH	Exposure, concentration	Toxicity index	Adverse effects	Reference
	(3) Amphipod (*Pontogammarus maeoticus*) (4) Crayfish (*Balanus improvisus*)						
Polychloro-biphenyls	Common seal	Sea	–	–	–	Study of effect of PCBs on seal reproduction	[80]
Diflubenzuron	Larval horseshoe crabs (*Limulus polyphemus*)	Sea	–	–	–	50 µg/l severe mortalities 5 µg/l slight delay in moulting	[63]
Endosulfan	Crab (*Oziotelphusa senex senex*)	Sea	–	0.2 – 18.62 mg/l	96 h LC$_{50}$ (sublethal) = 6.2 mg/l (lethal) 18.62 mg/l	Increase in body weight, haemolymph volume and hydration level	[49]
Kepone	Lamprey (*Petromyzon marinus*)	Sea	–	–	96 h LC$_{50}$ 444 µg/l (12 °C) 414 µg/l (20 °C) incipient lethal concentrations = 145 µg/l		[63,81]
Tetrachloro-1,2-benzoquinone (paper mill effluent constituent)	Fourhorn sculpin (*Myoxocephalus quadricornis*) also bleak (*Alburnus alburnus*) also Perch (*Perca fluviatilis*)	Gulf	–	0.1 – 0.5 mg/l (sublethal concentrations)	–	Skeletal abnormalities	[82]

Table 6.6. Organics in seawater (from author's own files)

Substance	Sample	Concentration, µg/l	Reference
$CHCl_3$	Seawater	0.026	[83,84]
CH_3CCl_3		0.046	[83,84]
CCl_4		< 0.005	[83,84]
$CHClCCl_2$		0.015	[83,84]
CCl_2CCl_2		0.005	[83,84]
$CHBr_3$		0.027	[85,86]
Total haloforms		0.119	[85,86]
2,6-Dinitrotoluene	Dobkai Bay, Japan	< 0.02 – 2.1	[87]
2,4-Dinitrotoluene	Dobkai Bay, Japan	0.13 – 28.3	
Azarenes	Dobkai Bay, Japan		
Quinoline		0.022	
Isoquinoline		0.013	
2-Methylquinoline		0.046	
1-Methylisoquinoline		0.043	
6-Methylquinoline		0.004	
4-Methylquinoline		0.003	
2,6-Dimethylquinoline		0.016	
2,4-Dimethylquinoline		0.055	
4-Azafluorene		0.006	
Benzo(*b*)quinoline		< 0.0001	
Acridine		0.009	
Phenanthridine or benzo(*b*)quinoline		0.002	
10-Azabenzo(*a*) pyrene		< 0.0001	
Dibenz(*c,b*)acridine		0.0007	
Dibenz *a,b*)acridine		0.003	
Dibenz(*a,z*)acridine		0.004	
Nonionic detergents exposed as $C_{12}H_{25}(C_6H_4O(C_2H_4O)_6H)$	Trieste Harbour	39 – 216	[88]
Total free amino acids	Open ocean	0 – 180	[89–94]
	North Sea	20 – 180	
	Baltic Ocean	4.8 – 84.5	[95]
	Mediterranean	5 – 92	[96]
Total combined amino acids	Open ocean	3 – 130	[97]
	Open ocean	10.5 – 87.5	[98]
	Open ocean	10 – 120	[99]
	Mediterranean	28 – 200	[97]
	Baltic Ocean	500	[96]
	North Sea	35 – 1350	[95]

Table 6.6. Continued

Substance	Depth (m)	Concentration, µg/l	Reference
(A) Total organic carbon filtered		0.035 – 1.22	[100]
seawater		0.57 – 1.74	[101]
		1.49 – 3.08	[103]
		0.74 – 2.44	[102]
		0.13 – 1.63	***
Scotian Shelf (filtered)	0	0.75 - 1.07	***
	0	0.81 – 0.93	
	25	0.73 – 1.14	
	50	0.62 – 0.97	
	100	0.62 – 0.85	
	150	0.71 – 0.81	
	200	0.56 – 0.76	
	400	0.51 – 0.72	
	500	0.53 – 1.64	
	0 – 500	0.77 – 0.88	
Halifax Harbour	1	1.04 – 1.29	***
(Filtered)	10	1.03 – 1.32	
(Unfiltered)	1	1.27 – 1.90	
	10	1.18 – 1.41	
Coastal area		1.12 – 1.31	***
(Filtered)			
(Unfiltered)		1.38 – 1.68	
Sargasso Sea		0.85	[104]
Vineyard Sound		1.07	
Santa Cruz		0.99	
Norwegian Fjord		1.00	
(B) Particulate organic carbon			
Surface seawater		0.025 – 0.2	
Deep seawater		0.003 – 0.015	
(C) Volatile organic carbon			
Gulf of St Lawrence	0 – 10	0.032	[105]
	10 – 50	0.036	
	50 – 100	0.030	
	100 – 250	0.030	
Scotian Shelf	0 – 10	0.041	[105]
	10 – 25	0.038	
	100 – 250	0.035	
	250 – 750	0.033	
	750 – 1500	0.026	

*** PD Goulden (private communication)

Table 6.6. Continued

Substance	Depth (m)	Concentration, µg/l	Reference
Central and North-Western Atlantic	0 – 10	0.03	[105]
	10 – 25	0.028	
	25 – 100	0.032	
	10 – 250	0.028	
	250 – 750	0.025	
	750 – 1500	0.026	
	1500 – 3000	0.024	
	3000 – 5000	0.026	
St Margaret's Bay, Nova Scotia	0 – 40	0.031	[105]
Halifax Harbour	0 – 10	0.033	[105]

Tetrachloro-1,2-Benzoquinone

Bengtsson [82] studies the effect of tetrachloro-1,2-benzoquinone pollution on skeletal parameters in fish, fourhorn sculpin (*Myoxocephalus quadricornis*), bleak (*Alburnus alburnus*) and perch (Perca fluviatilis) collected in the Gulf of Bosnia. Vertebral deformities were observed.

Cyclophosphamide, *N*-Methyl-*N*-Nitro-*N*-Nitroguanidine

Eggs (late blastula) or 406-day larvae of striped bass (*Morone saxatilus*) or sheepshead minnow (*Cyprinodon variegatus*), when exposed to cyclophosphamide or *N*-methyl-*N*-nitro-*N* nitroguanidine of concentrations of 1 – 1000 µm for 1 – 4 days, show a close dependent relationship between aberration frequency of chromosomes and the concentration of the toxicant in eggs and larvae of both species [107].

Polychlorobiphenyls

Reijnders [80] studied the effects of polychlorobiphenyls on seal reproduction.

Organochlorine Pesticides

Ferrando et al. [32] determined the toxicity of lindane and endosulfan to eels (*Anguilla anguilla*). LC_{50} values are tabulated. Toxicity is temperature-dependent.

6.3.2
Invertebrates

Hydrocarbon Oils, Diesel Oils

Various workers have studied the toxicity of hydrocarbon and diesel oils in oceanic waters on crustaceans [76–79] including mussels (*Mytilus edulis*)

[77, 108] and marine bivalves (*Venus verrucosa*) [84], and adverse effects including reduced growth rate and pumping activity of lateral cilia were observed.

Di(2-Ethylhexyl)Phthalate

Dragonfly larvae (*Odonta aeshna*) exposed to water and sediments containing di(2-ethylhexyl)phthalate (587 – 623 mg/kg di(2-ethylhexyl)phthalate in sediment) were shown after 40 days of exposure to contain 14.7 mg/kg di(2-ethylhexyl)phthalate in the tissue. This led to a reduction in predation efficiency of these organisms [60].

References

1. Dill DC, Murphy PG, Mayes MA (1987) *Bull Environ Contam Toxicol* **39**:869.
2. Wester PW, Canton JH, Dormans JAMA (1988) *Aqua Toxicol* **12**:323.
3. Wams TJ (1987) *Sci Total Environ* **66**:1
4. Russom CL, Drummond RA, Huffman AD (1988) *Bull Environ Contam Toxicol* **41**:589.
5. Carlson AR (1987) *Bull Environ Contam Toxicol* **38**:667.
6. Smith PD, Brockway DL, Stancil, Jr., FE (1987) *Environ Toxicol Chem* **6**:891.
7. Oikara A, Kukkonen J (1988) *Ecotoxicol Environ Safety* **15**:282.
8. McKim JM, Schmeider PK, Carlson RW, Hunt EP, Niemi GJ (1987) *Environ Toxicol Chem* **6**:295.
9. Cleland GB, McElroy PN, Sonstegard RA (1988) *Aqua Toxicol* **12**:141.
10. Call DJ, Poirier SH, Knuth ML, Harting SL, Lindberg CA (1987) *Bull Environ Contam Toxicol* **38**:352.
11. Tripathi G, Shukla SP (1988) *Ecotoxicol Environ Safety* **15**:277.
12. McKim JM, Schmieder PK, Niemi RJ, Carlson RW, Henry TR (1987) *Environ Toxicol Chem* **6**:313.
13. Khangarrot BS, Ray PK (1988) *Arch Hydrobiol* **113**:465.
14. Flores-Nava A, Vizcarra-Quiroz JJ (1988) *Aquaculture Fish Management* **19**:341.
15. Mitchell DG, Chapman PG, Long TJ (1987) *Bull Environ Contam Toxicol* **39**:1028.
16. Call DJ, Brooke LT, Kent RJ, Knuth ML, Poirier SH, Huot JM, Lima AR (1987) *Arch Environ Contam Toxicol* **16**:607.
17. Cossarini-Dunier M, Monod G, Demael A, Lepoot D (1987) *Ecotoxicol Environ Safety* **13**:339.
18. Ferrando MF, Almar MM, Andreu E (1988) *J Environ Sci Health* B **23**:45.
19. Heming TA, McGuiness EJ, George LM, Blumhagen KA (1988) *Bull Environ Contam Toxicol* **40**:764.
20. Palawski DU, Knowles CO (1986) *Environ Toxicol Chem* **5**:1039.
21. Newsted JL, Giesy JP (1987) *Environ Toxicol Chem* **6**:445.
22. Conti E (1987) *Aqua Toxicol* **10**:325.
23. Knowles CO, McKee MJ, Palawski DU (1987) *Environ Toxicol Chem* **6**:201.
24. Green DW, Williams KA, Hughes DRL, Shaik GAR, Pascoe D (1988) *Water Res* **22**:225.
25. Devilliers J (1988) *Sci Total Environ* **76**:79.
26. Gersich FM, Milazzo DP (1988) *Bull Environ Contam Toxicol* **40**:1.

27. Herring CO, Adams JA, Wilson BA, Pollard S (1988) *Bull Environ Contam Toxicol* **40**:35.
28. Bridgham SD (1988) *Arch Environ Contam Toxicol* **17**:731.
29. Rosenbaum EA, Caballero de Castro A, Guana L, Pelchen de D'Angelo AM (1988) *Arch Environ Contam Toxicol* **17**:831.
30. Saxena PK, Mani K (1988) *Environ Pollut* **55**:97.
31. Singh DK, Agarwal RA (1987) *Sci Total Environ* **67**:263.
32. Ferrando MD, Andreu-Moliner E, Almar MM, Cebrian C, Nunez A (1987) *Bull Environ Contam Toxicol* **39**:365.
33. Upadhyay OVB, Shukla GS (1986) *Environ Res* **41**:591.
34. Reddy MS, Rae KVR (1988) *Bull Environ Contam Toxicol* **40**:752.
35. Stephenson M, Mackie GL (1986) *Aqua Toxicol* **9**:243.
36. Dumpert K (1987) *Ecotoxicol Environ Safety* **13**:324.
37. Maas-Diepeveen JL, Van Leeuwen CJ (1988) *Bull Environ Contam Toxicol* **40**:517.
38. Baumann PC, Smith WD, Parland WK (1987) *Trans Am Fish Soc* **16**:79.
39. Van Hoogen G, Opperhuizen A (1988) *Environ Toxicol Chem* **7**:213.
40. Carlson AR, Kosian PA (1987) *Arch Environ Contam Toxicol* **16**:129.
41. Tana A (1988) *Water Sci Technol* **20**:77.
42. Rogers IH, Hall KJ (1987) *Water Pollut Res J Canada* **22**:197.
43. Reijnders P (1987) *Environmental Protection of the North Sea*, Paper No.5, Research Institute for Nature Management, London, UK.
44. Kawano M, Inoue T, Wada T, Hidaka H, Tatsukawa R (1988) *Environ Sci Technol* **22**:792.
45. Droy BF, Hinton DE (1988) *Mar Environ Res* **24**:259.
46. Matthiessen P, Whale GF, Rycroft RJ, Sheahan DA (1988) *Aqua Toxicol* **13**:61.
47. Keller AE, Dutton RJ, Crisman TL (1988) *Bull Environ Contam Toxicol* **41**:770.
48. Ward FJ, Northcote TG, Chapman MA (1987) *Water Air Soil Pollut* **32**:427.
49. Rajeswari K, Kalarani V, Reddy DC, Ramamurthi R (1988) *Bull Environ Contam Toxicol* **40**:212.
50. Thybaud E, Le Bras S (1988) *Bull Environ Contam Toxicol* **40**:731.
51. Takimoto Y, Ohshima M, Miyamoto J (1987) *Ecotoxicol Environ Safety* **13**:118.
52. Takimoto Y, Ohshima M, Miyamoto J (1987) *Ecotoxicol Environ Safety* **13**:126.
53. Van der Meer C, Teunissen C, Boog TFM (1988) *Bull Environ Contam Toxicol* **40**:204.
54. Clark JR, Borthwick PW, Goodman LR, Patrick JM Jr, Lores EM, Moore JC (1987) *Environ Toxicol Chem* **6**:151.
55. Malagrino W (1987) *Ambiente* **1**:37.
56. Tarazona JV, Nunez O (1987) *Bull Environ Contam Toxicol* **39**:1036.
57. Rice CP, White DS (1987) *Environ Toxicol Chem* **6**:259.
58. Van Leeuwen CJ, Niebeek G, Rijkeboer M (1987) *Ecotoxicol Environ Safety* **14**:1.
59. Le Blanc GA, Hilgenberg B, Cochrane BJ (1988) *Aqua Toxicol* **12**:147.
60. Woin P, Larsson P (1987) *Bull Environ Contam Toxicol* **38**:220.
61. Trim AH (1987) *Bull Environ Contam Toxicol* **38**:681.
62. Weis JS, Ma A (1987 *Bull Bull Environ Contam Toxicol* **39**:224.
63. Viyayakumari P, Reddy DC, Ramamurthi R (1987) *Bull Environ Contam Toxicol* **38**:742.
64. Ram RN, Sathanesan AG (1987) *Ecotoxicol Environ Safety* **13**:185.
65. Knezovich JP, Lawton MP, Harrison FL (1988) *Mar Environ Res* **24**:89.
66. Kukkonen J, Oikari A (1987) *Sci Total Environ* **62**:399.
67. Jong-Hwa L (1987) *Bull Fish Sci Inst* **3**:11.
68. Moles A, Babcock MM, Rice SD (1987) *Mar Environ Res* **21**:49.

69. Brule T (1987) *J Mar Biol Assoc* **67**:237.
70. Kocan RM, Von Westernhagen H, Landolt ML, Furstenberg G (1987) *Mar Environ Res* **23**:291.
71. Wan MT, Watts RG, Moul DJ (1988) *Bull Environ Contam Toxicol* **41**:609.
72. Seelye JG (1987) *N Am J Fish Management* **7**:598.
73. Mitchell DG, Chapman PM, Long TJ (1987) *Environ Toxicol Chem* **6**:875.
74. Wan MT, Moul DJ, Watts RG (1987) *Bull Environ Contam Toxicol* **39**:721.
75. Livingstone DR (1987) *Sci Total Environ* **65**:3.
76. Lawe DM, Pipe PK (1987) *Mar Environ Res* **22**:243.
77. Strömgren T (1987) *Mar Environ Res* **21**:239.
78. Axiak V, George JJ (1987) *Mar Biol* **94**:241.
79. Kasymov AG, Gasanov VM (1987) *Water Air Soil Pollut* **36**:9.
80. Reijnders PJH *Nature* **324**:456.
81. Mallatt J, Barron MG (1988) *Arch Environ Contam Toxicol* **17**:73.
82. Bengtsson BE (1988) *Water Sci Technol* **20**:87.
83. Eklund G, Josefsson B, Roos C (1978) *J High Res Chromatogr* **1**:34.
84. Eklund G, Josefsson B, Roos C (1977) *J Chromatogr A* **142**:575.
85. Hashimoto A, Sakino H, Yamagami E, Tateishi S (1980) *Analyst* **105**:787.
86. Hashimoto A, Kozima T, Shakino H, Akiyama T (1979) *Water Res* **13**:509.
87. Shinohara R, Kido A, Okamoto Y, Takeshita R (1983) *J Chromatogr A* **256**:81.
88. Favretto L, Stancher B, Tuinis R (1978) *Analyst* **103**:955.
89. Park K, Williams TD, Prescott JM, Hood DW (1962) *Science* **138**:531.
90. Riley JP, Segar DA (1970) *J Mar Biol Assoc* **50**:713.
91. Kawahara H, Maita Y (1971) *J Oceanogr Soc Japan* **27**:27.
92. Starikova ND, Korzhikova RI (1969) *Okeanogiya* **9**:509.
93. Lee C, Bada JL (1975) *Earth Planet Sci Lett* **26**:61.
94. Ziobin VS, Perlyuk MF, Orlova TA (1975) *Okeanogiya* **15**:643.
95. Garrasi C, Regens ET (1976) *Analytische Methoden zur säulenchromatographischen Bestimmung von Aminosäuren und zwischen Meerwasser und Sediment*, Bericht aus dem Projekt DFG-DE7413, Litoralforschung – Abwasser in Küstennähe, DFG Abschlußkolloquium, Bremerhaven, Germany.
96. Dawson R, Mopper K (1977) *Anal Biochem* **84**:191.
97. Daumas RA (1976) *Mar Chem* **4**:225.
98. Tatsumato M, Williams WT, Prescott JM, Hood DW (1961) *J Mar Res* **19**:89.
99. Rittenberg SC, Emery KO, Hulsemann J, Regens ET, Fay RS, Reuter JH, Grady JR, Richardson SH, Bray EE (1963) *J Sed Petrol* **33**:140.
100. Williams PM (1969) *Limnol Oceanogr* **14**:297.
101. Mackinnon MD *Mar Chem* (1978) **7**:17.
102. Gershey RM, Mackinnon MD, Williams PJ de B, Moore RH (1979) *Mar Chem* **7**:289.
103. Goulden PD, Brooksbank P (1975) *Anal Chem* **47**:1943.
104. Tam N, Takahashi Y (1985) *Int Lab* **Sept**:49.
105. Schwarzenbach RP, Bromiend RH, Gschwend PM, Zafiron OC (1978) *Org Geol Chem* **1**:93.
106. Tjeerdema RS, Crosby DG (1987) *Aqua Toxicol* **9**:305.
107. Means JC, Daniels CB, Baksi SM (1988) *Mar Environ Res* **24**:327.
108. Lowe DM, Pipe RK (1987) *Mar Environ Res* **22**:243.
109. White KL (1986) *J Environ Sci Health C* **4**:163.
110. McElroy AE, Farrington JW, Teal JM (1989) In: Varanasi U (ed) *Metabolism of Polycyclic Aromatic Hydrocarbons in the Aquatic Environment*, CRC Press, Boca Raton, FL, USA, p. 1–40.

111. Harvey RG (ed)(1985) *Polycyclic Hydrocarbons and Carcinogenesis*, American Chemical Society, Washington, DC, USA.

112. Lee ML, Novotny MV, Bartle KD (1982) *Analytical Chemistry of Polycyclic Compounds*, Academic, New York, USA, p. 462.

113. Grimmer S (1983) In: Grimmer G (ed) *Environmental Carcinogens, Polycyclic Aromatic Hydrocarbons: Chemistry, Occurrence, Biochemistry, Carcinogenicity*, CRC Press, Boca Raton, FL, USA.

114. EPA (1982) *EPA Test Method 610*, US Environmental Protection Agency, Washington, DC, USA.

115. Hennion M-C, Pichon V, Barcelo D (1984) *Trends Anal Chem* **13**:361.

7 Qualitative Toxicity Data for Organometallic Compounds, Fish and Invertebrates

Organometallic compounds can originate in one of two ways: by direct contamination of the water with organometallic compounds, or through the production of organometallic compounds in the water or sedimentary matter or living creatures by biomethylation of inorganic metals, as caused by various types of organisms.

A limited amount of work has been carried out on the adverse effects of various types of organometallic compounds in nonsaline waters and seawater on fish and on organisms other than fish.

7.1
Nonsaline Water

7.1.1
Fish

Organomercury Compounds

Adult and six-month-old teleost fish (*Channa punctatus*) were exposed to the organomercury fungicide Emison (methoxyethyl mercuric chloride). Examination of the fish after six months of exposure revealed liver abnormalities including hyperplasia and fatty necrosis, indicative of carcinogenesis. Severe physiometabolic dysfunction would lead to mortalities in teleost fish [1].

The toxic effect of methyl mercuric chloride, methoxyethyl mercuric chloride and mercuric chloride on the survival of the catfish (*Clarias batrachus L.*) has been examined. LC_{50} values of 0.43, 4.3 and 0.507 mg/l were obtained. Kidney damage was evident in exposed specimens [2].

Organoarsenic Compounds

Juvenile rainbow trout (*Salmo gairdneri*) fed for eight weeks on a diet containing arsenic trioxide (180 – 1477 µg/g As diet), disodium arsenate (137 – 1054 µg/g As diet), dimethylarsinic acid (163 – 1497 µg/g As diet) or arsanalic acid (193 – 1503 µg/g As diet) all underwent adverse effects on growth, food consumption and feeding behaviour when fed with inorganic arsenic compounds, but were unaffected by diets containing the organoarsenic compounds. In all cases, carcass arsenic concentrations were related to dietary arsenic concentration [4].

Organotin Compounds

A 24-hour LC_{50} value of $1.3\,\mu g/l$ has been reported for adult rainbow trout [3].

The concentrations of tributyl tin found in surface microlayers of natural waters were in the range $1.9 – 473\,\mu g/l$. Consequently, rainbow trout swimming near the surface could be at risk.

7.1.2
Invertebrates

Organomercury Compounds

Microtubes were unaffected upon exposure to $1\,mg/l$ of methyl mercury for $1 – 24$ hours, and severely disrupted upon exposure to $6\,mg/l$ of methyl mercury for $1 – 24$ hours [5].

Organotin Compounds

Exposure of adult fiddler crabs (*Uca pugilator*) to tributyltin concentrations as low as $0.5\,\mu g/l$ retarded limb regeneration and of ecolysis and produced morphological abnormalities in regenerated limbs [6].

Roberts [7] has reported on the acute toxicity of tributyltin to embryos and larvae of bivalve molluscs (*Crassostrea virginica* and *Mercenaria mercenaria*). Forty-eight-hour LC_{50} values of 1.30 and $3.96\,\mu g/l$ were obtained in *C. virginica* embryos and straight-hinge stage larvae, respectively, and 1.13 and $1.65\,\mu g/l$ in *M. mercenaria* embryos and larvae respectively. The 24-hour LC_{50} values for both species were greater than $1.3\,\mu g/l$ in embryos and $4.2\,\mu g/l$ in larvae. Evidence suggested that tributyltin causes delayed clam embryo development when present below the LC_{50} value. Tributyltin concentrations above $0.77\,\mu g/l$ in the water caused abnormal shell development.

The occurrence and concentrations of organotin compounds in the tissues of scallops (*Pecten maximus*), flame shells (*Lima hians*) [8], polychaetes, snails and bivalves [11], and mussels (*Mytilus edulis*) and oysters (*Crassostrea virginica*) [9] have been studied. Scallop, mussel and flame shell populations are adversely affected by organotin compounds [8]. High concentrations of tributyltin have been found in polychaetes, snails and bivalves living in marinas containing $2 – 646\,ng/l$ tributyltin [10]; i.e., levels above the Environmental Quality Target for tributyltin of $20\,ng/l$. San Diego Bay mussels exposed to $0.7\,\mu g/l$ organotin for 60 days sustained a 50% mortality rate in the case of mussels and a decline in condition in the case of oysters [11]. Various tissues in these organisms showed tin uptake within $0 – 30$ days.

Table 7.1. Organometallic compounds in freshwaters and rain (from author's own files)

Compound	Origin of sample	Concentration (μg/l)	Reference
$MeSn^{3+}$	River	0.001 – 0.04	[13]
Me_2Sn^{2+}		0.007 – 0.005	
Me_3Sn^+		0.0006 – 0.004	
Total Sn		0.005 – 0.58	
Me_3Sn^+	Rain	0.006	[14]
Bu_2Sn^{2+}	Rain	< 0.001	[15]
Bu_3Sn^+	Switzerland	< 0.001	
$BuSn^{3+}$	Rivers	0.05 – 0.050	[15]
$BuSn^{2+}$	Switzerland	0.010 – 0.040	
$BuSn^{3+}$		0.005 – 0.015	
MeHg	River Waal	0.31 – 1.15	[16]
MeHg	River	0.0059 – 0.012	[17]
MeHg	Rain	0.009	[17]
$PbEt_4$	Surface water	50 – 530	[18]

Organolead Compounds

From the limited data available, concentrations of organolead in creatures other than fish are appreciably lower than those that occur in fish (Table 2.18).

Data have been presented on the concentrations of ionic alkyl lead compounds in saltmarsh periwinkles (*Littorina irrorata*) collected in Maryland, Virginia. Male periwinkles accumulated higher concentrations of several alkyl lead species than females [12].

Table 7.1 shows some typical levels of organometallic compounds of tin, lead and mercury found in river and surface water and rain. The high levels of methyl mercury in the polluted River Waal and of tetraethyl lead originating from gasoline are notable.

7.2
Seawater

7.2.1
Fish and Invertebrates: Organotin Compounds

Zischke and Arthur [19] have determined 96-hour LC_{50} values of tributyltin compounds for mysids (*Mysidopsis bahia*). The age of the fish was an important factor in determining the sensitivity of juveniles to tributyltin compounds.

In chronic toxicity tests [20] carried out in the Chesapeake bay area on biota exposed to tributyltin, the survival of *Gammarus SP* was unaffected by 24-hour exposure to concentrations of up to 0.58 LC_{50}, although body weight was reduced by 64% relative to controls. Survival of *Brevoortia tyrannus* and

Table 7.2. Organometallic compounds in seawater (from author's own files)

	Concentration (μg/l)	Reference
Mercury		
MeHg in seawater	0.06	[23]
Arsenic		
Irish sea	2.49 – 2.65	[24]
Tin		
Gulf of Mexico	0.0022 – 0.062	[1]
Sn(IV)	< 0.00001 – 0.015	
Me$_2$Sn	0.00074 – 0.007	
Me$_3$Sn	< 0.00001 – 0.00098	
Total Sn	0.0036 – 0.085	
Old Tampa Bay		[1]
Sn(IV)	< 0.0003 – 0.0027	
MeSn	0.00086 – 0.0011	
Me$_2$Sn	0.0006 – 0.002	
Me$_3$Sn	< 0.00001 – 0.00095	
Total Sn	0.025 – 0.005	
Estuary		[1]
Sn(IV)	0.0003 – 0.020	
MeSn	< 0.00001 – 0.008	
Me$_2$Sn	0.00079 – 0.0022	
Me$_3$Sn	< 0.00001 – 0.0011	
Total Sn	0.0025 – 0.023	
Harbour water		[1]
Me$_2$Sn	< 0.01 – 0.02	
Me$_3$Sn	< 0.01 – 0.02	
SnH$_4$	0.2 – 20	
Me$_4$Sn	< 0.01 – 0.3	
BuSnH$_3$	< 0.05 – 0.3	
Surface water		[26]
Sn(IV)	0.001 – 0.009	
BuSnH$_3$	0.01 – 0.06	
Bu$_2$SnH$_2$	0.13 – 0.46	
Bu$_3$SnH	0.06 – 0.78	
Bottom water		[26]
Sn(IV)	0.003 – 0.005	
BuSnH$_3$	0.03 – 0.04	
Bu$_2$SnH$_2$	0.13	
Bu$_3$SnH	0.01 – 0.10	

Table 7.2. Continued

	Concentration (µg/l)	Reference
Estuary water		[27]
Bu₃Sn	0.08 – 0.19	
Bay samples		[1]
Sn(IV)	0.003 – 0.02	
MeSn	0.0007 – 0.008	
Me₂Sn	0.0008 – 0.002	
Me₃Sn	0.0003 – 0.001	
Total Sn	0.0002 – 0.023	
Lake Michigan, adjacent to coast		[1]
Sn(IV)	0.08 – 0.49	
MeSnCl₃	0.006 – 0.0013	
Me₂SnCl₂	< 0.0001 – 0.063	
BuSnCl₃	0.002 – 1.22	
Bu₂SnCl₂	0.01 – 1.6	
San Diego Bay, surface water		[1]
Sn(IV)	0.006 – 0.038	
MeSnCl₃	0.0002 – 0.0008	
Me₂SnCl₂	0.015 – 0.045	
BuSnCl₃	< 0.0001	
Bu₂SnCl₂	< 0.0001	
San Francisco Bay		[1]
Sn(IV)	0.0002 – 0.0003	
MeSnCl₃	< 0.0001	
Me₂SnCl₂	< 0.0001	
BuSnCl₃	< 0.0001	
Bu₂SnCl₂	< 0.0001	
Coast adjacent to San Francisco		[1]
Sn(IV)	0.0003 – 0.0008	
MeSnCl₃	< 0.0001	
Me₂SnCl₂	< 0.0001	
BuSnCl₃	< 0.0001	
Bu₂SnCl₂	< 0.0001	

larval *Menidia beryllina* was unaffected by 28 days of exposure to concentrations of tributyltin of up to 0.49 µg/l. Growth was reduced by 20 – 22% following exposure to 0.09 or 0.49 µg/l tributyltin. Noth and Kumar [21] have studied the effect of 13 months' exposure to butyltin-containing paints on the oyster *Crassostrea gigas*. Oyster weight, length and width were adversely affected. Embryonic and larval viability were unaffected. The toxicities of organometallic compounds in seawater are also discussed in Sect. 9.3.

Other Organometallic Compounds

Table 7.2 lists the information available on organometallic compounds that have been found in seawater. Traces of organically bound arsenic are ubiquitous. Organotin compounds are found only in certain coastal areas, where these compounds are used as antifoulants on boats and harbour works. Several governments have banned the use of organotin compounds in recreational craft, while other countries are debating the issue [22].

References

1. Ram RN, Sathyanesan AG (1987) *Environ Pollut* **47**:135.
2. Kiruagarum R, Joy KP (1988) *Ecotoxicol Environ Safety* **15**:171.
3. McGuire RJ, Tkacz RJ (1987) *Water Pollut Res J Canada* **22**:227.
4. Cockell KA, Hilton JW (1988) *Aqua Toxicol* **12**:73.
5. Czuba M, Seagull RW, Tran H, Cloutier L (1987) *Ecotoxicol Environ Safety* **14**:64.
6. Weis JS, Gottlieb J, Kwiatkowski J (1987) *Archiv Environ Contam Toxicol* **16**:321.
7. Roberts MH (1987) *Bull Environ Contam Toxicol* (1987) **39**:1012.
8. Minchin D, Duggan CB, King W (1987) *Mar Pollut Bull* **18**:604.
9. Chesler SN, Gump BH, Hertz HS, May WE, Wise SA (1978) *Anal Chem* **50**:805.
10. Langston WJ, Burt GR, Mingjiang Z (1987) *Mar Pollut Bull* **18**:634.
11. Pickwell GV, Steinert SA (1988) *Mar Environ Res* **24**:215.
12. Krishnan K, Marshall WD, Hatch WI (1988) *Environ Sci Technol* **22**:806.
13. Braman RS, Tompkins MA (1979) *Anal Chem* **51**:12.
14. Landy MP (1980) *Anal Chim Acta* **121**:39.
15. Muller MD (1987) *Anal Chem* **56**:617.
16. Kiemeneij AM, Kloosterboer JG (1976) *Anal Chem* **48**:575.
17. Minagawa K, Takizawa Y, Kifune I (1980) *Anal Chim Acta* **115**:103.
18. Potter HR, Jarview AW, Markell RN (1977) *Water Pollut Cont* **76**:123.
19. Zischke JA, Arthur JW (1987) *Archiv Environ Contam Toxicol* **16**:225.
20. Giudici MDN, Migliore L, Guarino SM, Gambardella C (1987) *Mar Pollut Bull* **18**:454.
21. Noth K, Kumar N (1988) *Chemosphere* **17**:465.
22. Laughlin RB, Linden O (1987) *Ambio* **16**:252.
23. Davies IM, Graham WC, Pirie JM (1979) *Mar Chem* **7**:111.
24. Hayward MG, Riley JP (1976) *Anal Chim Acta* **85**:219.
25. Jackson JA, Blair WR, Brinckman FE, Iveson WP (1982) *Environ Sci Technol* **16**:110.
26. Valkirs AO, Seliman PF, Stang PM, Homer V, Lieberman SH, Vafa G, Dooley CA (1986) *Mar Pollut Bull* **17**:319.
27. Ebdon L, Alonso JIG (1987) *Analyst* **112**:1551.

8 Effect of Toxicants on Phytoplankton, Algae and Weeds

8.1
Cations

8.1.1
Phytoplankton and Algae

Dallakyan et al. [1] studied the combined effect of zinc (100 – 1000 µg/l), chromium (100 – 1000 µg/l) and cadmium (100 – 500 µg/l) on the phytoplankton in a reservoir. Zinc and cadmium additions maximally inhibited phytoplankton production at the beginning of a blue-green bloom. Chromium had no such inhibiting effect.

Brand et al. [2] investigated the effect of copper and cadmium on the reproduction rates of 38 clones of marine phytoplankton. Cyanobacteria were the most sensitive to the copper toxicity and diatoms were the least sensitive.

Reproduction rates of cyanobacteria were reduced at cupric ion (i.e. copper(II)) activities above 10 nM (picomole), whereas eukaryotic algae still maintained maximal reproductive rates at 10 nM. Trends for divalent cadmium were the same as for copper. Concentrations of cadmium in natural seawater were not of significance in unpolluted water, but copper concentrations in upwelling water might affect cyanobacteria.

In a study of the toxic effect of total aluminium and copper concentrations on the green alga (*Scenedesmus*), it was observed that toxicity effects (mainly on growth rate) were due almost entirely to an increase in cupric ion activity as a result of indirect competition from aluminium in the growth media that displaced copper from chelators [3].

Claesson and Tornqvist [4] studied the toxicity of aluminium to two acido-tolerant green algae, chlorophycae (*Monoraphidium dybowskii* and *Stichoccus sp*). Exposure to 100 – 800 µg/l aluminium at pH 5 – 6 led to cell decomposition, even at 100 µg/l aluminium. Growth was also affected.

Growth of pure cultures of phytoplankton *Scenedesmus bijugatus* and *Nitzchia palea* in 10 – 50 µg/l and 20 – 40 µg/l cadmium, respectively, showed that the physiology of the algae was affected during the experimental growth phase, and the ratio of carbohydrate, protein and lipid was affected by cadmium [5].

Increased cyanide concentrations in the range of 100 – 700 µg/l inhibited the growth of the Nile water algae *Scenedesmus*, but had no effect on the growth of *Anabaena* [6].

Upon exposure to solutions of mercuric chloride and sodium chromate for 28 days, the aquatic macrophytes *Elchornia crassipes, Hydrilla verticillata* and the alga *Oedogonium aerolatum* accumulated more chromium than mercury. Exposure did not produce any significant changes in Hill activity, chlorophyll, protein, free amino acid, inorganic phosphorus, RNA, DNA, dry weight permeability or protease activity [7].

In studies of chronic exposure of algal periphyton, the communities were exposed to 50 – 1000 µg/l zinc for up to 30 days. Treatments as low as 50 µg/l zinc significantly changed algal community composition from diatoms to green or blue-green algae. A zinc concentration of 47 µg/l is the criterion of the Environmental Protection Agency for the 24 hour average of total recoverable zinc [8]. Starodub et al. [9] carried out short- and long-term studies on the individual and combined toxicities of copper, zinc and lead to *Scenedesmus quadricanda* freshwater green alga. Short-term exposure to the effect of combinations of 0 – 200 µg/l copper, 0 – 500 µg/l zinc and 0 – 6000 µg/l lead and the long-term effects of the single and combined metals on primary productivity were studied. Low concentrations of single metals had the greatest effect on primary productivity; copper was the most toxic and lead the least toxic in short- and long-term studies. The combined metals exhibited an antagonistic effect in short-term exposure and both synergistic and antagonistic effects in long-term experiments. Kuwabara [10] reviewed the physicochemical processes affecting the toxicity of copper, tin and zinc to algae.

8.1.2
Weeds

Nickel at the 100 µg/l level would reduce the growth rate of common duckweed (*Lemna minor*) by 30% in most surface waters and by 70% in very soft water [11]. The angiosperm *Cuseuta reflexa* undergoes a reduction in chlorophyll and protein content and percentage dry matter in biomass as well as an increase in tissue permeability in the presence of certain metals in the overlying water [12]. It is more sensitive to arsenic then to cadmium, lead, mercury and chromium in that order. Pick-up of aluminium, copper(II) and lead(II) from water by duckweed in amounts above a certain concentration will cause the plant to die [13]. The toxicity is believed to be due to the replacement of magnesium in chlorophyll and hence the loss of its normal activity.

Samples of coral (*Pocillopora damicornis*) in a coral reef adjacent to a tin smelter contained significantly higher concentrations of calcium, strontium, zinc, chromium, cobalt, molybdenum, nickel, magnesium, sodium and potassium than those found in coral from an uncontaminated site [14].

There was distinct evidence that these contaminants caused reduced coral growth rates and a low number of branching coral species.

8.2
Organic Compounds

8.2.1
Phytoplankton

Di-*n*-butylphthalate has a distinct adverse effect on the distribution and survival of marine phytoplankton. It also markedly affects growth and/or aggregation behaviour of algae and diatoms [15].

Rhee et al. [16] studied the long-term responses of phytoplankton (*Selanastrum capricornutum*) to 2,5,2',5'-tetrachlorobiphenyl in water. This compound caused a reduction in the percentage of fixed carbon incorporated into the cells, and this carbon was probably excreted.

Concentrations of permethrin between 0.75 and 1.5 µg/l in pond water caused a decline in populations of *Daphnia rosea*, and at 10 µg/l it caused the complete elimination of this species. *Acanthodiaptomus pacificus* behaved similarly. *Tropocylops praciuus* was slightly more tolerant [17].

8.2.2
Zooplankton

Ali et al. [18] obtained no evidence that very low concentrations of difluben-zuron had any adverse effects on zooplankton and benthic invertebrates in ponds which had been contaminated by this insect growth regulator present as an air drift from a nearby grove.

Day and Kaushik [19] studied the effect of short-term exposure to the synthetic pyrethroid fenvalerate in water on the rate of filtration and the rate of assimilation of *Chlamydomonas reinhardii* by three species of freshwater zooplankton, namely *Daphnia galeata mendotae*, *Ceriodaphnia cacustris* and *Diaptomus oregonensis*. Rates of filtration of *Chlamydomonas reinhardii* by all three species were decreased significantly at sublethal concentrations (0.05 µg/l) of fenvalerate in water after 24 hours' exposure. Rates of assimilation of algae by the three species were decreased at lethal concentrations of more than 0.05 µg/l fenvalerate. Changes in rates of filtration and assimilation can be used to monitor the effects of sublethal levels of toxicants.

Applications of 1000 µg/l of carbaryl insecticide to pondwater killed off all zooplankton but had no effect on phytoplankton, though changes in zooplankton densities affected phytoplankton community structures [20]. Lindane (γBHC) has no significant effect on natural zooplankton populations, but the population density of zooplankton was reduced, even at concentrations of lindane as low as 20 µg/l [21]. Rotifers and nauplii were particularly adversely affected.

Arthur [22] has studied the effects of pollution by diazinon in water (0.3 – 3.0 µg/l), chlorpyrifos (0.2 – 11 µg/l), pentachlorophenol (48 – 432 µg/l) on plankton and invertebrate communities and on survival, growth and reproduction.

8.2.3
Algae

Exposure of periphyton communities from brackish water mesocosinus to 1 – 10 µg/l 4,5,6-trichloroguiacol produced no evidence for adverse effects [23]. Exposure of natural periphyton communities to atrazine, alaclor, metolachlor and metribuzin reduced growth rates and rates of uptake of nutrients at least temporarily [24].

Minimum concentrations of terbutryn, diuron, monouron and atrazine for inhibiting the growth of microalgae have been reported as 1100 – 2800 µg/l (terbutryn and diuron) and 1100 – 17100 µg/l (monouron and atrazine) [25].

Hamilton et al. [26] has studied the effect of up to two years' exposure of lake periphyton communities to concentrations of atrazine in the range of 80 – 1500 µg/l. Chlorophyll-a, freshwater biomass, ash-free weight, cell numbers, species diversity, community carbon uptake and species-specific carbon uptake were measured. There was a shift from a chlorophyte- to a diatom-dominated community over the two-year period but *Cylindrospernum stagnate* and *Tetraspora cyclindrica* showed evidence of resistance to atrazine at 1560 µg/l.

Community productivity was reduced by 21% and 82% upon low and high exposure, respectively, returning to control levels in 21 days. The productivities of the larger algae were most affected. Reduced growth rates were obtained after exposure to the herbicide. Other workers have reported on a growth rate depression when green algae are exposed to atrazine [27].

Marine unicellular algae *Skeletonema costatum, Thalassiosira pseudonana* and *Chlorella sp.* have been exposed to water containing the brominated organic compounds decabromobiphenyloxide, pentabromomethyl benzene and pentabromomethyl benzene. The corresponding LC_{50} values were greater than 1.1, and 0.5 mg/l, respectively, the highest exposure concentrations tested [28].

The effect of atrazine (50 – 30000 µg/l) combined with either ethanol (0.1 – 3% v/v) or acetone (0.1 – 5%) on the growth of the green alga *Chlorella pyrenoidosa* has been studied [29]. Acetone arid atrazine interacted antagonistically, but only at solvent concentrations exceeding 4 – 5% with both solvents. Atrazine EC_{50} values (calculated using growth data in the additive solvent range) were between 50 and 80 µg/l.

The effect of 0 – 100 mg/l concentrations of lindane (γ-BHC) in freshwaters on the alga *S. obliquus* has been studied [30]. Daily samples were examined for algal growth, pigment content, accumulation and degradation.

The algal pigment content was affected at above 50 mg/l lindane in water. Accumulation was enhanced by exposure time and by vibration.

Walsh et al. [31] evaluated the effect of 21 pesticides in water on five different algal species by determining EC50 values.

The effects of the organophosphorus insecticide Phosalone on the sexual life cycle of the alga *Chlamydomonas reinhardii* have been examined [32]. The formation of gametes, young, mature zygotes and the meiotic division of mature zygotes were examined following two hours' exposure to 36.7 mg/l Phosalone. The formation of gametes and young zygotes was not affected by the treatment. Unlike control groups, the mature zygotes thus formed did not exhibit meiotic division ability in the first days of light exposure, but remained in the same state for five days and then underwent meiotic division on the sixth day of exposure. Stratton [33] has studied the inhibitory effects of from 0.1% to 14% of six organic solvents (methanol, acetone, hexane, ethanol, dimethyl sulfoxide and *N,N*-dimethyl formamide) in water towards five species of blue-green algae (*Anabaena sp., Anabaena cylindrica, Anabaena variablilis, Nostoc sp.* and *Anabaena inaequalis*). Acetone and dimethyl sulfoxide displayed intermediate toxicity in terms of growth inhibition (EC50 values 0.36% and 4.4%, respectively). Dimethyl sulfoxide and ethanol were highly toxic.

In 10 – 14-day growth experiments, methyl formamide and ethanol were confirmed as the most toxic organic solvents towards the green algae *Chlorella pyrenoidosca* [29] (EC_{50} of 0.84 and 1.18% v/v, respectively), followed by dimethyl sulfoxide, hexane, methanol and acetone (EC_{50} of 2.01%, 2.66%, 3.02% and 3.60% v/v, respectively).

Chlorella vulgaris cultures exposed to *p*-nitrophenol or *m*-nitrophenol in water at concentrations of between 5 and 20 mg/l for 20 – 30 days exhibited inhibited growth in the case of *p*-nitrophenol at 10 mg/l and stimulated growth in the case of *m*-nitrophenol at 5 μg/l during 20 – 30 days exposure, but inhibited growth at 15 mg/l during 15 days' exposure [34].

8.2.4
Weeds

Thorhang and Marcus [35] studied the effect of three oil dispersants (Corexit 9527, Arcochem D609 and Canco K(K)) on the subtropical/tropical sea grasses *Thalassia testudinum, Halodule wrightii* and *Syringodiumfiliforme*. At concentrations of below 1 ml dispersant per 10 ml oil in 100 litres of seawater, mortality rates were low even for long exposure times. At 10 ml dispersal per 100 ml oil in 100 litres of seawater, *Syringodium filiforme* and *Halodule wrightii* died. Conco K(K) was far more toxic than the other two dispersants.

8.2.5
Diatoms

Goutx et al. [36] studied the effects of 50 mg/l of 9,10-dihydroanthracene and its biodegradation products on the marine diatom *Phaeodactylum tricornatum*. Growth of the diatom was inhibited. Synergistic effects between 9,10-dihydroanthracene and its biodegradation products increased the toxicity of the hydrocarbon. Resistance to polychlorobiphenyls and cross-resistance to DDT were induced in a polychlorobiphenyl-resistant clone of *Ditylum brightwelli* by 30 days' exposure to 10 µg/l polychlorobiphenyl or polychlorobiphenyl concentrations which increased progressively from 10 to 30 µg/l over the 30 days [37].

Polychlorobiphenyl resistance persisted for two years. The polychlorobiphenyl-resistant *Ditylum brightwelli* exhibited greater tolerance to polychlorobiphenyl than the sensitive strain did under all environmental conditions which permitted its growth, even when the conditions of salinity, temperature and nitrogen availability were very different from those maintained during induction. Polychlorobiphenyl resistance decreased the tolerance of the strain to lower salinities and nitrogen limitation but increased its tolerance to lower temperatures.

References

1. Dallakyan GA, Korsak MN, Nikiforova EP (1988) *Water Resour* **15**:53.
2. Brand LE, Sunda WC, Guillard PRL (1986) *J Exp Mar Biol Ecol* **96**:225.
3. Rueter JG, O'Reilly KT, Petersen RR (1987) *Environ Sci Technol* **21**:435.
4. Claesson A, Tornqvist L (1988) *Water Res* **22**:977.
5. Sathya KS, Balakrishnan KP (1988) *Water Air Soil Pollut* **38**:283.
6. Shogata SA, Aboelela SI, Ali GH (1988) *Environ Technol Lett* **9**:1137.
7. Jana S (1988) *Water Air Soil Pollut* **38**:105.
8. Genter RB, Cherry DS, Smith EP, Cairns J (1987) *Hydrobiol* **153**:261.
9. Starodub ME, Wong PTS, Mayfield CI (1987) *Sci Total Environ* **63**:101.
10. Kuwabara JS (1986) *Stud Environ Sci* **28**:129.
11. Wang W (1987) *Environ Toxicol Chem* **6**:961.
12. Jana S, Dalal T, Barua B (1987) *Water Air Soil Pollut* **33**:23.
13. Mo SC, Choi DS, Robinson JW (1988) *J Environ Sci Health* **A23**:139.
14. Howard LS, Brown BE (1987) *Mar Pollut Bull* **18**:451.
15. Acey R, Healey, P, Unger TF, Ford CE, Hudson, DA (1987) *Bull Environ Contam Toxicol* **39**:1.
16. Rhee GY, Shane L, Denucci A (1988) *Appl Environ Microbiol* **54**:1394.
17. Yasuno M, Hanazato T, Iwakuma T, Takamura K, Ueno R, Takamura T (1988) *Hydrobiol* **159**:247.
18. Ali A, Nigg HN, Stamper JH, Kok-Yokaomi ML, Weaver M (1988) *Bull Environ Contam Toxicol* **41**:781.
19. Day K, Kaushik NK (1987) *Archiv Environ Contamin Toxicol* **16**:423.
20. Hanazato T, Yasuno M (1987) *Environ Pollut* **48**:145.
21. Lay JP, Müller A, Peichl L, Lang R, Kote F (1987) *Chemosphere* **16**:1527.

22. Arthur JW (1988) *Int J Environ Stud* **32**:97.
23. Molanda S, Blanch H (1988) *Water Sci Technol* **20**:193.
24. Krieger KA, Baker DB, Kramer JW (1988) *Archiv Environ Contamin Toxicol* **17**:299.
25. Paterson DM, Wright SJL (1988) *Lett Appl Microbiol* **7**:87.
26. Hamilton PB, Jackson GS, Kaushik NK, Solomon KR (1987) *Environ Pollut* **46**:83.
27. Hersh CM, Crumpton WG (1987) *Bull Environ Contam Toxicol* **39**:1041.
28. Walsh GE, Yoder MJ, McLaughlin LL, Lores LM (1987) *Ecotoxicol Environ Safety* **14**:215.
29. Stratton GW, Smith TM (1988) *Bull Environ Contam Toxicol* **40**:736.
30. Yi-Xiong L, Bo-Zen S (1987) *Hydrobiol* **153**:249.
31. Walsh GE, Deans CH, McLaughlin LL (1987) *Environ Contam Toxicol* **6**:767.
32. Pednekar MD, Gandhi S, Netrawate MS (1987) *Environ Int* **13**:219.
33. Stratton GW (1987) *Bull Environ Contam Toxicol* **38**:1012.
34. Megharaj M, Venkateswarlu K, Rao AS (1988) *Ecotoxicol Environ Safety* **15**: 320.
35. Thorhaug A, Marcus J (1987) *Mar Pollut Bull* **18**:124.
36. Goutx MM, Al-Mallah M, Bertrand JC (1987) *Mar Biol* **94**:111.
37. Cosper M, Snyder BJ, Arnold LM, Zaikowski LA, Wurster CF (1987) *Mar Environ Res* **23**:207.

9 Toxicity Index (LC$_{50}$), Mean (S$_x$) and Percentile (S$_{95}$) Concentrations of Toxicants

Whereas the data given in Chaps. 5 to 7 are generally of a more qualitative nature, discussing the effects on health and mortality of various pollutants on water-based creatures, the information given in this chapter is of a more quantitative nature: it focuses more on measurements of toxicity index (LC$_{50}$ or LE$_{50}$), mean concentrations (S$_x$) and percentile concentrations (S$_{95}$) of pollutants.

The toxicity data provided in this chapter are supplementary to those provided in Chapters 5–7.

This chapter reviews the information available on the toxicities of various types of pollutants—metallic, organometallic and organic—towards fish and creatures other than fish. Most of this information is concerned with the concentrations of these substances in the water to which the creatures are exposed, whether it is freshwater or seawater. Information has also been reported on the concentrations of toxicants found in tissues of creatures that are known to have been killed by pollutants (i.e. studies of acute exposures that resulted in fish being killed). Although there is a vast amount of literature that is available on the presence of metals in fish, it deals almost exclusively with levels in muscle tissue or in whole fish after chronic exposure. Van Hoof and Van Son [1] pointed out that investigating the causes of fish deaths by water analysis alone has serious drawbacks, since in many cases the causative agent may have been diluted, biodegraded or volatilised to a level that does not allow an unambiguous interpretation at the time of sampling. It may have eventually been displaced from the site where the fish were localised. The work of Van Hoof and Van Son [1] on copper, cadmium, zinc and chromium is discussed where relevant in this chapter.

Mount and Stephen [2] developed an autopsy technique for zinc-related fish mortalities and found that the ratio of opercle to gill zinc concentration gives valuable information for discriminating between acute and chronic exposure. Mount and Stephen [2] found that cadmium intoxications in the bluegill sunfish (*Lepomis macrochirus*) and catfish (*Ictalurus nebulosus*) could be demonstrated through the analysis of gill tissue. Martin et al. [3] and Kariya et al. [4] found similar results for copper in five different fish species. These findings were not confirmed by the work of Buings et al. [5], who found no significant differences between copper tissue levels in *Ictalurus nebulosus* exposed to acute lethal and subacute nonlethal concentrations.

9.1
Cations

The toxicities of metals towards fish and invertebrates are now discussed where the data are available; for comparison, each element is identified by mean (S$_x$) and 95% percentile (S$_{95}$) concentrations (discussed in Chaps. 10 and 11). Lethal LC$_{50}$ values for fish and invertebrates are given in Table 9.1 as a function of the type of water (nonsaline or saline), type of creature and exposure time.

9.1.1
Nonsaline Water

9.1.1.1
Aluminium

Fish

Rainbow trout (*Salmo gairdneri*), when exposed to aluminium in nonsaline waters, gave a four-day LC$_{50}$ value of 3800 µg/l. Depending on its concentration, the pH and the hardness of the water, aluminium can cause growth suppression [19], reduced survival rates [27] and delayed hatchings [19–26], as well as stress and embryo-larval mortalities [28] (see Table 5.7).

9.1.1.2
Ammonium

Fish

Rainbow trout (*Salmo gairdneri*), when exposed to nonsaline water containing ammonium ions, had a four-day LC$_{50}$ value of 530 µg/l [29]. Fathead minnow (*Pimephales promelas*) had a LC$_{50}$ value of 2170 µg/l under similar conditions [20]. Depending on its concentration and the exposure time, ammonium ions can cause liver and thyroid degeneration, hyperactivity and mortalities in fish [30] (see Table 5.7).

Invertebrates

Depending on its concentration, ammonium ions can reduce the survival rates [31] and growth [32] of invertebrates (see Table 5.8).

Table 9.1. Concentrations of elements in freshwater causing mortalities (LC_{50}) of salmonid and nonsalmonid fish (from author's own files) and invertebrates

Element	S_x (µg/l) (Table 10.5)	S_{95} (µg/l) (Table 10.5)	Fish		Invertebrates	
			Exposure time (days)	LC_{50} (µg/l)	Exposure time (days)	LC_{50} (µg/l)
Ni	220	900 (n/s, s)	Long-term	500 (s, n/s)	–	–
			100	2200 (s, n/s)		
			15	8000 (s, n/s)		
			4	35000 (s, n/s)		
Se	200	1300 (n/s)	4 (8)	2900 – 3060	2	1100 (as SeO_3) [6]
			10	300	2	5300 (as SeO_4) [6]
			10		2	680 (adult) [7] 750 (juvenile) (*Daphnia magna*)
V	100	1000 – 1600 (n/s)	7	2400 – 3000 (s) (saltwaters) 2900 – 5000 (s) (hardwaters)	–	
Cr	100	800 (s)	100	1150 (s)	3	30 – 80 (as Cr(IV)) (crustaceans)
			10	18300 (s)		
	100	1000 – 3000 (n/s)	4	3300 – 65000 (s)	3 – 5	1000
			60	200 (n/s)		
			4	25000 – 169000 (n/s)		
As	80	600 (n/s, s)	4	14400 [9]		
Ag	70	850 (n/s)	Short-term	10 – 10000 (as $AgNO_3$)	–	
Zn	23	200 (n/s)	500 – 1000	260 [10, 11] (softwaters)	4	70 (*Daphnia magna*)
			500 – 1000	1050 [10, 11] (hardwaters)	4	10000 (annelids, insect larvae)
			4	2600 (juvenile)	60	200 – 600 (snail, *Ancyclus fluviatus*) [12]
			4	2400 (adult) Rainbow trout (*Salmo gairdneri*)		
			4	13300 – 33000 (*Tilapia zilli*) [13]	60	2000 (amphipod *Allorchestes compressa*) [21]
			4	2600 – 52000 (*Clarias lazena*) [13]		
Pb	20	100 (n/s)	90	5500 [14]		Similar to fish
			40	900 [14]		
			4	1500 [14]		

Table 9.1. Continued

Element	S$_x$ (µg/l) (Table 10.5)	S$_{95}$ (µg/l) (Table 10.5)	Fish		Invertebrates	
			Exposure time (days)	LC$_{50}$ (µg/l)	Exposure time (days)	LC$_{50}$ (µg/l)
Cu	4	17 (s)	72	80 (s)	3	24 [21] (crustaceans)
			30	200 (s)		
			6	250 – 400 (s) [14]	4	400 – 2000 [21, 22] (molluscs)
			6		40	140 [17] (juvenile clams)
			6		8	5000 [17] (juvenile clams)
			6		4	480 (amphipod *Allorchestes compressa*)
Cd	4	16 (n/s)	100	180 (n/s)	4	680 (crustaceans)
			10	4000 (n/s)		
	2	6 (s)	700	2 (s)		
			4	< 10000 (*Notropis lutrensis*) [15]	4	780 (amphipod *Allorchestes compressa*) [23]
			4	< 10000 Fathead minnow (*Pimephales promelas*) [15]		
			4	12600 *Punctius conchonus* [16]		
			4	350000 (male) 371000 (female)		
			4	(*Herbistes reticulates*) [17]		
Hg	2	22 (n/s)	30	2 (as MeHg) [18]	3	0.2 (crayfish)
					2	110 (slipper limpet, *Cripidula fornicata*) [21]
Al	–	–	4	3800 Rainbow trout (*Salmo gairdneri*) [19]		
Fe	–	–	–	–	4	25610 – 43100 (isopod, *Asselus aquaticus*) [25]
NH$_4$	–	–	4	2170 Fathead minnow (*Pimephales promelas*) [20]	–	–

n/s: nonsalmonids
s: salmonids

9.1.1.3
Arsenic ($S_x = 80\,\mu g/l$, $S_{95} = 600\,\mu g/l$, Nonsalmonids)

Large amounts of arsenic enter the environment each year because of the use of arsenic compounds in agriculture and industry as pesticides, feed preservatives, herbicides, insecticides, feed additives and wood preservatives. Most of this is used as inorganic arsenic (arsenite, arsenate) and about 30% as organoarsenicals, such as monomethylarsinate and dimethylarsinate, used as agricultural chemicals. Arsenic is known to be relatively easily transformed between organic and inorganic forms in different oxidation states by biological and chemical action. Since the toxicities and biological activities of the different species vary considerably, information about the chemical form is of great importance in environmental analysis.

Of the two oxidation states, As(III) and As(V), the latter is more common in an oxidising environment and is more toxic.

Arsenic has some similar toxic properties to lead, mercury and cadmium in regards to bonding to sulfur and inhibiting the action of enzymes such as pyruvate dehydrogenase. The order of toxicity of arsenic compounds is arsines (As(III)) > arsenite (As(III)) > arsenate (As(V)) and arsenoorganic acids (As(V)). Arsenic, which is found mainly in the liver, kidneys, lungs and intestinal walls, is readily absorbed if water-soluble.

Fish

Exposure of fish to $4000\,\mu g/l$ arsenic for 30 days reduces fish growth [9]. Organoarsenic compounds are less toxic to fish than inorganic arsenic.

Invertebrates

Arsenic is relatively nontoxic to these creatures; $1000\,\mu g/l$ was required to cause mortalities in short-duration tests (Table 9.1). Insect larvae are the least sensitive and crustaceans are the most sensitive to arsenic.

9.1.1.4
Cadmium ($S_x = 4\,\mu g/l$, $S_{95} = 16\,\mu g/l$) (Nonsalmonids)
and $S_x = 2\,\mu g/l$, $S_{95} = 6\,\mu g/l$) (Salmonids)

Cadmium is increasingly being recognised as an important environmental pollutant with toxic effects on human and animal life at relatively low levels [33–35]. Environmental concentrations of cadmium are of serious concern, because cadmium accumulates in the human body throughout life, from $1\,\mu g$ body burden at birth to about $30\,mg$ in an adult, with about one third to be found in the kidneys [36]. Based on animal studies, cadmium is preferentially retained by the kidney and liver [38].

In view of the known accumulation of cadmium in biological tissues, a detailed study has been carried out to determine the rate of uptake in the common bluegill (*Lepomis macrochirus Raf.*) exposed to known amounts of cadmium in a carefully controlled aquatic environment. An important objective was to evaluate the relative rates of uptake in vital organs, including heart, skin, muscle, gut, gill, kidney, liver and/or bone.

Another study examined the chronic toxicity of cadmium (as well as copper and zinc) mixtures at sublethal concentrations to the fathead minnow using mortality, physical characteristics and reproduction as bioassay methods [37]. While these studies increase our understanding of the biological effects of relatively concentrated heavy metal pollutants in aquatic systems, they provide no evidence for actual rate of accumulation of toxic metals nor the distributions of these in vital organs. Moreover, only rarely in natural waters do the concentrations of toxic metals attain the levels used in most acute and chronic bioassay studies. Thus, experimental evidence for heavy metal accumulation and distribution in organisms exposed to environmentally unrealistic levels of heavy metal pollutants in natural waters should make more reliable and general predictions for the long-term effects of such pollutants possible.

In humans, cadmium accumulates in the liver and kidneys, the average level in wet kidney tissue being 25 – 50 µg/g. A level of 200 µg/g produces irreversible kidney damage.

A calcium-deficient diet enhances cadmium accumulation. The contributing factors to Itai-Itai disease were high cadmium intakes (>600 µg for most sufferers), a low-calcium diet and a lack of vitamin D. Itai-Itai disease is always accompanied by renal dysfunction. The levels of cadmium were high in the bones of Itai-Itai sufferers, 1.0 – 1.4% (ash weight). A low molecular weight protein in the liver, metallothionein (MW 7000), approximately one third of which is cysteine, bonds to heavy metals (especially cadmium and mercury) and protects against toxic metals. A sample of metallothionein was found to contain 4.2% cadmium.

Hypertension has been attributed to cadmium, although the topic is controversial. Respiratory and pulmonary damage is reported to occur upon breathing-in cadmium vapour or particles. Cadmium, unlike mercury and lead, does not affect the central nervous system—it cannot cross the placental membrane, and the mammary gland is an effective barrier.

Fish

A concentration of 4 – 13 µg/l cadmium for 30 to 60 days causes reduced growth in fish [37]. Nonsalmonid species are one tenth as sensitive to cadmium as salmonids [38–43].

Young life stages are more susceptible than adults. Impaired reproducibility of nonsalmonids occurs at a concentration of cadmium of 15 µg/l over 100 days (or 240 µg/l cadmium over ten days).

Depending on its concentration, cadmium can reduce the ability of the fish to withstand heat stress [44], cause branchial lesions and mucus secretion [45], and reduce alkaline phosphatase activity [45, 46] (see Table 5.7).

Four-day LC$_{50}$ values for cadmium in nonsaline water for red shiner (*Notropis lutrenis*), fathead minnow (*Pimephales promelas*) and rosy barb (*Puntius conchorinus*) were < 10,000 µg/l [15], 10,000 µg/l [47] and 12,600 µg/l [47], respectively. The corresponding values obtained for *Lebistes reticulatus* were 350,000 µg/l (males) and 371,000 µg/l (females) (see Table 5.7).

Invertebrates

Some species are more sensitive to cadmium than others. Thus *Daphnia magna* was adversely affected by 5 – 7 µg/l cadmium over 4 – 20 days [43,49], while the crustacean *Gammarin pulex* had a four-day LC$_{50}$ of 680 µg/l.

Crustacea are more sensitive to cadmium, and insect larvae the least sensitive. A rise in temperature increased toxicity, and increasing the pH reduced toxicity [49]. Table 9.2 presents results obtained in subacute and acute toxicity tests carried out by exposing rudd to water containing various levels of cadmium for various times.

One hundred per cent mortality occurred when the cadmium content of the water was somewhere in the range of 200 to 11,000 µg/l for up to 12 hours' exposure.

Exposure to increasing concentrations results in elevated cadmium levels in gill and kidney tissues, which are statistically significant against control levels at all exposure levels. Cadmium accumulation in gill tissue has also been observed by Mount and Stephen [2] in bluegills and brown bullheads after acute lethal exposure and by Sangalang and Freeman [50] in brook

Table 9.2. Toxicity of cadmium to rudd fish (*Scardinius erthropthalmus*). From author's own files

	Control fish	Subacute exposure	Acute exposure		
Concentration of water (µg/l)	3	250	1100	4000	11000
Composition of tissue (µg/l) (dry weight basis) Organ					
Muscle	0.3	0.41	0.6	0.5	3.2
Gill	2.6	2.5	3.9	10.4	87.9
Opercle	9.5	8.7	6.0	20.7	29.2
Liver	5.0	9.6	4.1	3.8	12.3
Kidney	4.2	13.7	14.4	12.8	28.2
Mortality (%) during an exposure time of	Nil	Nil	100	100	100
Weeks	> 10	3	–	–	–
Hours	–	–	< 12	< 12	< 12

trout after chronic sublethal exposure. Cadmium levels in gills from killed fish are significantly different from levels in exposed surviving fish.

The four-day LC$_{50}$ value for cadmium obtained for the amphipod *Allochestes compressa* in nonsaline water was 780 μg/l (see Table 5.8). *Daphnia magna* in nonsaline water had a 25-day LC$_{50}$ value of 10 μg/l. Depending on its concentration, cadmium can reduce the survival rates of invertebrates [51] (see Table 5.8).

9.1.1.5
Chromium (S$_x$ = 100 μg/l, S$_{95}$ = 800 μg/l (Salmonids), and S$_x$ = 100 μg/l, S$_{95}$ = 1000 – 3000 μg/l (Nonsalmonids)

Fish

The four-day LC$_{50}$ for salmonids is appreciably lower than that for non-salmonids, (Table 9.1); i.e. salmonids are more sensitive to chromium. Concentrations of chromium as low as 13 μg/l for 60 days adversely affect growth in fish, 720 μg/l for 60 days reduces growth of nonsalmonids, and 2300 μg/l for ten days or 100 μg/l for 100 days reduces growth in nonsalmonids. Chromium is more toxic to fish at lower pH values.

Table 9.3 presents results obtained in subacute and acute toxicity tests carried out by exposing rudd to water containing various levels of chromium (present as potassium dichromate) for various times. One hundred percent mortality of rudd occurred when the chromium content of the water was somewhere within the range 20 – 80 μg/l during a 12-hour exposure. Chromium levels were found in organs of rudd, and the most elevated values were detected in gill tissue in all fish killed. Chromium levels found

Table 9.3. Toxicity of chromium (as potassium dichromate). From author's own files

	Control fish	Subacute exposure	Acute exposure		
Composition of water (μg/l)	3	16	20	80	145
Composition of tissue (μg/l) (dry weight basis) Organ					
Muscle	< 0.2	< 2	0.5	0.8	0.6
Gill	< 0.2	< 2	4.9	48.2	30.6
Opercle	< 0.2	< 2	8.3	26.0	19.6
Liver	< 0.2	< 2	5.6	18.4	15.2
Kidney	< 0.2	< 2	10.3	23.8	27.8
Mortality (%) during an exposure time of	0	0	0	100	100
Weeks	> 10	3			
Hours			< 12	< 12	< 12

in the experiment using the highest chromium concentration (145 mg Cr/l) were lower than those found after exposure to 80 mg Cr/l, probably because of the shorter exposure time before death. Chromium levels in all organs of fish killed differed significantly from exposed fish that had survived, which had higher chromium concentrations in opercle, kidney and liver than in gill tissue. Similar results were reported by Van der Putte et al. [52] after exposing rainbow trout to hexavalent chromium.

Invertebrates

Insect larvae are least affected and crustacea most affected by chromium. Invertebrates are more sensitive than fish. Crustaceans are very sensitive, with a three-day LC_{50} of 30-80 µg/l as hexavalent chromium. Trivalent chromium is believed to be less toxic than the hexavalent form.

9.1.1.6
Cobalt

Invertebrates

Depending on its concentration, cobalt can decrease the muscle glycogen level in invertebrates [19] (see Table 5.8).

9.1.1.7
Copper ($S_x = 4$ µg/l, $S_{95} = 17$ µg/l, Salmonids)

The reported lowest adverse effect concentration for copper is 2 µg/l [53, 54]. The six-day LC_{50} value is 250 – 400 µg/l [55]. Doses for copper with salmonid fish range from 200 µg/l for 30 days to 80 µg/l for 72 days (lethal dose) (Table 9.1) to 100 µg/l for 30 days to 30 µg/l for 72 days (reduced growth). An increase in water temperature reduces the toxicity of copper to fish. Depending on the copper concentration, copper can cause decreased emergence success of fish eggs, reduced fish activity, necrosis, stress [10] and interference in nervous functions [57] (see Table 5.7).

Table 9.4 presents results obtained in subacute and acute toxicity tests carried out by exposing rudd to water containing various levels of copper for various times. One hundred percent mortalities occurred when the copper concentration of the water was somewhere in the range 50 – 250 µg/l for up to 12 hours' exposure.

In acute exposures where 100% mortalities of fish occurred in a few hours, the highest concentrations of copper were found in opercle and kidneys. The high concentration of copper found in the opercle may be partly due to adsorption.

Table 9.4. Toxicity of copper to rudd fish (*Scardinius erthropthalmus*). From author's own files

	Control fish	Subacute exposure	Acute exposure		
Composition of water (µg/l)	11	50	250	1200	1600
Composition of tissue (µg/l) (dry weight basis) Organ					
Muscle	0.7	1.6	2.3	2.2	4.0
Gill	5.5	8.9	22.9	29.3	43.2
Opercle	12.4	30.9	52.6	72.1	104
Liver	6.9	20.2	22.3	31.1	39.8
Kidney	6.0	28.5	30.4	39.0	100
Mortality (%) during an exposure time of	Nil	Nil	100	100	100
Weeks	< 30	3	–	–	–
Hours	–	–	< 12	< 12	< 12

Invertebrates

Crustaceans are the most sensitive to copper and molluscs the least sensitive [18, 56] (Table 9.1). The life stage of the invertebrate is important when determining the toxicity of copper [58]. Thus, adult clams had a one-day LC$_{50}$ of 500 µg/l while juveniles had an eight-day LC$_{50}$ of 5 µg/l and from 140 µg/l for 40 days to 30 µg/l for 72 days (impaired reproducibility).

The amphipod *Allochestes compressa* had a four-day LC$_{50}$ of 40 µg/l in nonsaline waters. Depending on its concentration, copper can reduce the survival rates of juvenile invertebrates [59] (see Table 5.8).

9.1.1.8
Iron

Invertebrates

The freshwater isopod *Asselus aquaticus*, when exposed to iron, had a four-day LC$_{50}$ value of 256,000 – 431,000 µg/l [25] (see Table 5.8).

9.1.1.9
Lead (S$_x$ = 20 µg/l, S$_{95}$ = 100 µg/l, Nonsalmonids)

Fish

LC$_{50}$ values differ little between salmonids and nonsalmonid fish. Impaired reproducibility of nonsalmonids occurs at 70 µg/l for 40 days' exposure or 400 µg/l for 90 days' exposure [60]. The four-day LC$_{50}$ values obtained for *Lebistes reticulatis* in nonsaline waters are 1,620,000 µg/l (males) and 1,630,000 µg/l females [17] (see Table 5.7).

Invertebrates

Crustaceans and gastropods have similar sensitivities to lead in long-term exposure tests.

9.1.1.10
Mercury ($S_x = 2 \, \mu g/l$, $S_{95} = 22 \, \mu g/l$, Nonsalmonids)

Mercury has recently been recognised as a toxic contaminant in the environment. The toxicity of mercury is related to its chemical form. Liquid mercury appears to have little effect, but mercury vapour is readily adsorbed, producing brain damage. Mercury I salts are relatively toxic compared to mercury II salts because of their low solubilities.

Mercury present in fish occurs almost entirely as methyl mercury. The WHO recommends a maximum daily intake of mercury by humans from all sources of $43 \, \mu g/day$, of which no more than $29 \, \mu g/day$ should be methyl mercury. In lakes and streams, mercury can collect in the bottom deposits, where it may remain for long periods of time.

Fish

Thirty days' exposure of fish to $1 \, \mu g/l$ inorganic mercury causes weight reduction and poor spawning. Three $\mu g/l$ mercury as methyl mercury chloride caused 88% mortality in fish [61]. The toxic effects of organomercury compounds are similar to those of inorganic mercury. Depending on its concentration and time of exposure, mercury will inhibit gonadal growth in fish (see Table 5.7).

Invertebrates

Most invertebrates are very sensitive to mercury; e.g. crayfish have a three-day LC_{50} of $0.2 \, \mu g/l$. Slipper limpets (*Crepidula fornicata*) have a two-day LC_{50} value of $1100 \, \mu g/l$ in nonsaline water (see Table 5.8).

9.1.1.11
Nickel ($S_x = 220 \, \mu g/l$, $S_{95} = 900 \, \mu g/l$, Nonsalmonids and Salmonids)
Fish

The long-term LC_{50} value ($500 \, \mu g/l$) is similar for both salmonid and non-salmonid fish. Impaired reproducibility of nonsalmonids occurs at $100 \, \mu g/l$ nickel for four days' exposure, $50 \, \mu g/l$ nickel for 15 days' exposure or $110 \, \mu g/l$ nickel for 100 days', compared with lethal concentrations of, respectively, 35,000, 8000 and $2200 \, \mu g/l$ (Table 9.1).

Invertebrates

Low concentrations of nickel ($15 \, \mu g/l$) impair invertabrate reproduction.

9.1.1.12
Silver (S$_x$ = 70 µg/l, S$_{95}$ = 850 µg/l, Nonsalmonids)
Fish

The toxicity of silver decreases depending on its chemical form, in the order nitrate, chloride, iodide, sulfide and thiosulfate. Thus eggs and larvae of *Pimephales promelas* exposed to 650 µg/l silver nitrate or 11,000 µg/l silver sulfide for 30 days caused 20% mortalities [62]. In general, nonsalmonids are more sensitive to silver than salmonids.

Invertebrates

Adverse effects have been observed at 10 µg/l silver [64].

9.1.1.13
Selenium (S$_x$ = 200 µg/l, S$_{95}$ = 1300 µg/l, Nonsalmonids)

In recent years, the physiological role of selenium as a trace element has created considerable speculation and some controversy. Selenium has been reported as having carcinogenic as well as toxic properties. Other authorities have presented evidence that selenium is highly beneficial as an essential nutrient [63,64].

Fish

Exposure of salmonid fish for 250 days to 40–50 µg/l selenium reduced blood volume [65]. Salmonids and nonsalmonids are equally sensitive to selenium. Depending on its concentration, selenium can cause mortalities in spawning, cataract development and reduced larval survival in fish (see Table 5.7).

Invertebrates

The toxicity of selenium depends on its chemical form. Thus the two-day LC$_{50}$ values for selenite and selenate are 1100 and 5300 µg/l [6], respectively. The two-day LC$_{50}$ for selenium in the case of *Daphnia magna* in nonsaline waters is 680 µg/l (adults) and 750 µg/l (juveniles) [7].

Depending on its concentration, selenium can have an adverse effect on growth and reproduction in invertebrates (see Table 5.8).

9.1.1.14
Vanadium (S$_x$ = 100 µg/l, S$_{95}$ = 1,000–1,600 µg/l, Nonsalmonids)
Fish

The lowest adverse effect concentration (LC$_{50}$) observed for this element is 1130 µg/l. Toxicity increases at higher pH values. The chemical form of vanadium also affects toxicity, vanadate being the most toxic and vanadium pentoxide the least [47].

9.1.1.15
Zinc ($S_x = 23\,\mu g/l$, $S_{95} = 200\,\mu g/l$, Nonsalmonids)

Fish

Exposure of fish to $210 - 520\,\mu g/l$ zinc for $30 - 140$ days affected nonsalmonid fish growth, while $200 - 300$ days' exposure to $180\,\mu g/l$ zinc affected reproduction. Salmonids are more sensitive to zinc than nonsalmonids upon short-term exposure but they have similar sensitivities upon long-term exposure. The four-day LC_{50} value in nonsaline water obtained for rainbow trout (*Salmo gairdneri*) exposed to zinc was $26,000\,\mu g/l$ (juveniles) and $24,000$ (adults) [13] (Table 5.7). Corresponding four-day LC_{50} values in nonsaline waters obtained for *Tilapia zilli* and *Clarias lazena* were $13,300 - 33,000\,\mu g/l$ and $26,000 - 52,000\,\mu g/l$ [14].

Table 9.5 presents results obtained in subacute and acute toxicity tests carried out by exposing rudd to water containing various levels of zinc for various times [14]. One hundred percent mortality of rudd occurred when the zinc level of the water was somewhere in the range $1600 - 7500\,\mu g/l$ for up to 12 hours' exposure.

The zinc values found in all tissue were significantly different from control values after exposure to $18,000\,\mu g/l$ zinc. In this case, all fish died within four hours and values found in gill tissues are clearly higher than in other tissues. These findings were not entirely confirmed after exposure to $7500\,\mu g/l$ zinc/l, in which case all fish died within 12 hours, although zinc levels in gills and opercle were significantly higher than control levels, and after exposure to $1600\,\mu g/l$ zinc/l which caused no deaths after 24 hours. In all experiments, kidney zinc levels were significantly higher than control values, suggesting

Table 9.5. Toxicity of zinc to rudd fish (*Scardinius erthropthalmus*). From author's own files

	Control fish	Subacute exposure	Acute exposure		
Concentration of water ($\mu g/l$)	180	800	1600	7500	18000
Composition of tissue ($\mu g/l$) (dry weight basis) Organ					
Muscle	16.4	22.4	10.5	6.6	11.2
Gill	47.9	101.9	38.6	51.2	647.2
Opercle	120.2	195.5	115.3	90.6	174.5
Liver	29.4	104.9	42.5	34.1	63.5
Kidney	57.0	151.7	154.6	92.2	216.1
Mortality (%) during an exposure time of					
Weeks	< 10	3	–	–	–
Hours	–	–	≥ 24	≤ 12	≤ 4

that this organ might in this case also give supplementary information about acute exposure.

After exposing rudd to 800 µg/l zinc for three weeks, there were higher levels of zinc in the opercle than in the kidney and gill tissue, suggesting that the opercle gives the most valuable information on nonlethal exposure. A study of the toxicity of zinc sulfate to rainbow trout dealt only with acute toxicity as measured by fish mortality [66].

Invertebrates

Four days' exposure to 70 µg/l zinc caused 50% mortality in *Daphnia magna*. The four-day LC$_{50}$ of annelids and insect larvae was 10,000 µg/l. Higher water hardness reduces the toxicity of zinc to some gastropods.

Sixty-day LC$_{50}$ values obtained for zinc in nonsaline waters were 200 – 600 µg/l in the case of *Ancylus ruviatis* snails, depending on snail size [12], and 2000 µg/l in the amphipod *Allorchestes compressa* [22] (see Table 5.8). Depending on its concentration, zinc can reduce the reproductive capacities of invertebrates (see Table 5.8).

9.1.2
Saline Waters (Estuaries, Bays, Coastal and Open Sea Waters)

The effect of metal concentrations on marine life, particularly invertebrates, has been examined by various workers and is summarised below.

9.1.2.1
Arsenic

The chemical form of arsenic in marine environmental samples is of interest from several standpoints. Marine organisms show widely varying concentrations of arsenic [67–69] and knowledge of the chemical forms in which the element occurs in tissues is relevant when interpretating these variable degrees of bioaccumulation and attempting to understand the biochemical mechanisms involved. Different arsenic species have different levels of toxicity [70] and bioavailability [71], and this is important in food chain processes, while physiochemical behaviour in processes such as adsorption onto sediments also varies with the species involved [72]. It has been shown that inorganic arsenic (III and V), monomethylarsenic (MMA) and dimethylarsenic (DMA) acids are present in natural waters [73], biological materials [74] and sediments [75]. Unpolluted seawater has a natural arsenic level of about 1 mg/kg.

The UK total diet survey suggests that at least 75% of total arsenic ingested originates from fish and shellfish. It is accepted that the arsenic in fish and shellfish is mainly organically bound; hence, if any of the more toxic inorganic arsenic is present it is of great interest. If the levels of total inorganic arsenic approach 1 mg/kg, the proportion of arsenic(III) relative to arsenic(V) also assumes importance, as the latter is considered to be more toxic than the former [77].

Crabs are susceptible to arsenic at the larval life stage [78]. Crustacea are the most sensitive to arsenic and annelids are the least sensitive. For instance, toxicity decreases in the order As(V) > organic arsenic > As(III). Fish are less susceptible (four-day LC_{50} = 15,000 – 28,000 µg/l) than invertebrates (four-day LC_{50} = 4000 µg/l).

The UK Arsenic in Food Regulations 1959 [76] state that foodstuffs must not contain more than 1 mg/kg of total arsenic. Certain exceptions are listed, which include fish and edible seaweed and their products, where arsenic contents of above 1 mg/kg may be accepted in certain circumstances.

9.1.2.2
Cadmium

Reduced salinity and higher water temperature both increase the toxicity of cadmium to marine invertebrates [79, 80]. Planktonic crustaceans have a four-day LC_{50} of 60 – 380 µg/l [83]. Young life stages of invertebrates are sensitive to cadmium [82]. Adult crustaceans are also susceptible to 60 µg/l of cadmium, causing 30% mortality in 60 days' exposure [80]. Fish are relatively resistant to cadmium, with a four-day LC_{50} of 6400 – 16,400 µg/l. Fiddler crabs *Uca annulipes* and *Uca triangularis* exposed to cadmium gave four-day LC_{50} values of 43,230 – 48,210 µg/l.

9.1.2.3
Chromium

The relative LC_{50} values of marine annelids, molluscs, crustacea and fish when exposed to trivalent chromium (exposure period not stated) are 2200 – 8000 µg/l, 14,000 – 105,000 µg/l, 2000 – 98,000 µg/l and 12,400 – 91,000 µg/l, respectively. Reduction in salinity from 35 to 15 g/kg reduces the four-day LC_{50} from 640,000 to 190,000 µg/l.

9.1.2.4
Copper

The importance of complexing agents in the mineral nutrition of phytoplankton and other marine organisms has been recognised for more than 30 years. Complexing agents have been held responsible for the solubilisation of iron and therefore its greater biological availability [83]. In contrast, complexing agents are assumed to reduce the biological availability of copper and to minimise its toxic effects. Experiments with pure cultures of phytoplankton in chemically defined media have demonstrated that copper toxicity is directly correlated to cupric ion in activity and independent of the total copper concentration. In these experiments, cupric ion (Cu^{2+})

concentrations can be varied in media containing a wide range of total concentrations through the use of artificial complexing agents. When Cu^{2+} concentration was calculated for earlier experiments with phytoplankton in defined media, it appeared that Cu^{2+} was toxic to a number of marine phytoplankton species in concentrations as low as 10^{-6} µmol/l. Since copper concentrations in the world's oceans typically range from 10^{-4} to 10^{-1} µmol/l, complexing agents and other materials affecting the solution chemistry of copper must maintain the Cu^{2+} activity at sublethal levels.

Copper may exist in particulate, colloidal and dissolved forms in seawater. In the absence of organic ligands or particulate and colloidal species, carbonate and hydroxide complexes account for more than 98% of the inorganic copper in seawater [84, 85].

The young life stages of crustacea and molluscs are more sensitive to copper (two-day LC$_{50}$ = 300 µg/l) than adults (two-day LC$_{50}$ = 30,000 µg/l). This also applies to fish [75]. Low salinity increases the toxicity of copper. Bivalve molluscs are the most copper-sensitive species yet examined, undergoing reduced growth in the presence of 3 – 10 µg/l copper for prolonged periods. Mortalities occurred beyond 500 days' exposure [86]. When exposed to copper, fiddler crabs *Uca annulipes* and *Uca triangularis* gave four-day LC$_{50}$ values of 12,820 – 14,810 µg/l.

9.1.2.5
Lead

Mollusc larvae are particularly sensitive to lead, with abnormal development occurring upon two days' exposure to 400 µg/l lead [78].

9.1.2.6
Mercury

The toxicities of organic and inorganic mercury to marine fauna are similar. Thus, fish embryos undergo damage when exposed to 67 µg/l mercury for four days [87] and poor hatching when exposed to 32 µg/l mercury for 32 days. Crustacea and molluscs are as sensitive to mercury as are fish (e.g. crab LC$_{50}$ = 8 µg/l) [88]. Fiddler crabs *Uca annulipes* and *Uca triangularis* gave a four-day) LC$_{50}$ value of 2750 – 2830 µg/l when exposed to mercury.

9.1.2.7
Nickel

Fish

This element is relatively nontoxic to fish. It is less toxic in saline water (four-day LC$_{50}$35,000 µg/l) than in nonsaline water (four-day LC$_{50}$ = 10,000µg/l). Toxicity is greater at higher water temperatures [89].

Invertebrates

Nickel is relatively nontoxic to marine organisms (four-day LC_{50} = 10,000 µg/l).

Planktonic crustacea and bivalve mollusc larvae are more sensitive (four-day LC_{50} 50 – 600 µg/l).

9.1.2.8
Selenium

In recent years, the physiological role of selenium as a trace element has created considerable speculation and some controversy. Selenium has been reported as having carcinogenic as well as toxic properties; other authorities have presented evidence that selenium is highly beneficial as an essential nutrient [63,64]. Its significance and involvement in the marine biosphere is not known. A review of the marine literature indicates that selenium occurs in seawater as selenite ions (SeO_3^{2-}), with a reported average of 0.2 µg/l [90].

Selenium is particularly toxic to marine invertebrates (four-day LC_{50} = 2900 to > 10,000 µg/l). The lowest observed adverse effect concentration was 200 µg/l.

9.1.2.9
Silver

Marine fish embryos and eggs are relatively insensitive to silver. A nine-day exposure to 90 µg/l silver had no adverse effect [91]. At 180 µg/l silver, growth deformities and 30% mortality were observed [92]. Increasing the salinity reduced the toxicity of the silver [93].

9.1.2.10
Vanadium

Vanadium has a tendency to concentrate in the environment for reasons not yet understood. Environmental mobilisation of vanadium and its compounds occurs in a number of ways during the net transport of vanadium into the oceans. Some of these transport processes include terrestrial run-off, industrial emissions, atmospheric wash-out (vanadium in the air comes only from industry, as there are no significant natural sources), river transport and oil spills, resulting in a complex ecological cycle. The possibility of vanadium deposition due to oil spillage has been discussed, but no evidence is yet available to confirm the release of vanadium from oil. Since crude oils are rather rich in vanadium (50 – 200 ppm), it is not inconceivable that some vanadium may be released upon the contact of oil with seawater. The LC_{50} of vanadium is greater than 10,000 µg/l.

9.1.2.11
Zinc

Marine life is relatively resistant to zinc at all life stages. Crustacea, bivalve molluscs and worms undergo damage or fatalities upon 1–2 weeks' exposure to 340 µg/l zinc [94,95]. Decreasing the salinity increased the toxicity of zinc to invertebrates and fish sevenfold [96–98]. Fiddler crabs *Uca annulipes* and *Uca triangularis* gave four-day LC$_{50}$ values of 66,420–76,950 µg/l when exposed to zinc.

9.1.3
Summary of Toxicity Data

Tables 9.6 and 9.7 compare the short-term four-day LC$_{50}$ values obtained for various metals in nonsaline and saline waters, respectively, with typical concentrations of these elements that have been found in natural waters. Such typical concentrations are summarised in Appendix 5.1 (nonsaline waters) and Appendix 5.2 (seawater). When the concentration of a metal giving a four-day LC$_{50}$ in an environmental water is lower than 50%, mortalities occur in this period.

The higher the four-day LC$_{50}$ value relative to the observed concentration in the environmental water, the fewer the mortalities. Thus, if the four-day LC$_{50}$ is 3000 µg/l and 5 µg/l are present in environmental water, then few or zero mortalities will occur. If 1000–2000 µg/l of the metal is present in environmental water then some mortalities (< 50%) and adverse effects will take place.

Nonsaline Freshwaters

Applying this treatment to the results in Table 9.6, it is apparent that in nonsaline waters the following creatures will undergo extensive mortalities (50–100%) upon short exposure to the quoted concentrations of metals:

Some types of fish exposed to 32 µg/l silver
Daphnia magna exposed to 630 µg/l zinc
Crustacea exposed to 200 µg/l copper
Crayfish exposed to 1.3 µg/l mercury.

If we consider the ratio of four-day LC$_{50}$ to environmental concentrations of metals in nonsaline waters, it is clear that some mortalities (< 50%) and certainly adverse effects are likely to occur when these elements occur at the higher end of their observed concentration range in the environment:

Crustacea in the presence of 23 µg/l chromium
Fish and amphipods in the presence of 200 µg/l copper
Fish in the presence of 3600 µg/l aluminium
Some types of fish in the presence of 630 µg/l zinc.

Table 9.6. Effect of short-term (four-day) exposure of creatures to typical concentrations of metals found in fresh waters (nonsaline). From author's own files

Element	Creature	4-day LC$_{50}^{a}$ (µg/l)	Typical concentration of metal (µg/l) in freshwater (see Appendix 5.1)	
			Maximum	Minimum
Nickel	Fish	3060 max	40	1.5
		2900 min	40	1.5
Vanadium	Fish	5000 max	24	0.1
		2900 min	24	0.1
Chromium	Fish	65000 max	23	0.05
		3300 min	23	0.05
	Crustacea	80 max	23	0.05
		30 min	23	0.05
Arsenic	Fish	14,400	490	0.42
Silver	Fish	6700 max	32	0.3
		7 min	32	0.3
Zinc	Fish	52000 max	630	0.86
		2400 min	630	0.86
	Daphnia magna	70	630	0.86
	Annelid	10000	630	0.86
Lead	Fish	1500	60	0.13
	Invertebrates	1500	60	0.13
Copper	Fish	400 max	200	0.48
		250 min	200	0.48
	Crustacea	24	200	0.48
	Mollusc	2000 max	200	0.48
		400 min	200	0.48
	Amphipod	480	200	0.48
Cadmium	Fish	371000 max	5	0.013
		< 10000 min	5	0.013
	Amphipod	780	5	0.013
Mercury	Crayfish	0.2	1.3	0.009
	Slipper limpet	1100	1.3	0.009
Aluminium	Fish	3800	3600	14
Iron	Isopod	43100 max	5000	1
		28610		

[a] Maximum to minimum range depending on creature type.

Table 9.7. Effect of short-term (four-day) exposure of creatures to typical concentrations of metals found in saline waters (from author's own files)

| Element | Creature | 4-day LC$_{50}^a$ (µg/l) | Typical concentration of metal (µg/l) in saline water (see Appendix 5.2) | | | |
| | | | Open seawater | | Coastal, bay and estuary waters | |
			Maxim.	Minim.	Maxim.	Minim.
Nickel	Fish	35,000	1.58	0.099	5.3	0.2
	Marine organisms	10,000	1.58	0.099	5.3	0.2
	Planktonic	600 max	1.58	0.099	5.3	0.2
	Crustacea	50 min	1.58	0.099	5.3	0.2
	Bivalve	600 max	1.58	0.099	5.3	0.2
	mollusc	50 min	1.58	0.099	5.3	0.2
Selenium	Invertebrates	> 10000 max	0.029	0.001	0.4	0.4
		2900 min	0.029	0.001	0.4	0.4
Vanadium	Fish	> 10,000	2.0	0.45	5.1	< 0.001
	Invertebrates	10,000	2.0	0.45	5.1	< 0.001
Chromium	Annelid	8000 max	1.26	0.005	3.3	0.15
		2200 min	1.26	0.005	3.3	0.15
	Mollusc	105,000 max	1.26	0.005	3.3	0.15
		14,000 min	1.26	0.005	3.3	0.15
	Crustacea	640,000 max	1.26	0.005	3.3	0.15
		2000 min	1.26	0.005	3.3	0.15
	Fish	190,000 max	1.26	0.005	3.3	0.15
		12,400 min	1.26	0.005	3.3	0.15
Arsenic	Fish	28,000 max	–	–	1.04	1.0
		15,000 min	–	–	1.04	1.0
	Invertebrates	4000	–	–	1.04	1.0
Zinc	Fiddler crab	76,950 max	10.9	0.05	250	0.007
		66,420 min	10.9	0.05	250	0.007
Copper	Crustacea (young)	300	8.6	0.006	20	0.065
	Mollusc (young)	300	8.6	0.006	20	0.065
	Crustacea (adult)	30,000	8.6	0.006	20	0.065
	Mollusc (adult)	30,000	8.6	0.006	20	0.065
Cadmium	Fish	16,400 max	0.3	0.01	5	0.013
		6400 min	0.3	0.01	5	0.013
	Fiddler crab	48,210 max	0.3	0.01	5	0.013
		42,230 min	0.3	0.01	5	0.013
Mercury	Crab	8		0.002	15.1	0.00002
	Fiddler crab	2830 max	0.078	0.002	15.1	0.00002
		2750 min	0.078	0.002	15.1	0.00002

[a] Maximum to minmum range depending on creature type
Greater than 50% mortality upon 4 days of exposure in the case of crab exposed to 15.1 µg/l mercury

Table 9.8. Effect of duration of exposure to metals in water on the mortalities of various creatures (from author's own files)

Element	Creature	4-day LC_{50} (µg/l)	Short-term exposure (4 days) Metal concentration in freshwater (µg/l) (see Appendix 5.1) Maxim.	Minim.	365 day LC_{50} (µg/l)[b]	Long-term exposure (1 year) Metal concentration in freshwater (µg/l) (see Appendix 5.2) Maxim.	Minim.
Cadmium	Fish (max)[a]	37,100	5	0.013	24,733	5	0.013
	Fish (min)[a]	< 10,000	5	0.013	< 666	5	0.013
	Amphipod	780	5	0.013	52	5	0.013
Zinc	Fish (max)[a]	52,000	630	0.86	3466	630	0.86
	Fish (min)[a]	2400	630	0.86	160	630	0.86
	Daphnia magna	70	630	0.86	4.6	630	0.86
	Annelid	10,000	630	0.86	666	630	0.86
Mercury	Crayfish	0.2	1.3	0.009	0.013	1.3	0.009
	Slipper limpet	1100	1.3	0.009	73	1.3	0.009

[a] Depending on type

[b] Assumed 365-day LC_{50} = 4-day $LC_{50}/15$, similar relationship assumed for mercury and zinc. Greater than 50% mortalities (i.e. LC_{50} < concentration of metal in freshwater): short-term (4-day) exposure: *Daphnia magna* exposed to 630 µg/l zinc, crayfish exposed to 1.3 µg/l mercury; long-term (1-year) exposure: fish and *Daphnia magna* exposed to 630 µg/l zinc, crayfish exposed to 1.3 µg/l mercury.

Thus, long-term exposure is more likely to produce mortalities at lower concentrations. This is illustrated in Table 9.8, where it is apparent, for example, that exposure to 780 µg/l and 52 µg/l cadmium for four days and 365 days, respectively, would kill 50% of amphipods. For environmental waters containing 5 µl/l cadmium, adverse effects on amphipods and possibly a small number of fatalities are more likely to occur during long-term exposure. In the case of zinc, short-term exposure under these conditions would lead to more than 50% mortalities in the case of *Daphnia magna* and long-term exposure would, in addition, have a similar effect on certain types of fish.

Saline Waters

Fewer types of creatures will undergo mortalities in saline waters due to the lower environmental concentrations of metals that occur compared to the concentrations of metals present in nonsaline inland waters. Thus, it is apparent from Table 9.7 that the only observed case where the environmental concentration exceeds the four-day LC_{50} value is that of crabs exposed to 15.1 µg/l mercury in seawater. Other cases where low percentage mortalities or adverse effects might occur (i.e. where the four-day LC_{50} environmental concentration ratio is low) include bivalve molluscs and plankton, as well as planktonic crustacea in the presence of 5.3 µg/l nickel and young crustacea and molluscs in the presence of 20 µg/l copper.

It will have been noted that in all of the above considerations only the four-day LC$_{50}$ test was discussed. This parameter gives the concentration of the test metal in the test water that will kill 50% of the creatures being tested in four days. Obviously, if the duration of the LC$_{50}$ test is increased, then a lower concentration of the test metal will be required over the extended period for 50% fatalities to occur. Thus, as shown in Fig. 10.1, the LC$_{50}$ value of 0.7 µg/l obtained for salmonid fish in nonsaline waters when exposed to cadmium for four days is approximately 15 times greater than the value of 0.05 µg/l obtained in a long-term 365-day exposure.

Exposure of the fish to various concentrations of cadmium in nonsaline waters would have the following results:

Environmental concentration of cadmium (µg/l)	Exposure time (days)	
	4	364
	% mortality	% mortality
0.005	Very low	< 50
0.05	Very low	50
0.5	< 50	> 50
0.7	50	> 50
5	> 50	> 50

9.2
Organic Compounds

9.2.1
Nonsaline Waters

It is seen in Tables 9.9 and 9.11 that LC$_{50}$ values of organic compounds cover a wide range, from as low as 0.01 – 0.1 mg/l (e.g. chlorophenols: very toxic compounds) to as high as 1000 mg/l (e.g. methylene chloride and alcohols: nontoxic compounds). Further information on the toxicities of organic compounds towards fish and creatures other than fish is given in Chap. 6.

Some particular examples where actual concentrations of contaminants in river waters and LC$_{50}$ values are known are quoted in Table 9.12.

9.2.2
Seawaters

Available data on seawaters is given in Table 9.10. Further information on the toxicities of organic compounds to seawater creatures is given below.

Table 9.9. LC_{50} values of organic compounds in nonsaline waters (from author's own files)

Compound	Organism	LC_{50} value	LC_{50} test duration (days)	Reference
Diethyl hexyl phosphate	Daphnia and fish	10 – 1000 mg/kg	4	[99, 100]
1-Octanol	Rainbow trout (*Salmo gairdneri*)	15.84 mg/l	4	[101]
Sodium decyl sulfonate	Lugworm	15.2 mg/l	4	[102]
Triton X-100	(*Arenicola*	15.2 mg/l	4	[102]
Sodium dodecyl benzene sulfonate	*marina*)	12.5 mg/l	4	[102]
Ethylene dibromide	*Hydra oligatis*	50 mg/l	3	
Methylene dichloride	Juvenile fathead minnows (*Pimphales promelas*)	502 mg/l	2	[112]
1,2,4-Trichlorobenzene	Fathead minnows (*Pimphales*	(a) 7.8 mg/l (b) 2.76 mg/l	4	[100]
1,4-Dichlorobenzene	*promelas*)	1.10 mg/l	4	[100]
1,2,3,4-Tetrachloro-benzene		4.2 mg/l	4	[100]
Pentachlorophenol	*Selenastrum capricornutum*	(a) 0.11 – 0.15 mg/l softwater	4	[104]
		(b) 0.76 mg/l hardwater	4	
	Roach (*Rutilus rutilus*)	0.028 mg/l	4	[105]
	Rainbow trout (*Salmo gairdneri*)	0.09 mg/l	4	[101]
2,4-Dichlorophenol	Rainbow trout (*Salmo gairdneri*)	4.64 mg/l	4	[101]
2,4,6-Trichlorophenol	Roach (*Rutilus rutilus*)	0.05 mg/l	4	[105]
2,3,4,6-Tetrachloro-phenol	Roach (*Rutilus rutilus*)	0.071 mg/l	4	[105]
Polychlorobiphenyl (Arochlor 1254)	Rainbow trout (*Salmo gairdneri*)	30 mg/l	4	[106]
Polycyclic aromatic hydrocarbons	*Daphnia magna*	LT measured		[107]
Picloram (4-amino-3,3,6-trichloro-picolinic acid)	Rainbow trout (*Salmo gairdneri*)	LC_{50} LC_{50}	96 192	[108]

Table 9.10. LC$_{50}$ values of organic compounds in seawater (from author's own files)

Compound	Organism	LC$_{50}$ value (mg/l)	LC$_{50}$ test duration (days)	Reference
Endosulfan	Crab (*Oziotelphusa senex senex*)	6.2 (sublethal) 18.62 (lethal)	4	[123,124]
Kepone	Lamprey (*Petroyzon marinus*)	414 – 444	4	[125]
Hexazinone	Juvenile pacific salmonid	276	4	[126]
Pronone 109	Fish	904	4	[126]
Vapar L	Fish	1686	4	[126]
3-Fluoro-methyl 4-Nitrophenol	Walleye (*Stizoastedion vitreum*)	LC$_{25}$ = 4.1 (gametes) LC$_{25}$ = 2.6 (eggs)	0.5 0.5	[127]
3-Fluoro-methyl 4-nitrophenol	Larval sea lamprey (*Petroyzon marinus*)	LC$_{99}$ = 1	8 h	[127]
Malathion	Teleosts *Channa punctatus* (Block)	1.73 mg/l	96	[109]
	Puntius sophore (Hamilton)	1.646 mg/l	96	[109]
Carbaryl	Catfish (*Clarius batrachus*)	24 – 61 mg/l 54 – 134 mg/l 49 – 123 mg/l 47 – 108 mg/l	24 48 72 96	[110]
Carbaryl	Lugworm (*Arenicola marina*)	7.2 mg/l	48	[102]
Parathion-ethyl		2.7 mg/l	48	[110]
Sodium dodecyl sulphate		15.2 mg/l	48	
Triton X-100		15.2 mg/l	48	
Malathion, endosulfan, Fenvalerate	Mummichog (*Fundulus heteroclitus*)	–	96	[111]
Ethylene dibromide	*Hydra oligactis*	50 mg/l	72	[112]
Methylene dichloride	Fathead minnow (*Pimephales promelas* Rafinesque)	502 mg/l 471 mg/l	48 192	[103]
Aniline	*Daphnia magna*	Maximum acceptable toxicant concentration (MATC) Aniline = 4.6 – 46.7 mg/l	21	[113]

Table 9.10. Continued

Compound	Organism	LC_{50} value (mg/l)	LC_{50} test duration (days)	Reference
2,4-Dichlorophenol		0.7 – 1.48 mg/l		[113]
Carbaryl	Lugworm (*Arenicola marina*)	7.2	3	[100]
	Catfish (*Clarius batrachus*)	46.9 – 107.7 mg/l	4	
Parathion-ethyl	Lugworm (*Arenicola marina*)	2.7 mg/l	3	[110]
Mirex	Rainbow trout (*Salmo gairdneri*)	5.0 mg/l	4	[106]
Malathion	Rainbow trout (*Salmo gairdneri*)	1.73 mg/l	4	[114]
Roundup herbicide	Rainbow trout (*Salmo gairdneri*) Chinook Coho salmon	7.4 – 12 mg/l	4	[115]
Rodeo herbicide	Rainbow trout (*Salmo gairdneri*) Chinook Coho salmon	580 mg/l	4	[115]
Bromacil	Fathead minnow (*Pimephales promelas*)	182 mg/l	4	[116]
Diuron	Fathead minnow (*Pimephales promelas*)	14.2 mg/l	4	[116]
Lindane	Teleost fish (*Anguilla anguilla*)	0.32 – 0.68 mg/l	4	[117]
Methylenebis thiocyanate	Chlorella pyrenozdosa	0.042 mg/l	4	[118]
	Guppy (*Poecilia reticulata*)	0.39 mg/l	4	
Cyanogen chloride	*Daphnia magna*	0.062 mg/l (adult) 0.029 mg/l (juvenile)	2	[119]
Acrylates, methacrylates	Juvenile Fathead minnow (*Pimephales promelas*)	0.38 – 2.1 mg/l	96	[113]
Diethyl hexyl phthalate	Review of toxic effects mammals	10 – 100 mg/l	–	[120]

Table 9.10. Continued

Compound	Organism	LC$_{50}$ value (mg/l)	LC$_{50}$ test duration (days)	Reference
1,2,4-Trichloro-benzene	Fathead minnow (*Pimephales promelas*)	2.76 mg/l	32	[121]
3,4-Dichloroaniline	Fathead minnow (*Pimephales promelas*)	6.99 – 8.06 mg/l	4	[122]

Table 9.11. Relative four-day LC$_{50}$ values for organic and organometallic compounds in nonsaline water creatures (from author's own files)

Most toxic LC$_{50}$ 0.01 – 10 µg/l	0.01 – 1 mg/l	1 – 10 mg/l	10 – 100 mg/l	Least toxic 100 – 1000 mg/l
Organotin compounds	Pentachlorophenol 2,4,6-Trichloro-phenol 2,3,4,6-Tetra-chlorophenol Lindane Methylene *bis* thiocyanate Cyanogen chloride	1,2,4-Trichloro-benzene 1,4-Dichlorobenzene 1,2,3,4-Tetra-chlorobenzene 2,4-Dichlorophenol Carbaryl Parathion Mirex Malathion Roundup 3,4-Dichloroaniline 3-Fluoro-4-methyl nitrophenol 3,4-Dichloroaniline	1-Octanol Sodium decyl sulfonate Triton X-100 Sodium dodecyl benzenesulfonate Ethylene dibromide Polychlorobiphenyls Endosulfan Carbaryl Diuron	Methylene dichloride Kepone Rodeo Bromacil Hexazinone Pronone 109 Varpar L

9.2.2.1
Polyaromatic Hydrocarbons

Many polyaromatic hydrocarbons have been shown to be directly carcino-genic to mammals when present in trace quantities (Table 9.13). These are attributed to particular materials that may be present in water samples and are also water-soluble to some extent, so that their occurrence in the environment has caused widespread concern. At least a hundred compounds of this type have been detected and characterised in environmental samples. The basic molecular structure consists of benzene rings either fused together or bridged by methylene side-chains. Alkyl substituents also occur.

These compounds can be produced by the biochemical degradation of other organic compounds under suitable conditions. They may occur in the environment due to the combustion of materials such as wood or leaves. Other sources of aromatic materials from which polyaromatic hydrocarbons may be derived include crude oil, which can contain 20% by weight of dicyclic and higher polyaromatic hydrocarbons and high-grade petrol, the

Table 9.12. Comparison of four-day LC$_{50}$ values for fish and other creatures and environmental concentrations of organic compounds in river waters (in order of increasing toxicity). From author's own files

Organic compound	Creature	4-day LC$_{50}$ (μg/l)	Environmental concentration (μg/l) (See Appendix 6.1)	
			Maxim.	Minim.
Most toxic				
Pentachlorophenol	Roach	28	250*	0.1
	Rainbow trout (*Salmo gairdneri*)	90	250*	0.1
	Selenastrum capricornutum	110 – 150	250*	0.1
Lindane	Teleost fish (*Anguilla anguilla*)	320 – 680	0.01	0.001
Malathion	Rainbow trout (*Salmo gairdneri*)	1730	0.032	0.027
Alkylbenzene sulfonate	Lugworm (*Arenicola marina*)	12,500	600	10
Nonionic detergents	Lugworm (*Arenicola marina*)	15,200	70	8
(Triton X-100)				
Least toxic				

* Environmental concentration exceeds 4-day LC$_{50}$, i.e., greater than 50% mortality of the test creature.

Table 9.13. PAHs commonly found in water

IUPAC	Molecular weight	Relative carcinogenicity	Abbreviation
Benzo(*ghi*)perylene	276	−	B(*ghi*)P
Chrysene	228	−	Ch
Fluoranthene	202	−	Fl
Indeno(1,2,3-*cd*)pyrene	276	+	IP
Phenanthrene	178	?	Ph
Perylene	252	−	Per
Pyrene	202	−	Pyr
Anthracene	178	?	An
Benzo(*a*)anthracene	228	+	B(a)A
Benzo(*b*)fluoranthene	252	++	B(b)F
Benzo(*j*)fluoranthene	252	++	B(j)F
Benzo(*k*)fluoranthene	252	−	B(k)F
Benzo(*a*)pyrene	252	+++	B(a)P
Benzo(*e*)pyrene	252	+	B(e)P

aromatic content of which is over 50%. Unsaturated fatty acids, terpenoids and steroids may also be potential polyaromatic hydrocarbon precursors.

The behaviour and effects of anthropogenic polycyclic aromatic hydrocarbons in aquatic biota in chronically and acutely polluted waterways have been intensely studied for many years [132–138]. Although molluscs have been shown to accumulate polyaromatic hydrocarbons, the question of whether the concentrations of potentially toxic and carcinogenic polyaromatic hydrocarbons are magnified through the food chain is not yet resolved.

The analytical chemistry of polyaromatic hydrocarbons in tissues can provide an important part of the answer to the biomagnification question, but it must be improved by new technology and the modification of existing analytical procedures to the point where unambiguous, detailed and reproducible data can be obtained on a routine basis. Furthermore, few papers dealing with the analytical methodology for determining polycyclic aromatic sulfur heterocycles and polycyclic aromatic nitrogen heterocycles in fish tissues are to be found in the literature. Considering that the heterocyclic fractions are at least as biologically active as the polyaromatic hydrocarbons [139–144], it is clearly desirable that techniques are developed that will provide accurate quantitative and qualitative data on the sulfur and nitrogen heterocycles in aquatic biota. Polyaromatic hydrocarbons have been shown to reduce the reproduction rate of plaice in seawater [146].

9.2.2.2
Chlorinated Insecticides

Persistent chlorinated hydrocarbons of agricultural and nonagricultural interest—such as 1,1,1-trichloro-2,2-bis-(p-chlorophenyl)ethane (DDT), polychlorinated biphenyls (PCBs) and hexachlorobenzene—now have a global distribution and can be detected in wildlife samples in variable amounts. PCBs, together with 1,1-dichloro-2,2-bis-(p-chlorophenyl)-ethylene (DDE), are the main types of chlorinated hydrocarbons found in Norwegian avian fauna and in fish along the Norwegian coast [147–149].

In Friefjorden, a fiord in south-east Norway, heavy local contamination with chlorinated hydrocarbons of industrial origin has been detected. The contaminants most often found in fish in this area are hexachlorobenzene, octachlorostyrene and decachlorobiphenyl. In addition, complex mixtures of PCBs and chlorinated naphthalenes have been detected [150, 151]. Decachlorobiphenyl has previously been found in arctic fox (*Alopex lagopus*) from Svalbard [152], and octachlorostyrene was first detected in birds in the Netherlands [153–155].

In a monitoring programme over the last six years, the above chlorinated hydrocarbons have been determined in samples from cod (*Gadus morhua*). Lindane and endosulfan and trichlorophon insecticides have been shown to cause erratic swimming behaviour, hyperventilation and mortalities in invertebrates [156].

9.2.2.3
Polychlorinated Paraffins

Polychlorinated paraffins are chlorination products of n-alkane mixtures which have been produced in technical amounts since the early 1930s [264–267]. Several chlorinated n-alkane fractions of petroleum from the range C_{10} – C_{30} are used mainly as additives to sealants and metal-cutting oils, as secondary plasticisers, and as flame retardants. The degree of chlorination varies between 10 and 72% depending on the application field. They are classified according to carbon chain length into short-chain (C_{10} – C_{30}), medium-chain (C_{14} – C_{17}) and long-chain (C_{17} – C_{30}) polychlorinated paraffins. Especially in the 1980s, after the ban of polychlorinated biphenyls, for which polychlorinated paraffins are good substitutes in some application fields, the production amounted to more than 300,000 t yearly, with demand from the United States alone making up a third of the world's consumption. The European production has been estimated at $\sim$ 140,000 t for 1991 [266]. Nowadays, it may be somewhat less than this because of the increased use of alternative, nonchlorinated products. The last German producer of polychlorinated paraffins, Hoechst, stopped production in 1995. Though the toxicity is rather low [268–270], bioconcentration factors are high, reaching values of nearly 1.4×10^5 in mussels with CP12:69 (polychlorinated dodecane with 69% chlorine content) [271].

9.2.2.4
Polychlorinated Biphenyls

Since their introduction, polychlorinated biphenyls have caused much ecological damage and have been shown to be harmful to humans. One aspect of these toxicants is that they have been shown to have a very severe adverse effect on wildlife by causing thin eggshells and consequently a poor reproductive rate in the laying season. Deleterious effects on seals have also been observed.

Polychorinated biphenyls have until recently been used extensively as cooling media in electrical transformers and also in railway engine repair shops. PCBs are marketed as Aroclors by Monsanto. All Aroclors are characterised by a four-digit number; the first two digits represent the type of molecule (e.g. 12 represents biphenyl, 54 terphenyl and 25 and 44 are mixtures of biphenyl and terphenyl), and the last two digits give the percentage by mass of chlorine (e.g. Aroclor 1260 is a carbon system with 60% m/m of chlorine). The compositions of two further Aroclors are given below:

Composition	Aroclor 1016 (16% mm chlorine)
Biphenyl	% w/w
2	0.03
4	1.1
2,4'	0.4
2,5,2'	12.7
4-4'	3.4

Aroclor 1254	(50% mm chlorine)
	% w/w
2,3,2',5'	5.3
2,5,2',5'	10.3
2,5,3',4'	3.3
2,4,5,2',5'	11.7
2,3,4,2',4',5'	4.9
2,4,5,2',4',5'	5.3

Polychlorinated biphenyls are sold under a variety of trade names, of which Aroclor is one. The following is a list of principal trade names used for PCB-based dielectric fluids which are usually classified as Apkarels: Aroclor (UK and USA), Pyoclor (UK), Inertren (USA), Pyanol (France), Clophen (Germany), Apirolio (Italy), Kaneclor (Japan), Solvol (USSR).

Aroclor causes severe weight reduction and liver degeneration in rainbow trout (*Salmo gairdneri*) [106] and inhibition of reproduction at the 50 – 100 µg/l level in nonsaline waters in *Daphnia puliccaria* [157] (see Tables 6.1 and 6.2).

Metabolites of polychlorinated biphenyls containing methylsulfonyl (MeSO$_2$) have been found to persist in biota. Based on tissue concentrations and pathological findings, a tentative suggestion has been made that persistent MeSO$_2$-CBs and MeSO$_2$-DDEs may be an important influence in a disease complex observed in Baltic seals [272]. The MeSO$_2$-CBs can exist as 2-, 3-, and 4-MeSO$_2$-substituted congeners; however, only the 3- and 4-substituted compounds have been found to persist in biota [273]. Similar to chlorobiphenyl precursors, MeSO$_2$-CBs congeners exhibit axial chirality if both of the phenyl rings have an asymmetric chlorine substitution pattern. The tri- and tetra-*ortho*-chlorine-substituted congeners may have hindered rotation about the phenyl–phenyl α bond at physiological and ambient temperatures. Thus, hindered congeners with axial chirality exist as atropisomeric (enantiomeric) pairs in the environment. There are 78 out of 209 theoretically possible polychlorinated biphenyls that exhibit axial chirality. Nineteen of the chiral polychlorinated biphenyls are predicted to form stable atropisomers under most environmental conditions [274], of which at least 12 (polychlorinated biphenyls 84, 88, 91, 95, 132, 136, 144, 149, 171, 174 and 183) have been detected in commercial polychlorinated biphenyl mixtures above 1% (w/w) [275]. Since the introduction of a MeSO$_2$ group at the 2- or 3-position will add an additional element of asymmetry, MeSO$_2$-

CBs may be chiral even if the parent polychlorinated biphenyl is not. Of these 837 theoretically possible MeSO$_2$-CBs, 456 are chiral [276]. Of these 456 congers, 170 may be environmentally stable due to tri- or tetra-*ortho*-substitution [276]. However, the number of environmentally relevant and chiral MeSO$_2$-CBs is lower still, since an apparent maximum of ~40 congeners can be derived from 20 precursor polychlorinated biphenyls with the correct structural features for sulfone formation [273].

9.2.2.5
Herbicides

The use of certain herbicides in or near to water will give rise to rapid decomposition of the affected vegetation, which in turn can cause deoxygenation of the water.

The most obvious method of entry of herbicides into river water is by their direct application to the water in order to control aquatic vegetation. When emergent vegetation is sprayed, some of the material may be sprayed directly onto the water surface and some may run off plants into the river. Any herbicide reaching the soil or the banks close to the water may or may not be available for leaching into the watercourse. If the herbicide remains in the plants after their death then it may enter the water when they decompose.

The above observations are also pertinent to field-applied herbicides, which may enter the water by spray drift, leaching from or erosion of the soil or via rotting vegetation or silage. The quantities reaching the water by leaching will depend upon the herbicide, rainfall and soil type. The terrain may also be important in that it will affect the pattern of leaching or run-off. The period that the herbicide persists in the soil is also important in that it will affect how long the pollution is likely to last. All of these factors apply at the same time, making each herbicide application an individual event, and so generalisation must be treated with caution.

The question of accidental spillage of a concentrate or a diluted spray into water must also be considered as well as malpractices such as the dumping of excess chemicals, washing out empty containers in ponds and rivers, and the improper disposal of containers. The herbicides may also be present in industrial or agricultural effluents.

Factors which will reduce the concentration of herbicides downstream and so must be taken into account are:

(1) The stability of the compound towards chemical and biological degradation, and its removal from the water by volatilisation.
(2) Its absorption into the mud at the bottom, into suspended material and into living organisms.

The herbicide 2,4-dinitrophenoxyacetic acid has been shown to cause death to macroinvertebrates during 12 months' exposure in nonsaline water [158] (see Table 6.2).

Diflubenzuron causes mortalities at 50 µg/l and moulting delay at 5 µg/l in larval horseshoe crabs (*Limulus polyphemus*) in seawater [159].

9.2.2.6
Polychlorodibenzo-*p*-Dioxins and Polychlorodibenzofurans

Polychlorinated dibenzo-*p*-dioxins, polychlorinated dibenzofurans and *ortho*-unsubstituted polychlorinated biphenyls (non-*ortho* polychlorinated biphenyls) are three structurally and toxicologically related families of anthropogenic chemical compounds that have in recent years been shown to have the potential to cause serious environmental contamination [160–164]. The substances are trace-level components or byproducts of several large-volume and widely used synthetic chemicals, principally polychlorinated biphenyls and chlorinated phenols [165] produced during combustion processes and by photolysis [166, 167].

In general, polychlorinated dibenzo-*p*-dioxins, polychlorinated dibenzofurans and non-*ortho* polychlorinated biphenyls are classified as highly toxic [168], although the toxicities are very dependent on the number and positions of the chlorine substituents [169]. About ten individual members of a total of 216 polychlorinated dibenzo-*p*-dioxins, polychlorinated dibenzofurans and non-*ortho* polychlorinated biphenyls are among the most toxic man-made or natural substances to a variety of animal species [155]. The toxic hazards posed by those chemicals are exacerbated by their propensity to persist in the environment [147] and to readily bioaccumulate [31,32,170] and although the rate of metabolism and elimination is strongly species-dependent [171–173], certain highly toxic isomers have been observed to persist in the human body for more than ten years [174].

The majority of scientific concern for the hazards of these compounds as been directed towards the disposition in the environment of the single most toxic isomer, 2,3,7,8-tetrachlorodibenzo-*p*-dioxin (2,3,7,8-TCDD) [106, 157–165]. More recently, however, investigations into the formation and occurrence of polychlorinated dibenzofurans suggest that this family of toxic compounds may also commonly occur at comparable or greater levels and could possibly pose a greater hazard than polychlorinated dibenzo-*p*-dioxins. Polychlorinated dibenzofurans are often found as cocontaminants in the dioxins and are more readily produced from pyrolysis of polychlorinated biphenyls [21,175–177]. Most importantly, polychlorinated dibenzofurans produced from the pyrolysis of polychlorinated biphenyls are predominantly the most toxic isomers, particularly those having a 2,3,7,8-chlorine substitution pattern [163]. A number of fires involving electrical transformers and capacitors have demonstrated the potential for the formation of hazardous levels of polychlorinated dibenzofurans from the pyrolysis of polychlorinated biphenyls [177–181]. In light of these findings, and because of a dearth of data pertaining to the occurrence of these compounds in the environment, polychlorinated dibenzofurans and non-*ortho* polychlorinated dibenzo-*p*-dioxins were included as target compounds in a survey of important US rivers and lakes for polychlorinated dibenzo-p-dioxins. The decision to include as many polychlorinated dibenzo-p-dioxins isomers as possible was based on several facts: (1) several other polychlorinated dibenzo-p-dioxins isomers are also extremely toxic; (2) pentachlorophenol, a large-volume fungicide and wood preservative, contains relatively high levels of hexa-, hepta- and octachlorodibenzodioxins and essentially no tetrachlorodibenzo-p-dioxins [106, 135, 136], and; (3) incineration of materials containing chlorophenols readily produces mixtures of polychlorinated dibenzo-p-dioxins, but 2,3,7,8-tetrachloro dibenzo-p-dioxins is a minor component. On the other hand, the highly toxic 1,2,3,7,8-pentachloroisomer is a major component of polychlorinated dibenzo-p-dioxin incineration products of pentachlorophenol.

Component-specific analyses can be a crucial link to the sources of contamination because different sources of polychlorinated dibenzo-p-dioxins and polychlorinated dibenzofurans usually produce mixtures with distinctly different relative component abundances [44]. On the other hand, the preferential accumulation of certain isomers in animals may prevent source identification from analyses of biological samples.

9.2.2.7
Nitrosamines

Many N-nitrosamines are toxic and carcinogenic, and furthermore the carcinogenic action exhibits a high degree of organ specificity. Nitrosamines are formed by the interaction between a nitrite and an amine with varying ease, depending on the nature of the amine and the prevailing conditions. The reaction is not restricted to secondary amines but also occurs with primary and tertiary amines and even quaternary ammonium salts. Thus, the precursors are widespread as naturally occurring compounds, and nitrosamines are generated in many commercial and industrial processes. It is therefore conceivable that trace amounts may be present in air and water in the vicinity of industrial sites.

Mills and Alexander [181] have discussed the factors affecting the formation of dimethylnitrosamine in samples of water and soil. Dimethylnitrosamine was formed as readily in sterilised samples as in nonsterile ones, indicating that although microorganisms can carry out an enzymatic nitrosation in some soils and waters, dimethylnitrosamine can be formed by a nonenzymatic reaction, even at near-neutral conditions. The presence of organic matter appears to be important for promoting nitrosation in the presence of the requisite precursors.

9.2.2.8
Other Insecticides

Other types of insecticides cause adverse effects in fish and invertebrates in nonsaline waters (see Tables 6.1 and 6.2).

Permethrin

This insecticide causes nerve poisoning and blocking of anaerobic and aerobic metabolism in the snail *Hymnaea acuminate* [182].

Phosphamidon

Phosphamidon causes glycogen depletion in muscles (i.e. reduced mobility) in freshwater prawn *Macrobmchum lamarrei* [183].

Fenitrothion

Fenitrothion causes growth abnormalities of follicle and epithelium in freshwater murrel *Channa punctatus* [184].

Carbaryl

The presence of carbaryl above certain concentration levels reduces the survival time of catfish (*Clarias leatrachus*) [110].

Malathion

Above certain concentration levels, malathion reduces the survival time of freshwater teleosts (*Channa punctatis*). It also produces mortalities upon five days of exposure at 44,000 μg/l in toad embryos (*Bufo arenarum*) [185].

Bromacil

Bromacil reduces the growth and survival time and deforms fry in fathead minnow (*Pimephales promelas*) [172].

Endosulfan

This insecticide in seawater increases the body weight, haemolymph volume and hydration at sublethal concentrations (6200 μg/l) and decreases these parameters in concentrations above the lethal level (18,600 μg/l) when the crab (*Oziotelphusa senex*) is exposed to endosulfan [116, 124].

9.2.2.9
Organic Esters

Di-2-Ethyl Hexylphthalate

This ester causes increased surfacing behaviour in *Daphnia magna* [99] as well as a deterioration of reproductive capacity and the immune system and carcinogenic activity in Daphnia and fish (see Table 9.10). Di-2-ethyl hexylphthalate causes mortalities of young baltic herring (*Clupea lapengus*) and Atlantic cod (*Gadus marina*) [186].

Acrylates and Methacrylates

These esters cause respiratory and metabolic inhibition and neurotoxicity in juvenile fathead minnow (*Pimphales promelas*) [113].

9.2.2.10
Surface-Active Agents

Sodium decyl sulfate, Triton X-100 nonionic detergents and sodium dodecyl benzene sulfonate surface-active agents have adverse effects on gills and epidermic receptors in the nonsaline water lugworm (*Arenicola marina L*) [102].

9.2.2.11
Phenol

The presence of phenol above certain concentrations in nonsaline waters causes immobilisation, paralysis and mortality in *Ascellus aquaticus* [187].

9.2.2.12
Pentachlorophenol

Pentachlorophenols above certain concentrations in nonsaline waters cause low survival rates (e.g. 31.6 h at 90 µg/l) in rainbow trout (*Salmo gairdneri*) [106].

9.2.2.13
Aniline

The presence of aniline above certain concentrations (1000 µg/l) causes inhibition of embryo development in the South African clawed toad (*Xenopus laevis*) [188].

9.2.2.14
p-Chloroaniline

Above certain concentrations (100,000 µg/l), *p*-chloroaniline kills embryos in the South African clawed toad (*Xenopus laevis*) [188].

9.2.2.15
Methyl Bromide

Methyl bromide at concentrations above 100,000 µg/l for a period of 1 – 3 months causes paralysis in the guppy (*Poecilia reticulata*) and at concentrations above 1800 µg/l for four days causes degenerative changes in the gills as well as oral mucosa [189].

9.2.2.16
Tetrachoro-1,2-Benzoquinone

This compound, when present in seawater, causes skeletal abnormalities in the fourhorn sculpin (*Myoxocephalus quadricornis*) [190].

9.2.2.17
3-Fluoro-Methyl-4-Nitrophenol-Lampricide

This compound in seawater damages the eggs and fry of walleye (*Stizostedium vitreum*) [190].

9.2.2.18
Linear Alkyl Benzene Sulfonates

In the past, linear alkyl benzene sulfonate bioaccumulation research has relied on quantifying on the total radioactivity without distinguishing between the parent surfactant and its biotransformation products [277]. Therefore, these data do not reflect the bioaccumulation potential of the parent surfactant [277]. Moreover, quantitative information on the biotransformation of linear alkyl benzene sulfonates is not available. However, it has been proposed that linear alkyl benzene sulfonates are transformed via ω-oxidation and subsequent β-oxidation steps to sulfophenylalkanoic acids. Since biotransformation results in a reduction of the concentration of linear alkyl benzene sulfonates in fish [278, 279], it is an important process that contributes to the overall linear alkyl benzene sulfonate bioaccumulation and is therefore an issue that deserves attention.

9.3
Organometallic Compounds

The four types of organometallic compounds that occur in the environment and that have been the most extensively studied are those of arsenic, lead, mercury and tin. These can originate in the ecosystem either as man-made pollutants or by microorganism-induced biomethylation of metals in sediments, fish or marine invertebrates. The toxic effects of organometallic compounds on fish and creatures other than fish are discussed in Chap. 9.

Organoarsenic Compounds

Organoarsenic species are known to vary considerably in their toxicity to humans and animals [127, 191]. Large fluxes of inorganic arsenic into the aquatic environment can be traced to geothermal systems [192], base metal smelter emissions, and localised arsenite treatments for aquatic weed control. The methylated arsenicals have entered the environment either directly as pesticides or by the biological transformation of the inorganic species [193, 194].

It has been shown that arsenic is incorporated into both marine and freshwater organisms in- the form of both water- and lipid-soluble arsenic compounds [195].

Studies to identify the chemical forms of these arsenic compounds have shown the presence of arsenite (As(III)), arsenate (As(V)), methylarsonic acid, dimethylarsinic acid and arsenobetaine [76]. Methyl arsenicals also appear in the urine and plasma of mammals, including humans, through the biotransformation of inorganic arsenic compounds [73].

The biological methylation of inorganic arsenic by microorganisms such as moulds and bacteria present in sediment sludges and muds has been established, although there is no unequivocal evidence for the proposed pathways [196–203].

Organoarsenical pesticides such as sodium methanearsonate and arsinic acid are used in agriculture as herbicides and fungicides. It is possible that these arsenicals enter soil, plants and consequently humans. On the other hand, arsenic is a ubiquitous element on Earth, and the presence of inorganic arsenic and several methylated forms of arsenic as monomethyl-, dimethyl- and trimethylarsenic compounds in the environment has been well-documented [204].

The occurrence of arsenic biomethylation in microorganisms [205], soil [206], animals and humans [207] has also been demonstrated. Therefore, further investigation of the fate of arsenicals in the physical environment and living organisms requires a knowledge of their complete speciation.

Organolead Compounds

The use of tetraalkyl leads as antiknock additives/octane enhancers for automotive gasolines has been reduced due to environmental considerations in several countries. However, the complete elimination of tetraalkyl lead additives is unlikely.

Organolead compounds are generally more toxic than inorganic lead compounds [208], and the toxicity of the alkylated lead compounds varies with the degree of alkylation, with tetraalkyl lead being the most toxic [209].

The highly polar dialkyl and trimethyl lead compounds in particular have a high toxicity to mammals [210] and are formed as a result of the degradation of tetraalkyl lead in aqueous medium [211].

Tetramethyl and tetraalkyl lead compounds are considerably more toxic than inorganic lead (1000 times) [212] or di- or trimethyl or triethyl lead compounds [228].

The high toxicity of tetraalkyl leads is attributed to their ability to undergo the following decomposition in the environment [211]:

$$R_4Pb \rightarrow R_3Pb^+ \rightarrow R_2Pb^{2+} \rightarrow Pb^{2+}$$

The formation of alkyl lead salts, probably associated with proteins, arising in tissues from rapid metabolic dealkylation of tetraalkyl lead compounds, is of toxicological importance when evaluating exposure to tetraalkyl leads.

The toxic effect of tetraalkyl leads on mammals has been attributed to the formation of trialkyl lead compounds in body fluids and tissues.

Wong et al. [213], Reisinger et al. [214] and others [182, 183] have demonstrated that microorganisms in lake sediment can transform inorganic and organic lead compounds into volatile tetraalkyl lead. The possibility of biomethylation of lead or organolead ionic species by microorganisms, reversing the decomposition mechanism given above, may add to the problem of lead toxicity already faced by humans, although the area is presently highly disputed [214].

Organically bound lead is a minor but important contribution to total lead intake by humans and animals. Alkyl lead salts such as trialkyl lead carbonates, nitrates and/or sulfates can be formed in tissues by the rapid metabolic dealkylation of tetraalkyl lead compounds.

An interest in the speciation of lead in environmental samples has resulted from several diverse lines of investigation. Organolead compounds have been detected in cod, lobster, mackerel and flounder meal (10 – 90% of the total lead burden [215], and in freshwater fish [216,217].

Fairly high concentrations of tetraethyl lead (30 ppm) have been detected in mussels collected at a buoy near the *SS Cavtat* incident, where a shipload of tetraethyl lead was sunk [215] in the Adriatic Sea. High organolead concentrations, mainly of tetraethyl lead, were also found in mussels in other parts of the Italian seas. The presence of tetraethyl lead in aquatic organisms may indicate that the alkyl lead compounds are not immediately metabolised by living organisms and may remain in their authentic forms in the living tissues for a long time. The occurrence of tetralkyl lead compounds in aquatic biota is highly significant because of the possibility of their incorporation into the food chain.

A steady input of organoleads into the environment results from the continued use of tetralkyl leads as antiknock additives. In addition, evidence for the chemical [198] and biological alkylation of organolead salts or of lead(II) salts has been obtained [213, 216–218].

Although organoleads may make only a small contribution to the total lead intake of an organism, it has been demonstrated that trialkyl lead salts arising in tissues from the degradation of tetralkyl leads are important in lead toxicity. The conversion of R$_4$Pb to R$_3$Pb$^+$ occurs rapidly in liver homogenates from rats to rabbits. Acute toxicities of tetralkyl leads and of trialkyl lead salts are similar and are at least an order of magnitude greater than dialkyl lead salts or inorganic lead salts. Relatively little is known of either the effect of chronic exposure to small amounts of such compounds or the levels of organic lead compounds, such as the tetralkyl leads, in biological and food material. Dialkyl lead salts cause toxic symptoms similar to those produced by inorganic lead salts, and they exhibit an affinity for thiol compounds. Triorganolead salts inhibit oxidative phosphorylation.

Speciation of alkyl lead compounds, including molecular and ionic, volatile and solvated forms, has become immensely important and in urgent demand in studies related to toxicity and environmental consequences.

The highly polar dialkyl and trialkyl forms in particular are more important species because of their high toxicity to mammals and the consequences of their formation as a result of the degradation of tetralkyl lead in aqueous medium [211].

Organomercury Compounds

Organomercury compounds are more toxic than metallic mercury [219], and, when present in the environment, inorganic mercury forms may cause serious illness in extremely polluted areas. Methyl mercury has been stated to be neurotoxic. Due to its chronic toxicity and its tendency to bioaccumulate, mercury is of prime interest. Being extremely volatile in the organic and elemental forms, mercury is well-dispersed in the atmosphere.

The interest in mercury contamination, and particularly in organomercury compounds, is a direct reflection of the toxicity of these compounds to humans. Some idea of the proliferation of work in this area can be derived from the reviews of Krenkel [220], Robinson and Scott [221] (460 references) and Uthe and Armstrong [222] (283 references).

All forms of mercury are potentially harmful to biota, but monomethyl and dimethyl mercury are particularly neurotoxic. The lipophilic nature of the latter compounds allows them to be concentrated to higher trophic levels, and the effects of this biomagnification can be catastrophic [223]. Certain species of microorganisms in contact with inorganic mercury produce methylmercury compounds [199]. Environmental factors influence the net amount of methylmercury in an ecosystem by shifting the equilibrium of the opposing methylation and demethylation processes. Methylation is the result of mercuric ion (Hg^{2+}) interference with biochemical C-1 transfer reactions [200]. Demethylation is brought about by nonspecific hydrolytic and reductive enzyme processes [224,225]. The biotic and abiotic influences that govern the rates at which these processes occur are not completely understood.

Although much of the early work on cycling of mercury pollutants has been performed in freshwater environments, estuaries are also subject to anthropogenic mercury pollution [220]. A strong negative correlation exists between the salinity of anaerobic sediments and their ability to form methylmercury from Hg^{2+}. As an explanation for this negative correlation, the theory was advanced that sulfide (derived by microbial reduction of sea and salt sulfate) interferes with Hg^{2+} methylation by forming mercuric sulfide, which is not readily methylated [213–216]. There are several reports in the literature on the methylation of Hg^{2+} by methylcobalamin [203,227,228].

The synthesis of methylmercury compounds from inorganic mercury by microorganisms, mould and enzymes in freshwater sediments has been investigated by some workers [98,191–198,203,218]. This biological methylation of mercury compounds provides an explanation for the fact that CH_3Hg^+ is found in fish, even if all known sources of mercury in the environment are in the form of inorganic mercury or phenyl mercury. The formation of the volatile CH_3HgCH_3 (bp 94 °C) may be a factor in the redistribution of mercury from aqueous industrial wastes. The process of methylation is fundamental to the knowledge of the turnover of mercury. It may be significant in the uptake and distribution of mercury in fish and in the mobilisation of mercury from deposits in bottom sediments into the general environment.

The organic mercury compounds produced, primarily dimethylmercury and methylmercury halides, are potentially more toxic than inorganic forms. Therefore, recent studies of environmental mercury have been concerned with its chemical speciation in order to determine not only the amounts of mercury present but also the chemical forms. More extensive data in this area will assist in determining the role of organic mercury in the global cycling of the element.

Andren and Harris [229] have reported a methylmercury concentration of 0.02 – 0.1 µg/kg mercury in unpolluted sediments. In two rockfish samples, the organic mercury concentration was 110 and 190 µg/kg (dry weight). This agrees quite well with the reported methylmercury concentration range of 70 – 200 µg/kg mercury in similar fish [229]. Matsunaga and Takahashi [8] found 0.2 – 0.4 µg/kg mercury in sediments.

It has been reported that organomercury compounds are significantly concentrated in fish [226], predominantly as methylmercury compounds. Fish in contact with water containing 0.01 µg/l and sediment containing 30 µg/kg of mercury have been found with 341 µg/kg in their flesh, i.e. a factor of 34,000 bioamplification in the flesh. At Minamata Bay, Japan, mercury levels in some fish attained 50 µg/kg wet weight, while levels of around 20 mg/kg were common. Experiments with brook trout have shown that over a period of nine months, the fish had accumulated 900, 2900 and 123,000 µg/kg of mercury from water containing 0.09, 0.29 and 0.93 µg/kg mercury, respectively, in their gonads.

Some of the toxic and adverse effects of organomercury compounds on fish and invertebrates in nonsaline waters are discussed in Chap. 10 and are summarised below.

Teleost	Liver abnormalities
Fish	Carcinogenesis [236]
Codfish	Kidney damage [231]
Microtubes	Severe disruption [237]

Four-day LC$_{50}$ values of 430, 4300 and 507 µg/l, respectively, were obtained for the catfish *Clarias batrachus L.* exposed to methyl mercuric chloride, hydroxyethyl mercuric chloride and mercuric chloride in nonsaline waters [230].

Organotin Compounds

These compounds have been the subject of environmental studies for two obvious reasons. The first is the increasing worldwide use of inorganic and organotin compounds in many industrial, chemical and agricultural areas, where very little is known about their environmental fate. Second, there is a great difference between the toxicities of the various organotin compounds; the toxicity depends on the organic moiety in the molecules.

Organic tin compounds have been applied in many fields (for instance as stabilisers for PVC, fungicides and miticides in agriculture and biocides, algicides, bactericides and molluscicides [232–235]) because their properties can be tailored by varying the type and the number of substituents in order to meet widely different requirements. Annual world production was estimated to be 33,000 tons in 1983, most of it dioctyltin maleate [233]. The toxicity and degradation of organotins in the environment depend strongly on the number and the nature of the substituents [232,238]. Organotin compounds with short alkyl chains or phenyl substituents generally exhibit considerable toxicity towards both aquatic organisms and mammals. Alkyltins with small alkyl chains degrade slowly in the environment [89, 193]; phenyltins are less stable and may, under certain conditions, rapidly lose the phenyl substituents. Organotin compounds may accumulate in sediments and aquatic organisms [90].

There is special interest in the biotic and abiotic methylation of tin compounds [239] and the fate of some organotins in aquatic ecosystems. One possible route is the eventual dealkylation of the trialkyltin species to $Sn(IV)$ and the microbial methylation of $Sn(IV)$ to the various methyltin species. Increasing methyltin concentrations with increasing anthropogenic tin influxes has been noted in Chesapeake Bay [240].

Methyltin species are ubiquitous in natural waters, although their concentrations are usually low (less than 1 ng/l) in waters relatively unimpacted by anthropogenic activity [241,242]. Mono- and dimethyltin are the dominant species [241–243], suggesting that methyltins, like methylmercury species, arise via stepwise methylation of the inorganic metal [244]. Not only are sediment slurries capable of methylating inorganic tin [243], but concentrations of methyltin species increase with estuarine surface-to-volume ratios [241]. Thus, tin methylation in aquatic environments is likely to occur in sediments.

Measurements of sediment methyltin concentrations show monomethyltin to be the dominant species in anoxic sediments, while trimethyltin is found in its highest concentrations in toxic sediments [245]. This suggests that tin methylation probably occurs in anaerobic sediments, while degradation of higher molecular weight organotins such as tributyltin, an antifouling agent, occurs in oxygenated environments. In recent studies of inorganic tin methylation, it has been confirmed that biomethylation occurs preferentially in anaerobic estuarine sediments [246]. Methyltins were produced to a maximum level of about 2 ng/l (dry weight) of sediment in 21 days [247]. Low concentrations of mono-, di- and trimethyltin compounds found in Baltimore Harbor sediments averaged at 8, 1 and 0.3 µg/kg dry weight of sediment, while sediment taken in a relatively unpolluted area had much lower organotin content (1.01 and 0.01 µg/kg).

Rapsonmanikis and Weber [227] examined the environmental implications of the methylation of tin(III) and methyltin(IV) ions in aqueous samples in the presence of manganese dioxide. Their studies were carried out with particular reference to the mechanisms involved and the role of dimethylcobalt complex carboanion donor, the carbocation donor iodomethane, and the oxidising agent manganese dioxide. The yields of the various methyltin ions were estimated, and some preliminary results were also presented on the further methylation of mono-, di- and trimethyltin, which indicated that the presence of a naturally occurring donor such as methylcobalamin would result in the formation of volatile tetraethyltin compounds.

Van Nguyen et al. [248] carried out an investigation of the fate in an aqueous environment of three organotin compounds (triphenyltin acetate, triphenyltin hydroxide and triphenyltin chloride) used in antifoulant paint compositions. The organotin compounds were leached from paint panels by shaking with distilled water for up to two weeks at room temperature, and the water and undissolved residues were then analysed. The results suggested that the organotin compounds ionised in aqueous media; a simple model was developed to explain the process.

Several investigators have reported ng – µg/l concentrations of organotin compounds in both freshwater and marine samples. Inorganic tin, methyltins and butyltins have been detected in marine and freshwater environmental samples [249–252]. The presence of inorganic tin, butyltin and methyltin species has been reported in Canadian lakes, rivers and harbours [253, 254]. Both organotins and inorganic tin were reported to be highly concentrated by factors of up to 10^4 in the surface microlayer relative to subsurface water [253, 254]. Inorganic tin, mono-, di- and trimethyltins have been detected at ng/l levels in saline, estuarine and freshwater samples [255, 256]. Methylation of tin compounds by biotic as well as abiotic processes has been proposed [257].

Possible anthropogenic sources of organotins have recently been suggested. Both polyvinylchloride and chlorinated polyvinylchloride have been shown to leach methyltin and dibutyltin compounds, respectively, into the environment [260].

Monobutyltin has been measured in marine sediments collected in areas associated with boating and shipping. Butyltin was not detected in areas free from exposure to maritime activity [261]. The use of organotin antifouling coatings in particular has stimulated interest in their environmental impact.

As discussed in greater detail in Sects. 7.1 and 7.2, the LC$_{50}$ values obtained for organotin compounds are extremely low, confirming the high toxicity of these compounds towards water-based creatures. Reported values for one- and two-day LC$_{50}$ are 1 – 4 µg/l, as seen in Table 9.14.

Table 9.14. LC$_{50}$ values obtained for organotin compounds (from author's own files)

Compound	Species	Type of water	LC$_{50}$ (µg/l)	Duration of LC$_{50}$ (test days)	Reference
Organotin compounds	Rainbow trout (*Salmo gairdneri*)	Nonsaline	1.3	1	[258]
Organotin compounds	Bivalve mollusc (*Crassostrea virginica*)	Nonsaline	1.3 (embryo) 3.96 (larvae)	2 2	[259]
Organotin compounds	Bivalve mollusc (*Mercenaria mercenaria*)	Nonsaline	1.13 (embryo) 1.65 (larvae)	2 2	[259]

Some adverse effects of the toxicity of organotin are summarised below (these are discussed further in Sects. 7.1 and 7.2):

Table 9.15. Some adverse effects of the presence of organotin in saline and nonsaline waters (from author's own files)

	Nonsaline waters		Saline waters
Fiddler crabs (*Uca pugilator*)	Retarded limb regeneration [263] Morphological abnormalities	*Gammarus GP* *Brevoorita tyrannus* and larval *Henidia berrylina*	Reduced bodyweight [51] Reduced growth rate
Bivalve molluscs (*Crassostrea virginica mercenaria*)	Acute toxicity to embryos and larvae, delayed clam embryo development [259]	Oyster *Crassostrea gigas*	Weight, length, width adversely affected [263]

References

1. Van Hoof F, Van San H (1981) *Chemosphere* **10**:1127.
2. Mount DI, Stephen CE (1967) *J Wildlife Managem* **31**:168.
3. Martin M, Stephenson MD, Martin JH (1977) *Calif Fish Game* **63**:95.
4. Kariya T, Haga Y, Hoga T, Tsuda T (1967) *Bull Japan Soc Sci Fish* **33**:818.
5. Buings WA, Leonard EN, McKim JM (1973) *J Fish Res Board Canada* **30**:583.
6. Schultz TW, Freeman SR, Dumont JN (1980) *Arch Environ Contam Toxicol* **9**:23.
7. Johnston PA (1987) *Aqua Toxicol* **10**:335.
8. Matsunaga K, Takahashi S (1976) *Anal Chim Acta* **87**:487.
9. Lima AR, Curtis C, Hammermeister DE, Markee TP, Northcott CE, Brooke LT (1984) *Arch Environ Contam Toxicol* **13**:595.
10. Hatakeyama S (1988) *Ecotoxicol Environ Safety* **16**:1.
11. Nemcsok JG, Hughes GM (1988) *Environ Pollut* **49**:77.
12. Meisner JD, Hum WQ (1987) *Bull Environ Contam Toxicol* **39**:898.
13. Hilmy AL, El-Domiaty NA, Daabees AY, Abdel Latife HA (1987) *Comp Biochem Phys C* **86**:263.
14. Vostal J (1972) *Mercury in the Environment*, CRC Press, Cleveland, OH, USA.
15. Carrier R, Beitinger TL (1988) *Water Res* **22**:511.
16. Gill TS, Pant JC, Tewari H (1988) *Ecotoxicol Environ Safety* **15**:153.
17. Schgal R, Savena AB (1987) *Int J Environ Studies* **29**:157.
18. Sheerbon EP (1979) *Hydrobiol J* **13**:75.
19. Thomsen A, Korsgaard B, Joensen J (1988) *Aqua Toxicol* **12**:291.
20. Lemby AD, Smith RJJ (1987) *Environ Toxicol Chem* **6**:225.
21. Harrison FL, Watness K, Nelson DA, Miller JE, Calabrese A (1987) *Estuaries* **10**:78.
22. Willis M (1988) *Arch Hydrobiol* **112**:299.
23. Ahsanullah M, Mohley MC, Rankin D (1988) *Aust J Mar Freshwater Res* **39**:33.
24. Bodar CWM, Leeuwen CJ, Voogt PA, Zandee PJ (1988) *Aqua Toxicol* **12**:301.
25. Maltby L, Snart JOH, Calow P (1987) *Environ Pollut* (1987) **43**:271.
26. Sadler K, Turnpenny AWH (1986) *Water Air Soil Pollut* **30**:593.
27. Zischke JA, Arthur JW (1987) *Archiv Environ Contam Toxicol* **16**:225.
28. Klanda RJ, Palmer RE, Lenkevich MJ (1987) *Estuaries* **10**:44.
29. Arthur JW, West CW, Allen KN, Hedtke SF (1987) *Bull Environ Contam Toxicol* **38**:324.
30. Clark JR, Borthwick PW, Goodman LR, Patrick JM Jr, Lores EM, Moore JC (1987) *Environ Toxicol Chem* **6**:151.
31. Isensee AR (1978) *Ecol Bull* **27**:255.
32. Decad GM, Birnbaum LS, Matthews HB (1982) In: Hutzinger O Frei RW, Merian E, Pocchiari F (eds) *Chlorinated Dioxins and Related Compounds: Impact on the Environment*, Pergamon, New York, NY, USA, p. 307–315.
33. Lee DHK (ed)(1972) *Metallic Contaminants and Human Health*, Academic, New York, NY, USA.
34. Friberg L, Piscator M, Nordberg G (1971) *Cadmium in the Environment*, CRC, Cleveland, OH, USA.
35. McCaull J (1971) *Environment* **13**:3.
36. Laurie DB, Joselow MM, Browder AA (1972) *Ann Int Med* **76**:307.
37. Eaton JG, McKim JM, Holcombe GW (1978) *Bull Environ Contam Toxicol* **19**:95
38. Alabaster JS, Lloyd R (eds)(1980) *EIFAC Water Quality Criteria for Freshwater Fish*, UN Food and Agriculture Organisation, Butterworths, London, UK.

39. Friberg L, Kjellstrom T, Nordberg G, Piscator M (1975) *Cadmium in the Environment*, EPA Report No.650 (3-75-049), EPA, Washington, DC, USA.

40. Reeder SW, Demayo A, Taylor MA (1979) *Guidelines for Surface Water Quality, Volume 1, Inorganic Chemical Substances, Cadmium*, Environment Canada, Ottawa, Canada.

41. Nriagu JO (ed)(1980) *Cadmium in the Environment, Part I: Ecological Recycling*, Wiley, New York, NY, USA.

42. Nriagu JO (ed)(1981) *Cadmium in the Environment, Part II: Health Effects*, Wiley, New York, NY, USA.

43. Canton JH, Slooff W (1982) *Ecotoxicol Environ Safety* **6**:113.

44. Rappe C, Buser HR, Bosshardt HP (1979) In: Nicholson WJ, Moore JA (eds) *Health Effects of Halogenated Aromatic Hydrocarbons*, New York Academy of Sciences, New York, NY, USA.

45. Greeve JC, Miller WE, Debacon M, Lang MA, Bartels CL (1988) *Environ Toxicol Chem* **7**:35.

46. Boge G, Bussiere D, Peres G (1988) *Water Res* **22**:441.

47. Knudtson BK (1979) *Bull Environ Contam Toxicol* **23**:95.

48. Attar EN, Maly EJ (1982) *Arch Environ Contam Toxicol* **11**:291.

49. Chapman PM, Farrell MA, Brinkhurst RO (1982) *Water Res* **16**:1405.

50. Sangalang GB, Freeman HC (1979) *Arch Environ Contam Toxicol* **8**:77.

51. Guidici MDN, Migliore SM, Guarino SM, Gambardella C (1987) *Mar Pollut Bull* **18**:454.

52. Van der Putte I, Lubbers J, Kolar Z (1981) *Aqua Toxicol* **1**:3.

53. Knittel MD (1980) Heavy Metal Stress and Increased Susceptibility of Steelhead Trout (*Salmo gairdneri*) to *Yersinia ruckeri* Infection. In: Eaton JG, Parrish PR, Hendricks AC (eds) *Aquatic Toxicology*, American Society for Testing Materials, Philadelphia, PA, USA.

54. Snarski VM (1982) *Environ Pollut Ser A* **28**:219.

55. Dixon DG, Hilton JW (1981) *J Fish Biol* **19**:509.

56. Ingersoll CG, Winner RW (1982) *Environ Contam Toxicol* **1**:321.

57. Sinley JR, Goettl JP, Davies PH (1974) *Bull Environ Contam Toxicol* **12**:193.

58. Nebecker AV, Cairns MA, Wise CM (1984) *Environ Contam Toxicol* **3**:151.

59. Guidici DNM, Migliore SM, Guarino SM (1987) *Hydrobiol* **146**:63.

60. Schmidt U, Huber F (1976) *Nature* **259**:157.

61. Dumas J-P, Pazdernik L, Belloneik S, Bouchard D, Vaillancourt G (1977) *Proc 12th Canadian Symp Water Pollution Res* **12**:91.

62. Holcombe GW, Phipps GL, Fiandt JT (1977) *Ecotoxicol Environ Safety* **7**:400.

63. Committee on Medical and Biological Effects of Environmental Pollution (1976) *Selenium*, National Academy of Sciences, Washington, DC, USA.

64. Luckey TD, Venugopal B (1976) *Chem Eng News* **54**:2.

65. Jaworski JF (1979) *Effects of Lead in the Environment 1978: Quantitative Aspects*, NRCC 16736, National Research Council, Ottawa, Canada.

66. Solbe JF de LG (1974) *Water Res* **8**:389.

67. Leatherhead YTM, Burton JD (1974) *J Mar Biol Assoc UK*, **54**:457.

68. Lunde G (1977) *Environ Health Persp* **19**:47.

69. Grieg RA, Wenzloff DR, Pearce JB (1976) *Mar Pollut Bull* **7**:185.

70. Luh M-D, Baker RA, Henley DE (1973) *Sci Total Environ* **2**:1.

71. Coulson EJ, Remington RE, Lynch KM (1935) *J Nutrition* **19**:225.

72. Jacobs LW, Syers JK, Keeney DR (1970) *Soil Sci Soc Ann Proc* **34**:750.

73. Andreae MO (1977) *Anal Chem* **49**:820.

74. Johnson DL, Braman RS (1975) *Deep Sea Res Oceanogr Abs* **22**:503.
75. Inverson DG, Anderson MA, Holm TR, Stanforth RR (1979) *Environ Sci Technol* **13**:1491.
76. HMSO (1959) *Arsenic in Food Regulations*, SI 1959, No. 1831, HMSO, London, UK.
77. Schroeder HA, Balassa JJ (1966) *J Chron Dis* **19**:85.
78. Martin M, Osborn KE, Billig P, Glickstein N (1981) *Mar Pollut Bull* **12**:305.
79. Phillips DJH (1980) Toxicity and Accumulation of Cadmium in Marine and Estuarine Biota. In: Nriagu JO (ed) *Cadmium in the Environment, Part I: Ecological Cycling*, Wiley, New York, NY, USA.
80. Taylor D (1981) A Review of the Lethal and Sublethal Effects of Cadmium on Aquatic Life. In: *Proc 3rd Int Cadmium Conference*, Miami, FL, USA, p. 75–81.
81. Thede H, Scholtz N, Fascher H (1979) *Mar Ecol Prog Ser* **1**:13.
82. Pesch GG, Stewart NE (1980) *Mar Environ Res* **3**:145.
83. Windom HL, Smith RG Jr (1979) *Mar Chem* **7**:157.
84. Zirino A, Yamamoto S (1972) *Limnol Oceanogr* **17**:661.
85. Turner DR, Whitfield M, Dickson AG (1981) *Geochim Cosmochim Acta* **45**:855.
86. Calabrese A, MacInnes JR, Nelson DA, Grieg RA, Yevich PP (1984) *Mar Environ Res* **11**:253.
87. Sharp JR, Neff JM (1980) *Mar Environ Res* **3**:195.
88. Maguire RJ, Carey JH, Hale EJ (1983) *J Agric Food Technol* **31**:1060.
89. Timourian H, Watchmaker G (1972) *J Exp Zool* **182**:379.
90. Riley JP, Skirrow G (1975) *Chemical Oceanography*, 2nd Edition, Volume I, Academic, New York, NY, USA, p.418
91. Voyer RA, Cardin JA, Heltshe JF, Hoffman GL (1982) *Aqua Toxicol* **2**:223.
92. Klein-MacPhee G, Cardin JA, Berry WJ (1984) *Trans Am Fish Society* **113**:247.
93. Coglianese MP (1982) *Archiv Environ Contam Toxicol* **11**:297.
94. Stromgren T (1982) *Mar Biol* **72**:69.
95. Calabrese A, MacInnes JR, Nelson DA, Miller JE (1977) *Mar Biol* **41**:179.
96. Bryant V, Newbery DM, McLusky DS, Campbell R (1985) *Mar Ecol Prog Series* **24**:139.
97. Jones MB (1975) *Mar Biol* **30**:13.
98. Herbert DWM, Wakeford AC (1964) *Int J Air Water Pollut* **8**:251.
99. Knowles CO, McKee MJ, Palawski DU (1987) *Environ Toxicol Chem* **6**:201.
100. Carlson AR, Kosian PA (1987) *Arch Environ Contam Toxicol* **16**:129.
101. McKim JM, Schmeider PK, Carlson RW, Hunt EP, Niemi GJ (1987) *Environ Toxicol Chem* **6**:295.
102. Conti E (1987) *Aqua Toxicol* **10**:325.
103. Dill DC, Murphy PG, Mayes MA (1987) *Bull Environ Contam Toxicol* **39**:869.
104. Smith PD, Brockway DL, Stancil FE Jr (1987) *Environ Toxicol Chem* **6**:891.
105. Oikari A, Kukkonen J (1988) *Ecotoxicol Environ Safety* **15**:282.
106. Cleland GB, McElroy PN, Sonstegard RA (1988) *Aqua Toxicol* **12**:141.
107. Newsted JL, Giesy JP (1987) *Environ Toxicol Chem* **6**:445.
108. Mayes MA, Hopkins DL, Dill DC (1987) *Bull Environ Contam Toxicol* **38**:653.
109. Khangarot BS, Ray PK (1988) *Arch Hydrobiol* **113**:465.
110. Tripathi G, Shukla SP (1988) *Ecotoxicol Environ Safety* **15**:277.
111. Trim AH (1987) *Bull Environ Contam Toxicol* **38**:681.
112. Herring CO, Adams JA, Wilson BA, Pollard S (1988) *Bull Environ Contam Toxicol* **40**:35.
113. Russom L, Drummond RA, Huffman AD (1988) *Bull Environ Contam Toxicol* **41**:589.

114. McKim JM, Schmeider PK, Niemi GJ, Carlson RW, Henry TR (1987) *Environ Toxicol Chem* **6**:313.
115. Mitchell DG, Chapman PM, Long TJ (1987) *Bull Environ Contam Toxicol* **39**:1028.
116. Call DJ, Brooke LT, Kent RJ, Knuth ML, Poirier SH, Huot JM, Lima AR (1987) *Archiv Environ Contam Toxicol* **16**:607.
117. Ferrando MD, Almar MM, Andreu-Moliner E (1988) *J Environ Sci Health* **B23**:45.
118. Maas-Diepeveen JL, Van Leeuwen CJ (1988) *Bull Environ Contam Toxicol* **40**:517.
119. Kononen DW (1988) *Bull Environ Contam Toxicol* **41**:371.
120. Wams TJ (1987) *Sci Total Environ* **66**:1.
121. Carlson AR (1987) *Bull Environ Contam Toxicol* **38**:667.
122. Call DJ, Poirier SH, Knuth ML, Harting SL, Lindberg CA (1987) *Bull Environ Contam Toxicol* **38**:352.
123. Rajeswari K, Kalarani V, Reddy DC, Ramamurthi R (1988) *Bull Environ Contam Toxicol* **40**:212.
124. Vijayakumari P, Reddy DC, Ramamurthi R (1987) *Bull Environ Contam Toxicol* **38**:742.
125. Mallatt J, Barron HC (1988) *Arch Environ Contam Toxicol* **17**:73.
126. Wan MT, Watts RG, Moul DJ (1988) *Bull Environ Contam Toxicol* **41**:609.
127. Seelye JG (1987) *N Am J Fish Mana* **7**:598.
128. Jong-Hwa L (1987) *Bull Natl Res Inst Fish Sci* **3**:11.
129. Moles A, Babcock MM, Rice SD (1987) *Mar Environ Res* **21**:49.
130. Axiak V, George JJ (1987) *Mar Biol* **94**:241.
131. Stromgren T (1987) *Mar Environ Pollut* **21**:239.
132. National Academy of Sciences (1975) *Petroleum in the Marine Environment*, National Academy of Sciences, Washington, DC, USA.
133. UN (1977) *Reports and Studies GESAMP No.6*, United Nations, New York, USA.
134. Neff JM (1979) *Polycyclic Aromatic Hydrocarbons in the Aquatic Environment, Sources, Fates and Biological Effects*, Applied Science, London, UK.
135. Farrington IW, Albaiges J, Burns KA, Dunn BP, Eaton P, Laseter JL, Parker PL, Wise S (1980) In: *The International Mussel Watch: Report of a Workshop Sponsored by the Environmental Studies Board Commission on Natural Resources National Research Council*, National Academy of Sciences, Washington, DC, USA, Chap. 2.
136. Howard JW, Fazio T (1980) *J AOAC* **63**:1077.
137. Connel DW, Miller GJ (1980) *CRC Crit Rev Environ Control* **11**:37.
138. Connel DW, Miller GJ (1980) *CRC Crit Rev Environ Control* **11**:105.
139. Davis KR, Schultz TW, Dumont JN (1981) *Arch Environ Contam Toxicol* **10**:371.
140. Dillon TM, Neff JM, Warner JS (1978) *Bull Environ Contam Toxicol* **20**:320.
141. Dumont JN, Schultz TW, Jones RD (1979) *Bull Environ Contam Toxicol* **22**:159.
142. Giddings JM *Bull Environ Contam Toxicol* (1979) **23**:360.
143. Parkhurst BR, Bradshaw AS, Forte JL, Wright GP *Bull Environ Contam Toxicol* (1979) **23**:349.
144. Southworth GR, Keffer CC, Beauchamp JJ (1980) *Environ Sci Technol* **14**:1529.
145. Wilson BW, Pelroy RA, Cresto JT (1980) *Mutat Res* **79**:193.
146. Brul T (1987) *J Mar Biol Assoc* **67**:237.
147. Holt G, Froslie A, Northeim G (1979) *Acta Vet Scand Suppl* **70**:28.
148. Brevik EM, Bjerk JE, Kveseth NJ (1978) *Bull Environ Contam Toxicol* **20**:715.
149. Kveseth NJ, Bjerk JE, Fimreite N, Stenersen J (1979) *Arch Environ Contam Toxicol* **8**:201.
150. Lunde G, Baumann Ofstad E (1976) *Fresen Z Anal Chem* **282**:395.
151. Baumann Ofstad E, Lunde G, Martinsen K, Rygg B (1978) *Sci Total Environ* **10**:219.

152. Norheim G (1978) *Acta Pharmacol Toxicol* **42**:7.
153. Koeman JH, Ten Noever de Brauw MC, de Vos RH (1969) *Nature* **221**:1126.
154. Ten Noever de Brauw MC, Koeman JH (1973) *Sci Total Environ* **1**:427.
155. Tucker RE, Young AL, Gray AP (eds)(1983) *Human and Environmental Risks of Chlorinated Dioxins and Related Compounds*, Plenum, New York, NY, USA.
156. Ferrando MD, Andreu-Moliner E, Alamar MM, Cebrian C, Nunez A (1987) *Bull Environ Contam Toxicol* **39**:365.
157. Bridgham SD (1988) *Archiv Environ Contam Toxicol* **17**:731.
158. Stephenson M, Mackie GL (1986) *Aqua Toxicol* **9**:243.
159. Weis JS, Ma A (1987) *Bull Environ Contam Toxicol* **39**:224.
160. Kimbrough RD (ed)(1980) *Halogenated Biphenyls, Terphenyls, Naphthalenes, Dibenzodioxins and Related Products*, Elsevier/North Holland Biomedical, Amsterdam, The Netherlands.
161. Hutzinger O, Frei RW, Merian E, Pocchiari F (eds)(1982) *Chlorinated Dioxins and Related Compounds: Impact on the Environment*, Pergamon, New York, USA.
162. Nicholson WJ, Moore JA (eds)(1979) *Health Effects of Halogenated Aromatic Hydrocarbons*, New York Academy of Sciences, New York, USA.
163. Lee DHK, Falk HL (eds)(1973) *Environ Health Persp* **5**:1.
164. Huff JR, Moore JA, Saracci DR, Tomatis L (1980) *Environ Health Persp* 36:221.
165. Esposito MP, Tiernan TO, Dryden FE (1980) *Dioxins*, US EPA Report No. EPA-600-80-197, US EPA, Washington, DC, USA.
166. Crosby DG, Wong AS (1976) *Chemosphere* **5**:327.
167. Lamparski LL, Stehl RH, Johnson RL (1980) *Environ Sci Technol* **14**:196.
168. McConnell EE (1980) In: Kimbrough RD (ed) *Halogenated Biphenyls, Terphenyls, Naphthalenes, Dibenzodioxins and Related Products*, Elsevier/North Holland Biomedical, Amsterdam, The Netherlands, p. 109–150.
169. Golstein JA (1980) In: Kimbrough RD (ed) *Halogenated Biphenyls, Terphenyls, Naphthalenes, Dibenzodioxins and Related Products*, Elsevier/North Holland Biomedical, Amsterdam, The Netherlands, p. 151–190.
170. Bickel MH, Muhelback S (1982) In: Hutzinger O Frei RW, Merian E, Pocchiari F (eds) *Chlorinated Dioxins and Related Compounds: Impact on the Environment*, Pergamon, New York, USA, p. 3036.
171. Di Domenico A, Viviano G, Zapponi G (1980) In: Kimbrough RD (ed) *Halogenated Biphenyls, Terphenyls, Naphthalenes, Dibenzodioxins and Related Products*, Elsevier/North Holland Biomedical, Amsterdam, The Netherlands, p. 105–114.
172. Ward CT, Matsumura F (1978) *Arch Environ Contam Toxicol* **7**:349.
173. Young AL (1979) In: Nicholson WJ, Moore JA (eds) *Health Effects of Halogenated Aromatic Hydrocarbons*, New York Academy of Sciences, New York, USA, p. 173–190.
174. Masuda I, Kuroki H (1980) In: Kimbrough RD (ed) *Halogenated Biphenyls, Terphenyls, Naphthalenes, Dibenzodioxins and Related Products*, Elsevier/North Holland Biomedical, Amsterdam, The Netherlands, p. 561–569.
175. Kuratsune M (1980) In: Kimbrough RD (ed) *Halogenated Biphenyls, Terphenyls, Naphthalenes, Dibenzodioxins and Related Products*, Elsevier/North Holland Biomedical, Amsterdam, The Netherlands.
176. Vos JG, Koeman JH, Van der Maas HL, Ten Noever de Brauw MC, de Vos RH (1970) *Food Cosmet Toxicol* **8**:625.
177. Buser HR, Bosshardt H-P, Rappe C (1978) *Chemosphere* **7**:109.
178. Smith RM, O'Keefe PW, Hilker DR, Jelus-Tyror BL, Aldous KM (1982) *Chemosphere* **11**:715.

179. Jansson B, Sundstrom G (1982) In: Hutzinger O Frei RW, Merian E, Pocchiari F (eds) *Chlorinated Dioxins and Related Compounds: Impact on the Environment*, Pergamon, New York, USA, p. 201–208.
180. Rappe C, Markland S, Bergqvist PA, Hansson M (1982) *Chem Scr* **20**:56.
181. Mills AI, Alexander M (1976) *J Environ Qual* **5**:437.
182. Singh DK, Agarwal RA (1987) *Sci Total Environ* **67**:263.
183. Upadhyaya OVB, Shukla GS (1986) *Environ Res* **41**:591.
184. Saxena PK, Mani K (1988) *Environ Pollut* **55**:97
185. Khangarot BS, Ray PK (1988) *Arch Hydrobiol* **113**:465.
186. Kocan RM, von Westernhagen H, Landolt ML, Furstenberg G (1987) *Mar Environ Res* **23**:291.
187. Green DWJ, Williams KA, Hughes DRL, Shaik GAR, Pascoe D (1988) *Water Res* **22**:225.
188. Dumpert K (1987) *Ecotoxicol Environ Safety* **13**:324.
189. Wester PW, Canton JH, Dormons JAMA (1988) *Aqua Toxicol* **12**:323.
190. Bengtssson BE (1988) *Water Sci Technol* **20**:87.
191. Webb JL (1966) *Enzyme and Metabolic Inhibitors, Vol. 3*, Academic, New York, USA, Chap. 6.
192. Staaffer RE, Ball JW, Jenne EA (1980) *Geological Survey Professional Paper 1044F*, US Government Printing Office, Washington, DC, USA
193. Getzendaner ME, Corbin HB (1972) *J Agric Food Chem* **20**:881.
194. Wong PTS, Chau IK, Luton L, Bengert GA (1977) Methylation of Arsenic in the Aquatic Environment. In: Hemphill DD (ed) *Proceedings of the Conference on Trace Substances in Environmental Health XI*, 7–9 June 1977, University of Missouri, Columbia, MO, USA.
195. Chapman AC (1926) *Analyst* **51**:548.
196. Challenger F (1945) *Chem Rev* **36**:315.
197. Vonk JW, Sisperstein AK (1973) *Antonie von Leeuwenhoeck* **39**:505.
198. Jensen S, Jernelov A (1968) *Nature* **223**:753.
199. McBride BC, Merilees H, Cullen WR, Picket W (1978) *ACS Symp Ser* **82**:94.
200. Wood JH, Kennedy FS, Rosen CG (1968) *Nature* **220**:173.
201. Wood JM (1974) *Science* **183**:1049.
202. Laudner L (1971) *Nature* **230**:452.
203. Bertilsson L, Neujahr HY (1971) *Biochem* **10**:2805.
204. Braman RS (1975) ACS Symp Ser **7**:108.
205. Cox DP (1975) ACS Symp Ser **7**:81.
206. Von Endt DW, Kearney PC, Kaufman DD (1968) *J Agric Food Chem* **16**:17.
207. Laskso JU, Peoples SA (1975) *J Agric Food Chem* **23**:674.
208. Wong PTS, Silverberg BA, Chau YK, Hodson PV (1978) In: Nriagu JO (ed) *Biogeochemistry of Lead in the Environment*, Elsevier, New York, USA, p. 270–342.
209. Muddock BG, Taylor D (1980) The Acute Toxicity and Bioaccumulation of some Lead Alkyl Compounds in Marine Animals. In: *Proceedings of the International Experts Discussion on Lead: Occurrence, Fate and Pollution in the Marine Environment*, Rovinia, Yugoslavia, Pergamon, New York, USA.
210. Granjean P, Neilson T (1979) *Res Rev* **72**:97.
211. Grove JR (1980) In: Branica M, Konrad Z (eds) *Lead in the Marine Environment*, Pergamon, Oxford, UK, p. 45–42.
212. Wong PTS, Chau JK, Kramer O, Bengert GA (1981) *Water Res* **15**:621.
213. Wong PTS, Chau YK, Luxon PL (1975) *Nature* **253**:263.
214. Reisinger PTS, Stoeppler M, Nurnberg HW (1981) *Nature* **291**:228.

215. Sirota GR, Uth JF (1977) *Anal Chem* **49**:823.
216. Chau YK, Wong PTS (1980) Lead in the Marine Environment. In: *Proceedings of the International Experts Discussion on Lead: Occurrence, Fate and Pollution in the Marine Environment*, Rovinia, Yugoslavia, Pergamon, New York, USA.
217. Reamer DC, Zoller WH, O'Haver TC (1978) *Anal Chem* **50**:1449.
218. Mor ED, Beccaria AM (1980) A Dehydration Method to Avoid Loss of Trace Elements in Biological Samples. In: *Proceedings of the International Experts Discussion on Lead: Occurrence, Fate and Pollution in the Marine Environment*, Rovinia, Yugoslavia, Pergamon, New York, USA.
219. Uth JF, Armstrong FAJ (1970) *J Fish Res Board Can* **27**:805.
220. Krenkel PA (1973) *Int Crit Rev Environ Cont* **3**:303.
221. Robinson S, Scott WB (1974) *A Selected Biogeography on Mercury in the Environment with Subject Listing*, Life Science, Miscellaneous Publication, Royal Ontario Museum, Toronto, Canada, p. 54.
222. Uthe JF, Armstrong FAJ (1974) *Toxicol Environ Chem Rev* **2**:45.
223. D'Itri PA, D'Itri FM (1978) *Environ Management* **2**:3.
224. Furukawa K, Tonomura K (1971) *Agric Biol Chem* **35**:604.
225. Furukawa K, Tonomura K (1972) *Agric Biol Chem* **36**:217.
226. Westoo G, Rydalv M (1969) *Var Foda* **21**:20.
227. Rapsomanikis S, Weber JH (1985) *Environ Sci Technol* **19**:352.
228. Imura N, Sukegawa E, Pan SK, Nagao K, Kim YJ, Kwan T, Ukita T (1971) *Science* **172**:1248.
229. Andren AW, Harriss RC (1973) *Nature* **245**:256.
230. Amemiya T, Takenchi M, Ito K, Ebara K, Harada H, Totani T (1975) *Ann Rep Tokyo Metropolitan Res Lab Pub Health* **26-1**:129.
231. Kirubagaran R, Joy KP (1988) *Ecotoxicol Environ Safety* **15**:171.
232. WHO Task Group, Sharrat M (Chairman) Vouk VB (Secretary) (1988) *Tin and Organotin Compounds*, Environ Health Criteria No. 15, WHO, Geneva, Switzerland, p. 1.
233. Zuckerman JJ, Reisdorf PR, Ellis III HE, Wilkinson RR (1978) ACS Symp Ser **28**:388.
234. Bock R (1981) *Res Rev* **79**:216.
235. Gächter R, Müller H (1979) *Handbuch der Kunststoff-Additive*, Hanser, Munich, Germany.
236. Ram RN, Sathyanesan AG (1987) *Environ Pollut* **47**:135.
237. Czuba M, Seagull RW, Tana H, Cloutier L (1987) *Ecotoxicol Environ Safety* **14**:64.
238. Laughlin, Jr., RB, French W, Johannsen RB, Guard HE, Brinckmam FE (1984) *Chemosphere* **13**:575.
239. Guard HE, Cobet AB, Coleman III WH (1981) *Science* **213**:770.
240. Jackson JAA, Blair WR, Brinckman FE, Iverson WP (1982) *Environ Sci Technol* **16**:110.
241. Byrd JT, Andreae MO (1982) *Science* **218**:565.
242. Braman RS, Tompkins MA (1979) *Anal Chem* **51**:12.
243. Hodge VF, Seidel SL, Goldberg ED (1979) *Anal Chem* **51**:1256.
244. Ridley WP, Dizcker LJ, Wood JM (1977) *Science* **197**:329.
245. Tugrul S, Balkas TI, Goldberg ED (1983) *Mar Pollut Bull* **14**:297.
246. Gilmour CC, Tuttle JH, Means JC (1985) In: Sigleo AC, Hattori A (eds) *Marine and Estuarine Geochemistry*, Lewis, Chelsea, MI, USA.
247. Gilmour CC, Tuttle JH, Means JC(1986) *Anal Chem* **58**:1848.
248. Van Nguyen V, Vasey IJ, Eng G (1984) *Water Air Soil Pollut* **23**:417.
249. Skopintsev BA (1976) *Oceanogr* **6**:361.

250. Williams PJ (1975) In: Riley JS, Skirrow G (eds) *Chemical Oceanography*, Vol. 3, 2nd Edition, Academic, New York, USA, p. 443–477.
251. Skopintsev BA (1986) *Oceanogr* **16**:630.
252. Wangersky PJ, Zika RG (1978) *The Analysis of Organic Compounds in Seawater*, Report No. 3, NRCC 16566, Marine Analytical Chemistry Standards Programme, National Research Council, Ottawa, Canada.
253. Van Hall CE, Stenger VH (1964) *Water Sewage Works* **111**:266.
254. Van Hall CE, Barth D, Stenger VA (1965) *Anal Chem* **37**:769.
255. Van Hall CE, Stenger VA (1967) *Anal Chem* **39**:503.
256. Van Hall CE, Safranko J, Stenger VA (1963) *Anal Chem* **35**:315.
257. Golterman HL (ed)(1969) *Methods for Chemical Analysis of Fresh Waters*, Blackwell, Oxford, UK, p. 133–145.
258. Maguire RJ, Tracz RJ (1987) *Water Pollut Res J Can* **22**:227.
259. Roberts MH (1987) *Bull Environ Contam Toxicol* **39**:1012.
260. HMSO (1972) *Notes on Water Pollution No. 59*, Water Pollution Laboratory, Stevenage, Hertfordshire, HMSO, London, UK.
261. Battori NS, Solar FR, Durich JO (1974) *Doc Inest Hidrol* **17**:303.
262. Weis JS, Gottlieb J, Kwiatkowski J (1987) *Arch Environ Contam Toxicol* **16**:321.
263. Noth K, Kumar N (1988) *Chemosphere* **17**:465.
264. Hardie DWF (1964) In: Standen A et al. (eds) *Kirk-Othmer Encyclopedia of Chemical Technology*, 2nd Edition, Interscience, New York, USA, Vol. 5, p. 231–240.
265. Zitko V (1980) Anthropogenic Compounds. In: Hutzinger O (ed) *The Handbook of Environmental Chemistry*, Volume 3, Part A, Springer, Berlin Heidelburg New York, p. 149–156.
266. BUA (1993) *Chlorparaffine: Paraffinwachse und Kohlenwasserstoffwachse, chloriert*, BUA-Stoffbericht 93, Beratergremium für umweltrelevante Altstoffe (BUA) der Gesellschaft Deutscher Chemiker, VCH, Weinheim, Germany.
267. Tomy GT, Fisk AT, Westmore JB, Muir DCG (1998) *Rev Environ Contam Toxicol* **158**:53.
268. Madley JR, Birtley RDN (1980) *Environ Sci Technol* **14**:1215.
269. Birtley RDN, Conning SDM, Daniel JW, Ferguson DM, Longstaff E, Swan AAB (1980) *Toxicol Appl Pharmacol* **54**:514.
270. Bucher JR, Alison RH, Montgomery CA, Huff J, Haseman JK, Farnell D, Thompson R, Prejean JD (1987) *Fund Appl Toxicol* **9**:454.
271. Renberg L, Tarkpea M, Sundstrom G (1984) *Ecotoxicol Environ Safety* **11**:361.
272. Olsson M, Karlsson B, Ahnland E (1994) *Sci Total Environ* **154**:217.
273. Letcher RJ, Norstrom RJ, Muir DCG (1998) *Environ Sci Technol* **32**:1656.
274. Kaiser KLE (1974) *Environ Pollut* **7**:93.
275. Frame GM, Wagner RE Carnahan JC, Brown JF Jr, May RJ, Smullen LA, Bedard DL (1996) *Chemosphere* **33**:603.
276. Nezel T, Müller-Plathe F, Müller MD, Buser H-R (1997) *Chemosphere* **35**:1895.
277. Tolls JJ, Kloepper-Sams P, Sijm DTHM (1994) *Chemosphere* **29**:693.
278. Newsome CS, Howes D, Marshall SJ, Van Egmond RA (1995) *Tenside Surf Deterg* **32**:498.
279. Comotto RM, Kimerle RA, Swisher RD (1979) In: Marking LL, Kimerle RA (eds) *Aquatic Toxicology and Hazard Assessment*, ASTM STP 667, American Society of Testing Materials, Philadelphia, PA, USA, p. 232–250.

10 Evaluating Toxicity via Water Analysis

10.1
Measurement of LC_{50}

The toxicity of a metal or an organic substance to fish, invertebrates, algae or bacteria is evaluated in toxicity tests, where the creatures are exposed under standard conditions to a range of concentrations of the test substance for a constant period of time. The number of creatures that either die or undergo a particular response (for example, growth reduction) during that period is counted at the end of the test period. A plot of log concentration of test substance versus log percentage of mortality or percentage of creatures undergoing a particular response enables one to read off the concentration of test substance that causes a 50% effect, as demonstrated in Fig. 10.1. Short-term tests are run for four days, although they are commonly run for 1, 10, 100 and 1000 days to obtain more information. From the graphs obtained, it is then possible (via interpolation) to read off the median lethal concentration (LC_{50}) or the median effect concentration (EC_{50}), i.e. the concentration which is calculated to cause, respectively, mortality or a particular response in 50% of the test population. It is also usual to identify the median concentration by the test duration, e.g. four-day LC_{50} or 100-day EC_{50}. Acute, chronic, lethal and sublethal tests can be performed in a similar manner.[1]

An example of this type of testing is shown in Fig. 10.2, which illustrates LC_{50} results obtained by exposing salmonid and nonsalmonid freshwater fish to cadmium in 1-, 10-, 100- and 1000-day tests. Both Figs. 10.1 and 10.2 illustrate that, as would be expected, LC_{50} decreases as the test duration is increased. Also, for any given test conditions, EC_{50} would always be less than LC_{50}. As might be imagined, the results obtained in such toxicity tests are affected by several factors, the most important of which are discussed next.

---

[1] *Acute toxicity* is the lethal response caused by a short exposure to a substance, at most a few days, and commonly four days.
   Chronic toxicity denotes the deleterious effects (not exclusively fatalities) resulting from prolonged exposure, i.e. more than a few days.
   Sublethal toxicity describes cases where deleterious effects are observed but not mortality.

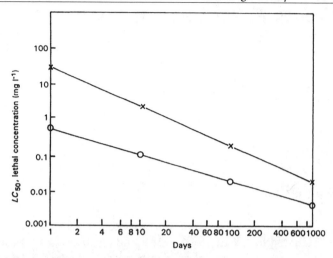

Figure 10.1. Effect of time of exposure to cadmium on LC_{50} for fish. x – nonsalmonid fish, o – salmonid fish. From author's own files

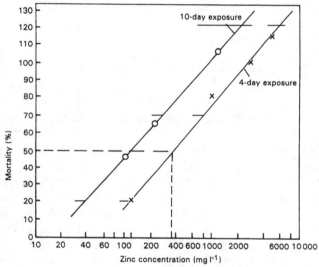

Figure 10.2. Method of obtaining LC_{50} by interpolation; toxicant: zinc. From author's own files

10.2
Factors Affecting LC$_{50}$

10.2.1
Factors Related to Tests

Space

Adequate space must be provided for the test creature, as overcrowding causes stress and consequently affects the sensitivities of the creatures to toxicants. A minimum of 2 l of sample is recommended per gram of biomass. Moreover, the sample should be changed every 24 hours and fresh toxicant added.

Water Flow Through the Test Chamber

The results obtained in static tests are of doubtful value in the case of creatures that normally inhabit flowing rivers. Flow-through tests are inherently more complicated, as they involve the provision of a dosing mechanism to maintain a constant concentration of toxicant in the flowing sample. They do, however, have the advantage of providing more constant chemical conditions through the duration of the test. In flow-through tests, 2 – 3 l of sample containing the controlled addition of toxicant should be passed through the test chamber per gram of biomass per day. This is equivalent to a 90% replacement of the test liquid per day.

In general, as illustrated in Table 10.1, freshwater creatures are more tolerant to toxic metals in static tests than in flow-through tests, all other test conditions being constant.

Table 10.1. Static tests versus flow-through tests in the measurement of LC$_{50}$ (from author's own files)

Element	Species	4 day LC$_{50}^{a}$ (mg/l)	
		Static test	Flow-through test
Cadmium	*Pimephales promelas*	31.0	4.3
Chromium	*Pimephales promelas*	36.2	36.8
Lead	*Salmo gairdneri*	471	8.0
Silver (as AgNO$_3$)	*Salmo gairdneri*	0.011	0.009
	Pimephales promelas	0.010	0.006
Zinc	*Pimephales promelas*	12.5	9.2

[a] pH, hardness and temperature are the same for static and flow-through tests for each element listed.

Temperature

Figure 10.3 shows the appreciable effect of sample temperature on four-day LC_{50} for the elements silver, cadmium and copper. It is clear, therefore, that when evaluating the toxicity of these three elements, the test temperature must be carefully controlled and reported with the test result.

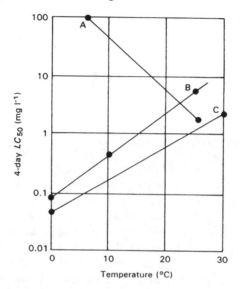

Figure 10.3. Effect of test temperature on four-day LC_{50}. (a) Silver (as silver nitrate) with salmonid fish, i.e. an increase in temperature increases toxicity. (b) Copper with salmonid fish. (c) Cadmium with nonsalmonid fish. In (b) and (c), an increase in temperature reduces toxicity. Mercury, cadmium, copper and zinc with freshwater invertebrates: an increase in temperature increases toxicity. From author's own files

Dissolved Oxygen Content of Sample

This should be controlled at a level exceeding 6 mg/l throughout the test.

Light

The light regime and duration of lighting should be controlled throughout the test. In general, the absence of light reduces stress, i.e. it decreases stress-linked mortality.

Chemical Form of the Toxicant Added to the Sample

Metals vary in terms of their solubility in water, depending on their chemical form, the sample pH and the presence in the sample of other chemicals such as phosphates and carbonates. Metal precipitation is expected at higher metal concentrations and higher pH or hardness. The greatest metal solubility is expected in soft acidic waters, and it is in these that the best correlation will be obtained between concentration of metal added to sample and LC_{50}.

10.2.2
Factors Related to the Sample

In addition to temperature, the following factors related to the sample should be as similar as possible in the test to those in the environmental water from which the samples were taken.

pH

As shown in Fig. 10.4, increasing the pH from 6.5 to 8.5 increases the four-day LC$_{50}$, i.e. it decreases the toxicity of chromium towards freshwater fish and invertebrates by a factor of approximately four. The effect of pH is most pronounced in hard waters and is related to an increase in metal solubility at higher pH [1–3]. Therefore, sample pH must always be reported when discussing toxicity data.

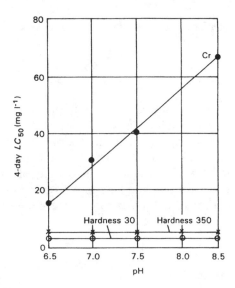

Figure 10.4. Effect of pH on LC$_{50}$ values of freshwater fish and invertebrates. Toxicant: chromium. From author's own files

Hardness

Increasing the sample hardness from 10 to 1000 increases the four-day LC$_{50}$ value, i.e. it decreases toxicity by a factor of between 15 (chromium and zinc) and 120 (nickel and copper; see Fig. 10.5) for freshwater fish, while sample hardness has no effect on LC$_{50}$ in the case of arsenic, vanadium and silver (as silver nitrate). Therefore, sample hardness should always be recorded when reporting toxicity data.

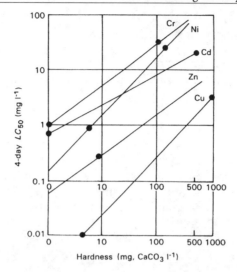

Figure 10.5. Effect of water hardness on LC$_{50}$. Toxicants: chromium, nickel, cadmium, zinc and copper. From author's own files

Salinity

As discussed above, in freshwater samples the hardness can significantly reduce the toxicity of a metal as derived from LC$_{50}$ measurements. In the case of estuary or seawater, where hardness is not as variable, it is salinity that, to a greater or lesser extent, determines the toxicity of the metal. An increase in sample salinity decreased the toxicity of chromium, copper, cadmium, lead, mercury, nickel and zinc to marine creatures. Typically, an increase in salinity from 10 to 35 g/kg can decrease the toxicity by a factor of 10.

10.2.3
Factors Related to the Creature

Age and Condition

Young life stages of fish and invertebrates are generally more sensitive to metals than are adults, particularly in the case of marine creatures. Test creatures should be disease-free.

Feeding

To minimise the toxic effects of animal waste, no feed should be given during short-term 24–48-hour tests. Feeding is necessary during long-term tests.

Acclimatisation

The exposure of creatures to toxicants prior to commencing the LC$_{50}$ test for periods of 1 to 1.5 weeks in some instances increases the tolerance of the creature to metals during subsequent LC$_{50}$ tests. For example, pre-exposure of American Flagfish (*Jordanella floridae*) embryos to zinc produced a considerably greater post-hatch tolerance to zinc compared to fry hatched without pre-exposure. After 30 days' [4] exposure to 0.14 µg/l zinc in the LC$_{50}$ test, pre-exposed creatures had nil mortality, while creatures which had not been pre-exposed had 100% mortality during the same period. The benefits of pre-exposure are generally transient.

Reported LC$_{50}$ values for metals obtained from a variety of sources for freshwater and marine fish and invertebrates are quite variable. Typically, a ten-day LC$_{50}$ value for chromium reported by various workers is in the range of 6 to 60 mg/l (i.e. $33 \pm 82\%$ for nonsalmonid fish), and in the range 1.5 to 35 mg/l (i.e. $18 \pm 93\%$) for salmonid fish. In view of the large number of variables discussed above, this is not surprising. Thus, considering hardness alone, it is clear from Fig. 10.5 that reported four-day LC$_{50}$ values for chromium range from 8 at 8 mg/l hardness to 100 at 500 mg/l hardness, i.e. a mean of $54 \pm 85\%$. Provided that parameters such as pH, hardness, salinity, test temperature and experimental parameters are reported with the LC$_{50}$ value, then data are meaningful and amenable to comparison with results obtained by other workers.

Reported LC$_{50}$ values obtained with fish for a range of elements are summarised in Table 10.2. LC$_{50}$ values obtained in tests of duration 1 – 100 days are included. For salmonid fish, mercury and cadmium are among the most toxic elements, while chromium and vanadium are the least toxic.

Table 10.2. Reported LC$_{50}$ values (mg/l) for various elements: salmonid and nonsalmonid fish (from author's own files)

Element		As	Zn	V	Ag	Se
Nonsalmonids						
Duration of	1	10 – 140	3.2 – 100	< 1 – 1000	15 – 100	19 – 152
toxicity test (d)	10	1.4 – 16	0.31 – 14	2.7 – 37	2.1 – 11	3 – 23
	100	0.16 – 2.3	0.04 – 1.4	0.1 – 1.4	0.19 – 1	0.35 – 2.3
	1000	0.05 – 0.85	0.007 – 0.16	–	–	
Salmonids						
Duration of	1	As above	As above	As above	As above	As above
toxicity test (d)	10					
	100					
	1000					

Continued on next page

Table 10.2. Continued

Element		Ni	Cd	Cr	Cu	Hg	Pb
Nonsalmonids							
Duration of toxicity test (d)	1	6.6 – 124	1 – 53	167 – 1670	40	0.35 – 1.2	1.9 – 35
	10	1.2 – 23	0.06 – 2.4	6 – 60	35	0.03 – 0.35	0.23 – 10
	100	0.23 – 4.3	0.005 – 0.35	0.15 – 1.7	5	0.002 – 0.02	0.03 – 0.66
	1000	0.05 – 0.81	0.0004 – 0.028	–	< 1	0.0002 – 0.002	0.005–0.1
Salmonids							
Duration of toxicity test (d)	1	As above	0.08 – 1	23 – 350	As above	As above	As above
	10		0.01 – 0.19	1.5 – 35			
	100		0.002 – 0.035	0.1 – 2.2			
	1000		0.001 – 0.005	0.007 – 0.015			

10.2.4
Cumulative LC$_{50}$ values

Only rarely does water that is toxic towards fish and invertebrates contain a single toxicant. If toxic impurities are present in any appreciable amount, then it is likely that several of them will adversely affect the fish. Assuming, as is generally the case, that no synergistic effects exist, then the effect of toxicants is additive. The following progressive dilution technique enables the cumulative effect of toxicants on fish to be assessed.

Polluted rivers have been assessed for their toxicity by performing toxicity tests in flowing water on the river bank using graded dilutions of river water. Caged creatures are exposed to the river water for a number of days and the mortality rate and pollutant concentrations are measured at daily intervals during this period. Simultaneously, caged creatures are exposed to a range of dilutions of river water and the same measurements repeated. From the results obtained, the dilution causing 50% mortality in two days is estimated from various fish species at each location.

Figures 10.6 (a) – (d) show test duration versus percentage mortality curves obtained in (A) polluted waters and (B) less polluted waters at zero dilution and ×1, ×2, ×5 and ×10 dilutions of river water. From these curves, the percentage of mortality occurring after two day's exposure for those polluted (A) and less polluted (Bb) river waters can be obtained. Plots of percentage mortality versus dilution enables the dilution corresponding to 50% mortality to be read off (Figs. 10.5 (a) and (b)). From these curves, it is clear (see Fig. 10.7 (a)) that for the more polluted water sample a 50% mortality rate results when the original river water sample has been diluted ×4 times, and for the relatively unpolluted water sample B (see Fig. 10.7 (b)) only ×2.8 times dilution is required to achieve the same effect. The results from these studies are presented not as a concentration of pollutants in the rivers but as cumulative fractions of the relevant laboratory-derived two-day LC$_{50}$ for each species and substance, the sum of which is compared with the toxicity observed at each location.

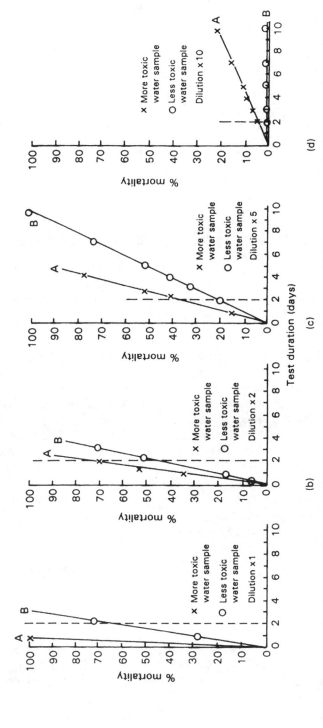

Figure 10.6. Test duration versus percentage mortality curves for a range of river water dilutions. x: a more toxic water, o: a less toxic water. From author's own files

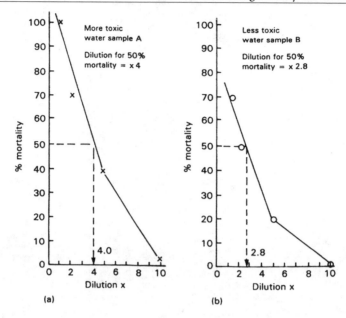

Figure 10.7. Dilution versus percentage mortality curves: (a) more polluted river water, (b) less polluted river water. From author's own files

Thus, considering a simple example, if a relatively toxic river water A before dilution contained 50 mg/l zinc and 10 mg/l copper, then the ×4 dilution of this, causing 50% mortality (Fig. 10.7 (a)), would contain 12.5 mg/l zinc and 2.5 mg/l copper, i.e. river-derived cumulative two-day $LC_{50} = 12.5 + 2.5 = 15$ mg/l. Similarly, if a relatively less toxic water B before ×2.8 dilution (Fig. 10.7 (b)) contained 8 and 1.5 mg/l of zinc and copper respectively, then the dilution would contain 2.8 and 0.5 mg/l zinc and copper, i.e., river-derived cumulative two-day $LC_{50} = 2.8 + 0.5 = 3.2$ mg/l. If the laboratory-derived two-day LC_{50} values for zinc and copper are, respectively, 12 and 6 mg/l, i.e. cumulative two-day LC_{50} is 18 mg/l, then the river-derived cumulative two-day LC_{50} as a fraction of the laboratory-derived two-day LC_{50} (i.e. the cumulative proportion of the laboratory-derived two-day LC_{50}) is given by:

Polluted water A = 2-day river-derived LC_{50}/2-day laboratory-derived LC_{50} = 15/18 = 0.83

Less polluted water B = 2-day river-derived LC_{50}/2-day laboratory-derived LC_{50} = 3.2/18 = 0.18

The difference observed between river-derived and laboratory-derived cumulative LC_{50} values can be ascribed to the effects of factors such as hardness, pH, temperature and dissolved oxygen prior to summation.

Table 10.3. Cumulative proportions of two-day LC_{50} (laboratory derived) values versus fishing status and water hardness (from author's own files)

Survey	Total hardness	Median cumulative 2-day LC_{50} ($\mu g/l$)		
		Fishless	Marginal	Fish present
1	11 – 0	0.45	0.42	0.05 – 0.25
2	100 – 170	0.1 – 0.2	0.16	0.13 – 0.16
3	100 – 300	> 0.28	–	< 0.28
4	134 – 292	> 0.1	–	< 0.1
5	500	> 0.32	0.25 – 0.32	< 0.25
6	70 – 745	0.32 – 2.95	0.3 – 0.37	0.005-0.02

This approach has been applied to an assessment of the fishery statuses of rivers, where it has been found that if the sum of the proportions of the two-day LC_{50} values exceeds about 0.3 then fish will not survive well enough to support fishing activities. Table 10.3 shows this effect for a range of river waters with different total hardnesses.

10.3
Continuous Exposure (S_x) and 95th Percentile Exposure (S_{95}) Concepts

Although the discussion below is concerned with toxic metals, similar considerations would apply in the case of organics.

Having defined a yardstick by which the toxicity of a given metal to a given creature, i.e. the LC_{50} value, can be evaluated, we now assess strategies for measuring the metal concentrations in freshwater and seawater. Clearly, a simple spot measurement of the concentration of a metal in a river water, for example, will not reflect the changes in concentration that occur over a period of time, and it is these that will dictate the long-term wellbeing or otherwise of creatures in that water.

Several environmental standard types of approach have been devised for assessing long-term water quality in terms of the metal concentration of the environmental water:

(1) A critical metal concentration in the water, which, if exceeded for any period of time, will cause damage to creatures; i.e. environmental change.
(2) A general reference value, reflecting relatively uncontaminated concentrations in creatures, for use when identifying areas receiving pollution inputs that may need control.
(3) A maximum safe concentration for continuous exposure, for use when calculating discharge limits for toxic metals; this could only be exceeded in the immediate vicinity of the discharge.

The application of standard (1) could offer short-term but not long-term protection to the receiving water. The application of (2) when calculating discharge limits would protect receiving water at a prohibitive cost (but may be necessary to ensure no mortalities of creatures at all). Type (3) standards should facilitate adequate discharges of toxicants at a reasonable cost, and this approach is the one that has been adopted by the EU and the UK to minimise the deleterious effects of toxicants discharged into receiving waters. Standards based on (1) would inevitably be lower than standards based on (3), making it necessary to have a much higher level of effluent control at a higher cost.

The toxic effect of a metal on fish is a consequence of not only the concentration but also the duration of the exposure, the adverse effect concentration becoming progressively lower as the period of exposure increases.

Consider, for example, the case of nickel. Various workers have reported LC_{50} values in the following ranges for nickel when nonsalmonid fish are subjected to toxicity tests of the stated durations (this variability encountered for a constant duration of toxicity test may be due to differences in hardness, pH, salinity, temperature, etc. in the various samples tested).

Duration of toxicity (d)	LC_{50} (mg/l)
1	6.6 – 124
10	1.24 – 23
100	0.23 – 4.3
1000	0.051 – 0.81

A curve of the type shown in Fig. 10.8 can be prepared from these data. When selecting potential values of the standard, a boundary line (dotted) drawn to enclose the lower limits of the reported adverse effect concentrations (i.e. conservative estimate) would describe a continuous standard in the form of an equation predicting the maximum acceptable concentration

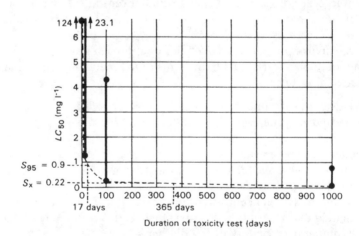

Figure 10.8. Test duration versus LC_{50} plot for nickel. From author's own files

of nickel (with no safety margin) permissible for a specified duration of time. Such real-time management of pollution control is rarely possible and an alternative approach, discussed below, is usually adopted.

Consideration of the relationship in Fig. 10.8 would enable most of the statistical values discussed above to be determined, if 100% of the time is assumed to be 365 days exposure, or longer. Thus, for continuous exposure, the 365+ day asymptote $S_x = 0.22$ mg/l (220 µg/l) nickel would represent the potential standard after the application of a suitable (probably small) safety factor. This long-term standard might be stated as the annual average concentration. However, adoption of this standard would allow higher concentrations to occur for shorter periods, and there is a potential risk that these excursions would be sufficiently great to cause damage to fish. To overcome this, the 95% percentile concept has been adopted, representing the concentration that could be safely exceeded for 5% of the year (i.e. on 17 days), and this value $S_{95} = 0.9$ mg/l (900 µg/l) nickel (excluding safety factors) can be found from Fig. 10.8. Adoption of this approach avoids the need for continuous daily monitoring of nickel concentration, since it states that the nickel content of the water only needs to be below $0.9 \, \text{mgl}^{-1}$ (900 µg/l) for 347 days of the year (365 daily samples), or for 49 weeks of the year (52 weekly samples).

We thus reach the conclusion that for nickel and nonsalmonid fish, the target standard that would enable fish to survive is that the nickel content of the water should not exceed 0.9 mg/l (900 µg/l) (or a slightly lower value if a safety factor is applied) for 95% of evenly spaced out (say, daily or weekly) samples of water taken during a year. It now remains to assess the actual 95th percentile and the arithmetic mean nickel concentrations for river or tidal water samples taken at a particular sampling point.

Ideally, the relationship between time and concentrations in river waters would be described by continuous water quality data, but this is rarely available for metals, being replaced by a series of discrete observations of concentrations. A population of samples of water quality may be summarised to estimate the frequency distribution of observed concentrations (or the probability density of each concentration) or, alternatively, the cumulative probability that the specified concentrations are not exceeded. It is assumed that the observed water concentrations of nickel adequately represent the natural distributions and therefore that the percentage of samples exceeding a specified concentration equate to the proportion of time for which that concentration will be exceeded in the water. On this basis, the hypothetical relationship between time and concentration of nickel enables the 95th percentile and the annual average concentration to be calculated.

Suppose that weekly analyses of river water over twelve months give the results and the distribution of results shown in Table 10.4. A plot of the percentage of time during which nickel contents are within the stipulated range versus the determined nickel content reveals that, for 5% of the time (i.e. 95th percentile), the nickel content is ≥ 1020 µg/l, and for 95% of the

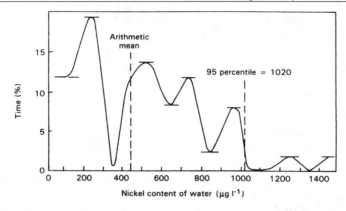

Figure 10.9. Ninety-five percentile: Determination of nickel in water over twelve months. From author's own files

time it is $\leq 1020\,\mu g/l$ (Fig. 10.9). The arithmetic mean is $448\,\mu g/l$ (Fig. 10.9). The 95th percentile value of $1020\,\mu g/l$ exceeds the target of $900\,\mu g/l$ nickel for 95% of the samples taken. Consequently, some mortality of nonsalmonid fish would be expected in the circumstances. If all of the concentrations quoted in Table 10.4 were halved, the 95th percentile value would decrease to $510\,\mu g/l$ (arithmetic mean $224\,\mu g/l$), which is less than the standard of $900\,\mu g/l$, and no adverse effect on nonsalmonid fish due to nickel would be expected.

Summarising, the available data for nickel ($\mu g/l$) are given below:

Standard			River samples		
Upper	Lower	Upper/lower	S_{95}	Arithmetic mean (m)	SS_{95}/m
S_{95}	S_x	S_{95}/S_x			
95% ile/900	220	4.09	95% ile/1020	448	2.28

It is the range of the ratios of the average to the 95percentile obtained for the river samples which determines the mode of expression of the standard that is most appropriate. Comparison of the observed ratios in rivers for nickel (or any other substance), i.e. S_{95}/arithmetic mean, with the ratio of the average standard and the 95th percentile standard, i.e. S_{95}/S_x, will indicate whether the upper or lower standard is appropriate. If the ratio of the standard S_{95}/S_x is less than or equal to the ratio in the river S_{95}/arithmetic mean, then the upper standard (95th percentile) should be selected, since the use of the lower or average concentration S_x as the standard would not guarantee against the short-term occurrence of high and damaging river concentrations of metal which exceed the 95th percentile standard. When, as in the case of nickel quoted above, the ratio of standards S_{95}/S_x ($= 4.09$) is substantially larger than the ratio in rivers S_{95}/arithmetic mean ($= 2.28$), then the use of the lower or average standard $S_x = 0.22$, i.e. the 365 day

Table 10.4. Weekly determinations of nickel (µg/l) in a river water over twelve months (from author's own files)

Week No	Nickel	Week No	Nickel	Week No	Nickel	Week No	Nickel	Week No	Nickel
1	10	11	90	21	420	31	710	41	780
2	120	12	210	22	210	32	60	42	940
3	500	13	900	23	630	33	210	43	160
4	620	14	1210	24	70	34	560	44	200
5	400	15	1410	25	100	35	720	45	220
6	810	16	50	26	150	36	410	46	520
7	420	17	210	27	210	37	510	47	170
8	210	18	430	28	70	38	910	48	400
9	610	19	560	29	420	39	200	49	500
10	700	20	700	30	520	40	200	50	600
								51	710
								52	910

Range of nickel (µg/l)	Number of samples in this range	Percentage of samples in this range, i.e percentage of time that nickel content is in the stated range
10 – 90	6	11.5
100 – 190	6	11.5
200 – 290	10	19.2
300 – 390	0	0
400 – 490	6	11.5
500 – 590	7	13.5
600 – 690	4	7.7
700 – 790	6	11.5
800 – 890	1	1.9
900 – 990	4	7.7
1000 – 1099	0	0
1100 – 1199	0	0
1200 – 1299	1	1.9
1300 – 1399	0	0
1400 – 1499	1	1.9

asymptote (see Fig. 10.8), will ensure that the 95th percentile is not transgressed, while providing adequate long-term protection.

The above discussion is concerned with nickel. However, similar considerations can apply to a range of other metals. Table 10.5 lists S_x and S_{95} values for a range of elements, from the least toxic (nickel, $S_x = 220$, $S_{95} = 900$ µg/l) to the most toxic (cadmium $S_x = 2$ µg/l, $S_{95} = 6$ µg/l and mercury $S_x = 2$ µg/l, $S_{95} = 22$ µg/l). These data are obtained by plotting the data shown in Table 10.5 in the same manner as is shown in the case of nickel. The maximum safe concentrations quoted in Table 10.5 are not amended by safety factors and have not been weighted for the effects of environmental factors such as water hardness, pH, temperature and, in saline waters, salinity. In practice, the available data do not permit this, and the effects of experimental factors demonstrated in short-term acute toxicity tests are extrapolated to long-term exposure.

Table 10.5. S_x and S_{95} values for metals in salmonid and nonsalmonid fish (from author's own files)

Fish species	Metal	Standard maximum safe concentration supporting fish life (µg/l), i.e.,	Standard S_{95} (µg/l) maximum metal concentration S_x (365 d) permitted for 17 days out of 365 days	S_{95}/S_x
Nonsalmonid	Ni	220	900	4.5
	Se	200	1300	6.5
	V	100	1000 – 1600	10 – 16
Salmonid	Cr	100	800	8
Nonsalmonid	As	80	600	4.8
Nonsalmonid	Ag	70	850	12.1
	Cr	100	1000 – 3000	10 – 30
	Zn	23	200	8.7
	Pb	20	100	5
	Cd	4	16	4
Salmonid	Cu	4	17	4.2
Nonsalmonid	Hg	2	22	11
Salmonid	Cd	2	6	3

10.4
Prediction of Fish Kills

If the initial number of fish in a given volume of water at the start of hour 1 is developed by I_i, and the percentage of fish killed by toxicants is P% per unit time (hours, days, weeks, etc.; hours are used in following calculations), then at the end of hours 1, 2, 3 and n, the number of fish surviving (I_n) are given by $I_n = I_i (1 - P/100)$, $I_i (1 - P/100)^2$, $I_i (1 - P/100)^3$ and $I_i (1 - P/100)^n$, respectively.

$$\text{i.e.} \quad I_n/I_i = (1 - P/100)^n \tag{10.1}$$

If, at the start of a period of time lasting n hours, there are I_i fish in a given volume of water, and at the end of that period of time there are I_n fish still alive, then $I_i - I_n$ fish have been killed in n hours. So the percentage of fish killed (F%) in that time is given by:

$$F = (I_i - I_n)100/I_i. \tag{10.2}$$

From Eq. (10.2):

$$FI_i = 100I_i - 100I_n$$
$$\therefore \quad 100I_n - 100I_i - FI_2 - I_i (100 - F)$$
$$\therefore \quad I_n/I_i = (100 - F)/100. \tag{10.3}$$

From Eqs. (10.1) and (10.3):

$$I_n/I_i = (100 - F)/100 = (1 - P/100)^n$$
$$\text{i.e.} \quad 1 - F/100 = (1 - P/100)^n$$
$$F = \{1 - (1 - P/100)^n\}100\% \tag{10.4}$$
$$\log(1 - P/100)^n = n\log(1 - P/100) = \log(1 - F/100)$$
$$\therefore \quad n = \log(1 - F/100)/\log(1 - P/100). \tag{10.5}$$

This equation can be used to calculate the hours of exposure (n) after which F% of the fish will have died for various assumed values of F%, i.e. the percentage of fish that are killed per hour of exposure. Conversely, if P and n are known it is possible to calculate values for F.

Table 10.6 gives the values of n when it is assumed that $P = 0.1$, 1 or 10% per hour and the percentage of fish killed F varies between 5% and 99%.

Table 10.6. Duration of toxicant exposure (n hours) at which F% of fish are killed for various assumed values of P%, the percentage of fish killed per hour (from author's own files)

P	F	$(1 - F/100)$	$(1 - P/100)$	$\log(1 - F/100)$	$\log(1 - P/100)$	$n = \dfrac{\log(1 - F/100)}{\log(1 - P/100)}$	Hours
0.1	5	0.95	0.999	−0.0327	−0.0004	81	
	10	0.90	0.999	−0.0458	−0.0004	114.5	
	15	0.85	0.999	−0.0706	−0.0004	176.5	
	20	0.80	0.999	−0.0969	−0.0004	242.0	
	50	0.50	0.999	−0.3010	−0.0004	752.0	50% fish killed (i.e., LD$_{50}$)
	99	0.1	0.999	−2.000	−0.0004	5000	99% fish killed
1	5	0.95	0.990	−0.0327	−0.0048	6.8	
	10	0.90	0.990	−0.0458	−0.0048	9.5	
	15	0.85	0.990	−0.0706	−0.0048	14.7	
	20	0.80	0.990	−0.0969	−0.0048	20.2	
	50	0.50	0.990	−0.3010	−0.0048	62.7	50% fish killed (i.e., LD$_{50}$)
	99	0.1	0.990	−2.000	−0.0048	417	99% fish killed
10	5	0.95	0.900	−0.0327	−0.0458	0.71	
	10	0.90	0.900	−0.0458	−0.0458	1.00	
	15	0.85	0.900	−0.0706	−0.0458	1.54	
	20	0.80	0.900	−0.0969	−0.0458	2.11	
	50	0.50	0.900	−0.3010	−0.0458	6.57	50% fish killed (i.e., LD$_{50}$)
	99	0.1	0.900	−2.000	−0.0458	43.7	99% fish killed

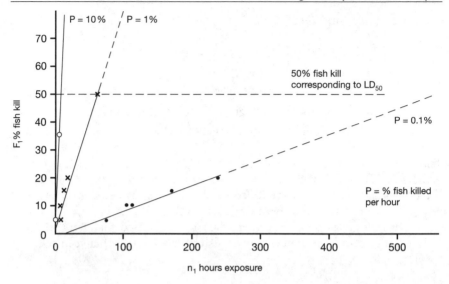

Figure 10.10. Plot of percentage fish kill (F%) versus exposure time (n hours). From author's own files

For example, for a fish kill of 50% in n hours (i.e. LD$_{50}$):

$$n = \log 0.5 / \log(1 - P/100) = -0.3010 / \log(1 - P/100).$$

Using this equation, it is possible to calculate the exposure time n (in hours) at which any specific value of fish kill (F%) will occur.

The data displayed in Table 10.6 are plotted graphically in Fig. 10.10.

If the fish kill (F$_1$%) for any given exposure time n_i (in hours) is known, it is possible to calculate the fish kill (F$_2$%) for any other exposure time n_2 hours.

Thus, from Eq. (10.5):

$$n_1 = \log(1 - F/100) / \log(1 - P/100),$$
$$n_2 = \log(1 - F_2/100) / (1 - P/100)$$
$$\text{i.e,} \quad \log(1 - F_2/100) = n_2/n_1 \log(1 - F_1/100). \tag{10.6}$$

If, for example, 50% (F$_2$%) fish are killed in five hours' exposure (n_2), then in two hours of exposure (n_1) F$_1$% of the fish will be killed.

$$\text{i.e,} \quad \log 0.5 = 5/2 \log(1 - F_1/100)$$
$$\text{i.e,} \quad \log(1 - F_1/100) = 0.4 \times \log 0.5 = -0.1204$$
$$\text{Thus} \quad 1 - F_1/100 = 0.758 \quad \text{i.e,} \quad F_1 = 24.12\%.$$

Or, if 50% (F_2) of fish are killed in twelve months (n_2), then in one month (n_1), F_i% of fish will be killed.

i.e $\log(1 - F_2/100) = 12 \log(1 - F_1/100)$

i.e $\log(1 - F_1/100) = \log 0.5/12 = -0.3010/12 = -0.0251$

$1 - F_1/100 = 0.944$

$F_1 = (1 - 0.944)100 = 5.6\%.$

If the % fish kill (F%) occurring during n hours is known, then the percentage fish kill per hour (P%) can be obtained:

$$\log(1 - P/100) = 1/n \log(1 - F/100)$$

From Eq. (10.6):

$$\log(1 - F_2/100) = n_2/n_1 \log(1 - F_1/100).$$

In other words, if we know the % mortality F_1 that occurs after n_1 days of exposure to a toxicant, we can calculate F_2, the % mortality that occurs during n_2 days of exposure to the toxicant at a particular concentration T mg/l of toxicant in the water.

A plot of % mortality (F) versus concentration of toxicant T yields the equation $F = m \log T + C$, where the slope m and the intercept C are found from the plot.

If F is known, one can calculate T from $\log T = F - C/m$, where T is the concentration of toxicant that causes % mortality F during n days of exposure.

References

1. Bowen HJM (1979) In: *Environmental Chemistry of the Elements*, Academic, London, UK.
2. Waldron HA (1980) In: Waldron HA (ed) *Metals in the Environment*. Academic, London, UK.
3. Smith AE (1973) *Analyst* 98:65.
4. Brown VM, Shurben DG, Shaw D (1970) *Water Res* 4:363.
5. Alabaster JS, Garland JHN, Hart IC, Solbe JF de LG (1972) *An Approach to the Problem of Pollution and Fisheries*, Symposium of the Zoological Society, London, UK, **29**:87.
6. Hart IC (1974) *The Toxicity to Fish of Some Rivers in the Yorkshire Ouse Basin*, WPR Report 1299, Water Pollution Research Laboratory, HMSO, London, UK.
7. Howells EJ, Howells ME, Alabaster JS (1983) *J Fish Biol* 22:447.
8. Brown VM (1968) *Water Res* 2:723.

9. HMSO (1985) *The Implementation of Directive 76/464 EEC on Pollution Caused by Certain Dangerous Substances Discharged in to the Aquatic Environment of the Community*, Department of the Environment Water and the Environment Circular 18/85, September 1985, HMSO London, UK

10. EC (1983) *Council of European Communities Directive of 26 September 1983 on Limit Values and Quality Objectives for Cadmium Discharges*, 83/513/EEC, OJL291, European Communities, Brussels.

11. EC (1982) *Council of European Communities, Directive on Limit Values and Quality Objectives for Mercury Discharges by the Chlor-alkali Electrolysis Industry*, 82/176/EEC, OJL81, European Communities, Brussels.

12. EC (1984) *Council of European Communities, Directive on Limit Values and Quality Objectives for Mercury Discharges by Sectors other than the Chlor-alkali Industry*, 84/156/EEC, OJL74, European Communities, Brussels.

11 Toxicity Evaluation Based on Animal Tissue Analysis

There are several reasons for monitoring the concentrations of toxic metals in creatures such as fish and shellfish.

11.1
Protection of Human Health

This applies to organisms that are harvested for food. Direct analysis of the organisms against accepted standards enables a decision to be made as to whether the organisms are acceptable for human consumption. In the UK for example [1], regulations exist concerning levels of zinc, chromium, copper, nickel and arsenic in fish and shellfish, and these are based on the maximum acceptable intakes of these foods for one week. It is stated that the 90th percentile consumption of fish should not exceed 0.79 kg per week, and for shellfish 0.26 kg per week. Table 11.1 shows weekly intakes of metals by consumers observing the recommendations that would result from the consumption of fish containing different levels of total metals. For example, the weekly recommended maximum intake of chromium from fish caught in coastal waters would be 0.237 mg, while that of arsenic would be 11.1 – 13.2 mg.

Table 11.1. Weekly intake of metals by consumers (from author's own files)

	Concentration of metal in organisms (mg/kg), dry weight						Weekly intake of metals by consumers	
	Cu	Ni	Zn	As	Cr	Total	Maximum intake (recom- mended, kg)	Weight (mg of total metals consumed)
Fish								
Coastal waters	0.5	0.7 – 1.4	4.6	14.1 – 16.7	0.3	20.2 – 23.5	0.79	17.3 – 18.5
Vicinity of municipal outfall	1 – 1.9	0.7 – 1.4	0.8 – 2.8	10	0.5 – 1.5	136.0 – 16.6		10.3 – 13.1
Remote area	1.2	0.2	2.4	0.5 – 1.5	0.2	4.5 – 5.5		3.5 – 4.3
Shellfish								
Vicinity of municipal outfall	0.4 – 2.4	2.2 – 6.5	0.3 – 0.9	10	1 – 10	13.9 – 29.8	0.26	3.6 – 7.7
Remote area	1	3	0.4	0.5 – 1.5	0.8	5.7 – 6.7		1.5 – 1.7

11.2
Protection of Animal Species

Biomagnification and bioaccumulation of metals and organics by fish and creatures other than fish (e.g., crustaceans, molluscs) in nonsaline and saline waters will now be considered.

Biomagnification is the increase in concentration of a toxicant through a food chain, and it has been observed for organochlorine pesticides [2], which occur at progressively higher concentrations along the food chain.

Two competing factors operate in bioaccumulation, namely the rate of uptake of metals or organics and their rate of loss, and these will govern whether there is a net decrease or an increase in toxicant content of the water-based creature [3–6].

11.2.1
Factors Affecting the Bioaccumulation of Cations

Bioaccumulation in fish and other creatures is greatest in the following circumstances:

(1) When the body weight is lowest, i.e. just after spawning, or in younger and smaller creatures
(2) During periods of low rate of growth
(3) In waters of low salinity
(4) In waters of higher temperature
(5) In the absence of competing metals, e.g. bioaccumulation is greater in soft waters than in hard waters
(6) When the species is close to the surface of the water.

Thus, rates of bioaccumulation are greater with creatures of low body weight and rate of growth in surface waters of low salinity and hardness which are at a relatively high temperature.

Since all of these factors have an influence on the extent of bioaccumulation, the ratio between the reported concentration of a metal in water (µg/l) and its concentration in animal or plant life (µg/kg dry weight) (i.e. bioaccumulation factor = µg/kg in plant or animal, µg/l in water) is by no means constant.

Because of bioaccumulation there is an increase in the concentration of a toxicant in a particular animal or plant species with time, and this has been extensively observed. Metals added to fresh or tidal water tend to be removed by absorption onto particulate matter or by chemical transformation into an insoluble form. Thus, sediment concentrations are normally higher than those of the overlying water. At the primary production level, macrophytes rooted in these metal-enriched sediments tend to have greater concentrations of metals than the sediment. This is also true for algae, whether attached or planktonic, as is illustrated in Table 11.2 for the

Table 11.2. Accumulation of metals from Humber and Severn estuary water into sediment (from author's own files)

Element	Accumulation factor = μg/kg dry weight in sediment/μg/l in water		
	Copper	Lead	Nickel
Severn Estuary	15,710 – 16,300	26,830 – 67,330	15,280 – 22,600
Humber Estuary	57,350 – 430,000	68,000 – 136,000	2,130 – 32,000
Element	Zinc	Arsenic	Cadmium
Severn Estuary	13,090 – 25,640	–	1,280 – 3,230
Humber Estuary	4,060 – 102,500	3,700 – 10,2500	800 – 4,000

case of the accumulation of metals in sediments in the Severn and Humber estuaries, UK [7].

Similar metal bioaccumulation phenomena have been observed in the case of fish and, indeed, bioaccumulation has been studied not only in the whole fish but also in individual fish organs, where appreciable differences have been reported between different organs. Van Hoof and Van Son [8] have reported on the extent of bioaccumulation occurring in five different organs taken from rudd (*Scardinius erythrophthalmus*) (muscle, gill, opercle, liver and kidney). Table 11.3 reports concentration factors for four metals (zinc, copper, cadmium and chromium) for organs taken from fish exposed to different levels of these metals for various exposure times of between four hours and greater than ten weeks. Of the various organs taken from this particular type of fish, it is seen in Table 11.4 that the highest concentration factors always occur in opercle tissue and the lowest in muscle, with other organs being intermediate.

It will be noted that higher concentration factors are obtained when the exposure time is extended from three to ten weeks, even though the metal concentrations in the water were lower in the ten-week test. Figure 11.1 plots concentration factors obtained from the opercle versus test duration and concentration of copper and zinc in the water (data from Table 11.3). It is apparent that the concentration factor increases linearly with increasing exposure time but seems to exhibit an exponential relationship with metal concentration in the water. A plot of the logarithm of the metal concentration and exposure time versus the concentration factor is linear, as shown in Table 11.5 and Fig. 11.2.

Thus,

$$n \log C_w = k C_f / C_w \quad \text{or} \quad \log C_w^n = k C_f / C_w,$$

where C_w is the concentration of the toxicant in the water in μg/l

C_f = concentration of the toxicant in the creature in μg/kg

n = duration of exposure of the fish to the toxicant

k = proportionality constant, the value of which depends on the toxicant.

Table 11.3. Interrelationship between concentration of metal in water, exposure time and concentration factor in rudd (*Scardinius erthropthalmus*) organs. From [8] and author's own files

Element	Exposure time	Water (µg/l)	Concentrations in organs (µg/kg)					Concentration factor in organ[a]				
			Muscle	Gill	Opercle	Liver	Kidney	Muscle	Gill	Opercle	Liver	Kidney
Zinc	3 weeks[b]	800[b]	24,400	101,900	195,000	104,900	151,700	30	127	244	131	189
	≤ 24 hours[b]	1600[b]	10,500	38,600	115,300	42,500	154,600	6.5	24	72	26	97
	≤ 12 hours[c]	7500[bc]	6,600	51,200	90,650	34,100	92,200	0.9	6.8	12.1	4.5	12
	≤ 4 hours[c]	18,000[bc]	11,200	647,200	174,500	63,500	216,100	0.6	36	9.7	3.5	12
Copper	> 10 weeks[b]	11[b]	700	5,500	12,400	6,900	6,000	64	500	1,127	627	545
	3 weeks[b]	50[b]	1,600	8,900	30,900	20,200	28,500	32	178	618	404	570
	< 12 hours[c]	250[c]	2,300	22,900	52,600	22,300	30,400	9.2	92	210	89	121
	< 12 hours[c]	1200[c]	2,200	29,300	72,100	31,100	39,000	1.8	24	60	26	32
	< 12 hours[c]	1600[c]	4,000	43,200	104,000	39,800	100,000	2.5	27	65	25	62
Cadmium	> 10 weeks[b]	3[b]	300	2,600	9,500	5,000	4,290	100	868	3,166	1,666	1,400
	3 weeks[b]	250[b]	410	2,500	8,700	9,600	15,700	1.6	0	35	38	55
	< 12 hours[c]	1100[c]	600	3,900	6,000	4,100	14,400	0.5	3.5	5.4	4.5	13
	< 12 hours[c]	4000[c]	500	10,400	20,700	3,800	12,800	0.12	2.6	5.2	0.95	3.2
	< 12 hours[c]	11000[c]	3,200	87,900	29,200	12,300	282,100	0.29	8.0	2.65	1.1	2.5
				27,800								
Chromium	¿ 10 weeks[b]	3	< 200	< 200	< 200	< 260	< 200	< 66	< 66	< 66	< 66	< 66
	3 weeks[b]	16	< 2000	< 2,000	< 2,000	< 2,000	< 2,000	< 125	< 125	< 125	< 125	< 125
	< 12 hours[c]	20	500	4,900	8,300	5,600	10,300	125	245	415	280	515
	< 12 hours[c]	80	800	48,200	26,000	18,400	23,800	10	602	325	230	297
	< 12 hours[c]	145	600	30,600	19,600	15,200	27,800	41	211	135	105	92

[a] Concentration factor = concentration in organ (µg/kg)/concentration in water (µg/l)

[b] Subacute toxicity tests, i.e. no fish mortalities

[c] Acute toxicity tests i.e. 100% fish mortality

Table 11.4. Summary of concentration factors obtained for different organs taken from rudd (*Scardinius erthropthalmus*) at different metal concentrations in water and different exposure times (from [8])

Exposure time (weeks)	Metal concentration in water (µg/l)		Highest concentration factor[a]	Lowest concentration factor[a]
3	Zn	800	244 (opercle)	28 (muscle)
	Cu	50	618 (opercle)	32 (muscle)
	Cd	250	55 (kidney)	1.6 (muscle)
10	Cu	11	1127 (opercle)	64 (muscle)
	Cd	3	3,166 (opercle)	100 (muscle)

[a] µg/l in tissue/µg/l in water

Table 11.5. Dependence of concentration factor obtained for the opercle on product of log(concentration factor) and exposure time (from author's own files)

Element	Exposure time of rudd (weeks)	Concentration of metal in water (µg/l)	(Log of concentration) × exposure time (a)	Observed concentration factor µg/kg/µg/l	Slope a/b
Zinc	3	800	8.71	244	0.036
	0.143 (1 day)	1,600	0.46	72	
	0.0715	7,500	0.28	12.1	
	0.024 (4 h)	18,000	0.10	9.7	
Copper	10	11	10.41	1127	0.0092
	3	50	5.10	618	0.0082
	0.0715 (0.5 day)	250	0.17	210	
	0.0715 (0.5 day)	1,200	0.22	60	
	0.0715 (0.5 day)	1,600	0.23	65	

This equation presents the relationship between the concentration of the toxicant in the water (C_w), the concentration in the fish organ (C_f) and the exposure time (n, in weeks).

Factor obtained for rudd opercle; from own files

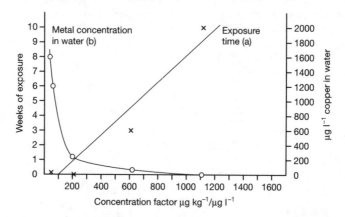

Figure 11.1. Relationships between **(a)** exposure time and copper concentration factor in water and **(b)** concentration of copper in water and concentration factor obtained for rudd opercle. From author's own files

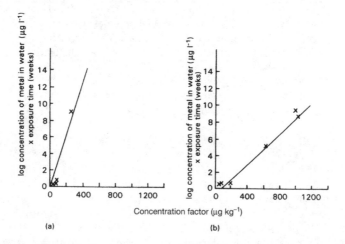

Figure 11.2. Linear relationship between product of log(concentration of metal in water) (µg/l) and exposure time (weeks) for **(a)** zinc, **(b)** copper. From author's own files

11.2.2
Bioaccumulation of Cations in Fish

The bioaccumulations of copper and zinc have been measured in the barnacle (*Balanus amphitrite*) in estuary water. At concentrations of $1 - 11\,\mu g/l$ copper in the water, between 39,700 and 625,700 µg/l of copper were found in the barnacle tissue, giving bioaccumulation factors of 3,609 and 625,700. At concentrations of $13 - 46\,\mu g/l$ zinc in the water, between 203,600 and

1,937,000 µg/kg zinc were found in barnacle tissue, giving bioaccumulation factors of between 18,509 and 1,937,000.

Langston and Zhan [10] studied the bioaccumulation of cadmium in the tellinid clam (*Macoma balthica*) taken at the coastline at Whitehaven, Cumbria. At 100 µg/l cadmium in water, the clam picked up 10.150 µg/kg cadmium during 29 days' exposure (0.35 µg Cd g/d), giving a bioaccumulation factor of 101. A bioaccumulation factor of 68,000 has been obtained for iron in kelp (*Ecklonia radiata*) taken in harbour water [11].

11.2.3
Bioaccumulation of Organic Compounds

Bioaccumulation factors of 15, 966 and 17 have been obtained upon the exposure of rainbow trout, channel catfish and bluegills, respectively, to carbendazin [12]. Seawater lampreys were exposed to water containing 50 – 485 µg/l Kegone for ten days and they gave an average bioaccumulation of 1900 [13]. Striped bass (*Morone saxatilis*) exposed to 100 µg/l Molinate (Ordram) in water for one day gave a bioaccumulation factor of 25.3 [14]. Upon exposure to 1000 µg/l fenitrothion for 1 – 3 days, the crustaceans *Daphnia pulex* and *Palaemon paucidens* gave, respectively, maximum bioaccumulation factors of 76 and 6 [14].

11.2.4
Bioaccumulation of Organometallic Compounds in Fish and Mussels

The data in Table 11.6 illustrate the bioaccumulation of tetramethyl lead present in water at a concentration of 3.46 µg/l into rainbow trout [15]. The concentration factor ranges from 124 after one day's exposure to 800 – 900 after seven days' exposure.

Table 11.6. Accumulation of tetramethyl lead in rainbow trout (*Oncorhynchus mykiss*). From [7]

Exposure (days)	Weight of fish (g)	Fish alive or dead	Water averaged (µg/l)	Fish wet (µg/kg) weight	Concentration factors[a]
1	0.1211	Dead	3.46	430	124
2	0.3661	Dead		1000	312
	0.7982	Dead		2000	578
3	0.4116	Dead		1320	382
	0.6300	Dead		2090	604
7	1.3045	Alive		2940	850
	1.5466	Alive		3230	934
	0.8100	Alive		2250	650
	0.4926	Alive		1730	500

[a] Concentration factor = concentration of Me_4Pb in fish (µg/l)/concentration of Me_4Pb in water (µg/l)

There are two reasons for spatially monitoring trends in the bioaccumulation in fresh and tidal waters:

(1) Macroscale, i.e. the identification of potentially unknown areas of elevated concentration and assessment of the extent of the zone of contamination.

(2) Monitoring of bioaccumulation in fresh and tidal waters as trends in time. These need to be maintained in order to identify trends in contamination, especially near effluent discharges so that stability, improvements or deteriorations in contaminant levels can be identified.

Spatial and time monitoring programs of the types discussed above will also provide information needed to assess the risk to top predators in a particular ecosystem.

The design of such a programme is typified by the US Mussel Watch Program [16], which takes into account the following factors:

- Species studied: *Mytilus edulis* mussel was used in this program as this creature had already been studied for factors affecting accumulation.
- Time of year: Late winter was chosen, as metal content is stable (i.e. avoiding post-spawning maximum).
- Size or age: Dominant size of population sampled to avoid effect of age and size.
- Position on shore: Collected on rocky shores to avoid contamination by soft sediments at level of shore exposed for approximately six hours each tidal cycle, i.e. 3 – 4 hours after high tide.
- Sample size: Minimum 25 animals to allow statistical assessment.
- Sampling: Transported alive in polyethylene bags regularly drained from free water. Placed in clean water for 24 hours prior to analysis to ensure gut contents are eliminated. Analysis of homogenised individual animals and shell dimensions recorded.

This scheme is designed to detect a 10% change in metal concentrations in *Mytilus* mussels with a confidence of 90%.

In one such study, mussels from a clean environment were suspended in cages at several locations in the Firth of Forth. A small number were removed periodically, homogenised and analysed for methylmercury. The rate of accumulation of methylmercury was determined and, by dividing this by mussel filtration rate, the total concentration of methylmercury in the seawater was calculated.

The methylmercury concentration in caged mussels increased from low levels (less than 0.01 µg/g) to 0.06 – 0.08 µg/g in 150 days (Figure 11.3), giving a mean uptake rate of 0.4 ng daily, i.e. a 10 g mussel accumulated 4 ng daily. The average percentage of total mercury in the form of methylmercury increased from less than 10% after 20 days to 33% after 150 days. This may be compared with analyses of natural intertidal mussels from the area in which the proportion of methylmercury was higher in mussels with lower (less than 10 µg/g) than in those with higher total mercury concentrations.

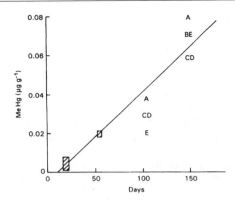

Figure 11.3. The increase with time of methylmercury concentrations in caged mussels at positions A – E. Methylmercury was not detectable (0.01 μg/g) after 20 days, and animals from all five stations contained 0.02 μg/g methylmercury after 55 days' exposure, as shown by the shaded rectangles. The case at position B was not sampled at 100 days' exposure. From author's own files

Davies, Graham and Pirie [17] calculated the total methylmercury concentration in the seawater as 0.06 μg/l, i.e. 0.1 – 0.3% of the total mercury concentration, as opposed to less than 5 – 32 ng/l methylmercury found in Minamata Bay, Japan. The bioaccumulation factor μg/kg/μg/l of methylmercury in mussels ranged from 17 (one day of exposure) to 1333 (150 days of exposure).

A potentially valuable consequence of this type of bioassay is that estimates of the relative abundance of methylmercury can be obtained at different sites through the exposure of 'standardised' mussels, as used in their experiment, in cages for controlled periods of time, and by comparing the resultant accumulations of methylmercury.

References

1. Mance G, Brown VM, Gardiner J, Yates J (1984) *Proposed Environmental Quality Standards for List II Substances in Water: Chromium*, Technical Report TR207, The Water Research Centre, Marlow, Buckinghamshire, UK.
2. Phillips DJH (1980) *Quantitative Aquatic Biological Indicators*, Applied Science, London, UK.
3. Bohn A (1975) *Mar Pollut Bull* **6**:87.
4. Leatherland M, Burton JD (1974) *J Mar Biol Assoc* **54**:457.
5. Prosi F (1977) *Schwermetallbelastung in den Sedimenten der Elsenz und ihre Auswirkung auf limnische Organismen*, Dissertation, University of Heidelberg, Germany.
6. Davis JJ, Perkins RW, Palmer RF, Hanson WC, Clive JF (1958) *Radioactive Materials in Aquatic and Terrestrial Organisms Exposed to Reactor Effluent Water*. In: 2nd UN International Conference on the Peaceful Uses of Atomic Energy, 1–13 Sept 1958, Geneva, Switzerland.

7. Chau YK, Wong PTS, Bengert GA, Kramer O (1979) *Anal Chem* **51**:186.
8. Van Hoof F, Van San M (1981) *Chemosphere* **10**:1127.
9. Anil AC, Wagh AB (1988) *Mar Pollut Bull* **19**:177.
10. Langston WJ, Zhon M (1987) *Mar Environ Res* **21**:225.
11. Higgins HW, Mackey DJ (1987) *Aust J Mar Freshwater Res* **38**:307.
12. Palawski DU, Knowles CO (1986) *Environ Toxicol Chem* **5**:1039.
13. Mallatt J, Barron NG (1988) *Arch Environ Contam Toxicol* **17**:73.
14. Tjeerdema RS, Crosby DG (1987) *Aqua Toxicol* **9**:305.
15. Takimoto Y, Oshima M, Miyamoto J (1987) *Ecotoxicol Environ Safety* **13**:126.
16. Baydon CE (1977) *J Mar Biol Assoc* **57**:675.
17. Davies IM, Graham WC, Pirie SM (1979) *Mar Chem* **7**:111.

Appendix 4.1
Concentration of Cations in Fish and Fish Organs

Element	Type	Concentration (mg/kg)	Reference
Arsenic	Herring	0.03	[1]
	Haddock	0.03	[1]
	Tuna	0.15	[1]
	Smelt	0.41 – 0.44	[2]
	Coho salmon	0.26 – 0.36	[2]
	Plaice	Total 24	[1]
		Inorganic 0.02 – 0.04	
	Herring	1.1 (total)	[1]
		0.02 – 0.04 (inorganic)	
	Haddock	2.6 (total)	[1]
		0.02 – 0.04 (inorganic)	
	Tuna	2.9 (total)	[1]
		0.12 – 0.020 (inorganic)	
	Dogfish (muscle)	18.7	[3]
Boron	Carp	Up to 1.5	[4]
Cadmium	*Fish*		
	Catfish	0.039	[5]
	Rainbow trout	0.10 – 0.11	[2]
	White bream	0.04	[6]
	Sardine	0.02	[6]
	Gilthead bream	0.03	[6]
	Grey mullet	0.09	[6]
	Carp	Up to 0.27	[4]
	Horse mackerel	0.17	[5]
	Striped mullet	0.02	[5]
	Crayfish	0.10	[5]
	Flathead	0.13	[5]
	Shark	0.08	[5]

Element	Type	Concentration (mg/kg)	Reference
Cadmium	*Organs*		
	Muscle	0.075 – 2.9	[7]
	Muscle	0.3	[8]
	Skin	0.14 – 10.9	[7]
	Kidney	3.1 – 5.6	[7]
	Kidney	7.1	[2]
	Kidney	4.2	[8]
	Gut	0.6 – 5.6	[7]
	Heart	1.5 – 5.6	[7]
	Bone	0.14 – 3.6	[7]
	Gill	0.038	[9]
	Gill	0.94 – 6.67	[7, 8]
	Liver	3.1 – 6.7	[2, 7]
	Liver (Perch)	0.17 – 0.90	[10]
	Liver (Pike)	0.17	[10]
	Liver	1.5 – 9.0	[10]
	Liver(White fish)	0.19 – 0.9	[10]
	Opercle	9.5	[8]
	Blue gill tissue		
	Kidney	5.6 – 13.1	[1]
	Gut		
	Heart		
	Liver		
	Muscle	0.14 – 1.7	[7]
	Skin		
	Bone		
Chromium	*Fish*		
	Coho jack	0.21	[2]
	Rainbow trout	2.2	[2]
	White bream	0.58	[6]
	Sardine	0.28	[6]
	Gilthead bream	0.49	[6]
	Grey mullet	0.10	[6]
	Horse mackerel	0.65	[6]
	Striped mullet	0.14	[6]
	Carp	Up to 2.2	[4]
Chromium	*Organs*		
	Muscle	0.5 – 0.8	[8]
	Gill	4.9 – 48.2	[8]
	Opercle	8.3 – 26.0	[8]
	Liver	5.6 – 18.4	[8]
	Kidney	10.3 – 27.8	[8]

Element	Type	Concentration (mg/kg)	Reference
Cobalt	*Fish*		
	Whale heart	0.07	[11]
	Whale meat	0.07	[11]
	Whale fat	0.38	[11]
	Trout	0.14	[11]
Copper	*Fish*		
	Rainbow trout	0.53 – 0.8	[2]
	Coho jack	1.06	[11]
	Whale heart	7.9	[11]
	Whale meat	2.29	[11]
	Whale fat	1.2	[11]
	Trout	2.6	[11]
	White bream	1.11	[6]
	Sardine	2.18	[6]
	Gilthead bream	1.20	[6]
	Grey mullet	1.70	[6]
	Horse mackerel	0.99	[6]
	Striped mullet	0.68	[6]
	Crayfish	3.46	[5]
	Flathead	0.39	[5]
	Shark	1.01	[5]
	Whale	1.2 – 7.6	[11]
	Organs		
	Perch liver	3.7 – 4.8	[10]
	White fish liver	24 – 62	[10]
	Pike liver	11.7	[10]
	Liver	6.9	[8]
	Liver	1.7	[12]
	Gill	5.5	[8]
	Gill	0.6	[12]
	Opercle	12.4	[8]
	Kidney	6.0	[8]
	Kidney	0.67	[12]
	Blood cell	0.27	[12]
	Blood serum	0.57	[12]
	Heart	3.0	[12]
	Spleen	3.0	[12]
	Gut	1.1	[12]
	Stomach	0.8	[12]
	Skin	0.64	[12]
	Muscle	0.22	[12]
	Bone	1.6	[12]

Element	Type	Concentration (mg/kg)	Reference
Lead	*Fish*		
	Rainbow trout	0.92 – 0.98	[2]
	Whale heart	0.62	[11]
	Whale meat	0.45	[11]
	Whale fat	1.37	[11]
	Carp	Up to 2.3	[4]
	Trout	0.89	[11]
	White bream	0.61	[6]
	Sardine	0.57	[6]
	Gilthead bream	0.68	[6]
	Grey mullet	1.36	[6]
	Horse mackerel	1.05	[6]
	Striped mullet	0.12	[6]
	Crayfish	0.48	[5]
	Flathead	0.92	[5]
	Shark	0.57	[5]
	Catfish	0.26	[9]
	Blue gill	0.32	[9]
	Miscellaneous fish	0.12 – 1.81	[13]
Lead	*Organs*		
	Liver	8.0	[2]
	Kidney	36.0	[2]
	Muscle	0.12 – 1.81	[13]
Manganese	White bream	0.51	[6]
	Sardine	1.63	[6]
	Grey mullet	0.33	[6]
	Horse mackerel	0.63	[6]
	Striped mullet	0.22	[6]
Mercury	Pickerel	0.24 – 1.11	[14]
	Carp	0.23 – 0.36	[15]
	Carp	0.22 – 2.4	[11]
	Carp	0.23 – 0.36	[15]
	Shiver	0.28 – 0.35	[11]
	Carp	Up to 2.9	[4]
	Chub	0.09 – 0.16	[11]
	Buffalo	0.12 – 0.41	[11]
	Blue cat	0.21 – 0.27	[11]
	Carp	1.5 – 2.7	[11]
	Carp	0.23 – 0.36	[16]
	Walleye	0.33 – 0.79	[11]
	Channel cat	0.26 – 0.55	[17]

Element	Type	Concentration (mg/kg)	Reference
	Channel cat	0.36 – 0.42	[15]
	Crappie	0.09 – 0.19	[11]
	Crappie	0.09 – 0.14	[16]
	Wallage	0.33 – 0.79	[15]
	Yellow perch	0.29 – 0.61	[15]
	Perch	0.51 – 0.53	[18]
	White bass	0.43 – 0.72	[15]
	Freshwater drin	0.30 – 0.67	[15]
	Coho salmon	0.51 – 0.69	[15]
	White sucker	0.35 – 0.56	[15]
	Gizzard shark	0.20 – 0.26	[15]
	Smallmouth bass	0.55	[15]
	Smelt	0.30	[11]
	Tuna	0.25 – 0.58	[19]
	Tuna	0.32 – 0.35	[18]
	Canned tuna	0.12 – 0.13	[18]
	Albacore tuna	0.93 – 0.94	[18]
	Barramundi	0.68	[18]
	Gemfish	0.32 – 0.29	[18]
	Miscellaneous fish	0.1 – 0.4	[18]
	Miscellaneous fish	0.11 – 4.01	[20]
	Miscellaneous fish	2.6 – 8.6	[21]
	Miscellaneous fish	2.06 – 7.23	[21]
Molybdenum	Carp	Up to 3.6	[4]
Nickel	*Fish*		
	Rainbow trout	0.15 – 0.20	[2]
	Whale heart	0.31	[11]
	Whale meat	0.17	[11]
	Whale fat	0.60	[11]
	Trout	0.34	[11]
	Carp	Up to 2.2	[4]
Nickel	*Organs*		
	Liver	0.92	[2]
	Kidney	1.9	[2]
Selenium	Miscellaneous fish	0.31 – 0.55	[22]
	Miscellaneous fish	0.4 – 6.6	[23]
	Smelt	0.31	[2]
	Coho salmon	0.38 – 0.55	[2]
	Crayfish	0.17 – 0.27	[5]
	Flathead	0.37	[5]

Element	Type	Concentration (mg/kg)	Reference
	Shark	0.19	[5]
	Carp	Up to 5.5	[4]
Silver	Whale heart	0.04	[11]
	Whale meat	0.02	[11]
	Whale fat	0.02	[11]
	Trout	0.04	[11]
Zinc	*Fish*		
	Rainbow trout	10.9 – 11.8	[2]
	Coho jack	24.6	[2]
	Whale heart	103	[11]
	Whale meat	42	[11]
	Whale fat	26	[11]
	Trout	39	[11]
	White bream	10.6	[6]
	Sardine	6.3	[6]
	Gilthead bream	9.5	[6]
	Grey mullet	12.2	[6]
	Horse mackerel	4.3	[6]
	Striped mullet	6.4	[6]
Zinc	*Organs*		
	Liver	12.6	[2]
	Liver	100 – 150	[10]
	Liver	29.4	[8]
	Muscle	16.4	[8]
	Gill	47.9	[8]
	Opercle	120	[6]
	Kidney	57.0	[8]
	Liver (perch)	107 – 120	[11]
	Liver (white fish)	463 – 487	[11]

References

1. Brooke PJ, Evans WH (1981) *Analyst* **106**:514.
2. Agemian H, Sturtevant DP, Austen KD (1980) *Analyst* **105**:125.
3. Beauchemin D, Bednas ME, BermanSS, McLaren JW, Siu KWM, Sturgeon RE (1988) *Anal Chem* **60**:2209.
4. Saiki MK, May TW (1988) *Sci Total Environ* **74**:199.
5. Adeloju SB, Bond AM, Hughes HC (1983) *Anal Chim Acta* **148**:59.
6. Ramelow G, Tugrul S, Ozkan MA, Tuncel G, Saydan C, Balkas TI (1978) *Int J Environ Anal Chem* **5**:125.
7. Blood ER, Grant GC (1975) *Anal Chem* **47**:1438.
8. Van Hoof F, Van San M (1981) *Chemosphere* **10**:1127.
9. Poldoski JE (1980) *Anal Chem* **52**:1147.
10. Borg H, Edin A, Holm K, Skold E (1981) *Water Res* **15**:1291.
11. Armannsson H (1979) *Anal Chim Acta* **110**:21.
12. Harvey BR (1978) *Anal Chem* **50**:1866.
13. Pagenkopf GK, Neumann DR, Woodriff R (1972) *Anal Chem* **44**:2248.
14. Davidson JW (1979) *Analyst* **104**:683.
15. Sivasankara Pillay KK, Thomas CC Jr, Sondel JA, Hyche CM (1971) *Anal Chem* **43**:1419.
16. Thomas RJ, Hagstrom RA, Kuchar EJ (1972) *Anal Chem* **44**:512.
17. Giam CS, Trujillo DA, Kira S, Hrung Y (1986) *Bull Environ Contam Toxicol* **25**:824.
18. Louie HW (1983) *Analyst* **108**:1313.
19. Holak W, Kruznitz B, Williams JC (1972) *J AOAC* **55**:741.
20. Uthe JF, Armstrong FAJ, Tam KC (1971) *J AOAC* **54**:866.
21. Jones P, Nickless J (1974) *J Chromatogr* A **89**:201.
22. Agemian H, Thomson R (1980) *Analyst* **105**:902.
23. Goulden PD, Anthony DHJ, Austen KD (1981) *Anal Chem* **53**:2027.

Appendix 4.2
Concentrations of Cations Found in Invertebrates

Cation	Type	Concentration (mg/kg)	Reference
Antimony	Oyster tissue	0.4	[1]
	Lobster tissue	0.071 – 0.089	[2]
Arsenic	Oyster tissue	13.4	[1]
	Lobster	11.9 – 15.9	[3]
	Canned crab	1.5 (total)	[4]
		0.06 – 0.10 (inorganic)	
	Whelk	3.2 (total)	[4]
		0.06 – 0.18 (inorganic)	
	Canned lobster	3.6 (total)	[4]
		0.06 – 0.08 (inorganic)	
	King prawn	14 (total)	[4]
		0.02 – 0.04 (inorganic)	
	Whelk	26 (total)	[4]
		0.10 – 0.18 (inorganic)	
	Lobster	24.6 – 25.5	[5]
	Scallops	7.0 – 7.8	[5]
	Mollusc	2 – 23.2	[6]
	Lobster hepato-pancreas	13.4	[7]
Bismuth	Mussel	0.0007 – 0.0023	[8]
	Oyster	0.0042	[8]
Bromine	Lobster	50.6 – 51.7	[3]
Cadmium	Oyster	0.0025	[18]
	Oyster	2.36 – 2.56	[18]
	Crab	0.71 – 0.83	[9]
	Crab	7.0	[10]
	Mussel (Mediterranean)	0.02 – 0.03	[11]

Cation	Type	Concentration (mg/kg)	Reference
	Mussel Port Phillip	0.5	[12]
	Mussel US West Coast	0.8 – 20.2	[13]
	Mussel	0.07 – 0.40	[11]
	Shrimp	0.07 – 0.24	[11]
	Crab	0.07 – 0.24	
	Oyster	0.07 – 0.24	
	Mussel	0.07 – 0.24	
	Clam	1.3	[14]
	Lobster	0.5 – 1.1	[10]
	Lobster hepato-pancreas	3.5	[7]
Chromium	Lobster	0.75	[3]
Cobalt	Lobster	0.34 – 0.44	[3]
Copper	Mussel		[10]
	Shrimp	0.75 – 2.65	
	Crab		
	Lobster		
	Lobster hepato-pancreas	63.0	[7]
Iron	Lobster	212 – 219	[3]
Lead	Crab	2.8	[9]
	Clam	0.83	[9]
	Mussel	0.43 – 0.61	[11]
	Lobster hepato-pancreas	2.5 – 12.0	[10]
	Lobster hepato-pancreas	12.4	[7]
	Oyster	0.48 – 0.61	[11]
	Crab	0.48 – 0.61	
	Shrimp	0.48 – 0.61	
	Mussel	0.48 – 0.61	
	Lobster	0.11 – 3.2	[10]
Manganese	Lobster hepato-pancreas	17.5	[17]
	Lobster	16, 57	[3]

Cation	Type	Concentration (mg/kg)	Reference
Mercury	Oyster	0.14 – 0.156	[14]
	Mussel (Mediterranean)	0.02 – 0.05	[10]
	Shrimp	0.02 – 0.05	[11]
	Crab	0.02 – 0.05	[11]
	Lobster	0.31	[5]
	Scallop	0.10	[5]
	Lobster	0.16	[3]
Nickel	Lobster	0.98	[7]
	Lobster hepato-pancreas	19.4	[7]
Plutonium	Mussel	0.3 – 13.9	
Selenium	Prawn	4.01	[15]
	Lobster	2.03 – 2.70	[17]
	Oyster	1.7	[10]
	Lobster	6.2 – 6.7	[5]
	Scallops	0.71 – 0.87	[5]
	Scallops	1.24	[15]
	Oyster	2.26	[17]
	Lobster	2.04 – 2.21	[3]
	Lobster	0.015	[3]
	Lobster	0.86 – 0.93	[3]
Strontium	Lobster hepato-pancreas	84.9	[7]
	Lobster	11.0	[3]
Vanadium	White shrimp	0.4 – 3.05	[16]
	Blue crab	1.09 – 1.84	[16]
	Oyster	0.53 – 1.42	[16]
Zinc	Lobster hepato-pancreas	852	[7]
	Lobster	548 – 888	[3]

References

1. De Oliveira E, McLaren JW, Berman SS (1983) *Anal Chem* **55**:2047.
2. Bertin KK, Dong Soo Lee (1983) In: Wong CS et al. (eds) *Trace Metals in Seawater, Proceedings of a NATO Advanced Research Institute on Trace Metals in Seawater,* Sicily, Italy, Plenum, New York, USA.
3. Chisela F, Gawlik D, Bratter P (1986) *Analyst* **111**:405.
4. Brooke PJ, Evans WH (1981) *Analyst* **106**:514.
5. Welz B, Melcher M (1985) *Anal Chem* **57**:427.
6. Maher WA (1981) *Anal Chim Acta* **126**:157.
7. Landsberger S, Davidson WF (1985) *Anal Chem* **57**:196.
8. Sivasankara Pillay KK, Thomas CC Jr, Sondel JA, Hyche CM (1971) *Anal Chem* **43**:1419.
9. McLaren JW, Berman SS (1981) *Appl Spectrosc* **35**:403.
10. Topping G (1982) *Report of the 6th ICES Trace Metal Intercomparison Exercise for Cadmium and Lead in Biological Tissue,* Cooperative Research Report No. 111, International Council for the Exploration of the Sea, Copenhagen.
11. Ramelow G, Ozkan MA, Tuncel G, Saydon C, Balkas TI (1978) *Int J Environ Chem Anal* **5**:125.
12. Ashworth MJ, Farthing RH (1981) *Int J Environ Chem Anal* **10**:35.
13. Goldberg ED, Bowen VT, Farrington JW, Harvey G, Martin JH, Parker PL, Risebrough RW, Robertson W, Schneider E, Gamble E (1978) *Environ Conserv* **5**:101.
14. Poldoski JE (1980) *Anal Chem* **52**:1147.
15. Maher WA (1985) *Mar Pollut Bull* **16**:33.
16. Blotcky AJ, Falcone C, Medina VA, Rack EP, Hobson DW (1979) *Anal Chem* **51**:178.
17. Bin Ahmad R, Hill JO, Magee RJ (1983) *Analyst* **108**:835.
18. Greenberry RR (1985) *Anal Chem* **57**:427.

Appendix 4.3
Types of Organic Compounds Found in Invertebrates

Organic Compound	Reference
Polyaromatic hydrocarbons	[1–6]
Phthalate esters	[7]
Volatile chloroaliphatics	[8,9]
Chlorinated insecticides	[10–18] ***
Polychlorinated biphenyls	[19–26]
Polychloroterphenyls	[27]
Organophosphorus insecticides	[28]
Organosulfur compounds	[29,30]
Polychlorodibenzofurans	[28]
Arsenobetaine	[31]
Coprostanol	[32]

*** Williams R and Holden AV, *National Institute of Oceanography, Wormley, Godalming, Surrey, UK,* Private communication

References

1. Chesler SN, Gump BH, Hertz H, May WE, Wise SA (1978) *Anal Chem* **50**:805.
2. Bjorseth BP, Knutzen J, Skei J (1979) *Sci Total Environ* **13**:71.
3. Dunn BP, Stich HF (1976) *J Fish Res Board Can* **33**:2040.
4. Kunte H (1967) *Arch Hyg Bakt* **151**:173.
5. Uthe JF, Musial CJ (1988) *J AOAC* **71**:363.
6. Giam CS, Trujillo DA, Kira S, Hrung Y (1986) *Bull Environ Contam Toxicol* **25**:824.
7. Giam CS, Chan HS, Neff GS (1975) *Anal Chem* **47**:2225.
8. Murray AJ, Riley JP (1973) *Anal Chim Acta* **65**:261.
9. Murray AJ, Riley JP (1973) *Nature* **242**:37.
10. Markin GP, Hawthorne JC, Collins HL, Ford JH (1974) *Pest Monit J* **7**:139.
11. Mills PA, Caley JF, Grithen RA (1963) *J AOAC* **46**:106.
12. Arias C, Vidal A, Vidal C, Maria J (1970) *An Bromat* **22**:273.
13. Kouyoumjian HH, Uglow RF (1974) *Environ Pollut* **7**:103.
14. Ernst W, Goerke H, Eder G, Schaefer RG (1976) *Bull Environ Contam Toxicol* **15**:55.

15. Ernst W, Schaefer RG, Goerke H, Eder GZ (1974) *Fresen Z Anal Chem* **272**:358.
16. US EPA (1974) *US Environmental Protection Agency Report No. EPA-600-4-74* 1974T 108, US Environmental Protection Agency, Washington, DC, USA.
17. Wilson AJ, Forester J, Knight J (1969) *US Fish Wildlife Circular 355 18-20*, Centre for Estuaries and Research, Gulf Breeze, FL, USA.
18. Neudorf S, Khan MAQ (1975) *Bull Environ Contam Toxicol* **13**:443.
19. Armour JA, Burke JA (1970) *J AOAC* **53**:761.
20. Gaul J, Cruze La Grange P (1971) *Separation of Mirex and PCBs in Fish*, Laboratory Information Bulletin, Food and Drug Administration, New Orleans, LA, USA.
21. Markin GP, Ford JH, Hawthorne JC, Spence JH, Davies J, Collins HL, Loftis CD (1972) Environmental Monitoring for the Insecticide Mirex, APHIS81-83, USDA, Washington, DC, USA.
22. Butler PA (1969) *Biol Sci* **19**:889.
23. McKenzie MD (1970) *Fluctuations in Abundance of the Blue Crab and Factors Affecting Mortalities*, Technical Report No. 1, South Carolina Wildlife Research Division, Charleston, SC, USA.
24. Mahood RK, McKenzie MD, Middlaugh DP, Bollar SJ, Davis JR, Spitzbergen D (1970) *A Report on the Cooperative Blue Crab Studies in South Atlantic States*, US Department of the Interior, Bureau of Commercial Fisheries (Project No. 2-79-R-1, 2-81-R-1, 2-82-R-1), Washington, DC, USA.
25. Teichman J, Bevenue H, Hylin JW (1978) *J Chromatogr* **151**:155.
26. Tanabe S, Tatsukawa R, Phillips DJH (1987) *Environ Pollut* **47**:41.
27. Freudenthal J, Greve PA (1973) *Bull Environ Contam Toxicol* **10**:108.
28. Deusch ME, Westlake WE, Gunther FA (1970) *J Agric Food Chem* **18**:178.
29. Kira S, Izumi T, Ogata M (1983) *Bull Environ Contam Toxicol* **31**:518.
30. Ogata M, Miyake Y (1979) *Water Res* **13**:1179.
31. Francesconi KA, Micks P, Stockten RA, Irgolie KJ (1985) *Chemosphere* **14**:1443.
32. Matusik JE, Hoskin GP, Spon JA (1988) *J AOAC* **71**:994.

Appendix 4.4
Concentrations of Metals Found in Nonsaline (Freshwater) Sediments

	River			Lake/Pond		
	Location	Concentration	Ref.	Location	Concentration	Ref.
Aluminium	–	46,200	[1]	–	26,200 – 63,800	[3]
		Total: 9,890 – 11,500	[2]	Lake Ontario, Canada	43,000	[4]
		Acid-extractable: 522 – 19,200	[2]			
Arsenic	River Edisto, USA	0.22 – 0.63	[5]	–	1.9 – 2.6	[3]
	–	1.9 – 7.1	[6]			
Antimony				–	0.01 – 2.9	[3]
Barium				–	163 – 175	[3]
				Lake Ontario, Canada	2700	[4]
Bromine				–	23 – 96	[3]
Cadmium	–	0.08 – 1.22	[7]	Lake Ontario, Canada	3.5 – 8.0	[3]
	River Arno, Italy	1.01 – 9.6	[8]	Lake Ontario, Canada	40.0	[4]
		Total: 0.06 – 27.5	[2]	Lake Ontario, Canada		
		Acid-extractable: 0.1 – 15.4	[2]			
Caesium				–	0.5 – 14.0	[3]
Calcium				–	12,300 – 40,000	[3]
Cerium				–	53 – 160	[3]

	River			Lake/Pond		
	Location	Concentration	Ref.	Location	Concentration	Ref.
Chlorine				–	20 – 609	[3]
Chromium	–	0.48 – 0.49	[9]	–	16 – 50	[3]
	River Susquehanna, USA	31.4 – 1143	[10]	Lake Ontario, Canada	110	[3]
		108	[1]			
	River Arno, Italy	450	[8]			
		Total: 3 – 368	–			
		Acid-extractable: 1.3 – 128	[2]			
Cobalt	River Arno, Italy	21.9	[8]	Lake Ontario, Canada	3.9 – 16.0	[3]
		57	[8]	Lake Ontario, Canada	200	[4]
		Total: 2.2 – 5.3	[2]			
		Acid-extractable: 1.8 – 48.9	[2]			
Copper	–	0.07	[11]	Lake Ontario, Canada	50	[4]
	River Rideau, Canada	4.2	[1]			
	River Arno, Italy	59.5 – 244	[8]			
		1.9 – 226	[12]			
		Total: 1 – 148				
		Acid-extractable 6.6 – 74	[2]			
Dysprosium					5.4 – 74.0	[3]
Europium					0.77 – 194	[3]
Gadolinium					6.4 – 22	[3]
Gold					0.25 – 19	[3]
Hafnium					1.7 – 12	[3]
Holmium					0.19 – 0.74	[3]

	River			Lake/Pond		
	Location	Concentration	Ref.	Location	Concentration	Ref.
Indium					5.3 – 19.0	[3]
Iridium					0.5 – 48	[3]
Iron		16.9 – 18.4	[11]	–	14,700 – 30,600	[3]
		31,000	[1]	Lake Ontario, Canada	30,000	[4]
		Total: 6960 – 15700	[2]			
		Acid-extractable: 1600 – 79800	[2]			
Lanthanum				–	28 – 73	[3]
Lead	–	0.11 – 0.13	[11]	–	20 – 180	[4]
	River Arno, Italy	60.7 – 170	[8]	Lake Ontario, Canada	100	[9]
		17 – 59	[7]			
		Total: 51 – 5060	[2]			
		Acid-extractable: 5 – 5160	[2]			
Lithium				Lake Ontario, Canada	50	[4]
Lutecium				–	0.52 – 1.20	[3]
				–	5900 – 16,800	[3]
Magnesium				Lake Ontario, Canada	16,000	[4]
					214 – 4500	[3]
Manganese	–	0.34	[11]	Lake Ontario, Canada	4500	[4]
	River Arno, Italy	553 – 704	[13]			
		582	[1]			
		5 – 3225				
		Total: 113 – 9,640	[12]			
		Acid-extractable: 37 – 9,600	[2]			

	River			Lake/Pond		
	Location	Concentration	Ref.	Location	Concentration	Ref.
Mercury	River Arno, Italy	0.91 – 4.4	[8]	Lake Erie, Canada	1.95 – 6.79	[16]
	River Loire, France	13.2 – 46.8	[14]			
	River inorganic	6.5 – 9.0	[15]			
		Total: 12 – 21.0				
Neodynium					15 – 137	[3]
					1 – 218	[3]
Nickel	River Arno, Italy	60.0 – 79.0	[8]	Lake Ontario, Canada	200	[4]
		72	[1]			
		Total: 7 – 238	[2]			
		Acid-extractable: 1.4 – 67.6	[2]			
Osmium					1 – 4.5	[3]
Phosphorus	–	Total: 675 – 1870	[9]		0.3 – 8.1	[3]
Platinum					5600 – 22,900	[3]
Potassium					19 – 49	[3]
Rubidium					45 – 500	[3]
Ruthenium					7.9 – 28.0	[3]
Samarium					3.3 – 9.2	[3]
Scandium					0.03 – 1.0	[3]
Selenium		0.09 – 0.93	[6]		0.1 – 1.0	[3]
Silver		1 – 5.53	[14]	Lake Moira, Canada	1.0 – 8.05	[17]
	River Arno, Italy	9.3	[8]			
Sodium					3000 – 9200	[3]

	River			Lake/Pond		
	Location	Concentration	Ref.	Location	Concentration	Ref.
Strontium					10 – 242	[3]
Tantalum					0.4 – 1.4	[3]
Terbium					0.95 – 2.4	[3]
Thorium					4.0 – 9.4	[3]
Titanium					800 – 3800	[3]
Uranium					0.78 – 4.3	[3]
Vanadium					28 – 68	[3]
Ytterbium					2.34 – 9.34	[3]
Zirconium					54 – 488	[3]

Radioactive Elements

[210]Lead	River sediment	[18–20]
[222]Radon	River sediment	[21]
[226]Radium	River sediment	[21, 22]
[228]Thorium	River sediment	[23]
[230]Thorium	River sediment	[23]
[232]Thorium	River sediment	[23]
[137]Caesium	River sediment	[18, 24]
[90]Strontium	River sediment	[24]
[237]Neptunium	River sediment	[25]
[238]Plutonium	River sediment	[25, 26]
[235]Uranium	Marine sediment	[27, 28]
[238]Uranium	Marine sediment	[13, 15]
[137]Caesium	Marine sediment	[29]
[144]Cerium	Marine sediment	[29]
[231]Palladium	Marine sediment	[24, 29]
[106]Ruthenium	Marine sediment	[29]
[204]Thallium	Marine sediment	[30]
[241]Americium	Marine sediment	[29]
[239]Plutonium	Marine sediment	[15, 24]
[240]Plutonium	Marine sediment	[15, 24]
[227]Actinium	Marine sediment	[13, 15]
[237]Actinium	Marine sediment	[15]
[210]Lead	Marine sediment	[22]

^{226}Radium	Marine sediment	[22, 27]
^{40}Potassium	Marine sediment	[27]
^{228}Thorium	Marine sediment	[13, 15, 27]
^{230}Thorium	Marine sediment	[13, 15, 27]
^{234}Thorium	Marine sediment	[13, 15]
^{232}Thorium	Marine sediment	[13, 15]
^{234}Uranium	Marine sediment	[27, 28]

References

1. Agemian H, Chau ASY (1977) *Arch Environ Contam Toxicol* **6**:69.
2. Malo BA (1977) *Environ Sci Technol* **11**:277.
3. Nadkarni RA, Morrison GH (1978) *Anal Chim Acta* **99**:133.
4. Agemian H, Chau ASY (1976) *Analyst* **101**:761.
5. Sandhu SS (1981) *Analyst* **106**:311.
6. Goulden PD, Anthony DHJ, Austen KD (1981) *Anal Chem* **53**:2027.
7. Sakata M, Shimoda O (1982) *Water Res* **16**:231.
8. Breder R (1982) *Fresen Z Anal Chem* **313**:395.
9. Aspila KI, Agemian H, Chau ASY (1976) *Analyst* **101**:187.
10. Pankow JF, Leta DP, Lin JW, Ohl SE, Shum WP, Janauer GE (1977) *Sci Total Environ* **7**:17.
11. Neilson Z (1977) *Vatten* **1**:14.
12. Lichtfusse R, Brümmer G (1978) *Chem Geol* **21**:51.
13. Anderson RF, Fleer AP (1982) *Anal Chem* **54**:1142.
14. Frenet-Robin M, Ottman F (1978) *Estuar Mar Coast Sci* **7**:425.
15. Anderson RF, Fleer AP (1982) *Anal Chem* **54**:1142.
16. Pillay KKS, Thomas CC Jr, Sondel JA, Hyche CM (1971) *Anal Chem* **43**:1419.
17. Lum KR, Edgar DG (1983) *Analyst* **108**:918.
18. Anderson RF, Schiff SL, Hesslein RH (1987) *Can J Fish Aqua Sci* **44**:231.
19. Binford MW, Brenner M (1986) *Limnol Oceanogr* **31**:584.
20. Appleby PG, Nolan PJ, Oldfield F, Richardson N, Higgit SR (1988) *Sci Total Environ* **69**:157.
21. HMSO (1986) *Methods for the Examination of Waters and Associated Materials. Measurement of Alpha and Beta Activity in Water and Sludge Samples. Determination of Radon-222 and Radon-226. The Determination of Uranium (including General X-Ray Fluorescence Spectrometric Analysis), 1985–1986*, HMSO, London, UK.
22. Harvey BR, Young AK (1988) *Sci Total Environ* **69**:13.
23. Joshi SR (1985) *Anal Chem* **57**:1023.
24. Grummitt WE, Lahaie G (1973) *Method for the Determination of Strontium-90 and Caesium-137 in River Sediments and Soils*, Report AECL 4365, Atomic Energy of Canada Ltd., Mississauga, Ontario, Canada.
25. Popplewell DS, Ham GJ (1987) *J Radioanal Nucl Chem* **115**:191.
26. Linsalata P, wren ME, Cohen N, Singh NP (1980) *Environ Sci Technol* **14**:1519.
27. Bojanowski B, Fukai R, Holm E (1987) *J Radioanal Nucl Chem* **110**:113.
28. Suttle AD Jr, O'Brien BC, Mueller DW (1969) *Anal Chem* **41**:1265.
29. Aston SR, Stanners DA (1982) *Estuar Coast Shelf Sci* **14**:687.
30. Matthews AD, Riley JP (1969) *Anal Chim Acta* **48**:25.

Appendix 4.5
Types of Organic Compounds Found in Sediments

Class	References		
	River	Lake	Marine
Aliphatic hydrocarbons	[39–44, 48]	–	[47–56]
Aromatic hydrocarbons	[46–48]	–	[57–60]
Polyaromatic hydrocarbons	[61–66]	[61]	[67–70]
Phenols	[71]	–	–
Fatty acids	[72]	–	–
Phthalate esters	[71, 74]	–	–
Carbohydrates	[38, 75]	–	–
Volatile chloroaliphatics	[14, 76–80]	–	–
Nonvolatile chloroaliphatics	[1, 2, 81–85]	–	[3]
Chlorophenols	[6–13]	–	[4]
Hexachlorobenzene		–	–
Chlorinated insecticides		–	[13]
Polychlorobiphenyls		–	[14, 15]
Nitrogen bases	–	–	[16]
Nitrogen-containing aromatics	–	–	[17]
Alkyl and aryl phosphates	–	[18]	–
Adenosine phosphates	[19]	–	–
Organophosphorus insecticides	[20–22]	–	–
Dioxins	[23]	–	–
Humic and fulvic acids	[24]	–	[25–28]
Herbicides	[29–31]	–	–
Inositol esters	–	[32]	–
Detergents	[33–37]	–	–
Priority pollutants (EPA)	–	–	[37]

References

1. Lee HB, Stokker YD, Chau ASY (1987) *J AOAC* **70**:1003.
2. Onuska FI, Terry KA (1985) *Anal Chem* **57**:801.
3. Beller HR, Simoneit BRT (1988) *Bull Environ Contam Toxicol* (1988) **41**:645.
4. Xie T-M (1983) *Chemosphere* **12**:1183.
5. Picer N, Picer M, Strohal P (1975) *Bull Environ Contam Toxicol* **14**:565.
6. Goerlitz DF, Law LM (1974) *J AOAC* **57**:176.
7. Kerkhoff MAT, de Vries A, Wegman RCC, Hofstee AWM (1982) *Chemosphere* **11**:165.
8. Kominar RJ, Onuska FL, Terry KA (1985) *J High Res Chromatogr* **8**:585.
9. Brown JF, Bedard DL, Brennan MJ, Carnation JC, Feng H, Wagner RE (1987) *Science* **236**:709.
10. Marin FA, Norooozian E, Otter RR, Van Dijck RCJM, De Jong CJJ, Brinkman T (1988) *J High Res Chromatogr* **11**:197.
11. Alford-Stevens AL, Budde WL, Bellar TA (1985) *Anal Chem* **57**:2452.
12. McMurtrey KD, Wildman NJ, Tai H (1983) *Bull Environ Contam Toxicol* **31**:734.
13. Lee H-B, Chau ASY (1987) *Analyst* **112**:37.
14. Teichman J, Bevenue A, Hylin JW (1978) *J Chromatogr A* **151**:155.
15. Jensen S, Renberg L, Reutergardh L (1977) *Anal Chem* **49**:316.
16. Kido A, Shinohara R, Eto S, Koga M, Hori T (1979) *Japan J Water Pollut Res* **2**:245.
17. Krone CA, Burrows DW, Brown DW, Robisch PA, Friedman AJ, Malins DC (1986) *Environ Sci Tech* **20**:1144.
18. Ishikawa S, Taketomi M, Shinohara R (1985) *Water Res* **19**:119.
19. Tobin SR, Ryan JF, Afghan BK (1978) *Water Res* **12**:783.
20. Rice JR, Dishberger HJ (1968) *J Agric Food Chem* **16**:867.
21. Deusch ME, Westlake WE, Gunther FA (1970) *J Agric Food Chem* **18**:178.
22. Kjolholt J (1985) *J Chromatogr A* **325**:231.
23. Smith LM, Stalling DL, Johnson JL (1984) *Anal Chem* **56**:1830.
24. Klenke T, Oskierski MW, Poll KG, Reichel B (1986) *Gas u Wasserfach (Wasser, Abwasser)* **127**:650.
25. Poutanen E-L, Morris RJ (1985) *Mar Chem* **17**:115.
26. Hayase K, Tsubota H (1985) *Geochim Cosmochim Acta* **49**:159.
27. Raspor B, Nurnberg HW, Valenta P, Branica M (1984) *Mar Chem* **15**:217.
28. Hayase K Tsubota H (1984) *J Chromatogr A* **295**:530.
29. Spengler D, Jumar A (1971) *Arch Pflanzenschutz* **7**:151.
30. Reeves RG, Woodham DW (1974) *J Agric Food Chem* **22**:76.
31. Wauchope RD, Myers RS (1985) *J Environ Qual* **14**:132.
32. Weimer WC, Armstrong DE (1977) *Anal Chim Acta* **94**:35.
33. Ambe Y, Hanya T (1972) *Japan Anal* **21**:252.
34. Longwell N, Maniece O (1955) *Anal Abstr* **2**:2244.
35. Sallee EM, Fairing JD, Hess RW, House R, Maxwell PM, Melpolder FW, Middleton FM, Ross J, Woelfel WC, Weaver PJ (1956) *Anal Chem* **28**:1822.
36. Ambe Y (1973) *Environ Sci Technol* **7**:542.
37. Ozretich RJ, Schroeder WP (1986) *Anal Chem* **58**:2041.
38. Pellenbarg R (1979) *Mar Pollut Bull* **10**:267.
39. Morgan NL (1975) *Bull Environ Contam Toxicol* **14**:309.
40. Meyers PA (1978) *Chemosphere* **7**:385.
41. Chesler SN, Gump BH, Hertz H, May WE, Wise SA (1978) *Anal Chem* **50**:805.
42. May WE, Chesler SN, Cram SP, Gump BH, Hertz HS, Enagonio DP, Dyszel SM (1975) *J Chromatogr Sci* **13**:535.

43. Chesler SN, Gump BH, Hertz HS, May WE, Dyszel SM, Enagonio DP (1976) *National Bureau of Standards (US) Technical Note No. 889*, Washington, DC, USA.
44. Berthou F, Gourmelun Y, Dreano Y, Friocourt MP (1981) *J Chromatogr A* **203**:279.
45. Mason RP (1987) *Mar Pollut Bull* **18**:528.
46. Vowles PD, Mantoura RFC (1987) *Chemosphere* **16**:109.
47. Blumer M, Sass J (1972) *Mar Pollut Bull* **3**:92.
48. Farrington JW, Quinn JG (1973) *Estuar Coast Mar Sci* **1**:71.
49. Walker JD, Colwell RR, Hamming MC, Ford HT (1975) *Environ Pollut* **9**:231.
50. Broman D, Colmsjo A, Ganning B, Naf C, Zebuhr Y, Ostman C (1987) *Mar Pollut Bull* **18**:380.
51. Mark HB Yu T-C, Mattson JS, Kolpack RL (1972) *Environ Sci Tech* **6**:833.
52. Zitko V, Carson WV (1970) *Tech Report Fish Res Board*, No. 217, Ottawa, Canada.
53. Scarratt DJ, Zitko VJ (1973) *J Fish Res Board Ottawa Canada*, **29**:1347.
54. MacLeod WD, Prohaska PG, Gennero DD, Brown DW (1982) *Anal Chem* **54**:386.
55. Hilpert LR, May WE, Wise SA, Chesler SN, Hertz HS (1978) *Anal Chem* **50**:458.
56. Albaiges J, Grimalt J (1987) *Int J Environ Anal Chem* **31**:281.
57. Hennig HF-KO (1979) *Mar Pollut Bull* **10**:234.
58. Hargrave BT, Phillips GA (1975) *Environ Pollut* **8**:193.
59. Takada H, Ishiwatari R (1985) *J Chromatogr A* **346**:281.
60. Krahn MM, Moore LK, Bogar RG, Wigren CA, Chau S-L, Brown DW (1988) *J Chromatogr A* **437**:161.
61. Giger W, Schaffner C (1978) *Anal Chem* **50**:243.
62. Tan YL (1979) *J Chromatogr A* **176**:316.
63. Bjorseth A, Knutzen J, Skei J (1979) *Sci Total Environ* **13**:71.
64. Garrigues P, Ewald M (1987) *Chemosphere* **16**:485.
65. De Leeuw JW, De Leer EWB, Damste JSS, Schuyl PJW (1986) *Anal Chem* **58**:1852.
66. Lee HK, Wright GJ, Swallow WH (1988) *Environ Pollut* **49**:167.
67. Readman JW, Preston MR, Mantoura RFC (1986) *Mar Pollut Bull* **17**:298.
68. Dunn BP, Stich HF (1976) *J Fish Res Board Ottawa, Canada* **33**:2040.
69. Dunn BP (1976) *Environ Sci Tech* **10**:1018.
70. Bates TS, Carpenter R (1979) *Anal Chem* **51**:551.
71. Goldberg MC, Weiner ER (1980) *Anal Chim Acta* **115**:373.
72. Farrington JW, Quinn JG (1971) *Geochim Cosmochim Acta* **35**:735.
73. Schwartz HW, Anzion GJM, Van Vleit HPM, Peerebooms JWC, Brinkman UAT (1979) *Int J Anal Chem* **6**:133.
74. Thuren A (1986) *Bull Environ Contam Toxicol* **36**:33.
75. McQuaker NR, Fung T (1975) *Anal Chem* **47**:1453.
76. Charles MJ, Simmons MS (1987) *Anal Chem* **59**:1217.
77. Murray AJ, Riley JP (1973) *Anal Chim Acta* **65**:261.
78. Murray AJ, Riley JP (1973) *Nature* **242**:37.
79. Novak J, Zluticky J, Kubelka V, Mostecky J (1973) *J Chromatogr A* **76**:45.
80. Amin TA, Narang RS (1985) *Anal Chem* **57**:648.
81. Bierl R (1988) *Fresen Z Anal Chem* **230**:4
82. Wegman H, (1985) *Deutsche Gewässerkund Mitteil* **29**:111.
83. Lee HB (1988) *J AOAC* **71**:803.
84. Wegman RCC, Greve PA (1977) *Sci Total Environ* **7**:235.
85. Schellenberg K, Leuenberger C, Schwartzenbach RP (1984) *Environ Sci Tech* **18**:652.

Appendix 5.1
Ranges of Metal Concentrations of Cations Found in Freshwaters

Element	Type of Water	Concentration (μg/l)	Reference
Aluminium	River	Total 1300 (pH 7.7)	[1]
		3600 (pH 4.9)	
		Labile 15 (pH 7.7)	
		520 (pH 4.9)	
		Total 200 (pH 4.6)	
		94 – 03 (pH 8.5)	
		73 (pH 8.1)	
		Labile 200 (pH 4.6)	
		39 – 42 (pH 8.5)	
		14.0 – 16.4 (pH 8.1)	
	Surface water	20 – 1430	[2]
	North Florida	210 – 260	
Antimony	River	1	[2]
	River Arve, Germany	0.32	[3]
	River Arne, Germany	0.066 – 0.14	[4]
	Lake	0.08 – 0.42	[5]
	Groundwater	0.77	[6, 7]
Arsenic	River	258 – 490	[8]
		210 – 240	[9]
		1.1 – 275	[10]
		2	[11]
	River Arve, Germany	2.8	[3]
	River Arne, Germany	0.42 – 0.69	[3]
	Groundwater	2.3	[6, 7]

Element	Type of Water	Concentration (μg/l)	Reference
Barium	River	10	[3]
		10 – 30	[4]
	River Arne, Germany	23	[4]
	Surface water	100 – 103	[11]
	Groundwater	41	[6,7]
Beryllium	River	0.4	[3]
	Surface water	< 0.01 – 0.31	[11]
		1	[2]
Bismuth	River	0.005	[3]
	Lake water	< 0.00015 (total and dissolved)	[12]
Caesium	Groundwater	0.006	[6,7]
Cadmium	River	0.013 – 0.29	[13]
		0.07 – 0.13	[4]
		0.03	[3]
	River Arne, Germany	5	[4]
	River Arne, Germany	0.88	[2]
	Surface water	4 – 130	[2]
	Groundwater	100 – 2600	[14]
Chromium	River	1	[3]
		16 – 23	[15]
	River Arne, Germany	0.05 – 0.25	[16]
	River Arve, Germany	1.44	[4]
	River Rhine, Germany	10	[4]
	Surface water	180	[2]
	North Florida	0.2 – 0.3	[16]
	Groundwater	1.0	[6,7]
Cobalt	River	0.2	[3]
	River Arne, Germany	0.013 – 0.09	[4]

Element	Type of Water	Concentration (μg/l)	Reference
	River Arve, Germany	0.12	[4]
	River Rhine, Germany	10	[4]
	Lakewater	54	[17]
	Groundwater	0.11	[6,7]
Copper	River	7	[3]
		0.51 – 6.5	[13]
		123 – 178	[15]
		0.48	[18]
	River Arve, Germany	14.8	[4]
	River Rhine, Germany	30	[4]
	River Thames, UK	30 – 200	[19]
	River Arne, Germany	0.53 – 2.35	[4]
	Surface water	14 – 15	[2]
		110	[2]
	Groundwater	3.7	[6,7]
Europium	River Arve, Germany	0.018	[4]
	River Arne, Germany	0.00008 – 0.0011	[4]
Gold	River (western USA and Alaska)	< 0.001 – 0.036	[20]
Iron	River	200 – 2950	[21]
		50 – 3925	[15]
		100	[2]
	River Arne, Germany	1 – 12.4	[3]
	River Arve, Germany	57	[3]
	Surface water (North Florida)	220 – 350	[16]
		150 – 5000	[2]
	Groundwater	0.15	[6,7]

Element	Type of Water	Concentration (µg/l)	Reference
Lead	River	0.9 – 1.0	[13]
		6 – 37	[15]
		2.1 – 34.8	[22]
	River Thames, UK	40 – 60	[19]
		0.13 – 0.15	[23]
	Surface water (North Florida)	17 – 42	[16]
Manganese	River	11 – 1835	[15]
		5.8 – 19.9	[24, 25]
		7	[3]
	River Arne, Germany	0.97 – 8.9	[4]
	River Arve, Germany	7.9	[4]
	River Thames, UK	200 – 1720	[19]
	Surface water	70 – 500	[2]
	Groundwater	3.2	[6, 7]
Mercury	River	0.51 – 1.3	[26, 27]
		0.07	[3]
	River Arne, Germany	0.017	[4]
	River Arve, Germany	0.009 – 0.047	[4]
	River Rhine, Germany	0.5	[4]
Molybdenum	River	1	[3]
	River Arne, Germany	0.74 – 1.12	[4]
	River Arve, Germany	4.08	[4]
Nickel	River	1.5	[3]
	River Thames, UK	20 – 40	[19]
	Surface water	10 – 40	[2]
	North Florida	8 – 10	[16]
Scandium	Groundwater	0.009	[6, 7]

Element	Type of Water	Concentration (μg/l)	Reference
Selenium	River	0.2	[3]
		0.2 – 0.9	[10]
	River Arne, Germany	0.0006 – 0.0023	[4]
	River Arve, Germany	0.031	[4]
		< 0.0002 → 50 (as selenite and selenite)	[28]
	Japan	0.005 – 0.012 (as elemental selenium)	[29]
		0.008 – 0.012 (as SeIV)	[30]
		< 0.002 – 0.016 (as SeIV)	–
		0.036 – 0.052 (as SeVI)	[30]
		0.003 – 0.020 (as SeVI)	–
		0.022 – 0.023 (as total Se)	[30]
		0.016 – 0.023 (as total Se)	–
	Groundwater	0.4	[6,7]
		0.02 – 0.7	[31]
Silver	River	0.3	[3]
		24 – 32	[15]
Titanium	River	3	[3]
	Surface water (North Florida)	24 – 31	[16]
Uranium	River Arve, Germany	1.36	[4]
	River Arne, Germany	0.37 – 0.49	[4]
Vanadium	River	0.9	[3]
		0.1 – 1	[33]
		24 (as V^{4+})	[32]
		23 (as V^{5+})	[32]
		21 (as total V)	[32]

Element	Type of Water	Concentration (µg/l)	Reference
	Surface water	4.5 – 5.2	[11]
	North Florida	3.9 – 24	[16]
	Groundwater	0.63	[6,7]
	Lake	0.1 – 1.5	[33]
Zinc	River	20	[3]
		14 – 202	[15]
	River Arne, Germany	0.86 – 5.13	[4]
	River Arve, Germany	630	[4]
	River Rhine, Germany	250	[4]
	Groundwater	8.9	[6,7]
	Surface water	10 – 250	[2]
	North Florida	2.5 – 48	[16]

Total nonmetallic elements

Element	Type of Water	Concentration (µg/l)	Reference
Bromine	Groundwater	78	[6,7]
Iodine	Groundwater	10	[6,7]
Nitrogen	Lakewater	1060 – 2940	[34]
	Surface water	1500 – 91000	[35]
Phosphorus	River	250 – 800	[36]
		20	[3]
Silicon		3000 – 5800	[36]
Sulfur	Lakewater	20	[37]

Anions

Element	Type of Water	Concentration (µg/l)	Reference
Borate	River	0.12 – 0.25	[38]
	Groundwater	44	[6,7]
Bromide	River	1.5 – 109.8	[39]
		< 500	[40]
	River Arne, Germany	0.7 – 4.7	[4]
	River Arve, Germany	4.7	[4]
	Surface water (North Florida)	40 – 140	[16]
	Groundwater	2000 – 280000	[4]
Fluoride	River	100 – 180	[42]
	Well water	600	[40]
Phosphate	River	160 – 550	[42]

Element	Type of Water	Concentration (μg/l)	Reference
Radioactive elements			
[14]Carbon	River	–	[59,65]
[214]Lead/ [214]Bismuth	River	–	[66]
[32]Phosphorus	River	–	[117,118]
[210]Polonium/ [210]Bismuth	River	–	[67,68]
[222]Radon	River	–	[71–78]
[231]Protoactinium	River	–	[70]
[223]Radium	River	–	[53,71–78]
[224]Radium	River	–	[53,71–78]
[226]Radium	River	–	[68,71,114]
[228]Radium	River	–	[53,71–78]
[40]Potassium	River	–	[69]
Tritium	River	–	[79–89]
[234]Uranium	River	–	[79–89]
[235]Uranium	River	–	[107–113]
[238]Uranium	River	–	[107–113]
[137]Caesium	River	–	[63,90–96]
[113]Cadmium	River	–	[97]
[141]Cerium	River	–	[98]
[144]Cerium	River	–	[98]
[60]Cobalt	River	–	[99]
[127]Iodine	River	–	[100]
[137]Iodine	River	–	[100]
[90]Strontium	River	–	[60,61,101–105]
[89]Strontium	River	–	[60,61]
[99]Technetium	River	–	[62,106]
[90]Yttrium	River	–	[63]
[95]Zirconium/ [95]Niobium	River	–	[64]
[222]Radon	Groundwater	–	[55,56]
[226]Radium	Groundwater	–	[53]
[90]Strontium	Groundwater	–	[52]
[214]Lead/ [214]Bismuth	Rain water	–	[44]
[22]Sodium	Rain water	–	[45,46]
[24]Sodium	Rain water	–	[45,46]
[237]Uranium	Rain water	–	[47]
[125]Antimony	Rain water	–	[48–52]
[140]Barium	Rain water	–	[48–52]
[137]Caesium	Rain water	–	[48–52]

Element	Type of Water	Concentration (μg/l)	Reference
[144]Cerium	Rain water	–	[48–52]
[127]Iodine	Rain water	–	[43]
[129]Iodine	Rain water	–	[43]
[131]Iodine	Rain water	–	[48–52]
[54]Manganese/ [85]Zinc	Rain water	–	[48–52]
[106]Ruthenium	Rain water	–	[48–52]
[89]Strontium	Rain water	–	[48–52]
[90]Strontium	Rain water	–	[48–52]
[238]Plutonium	Rain water	–	[48–52]
[240]Plutonium	Rain water	–	[48–52]
[226]Radium	Lake water	–	[115]
[228]Radium	Lake water	–	[115]
[238]Thorium	Lake water	–	[57, 58, 115]
[230]Thorium	Lake water	–	[57, 58]
[234]Thorium	Lake water	–	[57, 58]
[125]Antimony	Lake water	–	[115]
[7]Beryllium	Lake water	–	[115]
[137]Caesium	Lake water	–	[116]
[141]Cerium	Lake water	–	[115]
[144]Cerium	Lake water	–	[115]
[95]Niobium	Lake water	–	[115]
[103]Ruthenium	Lake water	–	[115]
[106]Ruthenium	Lake water	–	[115]
[95]Zirconium	Lake water	–	[115]

References

1. Zoltzer D, Schwedt G (1984) *Fresen Z Anal Chem* **317**:422.
2. Janssens E, Schutyser P, Dams R (1982) *Environ Tech Lett* **3**:35.
3. Thompson M, Ramsey MH, Pahlavanpour B (1982) *Analyst* **107**:1330.
4. Bart G, Von Gunter HR (1979) *Int J Eviron Anal Chem* **6**:25.
5. Abu-Hilal AH, Riley JP (1981) *Anal Chim Acta* **131**:175.
6. Kasuka Y, Tsuji H, Fujimoto Y, Ishida K, Fukai Y, Mamuro T, Matsumani I, Mizuhata A, Hurai S (1982) *J Radioanal Chem* **71**:7.
7. Kasuka JY, Tsuji H, Fujimoto Y, Ishida K, Mamuro I, Matsumani I, Mizohata A, Hirai S (1980) *Bull Inst Chem Res Kyoto Univ* **58**:171.
8. Chakraborti D, Irgolic KJ (1984) *Int J Eviron Anal Chem* **17**:241.
9. Subramanian KS, Leung PC, Meranger JC (1982) *Int J Eviron Anal Chem* **11**:121.
10. Thompson M, Pahlavanpour B, Walton SJ, Kirkbright GF (1978) *Analyst* **103**:568.
11. Lagas P (1978) *Anal Chim Acta* **98**:261.
12. Pillay KKS, Thomas CC Jr, Sondel JA, Hyche CM (1971) *Anal Chem* **43**:4119.
13. Poldoski JE, Glass GE (1978) *Anal Chim Acta* **101**:79.
14. Hasan MZ, Kumar A (1983) *Ind J Environ Health* **25**:161.
15. West MH, Molina JF, Yuan CL, Davis DG, Chauvin JV (1979) *Anal Chem* **51**:2370.
16. Tanaka S, Darzi M, Winchester JW (1981) *Environ Sci Tech* **15**:354.
17. McMahon JW, Docherty KE, Judd JM (1985) *Hydrobiol* **126**:103.
18. Yoshimura K, Nigo S, Tarutani T (1982) *Talanta* **29**:173.
19. Edward-Inatimi EB (1983) *J Chromatogr* **256**:253.
20. McHugh JB (1984) *J Geochem Explor* **20**:303.
21. Gine MF, Bergamin FH, Zagatto EAG, Reis BF (1980) *Anal Chim Acta* **114**:191.
22. Beneš P, Koč J, Štulik K (1979) *Water Res* **13**:967.
23. Ball JW, Thompson JM, Jenne EA (1978) *Anal Chim Acta* **98**:67.
24. Hadjiioannou TP, Hadjiioannou SI, Avery J, Malmstadt HV (1977) *Anal Chim Acta* **89**:231.
25. Hadjiioannou TP, Hadjiioannou SI, Avery J, Malmstadt HV (1976) *Clin Chem* **22**:802.
26. Mercury Analysis Working Party of the Bureau International Technique du Chlore (1979) *Anal Chim Acta* **109**:209.
27. Kopp JF, Longbottom HC, Lohring LB (1972) *J Am Water Works Assoc* **64**:20.
28. Cheam V, Agemian H (1980) *Anal Chim Acta* **113**:237.
29. Uchida H, Shimoishi Y, Toei K (1980) *Environ Sci Tech* **14**:541.
30. Shimoishi Y, Toei K (1978) *Anal Chim Acta* **100**:65.
31. Roden DR, Tallman DE (1982) *Anal Chem* **54**:307.
32. Orvini E, Lodola L, Sabbioni E, Pietra R, Goetz L (1979) *Sci Total Environ* **13**:195.
33. Meinecke G (1973) *Zur Geochemie des Vanadiums, Clausthaler Hefte zur Lagerstätten-Kunde und Geochemie der mineralischen Rohstoffe 2*, Springer, Heidelberg.
34. Simeonov V, Andrew G, Stoianov A (1979) *Fresen Z Anal Chem* **297**:418.
35. Johansen HS, Middelboe V (1976) *Int J Appl Rad Isot* **27**:591.
36. Urasa IT (1984) *Anal Chem* **56**:904.
37. Landers DH, David MB, Mitchell MJ (1983) *Int J Environ Anal Chem* **14**:245.
38. Aznarez J, Bonilla A, Vidal JC (1983) *Analyst* **108**:368.
39. Morrow CM, Minear RA (1984) *Water Res* **18**:1165.
40. Mosko JA (1984) *Anal Chem* **56**:269.
41. Maaschelein WJ, Denis M (1981) *Water Res* **15**:857.

42. Zelensky I, Zelenska V, Kaniansky D, Havassi P, Lednarova V (1984) *J Chromatogr* **294**:317.
43. Muramatsu Y, Ohmomo Y, Christoffers D (1984) *J Radioanal Nucl Chem* **83**:353.
44. Michaelis ML (1969) *Chemiker Zeitung Chem Apparat* **93**:883.
45. Yasulenis RY, Luyanas VY, Kekite VP (1972) *Soviet Biochem* **14**:673.
46. Burden BA (1968) *Analyst* **93**:715.
47. Suzuki T, Sotobayashi I, Koyame S, Kanda Y (1968) *J Chem Soc Japan Pure Chem* **89**:1084.
48. Kimura T, Hamada T (1980) *Anal Chim Acta* **120**:419.
49. Eichholz GG, Galli AN (1970) *Radiochem Radioanal Lett* **4**:315.
50. Perkins RW (1969) *Procedure for the Continuous Separation and Subsequent Direct Counting of Short-Lived Cosmic Ray Produced Radio Nuclides in Rain Water*, Report BNWL-1051, Pt 2, US Atomic Energy Commission, Washington, DC, USA.
51. Cambray OH, Fisher EM, Salmon L (1970) *Methods of Collection and Analysis of Radioactivity from Distant Nuclear Test Explosions*, Report AERE-R5898, Atomic Energy Authority, HMSO, London, UK.
52. Ardisson G (1982) *Trends Anal Chem* **1**:281.
53. Higuchi H, Uesugi M, Satoh K, Ohaski N, Noguchi M (1984) *Anal Chem* **56**:761.
54. Yar A (1972) *Gidrokhim Mater* **53**:163.
55. Kobal I, Kristan J (1972) *Radiochem Radioanal Lett* **10**:291.
56. Kobal I, Kristan J (1973) *Mikrochim Acta* **61**:219.
57. De Jong IG, Wiles DR (1984) *Water Air Soil Pollut* **23**:197.
58. De Jong IG, Wiles DR (1984) *J Radioanal Chem* **82**:120.
59. Fedorak PM, Foght JM, Westlake DWS (1982) *Water Res* **16**:1285.
60. Gregory LP (1972) *Anal Chem* **44**:2113.
61. Lapid J, Munster M, Farhi S, Eini M, La Louche L (1984) *J Radioanal Nucl Lett* **86**:231.
62. Robb P, Warwick P, Malcombe-Lawes DJ (1985) *J Radioanal Nucl Chem* **89**:323.
63. Mundschenk H (1974) *Deutsche Gewässerkund Mitteil* **18**:72.
64. Linsalata P, Cohen N (1982) *Health Physics* **43**:742.
65. Foyn YE, Hanneborg S (1971) *Mar Biol* **8**:57.
66. Hashimoto T, Satoh K, Aoyagi M (1985) *J Radioanal Nucl Chem* **92**:407.
67. MacKenzie AB, Scott RD (1979) *Analyst* **104**:1151.
68. Furnica G, Toader M (1969) *Igiena (Bucharest)* **18**:227.
69. Gertner A, Grdinic V (1969) *Mikrochim Acta* **57**:25.
70. Sill CW (1978) *Anal Chem* **50**:1559.
71. Kelkar DN, Joshi PV (1969) *Health Phys* **17**:253.
72. Darrall KG, Richardson PJ, Tyler JFC (1973) *Analyst* **98**:610.
73. Benes P, Sedlacek J, Sebesta F, Sandrik R, John J (1981) *Water Res* **15**:1299.
74. Johnson JO (1971) *US Geological Survey Water Supply Paper* 1696G.
75. Michel J, Moor WS, King PT (1981) *Anal Chem* **53**:1885.
76. Zeigelheim CJ, Busigin A, Phillips CR (1982) *Health Phys* **42**:317.
77. Pritchard HM, Gesell TF (1977) *Health Phys* **33**:577.
78. Perkins RW (1969), *Radium and Radiobarium Measurement in Seawater and Freshwater by Sorption and Direct Multidimensional X-Ray Spectrometry*, Report BNWL 1051, Atomic Energy Commission, Washington, DC, USA.
79. Noakes JE, Neary MP, Spaulding JD (1973) *Nucl Instrum Meth* **109**:177.
80. Files P (1973) *Radiochem Radioanal Lett* **15**:213.
81. Povinec P, Chudy M, Seliga M, Sarao S, Szarka J (1972) *J Radioanal Chem* **12**:513.
82. Tistchenko S, Dirian G (1970) *Bull Soc Chim Fr* **1**:16.

83. Knowles FE, Baratta EJ (1971) *Radiol Health Data* **12**:405.
84. Theodorsson P (1974) *Int J Appl Rad Isot* **25**:97.
85. Florkowski T, Payne BR, Sauzay G (1970) *Int J Appl Rad Isot* **21**:453.
86. Froehlich K, Herbert D, Andreev A (1972) *Isotopenpraxis* **8**:130.
87. Wolf M (1973) *Int J Appl Rad Isot* **24**:299.
88. Vinogradov AP, Devirts AL, Dobkina EI (1968) *Geokhimiya* **10**:1147.
89. Schell WR, Nevissi A, Huntamer D (1978) *Mar Chem* **6**:143.
90. Rebak W, Ubl G (1970) *Report Staatliche Zentrale für Strahlenschutz s25-1/70.*
91. Stewart ML, Pendeton RC, Lords JL (1972) *Int J Appl Rad Isot* **23**:345.
92. Senegacnik M, Paljk S (1969) *Z Anal Chem* **244**:306.
93. Senegacnik M, Paljk S (1969) *Z Anal Chem* **244**:375.
94. Haberer K, Stürzer U (1970) *Gas u Wasserfach (Wasser, Abwasser)* **111**:29.
95. Tereda K, Hayakawa H, Sawada K, Kiba T (1970) *Talanta* **17**:955.
96. Kapustin VK, Egorov AI, Eeonov VV (1981) *Soviet J Water Chem Technol* **3**:119.
97. Palogyi S, Larsen RP, Tisue GI (1985) *J Radioanal Nucl Lett* **96**:161.
98. Haberer K, Stürzer U (1972) *Gas u Wasserfach (Wasser, Abwasser)* **113**:122.
99. Claassen HC (1970) *Anal Chim Acta* **52**:229.
100. Haberer K, Stürzer U (1968) *Gas u Wasserfach (Wasser, Abwasser)* **109**:1287.
101. Gusev NG, Ya U, Margulis AN, Marei A, et al. (1966) *Dosimetric and Radiometric Methods*, Atomizdat, Moscow, Russia, p. 44 (in Russian).
102. Baratta EJ, Knowles FR (1986) *J AOAC* **69**:540.
103. Ankudina MM, Mordberg EL, Nekhorossheva MP, et al. (1970) *The Isolation of Strontium-90, Collection: Collection of Radiometric and Gamma Spectrometric Methods of Analysing Materials in the Environment*, Leningrad, p. 26 (in Russian).
104. Haberer K, Stürzer U (1971) *Gas u Wasserfach (Wasser, Abwasser)* **121**:186.
105. Tesetmale G, Leredde JL (1970) *Report CEA R-3908*, Centre of Nuclear Studies, Fontenay-aux-Roses, France.
106. Golchert NW, Sedlet J (1969) *Anal Chem* **41**:669.
107. Szy D, Sebessy L, Balint G (1971) *J Radioanal Chem* **7**:57.
108. Gorbushina LV, Zhil'tsova LY, Mtveeva EN, Surganova NP, Tenyaev VG, Tyminskii VG (1972) *J Radioanal Chem* **10**:165.
109. Takebayashi T, Matsuda H, Umemeoto S (1973) *Talanta* **20**:892.
110. Bertine KK, Chau LH, Turekian KK (1970) *Geochim Cosmochim Acta* **34**:641.
111. Doerschel E, Stolz W (1970) *Radiochem Radioanal Lett* **4**:277.
112. Fleischer RL, Lovett DB (1968) *Geochim Cosmochim Acta* **32**:1126.
113. Gladney ES, Peters RJ, Perrin DR (1983) *Anal Chem* **55**:967.
114. Elsinger RJ, King PT, Moore WS (1982) *Anal Chim Acta* **144**:277.
115. Durham RW, Joshi SR (1981) *Water Res* **15**:83.
116. Hashimoto T (1971) *Anal Chim Acta* **56**:347.
117. Milham RC (1973) Report DP US-73-35, US Atomic Energy Commission, Washington, DC, USA.
118. Furnica G, Ionescu H (1969) *Igiena (Bucharest)* **18**:105.

Appendix 5.2
Range of Cation Concentrations
Found in Open Sea Water and Estuary Waters

(a) Open Sea Waters

Element	Location	Concentration (μg/l)	Consensus value (μg/l)	Ref.
Aluminium	Open seawater surface	0.1		[1]
	Open seawater 3 km depth	0.6		[1]
Bismuth	Pacific, surface	< 0.00005		[2]
	Pacific, 2500 m depth	< 0.000003		[3]
Cadmium	Open ocean, salinity 35%	0.03		[4]
	Arctic Sea	0.010 – 0.045		[5]
	Arctic Sea, surface	0.0127		[6]
	Arctic Sea, 2000 m depth	0.023		[6]
	Arctic Sea	0.018		[5]
	Pacific	0.02 – 0.04		[7]
	Kattergat/Skaggerat	0.022		[8]
	Norwegian Sea	0.02 – 0.025 (surface)		[9]
		0.02 – 0.025 (3000 m)		
	Sargasso Sea	0.035 – 0.042 (216 m)		[10]
		0.109 – 0.126 (4926 m)		[10]
	Baltic Sea	0.03 – 0.06		[11]
	Open sea	0.03		[4]
	Open sea	0.079		[12]
	Open sea	0.12 – 0.30		[13]
	Open sea	0.03 – 0.17		[14]

Element	Location	Concentration (µg/l)	Consensus value (µg/l)	Ref.
Chromium	Pacific	Cr(III) 0.005 – 0.52		[15, 16]
		Cr(IV) 0.03 – 0.96		[15, 16]
		Organic Cr 0.07 – 0.32		[15, 16]
		Total Cr 0.06 – 1.26	0.03	[15, 16]
	Mediterranean	Cr(III) 0.02 – 0.05		[17]
		Cr(IV) 0.05 – 0.38		[17]
	Open ocean	Total Cr 0.07 – 0.97	0.03	[18]
		Cr(III) 0.08 – 0.22		[19]
		Cr(IV) 0.13 – 0.68		[19]
		Total Cr 0.18 – 0.19	0.03	[4]
Cobalt	North Sea	0.07 – 0.16	0.005	[20]
	Open ocean, salinity 35%	0.003		[4]
	Open sea	0.04		[21]
	Open sea	0.003		[4]
	Open sea	0.15 – 0.16		[13]
Copper	Pacific	0.3 – 2.8	0.05	[7]
	Open ocean, salinity 35%	0.121		[4]
	Good quality seawater	0.36 – 8.6		[22, 23]
	Sargasso Sea	0.072 – 0.081 (216 m)		[10]
	Sargasso Sea	0.26 – 0.33 (4926 m)		[10]
	Baltic Sea	0.59 – 0.99		[8]
	Baltic Sea	0.6 – 1.0		[10]
	Baltic Sea	0.0063 – 0.0252 (organic)		[24]
		0.6 – 0.751 (total)		[24]
	North Sea	0.0208		[8]
	Norwegian Sea	0.08 – 0.10 (surface)		[9]
	Norwegian Sea	0.08 – 0.10 (3000 m)		[9]
	Danish sound	0.48		[8]
	Arctic Sea	0.097		[24]
	Open sea	0.341		[12]
	Open sea	0.48 – 1.51		[13]
Iron	Pacific	140 – 320	0.2	[7]
	Open ocean, salinity 35%	0.2		[4]

Element	Location	Concentration (µg/l)	Consensus value (µg/l)	Ref.
	Open seawater	2.1		[21]
	Open seawater	3.25		[12]
	Pacific	< 0.01 – 0.7		[7]
Lead	Pacific	0.6 – 0.8		[7]
	Arctic Sea	0.01 – 28		[4]
	Arctic Sea	0.019 – 0.021		[5]
	Open ocean, salinity 35%	0.095		[4]
	Sargasso Sea	0.000041 (surface) 0.0083 – 0.012 (4800 m)		[25]
	West North Atlantic	0.00017 – 0.0003		[26]
	Norwegian Sea	< 0.0002 (3000 m) 0.025 – 0.065 (surface)		[9]
	Open sea	0.095		[4]
	Arctic Sea	0.015		
	Open sea	0.0083		
	Open sea	0.03 – 9.0		
	Open sea	< 0.04 – 0.28		
Manganese	Open ocean, salinity 35%	0.018	0.02	[4]
Mercury	Atlantic, open sea	0.021 – 0.078	< 0.2	[27]
	Open ocean	0.002 – 0.011		[27]
	Off Iceland	0.04		[28]
Molybdenum	Pacific	11.2 – 12.0		[7]
	Noncentral Pacific	3.2		[29]
	Seawater, Japan	11.5		[30]
	Open sea	5.3		[21]
Nickel	Pacific (4000 m)	0.45 – 0.84	0.17	[31]
	Pacific	0.15 – 0.93		[7]
	Pacific (surface)	0.16 – 0.29		[31]
	Open ocean, salinity 35%	0.341 – 0.608		[12]
	Open ocean	0.38 – 0.46		[32]
	Open ocean	0.27		[4]

Element	Location	Concentration (µg/l)	Consensus value (µg/l)	Ref.
	Norwegian Sea	0.175 – 0.20 (surface)		[9]
		0.175 – 0.20 (3000 m)		[9]
	Sargasso Sea	0.26 – 0.27 (216 m)		[10]
		0.45 – 0.47 (4926 m)		[10]
	Baltic Sea	0.6 – 0.9		[11]
	Arctic Sea	0.099		[24]
	Open sea	0.545		[12]
	Open sea	0.76 – 1.58		[13]
Rare earths	North Atlantic below mixed layer	La 13.0×10^{-12} mole/kg Ce 16.8 mol/kg Nd 12.8 mol/kg Sm 2.67 mol/kg Eu 0.644 mol/kg Gd 3.4 mol/kg Dy 4.78 mol/kg Er 4.07 mol/kg Yb 3.55 mol/kg		[33]
Rhenium	Atlantic	6 – 8		[34]
Selenium	Seawater	0.021 – 0.029		[35]
	Open ocean	0.00095		[21]
Silver	Open sea	0.08		[13]
Thorium	Open sea	< 0.0002		[21]
Tin	Open sea	Sn(IV) 0.02		[36]
		Sn(II) 0.05		[36]
Uranium	Seawater	1.9		[37]
		2.6		[32]
Vanadium	Pacific	1.73 – 2.00	2.5	[38]
		1.29 – 1.87		[7]
	Adriatic	1.64 – 1.73		[38]
	Open sea	0.45		[21]
Zinc	Pacific	1.9 – 3.0	0.49	[7]
	Arctic sea	0.125 – 0.16		[5]
		0.05 – 0.34		[5]

Element	Location	Concentration (μg/l)	Consensus value (μg/l)	Ref.
	Open sea, salinity 35%	0.28		[4]
		4.9		[21]
	Norwegian Sea	0.08 – 0.30 (surface)		[9]
		0.10 – 0.18 (3000 m)		[9]
	Open sea	0.074		[12]
		0.3 – 10.9		[14]
		2.6 – 10.1		[13]

Radioactive Metals

Element	Location	Reference
^{210}Polonium	Seawater	[113]
^{210}Polonium/^{210}Lead	Seawater	[5–52]
^{222}Radon	Seawater	[67,74,77,79–81,114]
^{226}Radium	Seawater	[67–78,114]
^{228}Radium	Seawater	[67–78,114]
^{40}Potassium	Seawater	[50]
^{228}Thorium	Seawater	[59,62–66]
^{230}Thorium	Seawater	[59,62–66]
^{234}Thorium	Seawater	[59,62–66]
^{234}Uranium	Seawater	[39–50]
^{235}Uranium	Seawater	[39–50,113]
^{237}Uranium	Seawater	[39–50,113]
^{238}Uranium	Seawater	[39–50,113]
^{137}Caesium	Seawater	[100–107]
^{60}Cobalt	Seawater	[109,110]
^{55}Iron	Seawater	[98]
^{54}Manganese/^{85}Zinc	Seawater	[111,112]
^{106}Ruthenium	Seawater	[108]
^{90}Strontium	Seawater	[102]
^{236}Plutonium, ^{238}Plutonium, ^{239}Plutonium, ^{240}Plutonium and ^{242}Plutonium	Seawater	[82–100]

(b) Coastal, Bay and Estuary Waters

Element	Location	Concentration (µg/l)	Ref.
Aluminium	Seto Upland Sea, Japan and Pacific Ocean	6.4 – 63	[115]
Antimony	North Sea	0.3 – 0.82	[116]
Arsenic	North Sea (soluble metals)	1.0	[117]
	North Sea (coastal water)	1.04	[118]
Barium	Kwangyana Bay, Korea	4.8	[119]
Bismuth	Seawater	0.02 – 0.11	[120]
	Kattegat	0.0015 – 0.003	[121]
	San Diego Bay	0.000 05 – 0.000 06 (dissolved)	[3]
		0.00013 – 0.002 (total)	
	North Sea	0.2 – 0.68	[116]
Cadmium	Near shore seawater (salinity 29‰)	0.02 – 0.0025	[4]
	Sandy Cove, USA	0.04 – 0.05	[122]
		0.24 – 0.28	[123]
	Bermuda	0.029	[122]
	Kwangyana Bay, Korea	0.20	[119]
	Coastal seawater	0.05 – 0.2	[124]
		0.020 – 0.28	[124]
	Seawater	0.056 – 0.08	[12]
		0.053	[125]
		0.053 – 0.07	[32]
		0.2	[126]
	North Sea (soluble metals)	0.02	[117]
	Straits of Gibraltar	< 2.8	[127]
	Heligoland Bight	0.02 – 0.07	[128, 129]
	Sea off California	0.015 – 0.016 (surface)	[130]
		0.94 – 0.099 (2950 metres)	[130]
	German Bight	0.024 – 0.768	[131]
	Danish coastal water	0.06 – 0.80	[132]

Element	Location	Concentration (µg/l)				Ref.
	Clyde coastal water	0.11 – 0.25				[14]
	Chesapeake Bay	0.05				[133]
	Canadian coastal water	0.035 – 0.048				[134]
	Danish coastal water	0.2 – 5.0				[132]
	Mediterranean	< 5.4				[127]
	Southampton Water	< 0.1 – 0.35				[13]
	Cape San Blay	0.013 (5 m)				[135]
		0.0045 (70 m)				[135]
	North Sea	0.2 – 0.4				[121]
	Near shore water	0.02 – 0.025				[4]
	Coastal water	0.3 – 1.0				[136]
	Estuary water (salinity 10‰)	0.5				[137]
	Estuary water (salinity 24.1‰)	2.1				[137]
	Coastal water	0.1				[127]
	Coastal water	0.05 – 0.07				[138]
	Gota River Estuary (salinity 0.5‰)	0.02				[139]
	Gota River Estuary (salinity 32‰)	0.02				[139]
Cerium	Kwangyana Bay, Korea	16.7				[119]
Chromium		Cr(III)	Cr(VI)	Organic Cr	Total Cr	
	North Sea soluble metals				0.4	[117]
	Sea of Japan	0.57 – 0.093	0.088 – 0.15	0.18 – 0.32	0.37 – 0.50	[140]
	Japan Coast	0.04 – 0.06				[141]
	Sandy Cove, USA				0.84	[123]
	Kwangyana Bay, Korea				2.33	[119]
	Port Hacking, Australia	0.27	0.49	0.56		[142]
	Drummoyne Bay, Australia	0.32	0.95	0.69	1.96	[142]

Element	Location	Concentration (μg/l)				Ref.
		Cr(III)	Cr(VI)	Organic Cr	Total Cr	
	Botany Bay, Australia	0.45	1.26	0.71	2.41	[142]
	Coastal seawater UK			0.095 – 0.100		[18]
		0.46	0.60			[144]
					3.3	[37]
	Kwangyana Bay, Korea				2.3	[140]
	Canadian coastal water				0.15 – 0.5	[134]
	Coastal water				0.25	[127]
	Coastal water				0.25 – 0.29	[138]
	Estuary water (salinity 10‰)				0.9	[137]
	Estuary water (salinity 24.1‰)				0.5	[137]
Cobalt	Sandy Cove USA	0.02				[123]
	Coastal seawater	0.018 – 0.02				[124]
		0.044				[37]
	(salinity 29.5‰)	0.017 – 0.018				[38]
		0.015 – 0.028				[32]
	Shitukawa Bay, Japan	0.07 – 0.16				[29]
	North Central Pacific	0.24				[29]
	Botany Bay, Australia	0.25				[29]
	North-west coast, USA	0.13				[29]
	Port Hacking, Australia	0.25				[29]
	Cronhulla Beach, Australia	0.21				[133]
	Chesapeake Bay	< 0.1				[133]
	Southampton Water	< 0.1 – 0.16				[13]
	Menai Straits	0.07				[29]
	Shore seawater	0.017 – 0.018				[4]
	Canadian coastal water	0.01				[134]
	Coastal Water	< 0.1				[127]
	Coastal Water	0.015 – 0.028				[138]

Element	Location	Concentration (µg/l)	Ref.
Copper	North Sea, soluble metals	0.2	[117]
	Kwangyana Bay, Korea	1.1	[119]
	Sandy Cove, USA	0.6 – 0.7	[123]
	Chirihaua, Japan	20	[144]
	Gironde Estuary	3.7	[145]
	Estuary water	2.0 – 2.01	[4]
	Near shore water (salinity 29.5‰)	0.17 – 1.03	[4]
	Coastal seawater	0.6 – 0.7	[124]
	Sunlace Water, North Pacific	0.64	[146]
	Seawater	0.66 – 0.72	[165]
		0.50 – 0.73	[32]
		0.16 – 0.34	[31]
		0.2	[146]
	Poor-quality seawater	6.8 – 15.8	[17]
	Chesapeake Bay	2.0	[133]
	Osaka Bay, Japan	0.89 – 2.66	[148]
	Delaware Bay	0.83 – 2.18 (surface)	[65]
		0.73 – 0.91 (16 m)	[65]
	North Sea	2.82 – 9.7	[116]
	Canadian Coast	1.1 – 1.2	[50]
	Cape San Blas	0.123 (5 m)	[51]
		0.065 (70 m)	[51]
	Southampton Water	0.48 – 2.6	[13]
	Heligoland Bight	0.3 – 2.04	[128, 129]
	Sandy Cove, USA	0.6 – 0.7	[130]
	Sea off California	0.069 – 0.105 (surface)	[130]
		0.098 – 0.24 (2950 m)	
	Kwangyana Bay, Korea	1.1	[119]
	Coastal water	0.5 – 0.73	[138]
	Coastal water	1.0	[127]
	Coastal water	0.6 – 3.4	[136]
	Gota River Estuary (salinity 0.5‰)	1.2	[139]
	Gota River Estuary (salinity 32‰)	0.3	[139]

Element	Location	Concentration (µg/l)	Ref.
Iron	Sandy Cove, USA	1.4 – 1.5	[123]
	Kwangyana Bay, Korea	250	[119]
	Coastal seawater	1.4 – 1.6	[124]
	(salinity 29.5‰)	1.0 – 7.2	[4]
	Estuary water	2.1	[37]
	Seawater	3.2 – 3.4	[32]
	Delaware Bay	2.46 – 35.1 (surface)	[149]
		2.1 – 5.2 (10 m)	[149]
	Heligoland Bight	1.13	[129]
	Osaka Bay, Japan	15.4 – 65.5	[148]
	Canadian coastal water	3.5 – 4.2	[134]
	Chesapeake Bay	2.1	[133]
	Coastal water	5.0	[127]
	Coastal water	3.2 – 3.7	[138]
	Gota River Estuary (salinity 0.5‰)	170	[139]
	Gota River Estuary (salinity 32‰)	16	[139]
Lanthanium	Kwangyana Bay, Korea	0.72	[119]
Lead	Guanalana Bay	0.07 – 0.55	[150]
	Sandy Cove, USA	0.22 – 0.35	[123]
	Near shore seawater (salinity 29.5‰)	0.14 – 0.22	[4]
	Coastal seawater	0.22 – 0.35	[124]
	Seawater	0.2 – 0.3	[147]
		7.1	[126, 165]
		0.038 – 0.29	[12]
		0.06 – 0.11	[32]
		0.51 – 0.65	[151, 152]
	North Sea, soluble metals	0.05	[117]
	Canadian coastal water	0.34 – 0.36	[134]
	North Sea	1.8 – 7.44	[116]
	Chesapeake Bay	0.3	[133]
	Danish Coastal water	0.8 – 80	[132]
	Clyde water	0.02 – 0.36	[14]

Element	Location	Concentration (μg/l)	Ref.
	Heligoland Bight	0.07	[129]
	Southampton Water	< 0.1 – 0.6	[13]
	Danish coastal water	4.5 – 200	[13]
	Coastal water	0.5 – 2.4	[136]
	Coastal water	0.06 – 0.11	[138]
	Coastal water	3.1 – 12	[153]
	Coastal water	0.25	[127]
	Near shore water	0.22	[4]
	Gota River Estuary (salinity 0.5‰)	0.30 – 0.36	[139]
	Gota River Estuary (salinity 32‰)	0.06 – 0.07	[139]
Manganese	Sandy Cove, USA	1.4 – 1.8	[123]
	Kwangyana Bay, Korea	1.5	[119]
	Chirihama Bay, Japan	60	[144]
	Near shore seawater (salinity 29.5‰)	0.71 – 1.06	[4]
	Tamar Estuary, UK	20 – 250	[154]
	Coastal seawater	1.4 – 1.6	[31]
	Estuary and seawater	1.89 – 2.0	
	South-west Bermuda	1.4 – 1.8	[155]
	Canadian coastal water	0.78 – 0.95	[134]
	Osaka Bay, Japan	11.1 – 30.6	[148]
	Chesapeake Bay	2.0	[133]
	Heligoland Bight	0.35	[129]
	Near shore seawater	0.7 – 1.06	[4]
	Coastal water	1.4 – 75.1	[28]
	Coastal water	4.0	[127]
	Coastal water	1.9 – 2.5	[138]
Mercury	Seawater	0.000018 – 0.000026	[165]
	Seawater	0.01	[2]
	Coastal samples	0.05	[2]

Element	Location	Concentration (μg/l)	Ref.
	River Loire Estuary (salinity 20 – 30‰)	0.6 – 1.1	[156]
	River Loire 0 – 10 km upstream of estuary (salinity 10 – 20‰)	1.4 – 11.6	[156]
	River Loire 10 – 15 km upstream of estuary (salinity 1 – 10‰)	1.0 – 7.0	[156]
	River Loire 15 – 30 km upstream of estuary (salinity < 1‰)	1 – 15.1	[156]
	North Sea soluble metals	0.002	[117]
Molybdenum	Kagoshaima Bay, Japan	8.16 – 9.7	[157]
	Estuary water	5.3	[37]
	Seawater	2.1 – 18.8	[158]
	Coastal water	7 – 200	[153]
	Coastal water	10.1 – 10.3	[118]
Nickel	Sandy Cove USA	0.33 – 0.40	[123]
	Profile to 1200 m in Santa Catalina	0.3 – 0.6	[159]
	Near shore seawater (salinity 29.5‰)	0.33 – 0.39	[4]
	Coastal seawater	0.33 – 0.4	[123]
	Estuary water	1.2 – 1.3	[37]
	Seawater	0.341 – 0.608	[32]
	North Sea soluble metals	0.25	[117]
	Southampton water	0.50 – 1.58	[13]
	Cronhulla Beach, Australia	2.5	[29]
	Chesapeake Bay	1.2	[133]
	Heligoland Bight	0.2 – 1.2	[128, 138]
	Osaka Bay, Japan	2.41 – 5.33	[148]

Element	Location	Concentration (μg/l)	Ref.
	North-West Coast, USA	1.1	[28]
	Canadian Coastal water	0.37 – 0.43	[134]
	Port Hastings, Australia	2.9	[29]
	Botany Bay, Australia	3.8	[29]
	Sea off California	0.22 – 0.3 (surface)	[134]
		0.60 – 0.67 (2950 m)	[134]
	Menai Straits	1.9	[28]
	Coastal water	0.58	[127]
	Gota River Estuary (salinity 0.5‰)	1.2 – 1.3	[139]
	Gota River Estuary (salinity 32‰)	0.4	[139]
Thorium	Estuary and seawater	≤ 0.0002	[37]
Uranium	Estuary and seawater	1.90	[37]
	Kwangyana Bay, Korea	1.36 – 1.86	[119]
	Coastal water	3.08 – 3.1	[118]
Rare earths	Kwangyana Bay, Korea	Ce 16.7	[119]
		La 0.72	[119]
Scandium	Kwangyana Bay, Korea	0.098	[119]
	Estuary water	0.00095	[37]
Selenium	Seawater	0.4	[160]
Silver	Southampton water	< 0.01 – 0.08	[13]
Vanadium	Kwangyana Bay, Korea	2.14	[119]
	Estuary and seawater	0.45	[37]
	Osaka Bay, Japan	0.23 – 0.88	[148]
	Coastal water	1.22 – 1.23	[118]
	Coastal water	< 0.01 – 5.1	[153]

Element	Location	Concentration (μg/l)	Ref.
Zinc	North Sea soluble metals	1.0	[123]
	Sandy Cove, USA	1.5 – 1.9	[123]
	Kwangyana Bay, Korea	45.9	[119]
	Near shore seawater (salinity 29.5‰)	0.29 – 0.44	[4]
	Estuary and seawater	4.5 – 4.9	[37]
	Cape San Blas	0.055 (5 m)	[135]
		0.030 (70 m)	[135]
	Heligoland Bight	1.3 – 6.6	[128, 129]
	Osaka Bay, Japan	5.3 – 29.1	[149]
	Chesapeake Bay	4.8	[133]
	Danish coastal water	0.5 – 250	[132]
	Clyde water	2.0 – 23.0	[14]
	North Sea	7.0 – 22.0	[116]
	Southampton Water	1.9 – 13.2	[13]
	Sea off California	0.007 (surface)	[130]
		0.60 – 0.65 (2950 m)	[130]
	Coastal water	0.29 – 0.44	[4]
	Coastal water	3.28	[127]
	Coastal water	1.6 – 2.0	[138]
	Coastal seawater	1.5 – 1.9	[124]
	Seawater	1.6 – 1.9	[32]
		4.1	[126]
		0.72 – 0.84	[31]
	Gota River Estuary (salinity 0.5‰)	7.6 – 8.4	[139]
	Gota River Estuary (salinity 32‰)	0.5	[139]
Anions in seawater			
	Phosphate, Japanese inland waters	12.8 – 46.0	[161, 162]
	Fluoride, Port Lonsdale, Victoria Australia (salinity 35‰)	1280 – 1430	[163]
	Iodate	30 – 60	[164]
	Iodide	0 – 20	[164]
	Organic iodine	< 5	[164]

Note: The North sea and Mediterranean are included in this list as these are both subject to a high degree of metal contamination originating from surrounding coastal areas.

References

1. Moore RM (1983) In: Wong CS, Bruland KW, Boyle E, Burton E (eds) *Trace Metals in Seawater*, Proceedings of a NATO Advanced Research Institute on Trace Metals in Seawater, Sicily, Italy, 1981, Plenum, New York, USA.
2. Fitzgerald WF (1975) Mercury Analysis in Seawater using Cold Trap Preconcentration and Gas Phase Detection. In: Gibb TRP Jr (ed) *Analytical Methods in Oceanography*, American Chemical Society, Washington, DC, USA.
3. Pillay KKS, Thomas CC Jr, Sondel JA, Hyche CM (1971) *Anal Chem* **43**:1419.
4. Sturgeon RE, Berman SS, Willie SN, Desaulniers JAH (1981) *Anal Chem* **53**:2337.
5. Jagner D, Josefson M, Westerlund S (1981) *Anal Chim Acta* **129**:153.
6. Danielsson L-G, Magnusson B, Westerlund S (1978) *Anal Chim Acta* **98**:47.
7. Miyazaki A, Kimura A, Bansho K, Umezaki Y (1981) *Anal Chim Acta* **144**:213.
8. Magnusson B, Westerland A (1983) In: Wong CS, Bruland KW, Boyle E, Burton E (eds) *Trace Metals in Seawater*, Proceedings of a NATO Advanced Research Institute on Trace Metals in Seawater, Sicily, Italy, 1981, Plenum, New York, USA.
9. Brugmann L, Danielsson L-G, Magnusson B, Westerlund S (1983) *Mar Chem* **13**:327.
10. Boyle EA, Edmond JM (1977) *Anal Chim Acta* **91**:189.
11. Magnusson B, Westerlund S (1980) *Mar Chem* **8**:231.
12. Stukas VJ, Wong CS (1983) In: Wong CS, Bruland KW, Boyle E, Burton E (eds) *Trace Metals in Seawater*, Proceedings of a NATO Advanced Research Institute on Trace Metals in Seawater, Sicily, Italy, 1981, Plenum, New York, USA.
13. Armannsson H (1979) *Anal Chim Acta* **110**:21.
14. Campbell WC, Ottaway JM (1977) *Analyst* **102**:495.
15. Grimaud D, Michard G (1974) *Mar Chem* **2**:229.
16. Kuwamoto T, Murai S (1970) *Preliminary Report of the Hakuho-Maru Cruise K11-68-4*, Ocean Research Institute University, Tokyo, Japan, p. 72.
17. Fukai R (1967) *Nature* **213**:901.
18. Willie SN, Sturgeon RE, Berman SS, (1983) *Anal Chem* **55**:981.
19. Batley GE, Matousek JP (1980) *Anal Chem* **52**:1570.
20. Motomizu S (1973) *Anal Chim Acta* **64**:217.
21. Greenberg RR, Kingston HM (1982) *J Radioanal Chem* **71**:147.
22. Sheffrin N, Williams EE, Weller EC (1982) *Anal Proc* **19**:483.
23. Krembling KP (1983) In: *Trace Metals in Seawater* Wong CS, Bruland KW, Boyle E, Burton E (eds) Proceedings of a NATO Advanced Research Institute on Trace Metals in Seawater, Sicily, Italy, 1981, Plenum Press, New York, USA.
24. Mort L, Nurnberg HW, Dyrssen D (1983) In: Wong CS, Bruland KW, Boyle E, Burton E (eds) *Trace Metals in Seawater*, Proceedings of a NATO Advanced Research Institute on Trace Metals in Seawater, Sicily, Italy, 1981, Plenum, New York, USA.
25. Shaube BK, Pattrson CC (1983) In: Wong CS, Bruland KW, Boyle E, Burton E (eds) *Trace Metals in Seawater*, Proceedings of a NATO Advanced Research Institute on Trace Metals in Seawater, Sicily, Italy, 1981, Plenum, New York, USA.
26. Lyte E, Huested SW (1983) In: Wong CS, Bruland KW, Boyle E, Burton E (eds) *Trace Metals in Seawater*, Proceedings of a NATO Advanced Research Institute on Trace Metals in Seawater, Sicily, Italy, 1981, Plenum, New York, USA.

27. Fitzgerald WF, Lyons WB, Hunt CD (1974) *Anal Chem* **46**:1882.
28. Olafsson J (1983) In: Wong CS, Bruland KW, Boyle E, Burton E (eds) *Trace Metals in Seawater*, Proceedings of a NATO Advanced Research Institute on Trace Metals in Seawater, Sicily, Italy, 1981, Plenum, New York, USA.
29. Batley GE, Matousek JP (1977) *Anal Chem* **49**:2031.
30. Shriadah MMA, Kataoka M, Ohzeki K (1985) *Analyst* **110**:125.
31. Pihlar B, Valenta P, Nurnberg HW (1981) *Fresen Z Anal Chem* **307**:337.
32. Mykytiuk AP, Russell DS, Sturgeon RE (1980) *Anal Chem* **52**:1281.
33. Elderfield H, Greaves HJ (1983) In: Wong CS, Bruland KW, Boyle E, Burton E (eds) *Trace Metals in Seawater*, Proceedings of a NATO Advanced Research Institute on Trace Metals in Seawater, Sicily, Italy, 1981, Plenum, New York, USA.
34. Matthews AD, Riley JP (1970) *Anal Chim Acta* **51**:483.
35. Willie SN, Sturgeon RE, Berman SS, (1986) *Anal Chem* **58**:1140.
36. Brinkmann FE (1983) In: Wong CS, Bruland KW, Boyle E, Burton E (eds) *Trace Metals in Seawater*, Proceedings of a NATO Advanced Research Institute on Trace Metals in Seawater, Sicily, Italy, 1981, Plenum, New York, USA.
37. Colella MB, Siggia S, Barnes RM (1980) *Anal Chem* **52**:2347.
38. Van den Berg CMG, Huang ZQ (1984) *Anal Chem* **56**:2382.
39. Spence R (1968) *Talanta* **15**:1307.
40. Bertine KK, Chan LH, Turekian KK (1970) *Geochim Cosmochim Acta* **34**:641.
41. Hashimoto T (1971) *Anal Chim Acta* **56**:347.
42. Kim YS, Zeitlin H (1971) *Anal Chem* **43**:1390.
43. Williams WJ, Gillam AH (1979) *Analyst* **103**:1239.
44. Smith J, Grimaldi O (1957) *Bull US Geol Survey* **1006**:125.
45. Leung G, Kim YS, Zeitlin H (1972) *Anal Chim Acta* **60**:229.
46. Kim YS, Zeitlin H (1972) *Anal Abstr* **22**:4571.
47. Korkish J, Koch W (1973) *Mikrochim Acta* **1**:157.
48. Babano PG, Rigali L (1978) *Anal Chim Acta* **96**:199.
49. Kim KH, Burnett WC (1983) *Anal Chem* **55**:1796.
50. Bowie SHU, Clayton CG (1972) *Trans Inst Miner Metals* **B81**:215.
51. Shannon LV, Orren MJ (1970) *Anal Chim Acta* **52**:166.
52. Nozaki Y, Tsunogai S (1973) *Anal Chim Acta* **64**:209.
53. Cowen JP, Hodge VF, Folsom TE (1977) *Anal Chem* **49**:494.
54. Tsunogai S, Nozaki Y (1971) *Geochem J* **5**:165.
55. Shannon LV, Cherry RD, Orren MJ (1970) *Geochim Cosmochim Acta* **34**:701.
56. Hodge VF, Hoffman FL, Folsom TR (1974) *Health Phys* **27**:29.
57. Folsom TR, Hodge VR (1975) *Mar Sci Commun* **1**:213.
58. Folsom TR, Hodge VR, Gurney ME (1975) *Mar Sci Commun* **1**:39.
59. Goldberg ED, Koide M, Hodge VF (1976) *Determination of Thorium Isotopes in Seawater*, Scripps Institute of Oceanography, La Jolla, CA, USA
60. Flynn A (1970) *Anal Abstr* **18**:1624.
61. Reid DF, Key RM, Schink DR (1979) *Earth Planet Sci Lett* **43**:223.
62. Huh CA, Bacon MP (1985) *Anal Chem* **57**:2138.
63. Bacon MP, Anderson RF (1983) In: Wong CS, Bruland KW, Boyle E, Burton E (eds) *Trace Metals in Seawater*, Proceedings of a NATO Advanced Research Institute on Trace Metals in Seawater, Sicily, Italy, 1981, Plenum, New York, USA.
64. Spencer DW, Sachs PL (1970) *Mar Geol* **9**:117.
65. Krishnaswami S, Lal D, Somayajulu BLK, Weiss RF, Craig H (1976) *Earth Planet Sci Lett* **32**:420.

66. Anderson RF (1981) *The Marine Geochemistry of Thorium and Protoactinium*, WH01-81-1, PhD Dissertation, Massachusetts Institute of Technology/Woods Hole Oceanographic Institute, Cambridge, MA, USA.
67. Moore WS, Reid DF (1973) *J Geophys Res* **78**:8880.
68. Moore WS (1976) *Deep Sea Res Oceanogr Abst* **23**:647.
69. Moore WS, Cook LM (1975) *Nature* **253**:262.
70. US EPA (1975) *Radiochemical Methodology for Drinking Water Regulations*, EPA 600/4-75-005, US Environmental Protection Agency, Washington, DC, USA.
71. Ku TL, Huh CA, Chen PS (1980) *Earth Planet Sci Lett* **49**:293.
72. Chung Y (1980) *Earth Planet Sci Lett* **49**:319.
73. Moore WS (1982) *Estuar Coast Shelf Sci* **12**:713.
74. Michel J, Moore WS, King PT (1981) *Anal Chem* **53**:1885.
75. US EPA (1976) *National Interim Primary Drinking Water Regulations* EPA-57019-79-003, US Environmental Protection Agency, Washington, DC, USA.
76. Key RM, Brewer RL, Stockwell JH, Guinasso NL Jr, Schink DR (1979) *Mar Chem* **7**:251.
77. Broecker WS (1965) An Application of Natural Radon to Problems in Oceanic Circulations. In: *Proceedings of the Symposium on Diffusion in the Oceans and Fresh Waters*, Lamont Geological Observatory, New York, USA, p. 116–145.
78. Reid DF, Key RM, Schink DR (1974) *EOS Trans Am Geophys Union*, December 1974.
79. Schink D, Guinasso N Jr, Charnell R, Sigalove J (1970) *IEEE Trans Nucl Sci* **NS-17**:184.
80. Chung Y (1971) *Pacific Deep and Bottom Water Studies Based on Temperature, Radium and Excess Radon Measurements*, Dissertation, University of California, San Diego, CA, USA.
81. Lucas HF (1957) *Rev Sci Instrum* **28**:680.
82. Comar CL (1976) *Plutonium, Facts and Interferences*, EPRI EA-43-SR, Electric Power Research Institute, Palo Alto, Ca, USA.
83. Livingston HD, Mann DR, Bowen VT (1975) In: Gibb TRP Jr (ed) *Analytical Methods in Oceanography*, Advances in Chemistry Series No. 147, ACS, Washington, DC, USA.
84. Wong KM (1971) *Anal Chim Acta* **56**:355.
85. Pillai KC, Smith RC, Folsom TR (1964) *Nature* **203**:568.
86. Ballestra S, Holm E, Fukai R (1978) In: *Proceedings of the Symposium on the Determination of Radionuclides in Environmental and Radiological Materials*, Central Electricity Generating Board, London, UK.
87. Holm E, Fukai R (1977) *Talanta* **24**:659.
88. Sakanous M, Nakamura M, Imai T (1971) Rapid Methods for Measuring Radioactivity in the Environment. In: *Proceedings of the Symposium*, Neuherberg, IAEA, Vienna, Austria, p. 171.
89. Statham C, Murray CN (1976) *Report of the International Committee of the Mediterranean Ocean* **23**:163.
90. Hampson BL, Tennant D (1973) *Analyst* **98**:873.
91. Levine H, Lamanna A (1965) *Health Phys* **11**:117.
92. Aakrog A (1975) *Reference Methods of Marine Radioactivity Studies II*, Technical Report Services No. 169, IAEA, Vienna, Austria.
93. Chu A (1972) *Anal Abst* **22**:427.
94. Livingston HD, Mann DR, Bowen UJ (1972) *Report of the Atomic Energy Commission*, US COO-3563-12, Woods Hole Oceanographic Institute, Woods Hole, MA, USA.

95. Delle Site A, Marchhionni V, Testa C, Triulzi C (1980) *Anal Chim Acta* **117**:217.
96. Testa C, Delle Site A (1976) *J Radioanal Chem* **34**:121.
97. Anderson RF, Fleer AP (1982) *Anal Chem* **54**:1142.
98. Testa C, Staccioli L (1972) *Analyst* **97**:527.
99. Hirose K, Sugimura YJ (1985) *J Radioanal Nucl Chem* **92**:363.
100. Silant'ev AN, Chumichv UB, Vakulouski SM (1970) *Trudy Inst Eskp Met Glav Uprav Gidromet, Sluzhty Sov Minist SSSR* **15**:2 [Zhur Khim 19GD (1) Abstract No.1, G209].
101. Gordon CM, Larson RE (1970) *Radiochem Radioanal Lett* **5**:369.
102. Dutton JRW (1970) *Report of the Fisheries and Radiobiological Laboratory*, FRL 6, Ministry of Agriculture, Fisheries and Food, London, UK.
103. Lewis SR, Shafrir NH (1971) *Nucl Instrum Method* **93**:317.
104. Janzer VJ (1973) *J US Geol Survey* **1**:113.
105. Yamamoto O (1967) *Anal Abstr* **14**:6669.
106. Mason WJ (1974) *Radiochem Radioanal Lett* **16**:237.
107. Morgan A, Arkell GM (1963) *Health Phys* **9**:857.
108. Kiba T, Terada K, Kiba T, Suzuki K (1972) *Talanta* **19**:451.
109. Hiraide M, Sakurai K, Mizuike A (1984) *Anal Chem* **56**:2851.
110. Tseng CL, Hsieh YS, Yang MH (1985) *J Radioanal Nucl Chem* **95**:359.
111. Flynn WW (1973) *Anal Chim Acta* **67**:119.
112. Stah SM, Rao SR (1972) *Curr Sci (Bombay)* **41**:659.
113. Cowen JP, Hodge VF, Folsom TR (1977) *Anal Chem* **49**:494.
114. Perkins RW (1969) *Radium and Radiobarium Measurement in Seawater and Freshwater by Sorption and Multi-dimensional X-ray Spectrometry*, Report BNWL 1051, Atomic Energy Commission, Washington, DC, USA.
115. Korenaga T, Motomizu S, Toei K (1980) *Analyst* **105**:328.
116. Gillain G, Duyckaerts G, Disteche A (1979) *Anal Chim Acta* **106**:23.
117. Hill JM, O'Donnell AR, Mance G (1984) *The Quantities of Some Heavy Metals Entering the North Sea*, Technical Report TR205, Water Research Centre, Stevenage, UK.
118. Schreedhara Murthy RS, Ryan DE (1983) *Anal Chem* **55**:682.
119. Lee C, Kim NB, Lee IC, Chung KS (1977) *Talanta* **24**:241.
120. Florence TM (1974) *J Electroanal Chem* **49**:255.
121. Eskilsson H, Jagner D (1982) *Anal Chim Acta* **138**:27.
122. Pruszkowska E, Carnrick GR, Slavin W (1983) *Anal Chem* **55**:182.
123. Berman SS, McLaren JW, Willie SN (1980) *Anal Chem* **52**:488.
124. Sturgeon RE, Berman SS, Willie SN, Desaulniers JAH, Mykytiuk AP, McLaren JW, Russell DS (1980) *Anal Chem* **52**:1585.
125. Guevremont R, Sturgeon RE, Berman SS (1980) *Anal Chim Acta* **115**:163.
126. Jagner D (1978) *Anal Chem* **50**:1924.
127. Sturgeon RE, Berman SS, Willie SN, Desaulniers JAH, Russell DS (1979) *Anal Chem* **51**:2364.
128. Schmidt D (1980) *Heligolander Meeresunter-Suchungen* **33**:576.
129. Lo JM, Yu JC, Hutchinson FI, Wai CM (1982) *Anal Chem* **54**:2536.
130. Bruland KW, Franks RP, Knauer GA, Martin JH (1979) *Anal Chim Acta* **105**:233.
131. Sperling RK (1982) *Fresen Z Anal Chem* **310**:254.
132. Drabach I, Pheiffer Madsen P, Sorensen J (1983) *Int J Environ Anal Chem* **15**:153.
133. Kingston HM, Barnes IL, Brady TJ, Rains TC, Champ MA (1978) *Anal Chem* **50**:2064.
134. Wan CC, Chiang S, Corsini A (1985) *Anal Chem* **57**:719.
135. Peotrowicz SR (1983) In: Wong CS, Bruland KW, Boyle E, Burton E (eds) *Trace Metals in Seawater*, Proceedings of a NATO Advanced Research Institute on Trace Metals in Seawater, Sicily, Italy, 1981, Plenum, New York, USA.

136. Scarponi G, Capodaglio G, Cescon P, Cosma B, Frache R (1982) *Anal Chim Acta* **135**:263.
137. Stein VB, Canelli E, Richards H (1980) *Int J Environ Anal Chem* **8**:99.
138. Slavin W (1980) *Atom Spectrosc* **1**:66.
139. Danielsson L-G, Magnusson B, Westerlund S, Zhang K (1982) *Anal Chim Acta* **144**:183.
140. Nakayama E, Kuwamoto T, Tokoro H, Fujinaga T (1981) *Anal Chim Acta* **131**:247.
141. Ishibashi M, Shigematsu T (1950) *Bull Inst Chem Res Kyoto Univ* **23**:59.
142. Mullins TL (1984) *Anal Chim Acta* **165**:97.
143. Cheucas L, Riley JP (1966) *Anal Chim Acta* **35**:240.
144. Murata M, Omatsu M, Mushimoto S (1984) *X-Ray Spectrosc* **13**:83.
145. Berger P, Ewald M, Liu D, Weber JH (1984) *Mar Chem* **14**:289.
146. Yoshimura K, Nigo S, Tarutani T (1982) *Talanta* **29**:173.
147. Shreedhara Murthy RS, Ryan DE (1982) *Anal Chim Acta* **140**:163.
148. Sugimae A (1980) *Anal Chim Acta* **121**:331.
149. Pellenberg RE, Church TM (1978) *Anal Chim Acta* **97**:81.
150. Acebal SA, De Luca Rebello A (1983) *Anal Chim Acta* **148**:71.
151. Torsi G (1977) *Anal Chem (Rome)* **67**:557.
152. Torsi G, Desimoni E, Palmisano F, Sabbatini L (1981) *Anal Chim Acta* **124**:143.
153. Tominaga M, Bansho K, Umezaki Y (1985) *Anal Chim Acta* **169**:171.
154. Knox S, Turner DR (1980) *Estuar Coastal Mar Sci* **10**:317.
155. Carnrick GR, Slavin W, Manning DC (1981) *Anal Chem* **53**:1866.
156. Frenet-Robin M, Ottmann F (1978) *Estuar Coast Mar Sci* **7**:425.
157. Kiriyama T, Kuroda R (1984) *Talanta* **31**:472.
158. Nakahara T, Chakrabarti CL (1979) *Anal Chim Acta* **104**:99.
159. Lee DS (1982) *Anal Chem* **54**:1 82.
160. Tzeng J-H, Zeitlin H (1978) *Anal Chim Acta* **101**:71.
161. Paulson AJ (1986) *Anal Chem* **58**:183.
162. Motomizu S, Wakimoto T, Toei K (1982) *Anal Chim Acta* **138**:329.
163. Rix CJ, Bond AM, Smith JD (1976) *Anal Chem* **48**:1236.
164. Truesdale VW (1978) *Mar Chem* **6**:253.
165. Murthy RSS, Ryan DE (1982) *Anal Chem Acta* **140**:163.

Appendix 6.1
Ranges of Concentrations of Organic Compounds Found in Fresh Waters

		Concentration (μg/l)	Reference
Haloforms			
$CHCH_3$	River	0.56 – 0.75	[1]
		0.20 – 0.67	[2]
$BrCl_2CH$	Lake water	54.6 – 59.1	[3]
	River	< 0.1	[4]
		0.06	[1]
		0.1 – 7.6	[2]
Br_2ClCH	River	< 0.1	[4]
		0.08	[1]
		4.66	[2]
Br_3CH	River	< 0.1	[4]
		0.15 – 0.21	[1]
		0.51	[2]
CCl_4	River	0.02 – 0.12	[1]
	Lake water	11.8 – 14.3	[3]
Cl_2CHCH_2Cl	River	0.05 – 0.09	[1]
	Lake water	7.8 – 11.4	[3]
Cl_2CHCH_2Cl	Lake water	8 – 20	[5]
$Cl_2CHCHCl_2$	Lake water	2 – 5	[5]
Total haloforms	River	0.92 – 1.31	[1]
	River	5.47 – 13.44	[2]
	River	11.8 – 14.3	[4]
	Lake	62.4 – 70.5	[3]

		Concentration (μg/l)	Reference
Polyaromatic hydrocarbons			
Benzo(*a*)pyrene	River	0.032 – 0.038	[6]
Fluoranthrene		0.02 – 1.1	[7]
Benzo(*k*)fluoranthene		0.03 – 049	
Benzo(*a*)pyrene		0.10 – 0.65	
Perylene		0.03 – 0.20	
Indene(1,2,3-ed)pyrene		0.4 – 0.32	
Benzo(*ghi*)perylene		0.04 – 0.12	
Pyrene		0.05 – 0.43	
Benzo(*a*)anthracene/ Chrysene		0.14 – 0.53	
Benzo(*b*)fluoranthrene		0.13 – 0.57	
Total PAH		< 0.1 – 4.3	
Chlorinated insecticides			
α-BHC	River	0.003	[8]
α-BHC	River	0.002	[9]
α- and γ-BHC plus hexachlorobenzene	River	0.018	[10]
β-BHC	River	0.0004	[8]
	River	0.013	[9]
	River	0.023	[10]
	Surface water	0.006 – 0.078	[11]
γ-BHC	River	0.69	[10]
	River	0.006	[9]
	Surface water	0.006 – 0.078	[11]

		Concentration (μg/l)	Reference
γ-BHC	River	0.69	[10]
	River	0.006	[9]
	Surface water	0.004 – 0.02	[11]
δ-BHC	River	0.016	[9]
DDT	River	0.042	[10]
p,p′-DDT	River	0.051	[12]
	Surface water	0.009 – 0.037	[11]
o,p′-DDT	Surface water	0.005 – 0.025	[11]
DDE	River	0.022	[10]
p,p′-DDE	Surface water	0.002 – 0.010	[11]
Lindane	River	0.01	[12]
	River	0.001	[8]
Dieldrin	River	0.031	[12]
Aldrin	River	0.02	[12]
Endrin	River	0.038	[12]
γ-Chlordane	River	0.03	[12]
Methoxychlor	River	0.12	[12]
Endosulfan	River	0.028 – 0.28	[13]
Heptachlor	Surface water	0.001 – 0.007	[11]
Hexachlorobenzene	Surface water	0.002 – 0.008	[11]
Total chlorinated insecticides	River	0.0034	[5]
	River	0.037	[5]
	River	0.761	[8]
	River	0.300	[12]
	Surface water	0.029 – 0.185	[11]

		Concentration (μg/l)	Reference
Other types of insecticides			
Ronnel		0.002 – 0.022	[14]
Dursban		0.030 – 0.043	
Diazinon		0.020 – 0.037	
Malathion		0.027 – 0.032	
Parathion		0.037 – 0.039	
Parathion-methyl		0.021 – 0.038	
Polychlorinated biphenyls	Ground water	0.0001 – 0.0002 (as Aroclor 1016)	***
Pentachlorophenol	River	10 – 250	[15]
	Well water	0.1	[16]
Dibutyl phosphate	River	< 0.1 – 1.0	[17, 18]
	River	45	[19]
	River Meuse	< 0.1 – 0.9	[17, 20]
Di-2-ethylhexyl phthalate	River	0.4 – 4.2	[17, 18]
	River	10	[17, 18, 21]
	River Meuse	0.1 – 1.1	[17, 18]
Nonionic detergents (phenol polyoxyalkylene condensates)	River	8 – 70	[20]
Alkyl benzene sulfonates	River	270 – 600	[22]
	River	10 – 600	[22]
	2-C_{10}	4400 – 4800	[23]
	3-C_{10}	3700 – 6200	
	4,5-C_{10}	1200 – 16,200	
	2-C_{11}	3900 – 12,200	
	3-C_{11}	8200 – 8300	
	4,5,6-C_{11}	42,700 – 43,700	
	2-C_{12}	< 200 – 4200	
	3-C_{12}	1200 – 3400	
	4,5,6-C_{12}	13,700 – 17,100	

		Concentration (μg/l)	Reference
	2-C_{13}	< 200 – 800	
	3-C_{13}	< 200 – 800	
	4,5,6-C_{13}	< 200 – 6200	
Methylene blue active substances	Ground water	20 – 22	[24]
Fluorescent whitening agents	River	5 – 7	[22]
	River	1 – 7	[25]
Fatty acids	River (total C_{10}-C_{19}FA)	4.13 – 527	[26]
Nitriloacetic acid	River	0.4	[27]
Dissolved organic carbon	River	6000 – 10,000	[28]
	Lake water	1500 – 3080	[29]
	Ground water	300 – 6300	[24]
Dissolved inorganic carbon	Lake water	1060 – 6190	[30]
	Ground water	1000	[24]
Gaseous organic carbon	Lake water	1900 – 2310	[31]
Particulate organic carbon	Lake water	100 – 300	[3]

*** LeBel GL, Williams DT, private communication

References

1. Von Rensberg JFJ, Van Huysteen JJ, Hassett AJ (1978) *Water Res* **12**:127.
2. Kirschen NA (1980) *Varian Instrum Appl* **14**:10.
3. Dietz EA, Singley KF (1979) *Anal Chem* **51**:1809.
4. Fielding M, McLouglin K, Steel C (1977) *Water Research Centre Enquiry Report ER532*, Water Research Centre, Stevenage Laboratory, Stevenage, UK.
5. Hagenmaier H, Werner G, Janer W (1982) *Z Wasser Abwass Forsch* **15**:195.
6. Monarca S, Causey BS, Kirkbright GF (1979) *Water Res* **13**:503.
7. Acheson MA, Harrison RM, Perry R, Wellings RA (1976) *Water Res* **10**:207.
8. Stachel B, Baetjer K, Cetinkaya M, Deuszeln J, Gable B, Kozicki R, Lahl U, Lierse K, Podbielski A, Thiemann W (1981) *Anal Chem* **53**:1469.
9. Suzuki M, Yamoto Y, Watanabe T (1977) *Environ Sci Technol* **11**:1109.
10. Sackmauerova M, Pal'usova O, Szokolay A (1977) *Water Res* **11**:551.
11. Leoni V, Puccetti G, Grella A (1975) *J Chromatogr* **106**:119.
12. Aspila KI, Carron JM, Chau AS (1977) *J Assoc Anal Chem* **60**:1097.
13. Bergemann H, Hellman H (1981) *Deutsch Gewässerkund Mitteil* **24**:31.
14. McIntyre AE, Perry R, Lester JN (1980) *Environ Tech Lett* **1**:157.
15. Ervin HE, McGinnis GD (1980) *J Chromatogr* **190**:203.
16. Morgade C, Barquiet A, Pfaffenberger CD (1980) *Bull Environ Contam Toxicol* **24**:257.
17. Schouten MJ, Copius Peereboom JW, Brinckman UAT (1979) *Int J Environ Anal Chem* **7**:13.
18. Schouten MJ, Schwartz H, Anzion CJM, Van Vleits HPM, Copius Peereboom JW, Brinckman UAT (1979) *Int J Environ Anal Chem* **6**:133.
19. Mori S (1976) *J Chromatogr* **129**:53.
20. Jones P, Nickless G (1978) *J Chromatogr* **156**:99.
21. Ton N, Takehashi Y (1985) *Int Lab* **September**:49.
22. Uchiyama M (1979) *Water Res* **13**:847.
23. Hon-Nami H, Hanya T (1978) *J Chromatogr* **161**:205.
24. Hughes JL, Eccles LA, Malcolm RL (1974) *Ground Water* **12**:283.
25. Schwartzenbach RP, Bromund RH, Gschwend PM, Zafirou OC (1978) *Organic Geochem* **1**:93.
26. Hullett DA, Eisenreich SJ (1979) *Anal Chem* **51**:1953.
27. Aue WA, Hastings CR, Gerhardt KO, Pierce II JO, Hill HH, Moseman RF (1972) *J Chromatogr* **72**:259.
28. Baker CD, Bartlett PD, Farr IS, Williams GI (1974) *Freshwater Biol* **4**:467.
29. Goulden PD, Brooksbank P (1975) *Anal Chem* **47**:1943.
30. Games LM, Hayes JM (1976) *Anal Chem* **48**:130.
31. Kraubeck HJ, Lampert W, Bredie H (1981) *Fachzeitschrift Lab* **25**:2009.

Index